# Botany

Harper International Edition
HARPER & ROW, PUBLISHERS
New York, Hagerstown, San Francisco, London

# Botany

## BASIC CONCEPTS IN PLANT BIOLOGY

TERRY L. HUFFORD

GEORGE WASHINGTON UNIVERSITY

*Sponsoring Editor: Jeffrey K. Smith*
*Project Editor: Karla B. Philip*
*Designer: Emily Harste*
*Production Supervisor: Kewal K. Sharma*
*Compositor: The Clarinda Company*
*Printer and Binder: Halliday Lithograph Corporation*
*Art Studio: J & R Technical Services Inc.*
*Cover Photo: Emily Harste*

**BOTANY:** *Basic Concepts in Plant Biology*



*Library of Congress Cataloging in Publication Data*

*Hufford, Terry L*
*Botany: basic concepts in plant biology.*

*Includes index.*
*1. Botany. I. Title.*
*QK47.H84 581 77-18028*
*ISBN 0-06-042972-0*

# Contents

*Chapter 11*
**THE VASCULAR PLANTS—A SURVEY 257**



*Chapter 12*
**MOVEMENT OF MATERIALS THROUGH PLANT CELLS AND ORGANS 294**



*Chapter 13*
**PLANT FOODS, FOOD SYNTHESIS, AND DIGESTION 318**

## Chapter 14
## ROLE OF FOODS IN PLANTS—I. RESPIRATION 350



## Chapter 15
## ROLE OF FOODS IN PLANTS—II. ASSIMILATION, GROWTH, AND DIFFERENTIATION 369



## Chapter 16
## PLANT REPRODUCTION 405

# Preface

Botany, as any science, is based upon the observations of people. The questions that arise as a result of these observations, the posing of hypotheses to answer these questions, the efforts to test the validity of these hypotheses through carefully constructed experiments, and, from these experiments, the formulation of "best inferences" to explain the aforementioned observations are all part of the ever-changing nature of botany.

*Botany: Basic Concepts in Plant Biology* represents my views of the materials and procedures necessary to effectively incorporate botanical inferences and their applicability to human interests and welfare. Because new botanical information becomes available every day, existing inferences are in a constant state of flux, with some discarded, others modified, some strengthened, and others newly proposed. Thus, to approach botany in a static or standardized way is to discount the dynamic nature of the science. Moreover, it must be remembered that general botany is a subject for all students, not just those interested in a career in some aspect of the botanical sciences. We cannot meet our obligations to those students by insisting that they memorize what scientists have discovered or what we tell them are the "facts," and then giving them these details in such breadth and depth that the memorization of them becomes an end unto itself rather than a basis for understanding the broader concepts. For the study to become meaningful to the students, it must be experienced, and this should be the purpose of classroom procedure.

This textbook assumes a pedagogy known by some as the "Socratic method" and by others as the "observation-discussion method," whereby the instructor assumes the role of questioner and, at times, devil's advocate. Thus, the student is led, by question and discussion, to make critical observations; and from these observations, to arrive at inferences based upon causal relationships. It is hoped that the student will thereby become aware of the necessity to explain botanical phenomena on the basis of consequences of preceding physical or physiological events rather than on the supposed needs or desires of the organism. This textbook is intended primarily to supplement what is observed and discussed in the classroom, and I assume that textual material will *not* be read until the students have had the opportunity to make their own observations and draw their own conclusions, insofar as possible, as a result of class experiences.

Since this textbook is designed to serve primarily to supplement, enlarge upon, and perhaps offer alternative viewpoints to material discussed in class, and not as a primary teaching tool, the Socratic philosophy is sparingly used in the text. Nevertheless, an attempt has been made to present certain subject matter in an observation-hypothesis-experimentation-results-inferences approach. In addition, each chapter is organized in a way that general concepts are presented first, followed by specific details or in-depth coverage of these concepts. Thus, the instructor of a nonmajors' course may wish to hold the student responsible for only the conceptual aspects, while those teaching a majors' course may wish to assign a portion or all of the in-depth coverage. I hope that students having a basis for understanding such details will be encouraged to study them carefully. For students who do not have such a basis, the details simply provide additional insight into the complexity of the subject. If these latter students attempt to memorize all the information given in this text, much of it will be simply that, memorization—and memorization without understanding is empty knowledge.

It is my belief that a basic body of knowledge exists which is as important to the botany and biology major as to the nonmajor. It is important to establish both conceptual knowledge and a process of reasoned approach to interpretation of botanical data. Thus, some materials in the text may seem elementary or obvious, but it is important to remember that not all students have the same background, and even some of the more sophisticated and learned of students, while they may not admit it, were so busy memorizing facts in their high school biology courses that they missed very basic conceptual ideas.

Depth of coverage of what may be considered the basic principles of botany will be variable. This reflects, in part, the interests and bias of the writer. For example, discussion of the plant kingdom per se is minimal, as it is my belief that while the salient features that distinguish one group of plants from another and the economic and ecological importance of each group are important to an understanding of basic botanical concepts, *detailed* aspects of plant diversity are not. Such discussion should be left to a second course such as one titled "The Plant Kingdom." In addition to some variability in depth of coverage, certain topics will be discussed in several chapters; thus, the students are encouraged to note interrelationships and to keep enlarging and modifying their basic concepts as they progress through the course. At times, students may be allowed to arrive at incorrect inferences, providing those inferences appear valid for the information given. Later, additional information should provide a basis for the student to realize the error of a previous inference and change it accordingly. The danger of drawing conclusions based upon limited information is thus demonstrated. At times, the author may draw inferences at variance with those more generally accepted. This is done for several reasons: 1) to indicate that different inferences, often equally valid, can be drawn given the same data; 2) to bring attention to the historic fact that in botanical investigations, the majority view has *not* always been correct; and 3) to provide a basis for discussion among students and between student and instructor.

A major purpose of this text is to help students arrive at concepts which are both reasonable and usable rather than to encourage memorization of definitions and dogma. The students should understand that our knowledge of plants is really just a compilation of the "best inferences" we can make, based upon current information. How much the students can learn is limited only by their capacity for critical observation, their knowledge of what can and cannot be done with experimental information, their ability to become accustomed to the inductive and deductive reasoning embodied in a scientific approach, and the time available for considering alternative points of view.

This textbook of botany does not represent the unique ideas of one individual but rather the synthesis of experiences and knowledge gained from many good instructors and from students. Although I had many good teachers during my undergraduate and graduate days, three profoundly influenced my thinking concerning both the subject matter and the art of teaching. These are Drs. Richard A. Popham, Clarence Taft, and Jacob Verduin.

The author is deeply indebted to Drs. J. M. Herr, Jr., Lois A. Pfiester, and Milton R. Sommerfield for their sympathetic and helpful criticisms of the manuscript. The author is also indebted to many unnamed others who reviewed earlier stages of manuscript preparation. Their comments aided greatly the development of the final manuscript.

Illustrations have been secured from many sources, and these are properly credited in each instance. The fine cooperation received from these many contributors is greatly appreciated. David R. Williams and Robert G. Trumbull III are deserving of special thanks for their many hours spent in taking photographs and for their assistance and support during manuscript and art preparation.

Without the assistance of William Bangham, however, this work would probably not have been completed. As friend, critic, and confidant, he aided in all stages of manuscript preparation. Without his help and encouragement, the task would have been formidable indeed.

Finally, I would like to express my apologies to my family for all the time I took away from them and my great appreciation for their patient understanding. My wife, Jan, deserves special thanks for proofreading and typing the final manuscript.

*Terry L. Hufford*

# To the Student

*Botany: Basic Concepts in Plant Biology* is intended to supplement your classroom experiences. It will expand your concepts discussed in class and will often offer alternative points of view. The detailed aspects given in the latter part of many chapters are intended as insight into the complexity involved and as general information, not as information to be memorized without understanding and regurgitated on a test.

As with any other human endeavor, botanists do not always agree on certain aspects. Additional study and research may tend to support one point of view over another, but frequently both views may be found to be correct. Different plants may simply have different ways of doing things. It must be understood that certain botanical information may be transitory; what we believe to be true today may be disproven tomorrow. Many of the basic concepts discussed in this book have stood the test of time, however, it is important that you approach this subject as you would any other—with an open mind, remembering that there is room for differences of opinion, but aware also that not everything you hear or read is necessarily true. Even botany teachers (or textbook writers) can sometimes be misinformed or can misinterpret certain points. You should attempt to evaluate critically what you hear or read. Obviously, such evaluation requires information. Part of that body of information should come from your own experiences, part from your in-class experiences, and part from your textbook, but you should add to this information by doing as much outside reading as time allows. Remember that inferences drawn upon limited information may be just as incorrect as those drawn upon faulty information.

I hope that as you proceed through the course, you will be impressed with the need to make critical observations. I further hope that you will use the information available to you to arrive at inferences based upon causal relationships. In other words, for any botanical phenomenon there are underlying physical or physiological events that account for that phenomenon. There is no need to explain plant behavior on the basis of some mystical or supernatural event, nor do we need to associate it with some supposed need or desire on the part of the plant. There is much we do know about plant activities; there is much we do not know. This does not mean that the currently unexplained is unexplainable; we simply have not yet discovered the explanation. As scientists, we have faith that there is a logical and reasonable answer if we but ask the right question, discover the right means for extracting pertinent data, or have that flash of insight which will provide the answer.

There is, of course, little question concerning the necessity and importance of plants in your life, as well as the certain pleasures you may draw from them. I hope this text will not only prove to be informative and make you think about plants in a way you may not have previously, but that it will also be enjoyable reading.

*T.L.H.*

# Chapter 1

## INTRODUCTION

Plant biology, or botany as it is more commonly known, involves the study of plants. The farmer, gardener, forester, florist, nurseryman, orchardist, pharmacologist, physician, environmentalist, geologist, geographer, and others find themselves to a greater or lesser degree confronted by an array of problems that lie within the field of botany, as does the student or anyone else who wishes to exercise his or her green thumb in planting a garden or raising some house plants. Some of the answers to their problems have been arrived at through trial and error, some through insight, but many have been or will be arrived at through the experimentation of trained botanists.

## ORIGIN OF PLANT LIFE

Before looking at aspects of plants per se, it might be a good idea to consider how plants came to be in the first place. The easiest and simplest way to answer this might be to say, "We really do not know," or "It was by Divine creation." But geochemical, geophysical, and paleontological evidence and experimentation dealing with the formation of organic entities allow us to formulate a somewhat better answer. By offering these theories on the origin of life, botanists are not necessarily discounting the hand of God in the scheme of things. We still have the even more basic question left unanswered: how did the universe come into being?

Discussion of the origin of life is somewhat premature at this point as certain aspects involve cell structures and physiological processes which will be discussed in later chapters. But let us give it a try anyhow.

The earth itself seems to be some 4.7 billion years old by most accounts. It formed apparently by the accretion of gases and dust swirling in orbit around the star that is our sun. As the earth formed and its outer portions cooled, it was surrounded by water vapor and other gases such as methane, ammonia, hydrogen, nitrogen, and carbon dioxide. Geological evidence from the oldest known sedimentary rocks (dated at approximately 3.75 billion years) seems to indicate that the early atmosphere, while oxygen deficient, was not entirely devoid of free oxygen. This oxygen was presumably produced nonbiologically. As there was no ozone layer to screen out the ultraviolet light, its intensity could have induced the photodissociation of water vapor, thus producing oxygen. In the highly reducing atmosphere of the time, the small amounts of oxygen were consumed rapidly through reaction with previously unoxidized material. Thus, any life forms evolving under such circumstances had to be anaerobic (capable of living without free oxygen).

The combination of intense heat created by the sun and by the molten interior of the earth caused water on the earth to vaporize. When this water vapor reached the upper atmosphere, it cooled and fell back to earth as rain. Thus, this early stage of the earth's development was probably characterized by a long period of violent storms. The combination of lightning, ultraviolet radiation, and radioactivity from the earth's crust broke apart the simple gases of the atmosphere and re-formed them into more complex molecules. These were washed out of the atmosphere by the rain and collected in the oceans, which became larger as the earth cooled. Consequently, the oceans, which were not saline like those of the present, became a sort of "melting pot" of organic and inorganic molecules. Aggregation and formation of more complex molecules in this so-called primordial soup led eventually to living cells, called the *protoorganisms* by some.

It is postulated that these first primitive organisms lacked a genetic system and thus the ability to reproduce. Obviously, if life was to persist and become widespread, a reliable mechanism for self-replication had to develop. Somehow, these protoorganisms did develop this intricate genetic apparatus, and from that point, life probably evolved rather rapidly. Once a mechanism for the coding of proteins was established, more and more enzymes could be formed, and inevitably, energy-generating systems could be evolved. The source of energy for these early forms of life was undoubtedly the simple organic molecules present in the primordial soup; thus, these organisms were *heterotrophs* (organisms which obtain food "other," *hetero*, than from self.)

According to Schopf (1974), a possible course of evolutionary progression might have been as follows:

1. Biological systems initially lacked porphyrin-containing molecules (porphyrin is a compound contained in certain pigments). The most primitive organisms were anaerobic (*an*, "without"; *aer*, "air"; *bi*, "manner of living") heterotrophs, dependent entirely upon *fermentation* (the breakdown of foods for energy under anaerobic conditions) of organic compounds abiologically (*a*, "without"; *bios*, "life") synthesized in the essentially anoxic (without oxygen) environment.
2. As progressive depletion of the supply of relatively simple organic materials available from the aqueous environment occurred, organisms appeared which possessed iron-containing porphyrins similar functionally to modern cytochromes (electron transport enzymes). These organisms were able to carry out energy-forming reactions anaerobically, with sulfate, carbon dioxide, nitrate, or organic compounds such as fumarate acting as terminal or final electron acceptors. They were not only able to utilize nutrients not available via fermentative pathways, but they obtained a greater energy yield from these nutrients.
3. Further evolution resulted in the appearance of enzymes that extended the previously established biosynthetic pathway to metalloporphyrin, yielding porphyrins complexed with magnesium

rather than iron. This resulted in the formation of chlorophylls and the establishment of anaerobic photoheterotrophy: sunlight was now the energy source rather than organic compounds, although organic nutrients were still required from outside the cell. Freed of the need to degrade large amounts of organic matter for energy, those photosynthetic organisms were able to invade habitats containing only enough organic material to supply the synthetic requirements of the cell. Such habitats were too dilute in organic substances for organisms dependent upon them for energy as well.

4. With the addition of carbon-dioxide—fixing pathways similar to those of living autotrophs (organisms which produce their own food) the photoassimilators (anaerobic photoheterotrophs) gave rise to anaerobic photoautotrophs (organisms capable of utilizing the energy of sunlight and forming their own organic materials).
5. With the development of phycobilins (a type of water soluble pigment) and the establishment of linkage between photosystem I and photosystem II of photosynthesis, hydroxyl ions from water could be used as electron donors, with the resulting introduction of free oxygen into the environment.

As the result of the latter development, the amount of oxygen in the atmosphere increased. Some of it was converted to ozone, which acts to screen out some of the ultraviolet rays. As ultraviolet radiation is destructive to living systems, living organisms could now move out of the deeper waters and eventually onto land. A second effect of free oxygen in the atmosphere was to open the way to a much more efficient utilization of food for energy. As we will see in Chapter 14, aerobic (with free oxygen) respiration yields far more energy than does fermentation.

Although there is obviously no living example of the so-called first cell, are there organisms existing today which some believe may be similar to the organisms represented by Schopf's steps 1, 2, 3, 4, and 5? The porphyrin-lacking organism of step 1 may have been quite similar to a group of present-day organisms called *mycoplasmas.* These are the simplest living cells known (Figure 1.1). They are very tiny, about 0.15 $\mu$m (micrometer, equals to 1/1,000,000 of a meter), are enclosed by a delicate *plasma membrane* but lack a *cell wall.* Organisms similar to those of step 2 are found in present-day *anaerobic bacteria.* The present-day *blue-green algae* may be quite similar to the organisms described in step 5. Thus, in existing organisms, we can find examples of the possible evolution of oxygen-producing, photosynthetic plants.

Note that all of the cells discussed above are what we term *prokaryotic* cells. Yet all plant groups existing today, with the exception of the bacteria and blue-green algae, are composed of organisms having *eukaryotic* cells. True multicellularity does not exist in the prokaryotes; thus, the multicel-

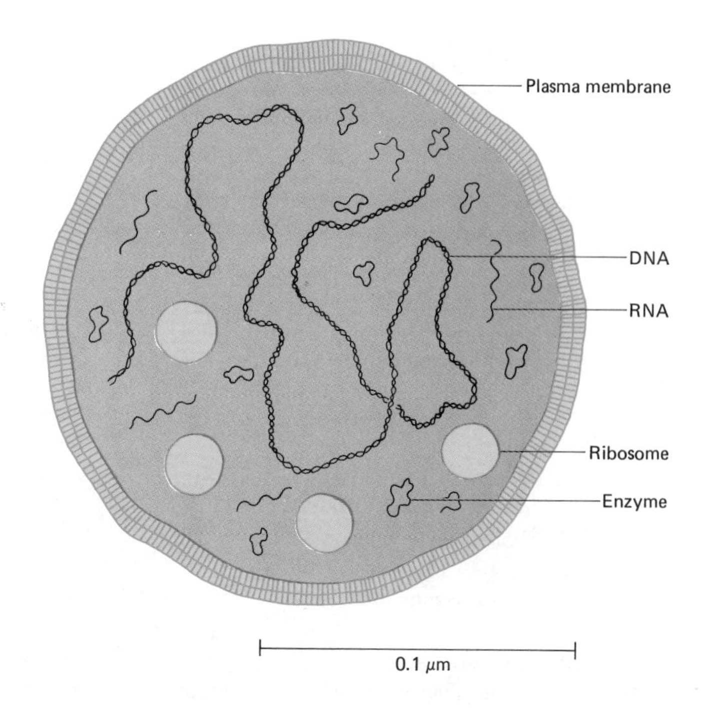


*Figure 1.1*
Diagrammatic representation of a mycoplasm cell. [Redrawn from H. J. Morowitz and M. E. Tourtellotte, "The Smallest Living Cells." Copyright © 1962 by Scientific American, Inc. All rights reserved].

lular plant body did not evolve until the advent of the eukaryotic cell. What is the difference between a prokaryotic cell and a eukaryotic cell? The major differences are summarized in Table 1.1.

How then did the eukaryotic cell evolve? We really do not know how, nor do we know when. In 1965, Barghoorn and Schopf reported finding eukaryotes in bedded black cherts of the approximately 900-million-year-old Bitter Springs formation of central Australia, and later, Schopf (1975) reported finding such structures in the 1500-million-year-old Bungle Bungle Dolomite of northern Australia. But in 1975, Knoll and Barghoorn disavowed the 1965 report, stating that more comprehensive studies had provided no good evidence for the presence of eukaryotes in Bitter Springs cherts. They also believe that all reports of earlier eukaryotes are not conclusive. However, most investigators agree that multicellular eukaryotic organisms existed about 600 million years ago.

How did such cells arise? Currently, there are two major hypotheses: an endosymbiotic, or serial symbiosis, origin; a linear development by a process of evolutionary change; the classical, or traditional, view. A plausible phylogeny (developmental history) of certain prokaryotic and eukaryotic organisms has been expressed for the former hypothesis by Margulis, for the latter hypothesis by Uzzell and Spolsky.

In short, the symbiotic hypothesis postulates that the common ancestor of all eukaryotes was the amoeboflagellate heterotroph. These ancestral

*Table 1.1*
MAJOR DIFFERENCES BETWEEN PROKARYOTES AND EUKARYOTES

| PROKARYOTES | EUKARYOTES |
|---|---|
| Mostly small cells (1 – 10 μm). All are microbes. The most morphologically complex are filamentous or mycelial with fruiting bodies (e.g., actinomycetes, myxobacteria, blue-green algae) | Mostly large cells (10 – 100 μm). Some are microbes, most are large organisms. The most morphologically complex are the vertebrates and the flowering plants. |
| Nucleoid, not membrane bounded | Membrane-bounded nucleus |
| Cell division direct, mostly by "binary fission." Chromatin body which contains DNA, but no protein; does not stain by the Feulgen technique. No centrioles, mitotic spindle, or microtubules | Cell division by various forms of mitosis. Many chromosomes containing DNA, RNA, and proteins; stain bright red by the Feulgen technique. Centrioles present in many; mitotic spindle (or at least some arrangement of microtubules) occurs |
| Sexual systems absent in most forms; when present, unidirectional transfer of genetic material from donor to host takes place | Sexual systems present in most forms; equal genetic participation of both partners (male and female) in fertilization |
| Multicellular organisms never develop from diploid zygotes. No tissue differentiation | Meiosis produces haploid forms, diploids develop from zygotes. Multicellular organisms show extensive development of tissues |
| Includes strict anaerobes (these are killed by $O_2$) and facultatively anaerobic, microaerophilic, and aerobic forms | All forms are aerobic (these need $O_2$ to live); exceptions are clearly secondary modifications |
| Enormous variations in the metabolic patterns of the group as a whole. Mitochondria absent; enzymes for oxidation of organic molecules are bound to cell membranes, that is not packaged separately | Same metabolic patterns of oxidation within the group (i.e., Embden-Meyerhof glucose metabolism, Krebs cycle oxidations, molecular oxygen combines with hydrogens from foodstuffs, and catalyzed by cytochrome, water is produced) |
| | Enzymes for oxidation of 3-carbon organic acids are packaged within membrane bounded sacs (mitochondria) |
| Simple bacterial flagella if flagellated, flagellin proteins | Complex (9 + 2) flagella or cilia if flagellated or ciliated, tubulin proteins |
| If photosynthetic, enzymes for photosynthesis bound to cell membrane (chromatophores), not packed separately. Anaerobic and aerobic photosynthesis, sulfur deposition, and $O_2$ elimination | If photosynthetic, enzymes for photosynthesis packaged in membrane-bounded chloroplasts. $O_2$-eliminating photosynthesis |
| No intracellular movements | Intracellular movements: phagocytosis, pinocytosis, streaming, and others |

*Source:* Adapted from Margulis, L., "The Origin of Plant and Animal Cells," *Amer. Sci.*, 1971, 59.

cells were the products of two endocellular symbioses: the protoameboid host plus the protomitochondrial and protoflagellar symbionts. Some eukaryotes acquired photosynthesis by the incorporation of protoplastid symbionts, thus forming the green plants. It is believed that a primitive aerobic bacterium constituted the protomitochondrian, a primitive mobile bacterium called a spirochaeta constituted the protoflagellum, while a primitive blue-green alga constituted the protoplastid. Presentation of all of the evidence utilized to support this view is beyond the scope of this text, but in general, it is based upon the prokaryotic-eukaryotic discontinuity in the fossil record (i.e., the fact that no intermediate forms are found) and upon biochemical and morphological data.

The classical hypothesis, on the other hand, postulates that many eukaryotic features evolved by fixation of mutations in a lineage leading to eukaryotes from the last common ancestor of eukaryotes and prokaryotes. According to Stanier (1970), the basic evolutionary step in the differentiation of eukaryotic from prokaryotic cells was invagination of the plasma membrane. This results in the formation of the double membranes associated with the organelles of eukaryotic cells. The initial step in such a scheme was the duplication of genetic material without cell division. The plasma membrane then invaginated near the attachment sites of this genetic material to the plasma membrane, forming double-membrane structures, with each set of genetic material attached to the inner surface of the inner membrane of these structures. Initially, these structures were probably very similar, with each having all the genetic material of the cell and the membrane-bound respiratory and photosynthetic activities. With time, differentiation of function occurred: the nuclear membranes lost photosynthetic and respiratory activities, photosynthetic activity becoming sequestered in one organelle, respiratory activity in another. The organellar genetic material lost many of its duplicated characters, while the nuclear genetic material, through additional duplication, increased in size and differentiated tremendously. What evidence is available to support this view? Suprisingly, much of the same biochemical and morphological data used to support the contrasting view.

The controversy regarding the origin of eukaryotic cells illustrates a fairly common phenomenon in science. That is, different investigators utilizing the same data will interpret these data differently. Which view is the correct one? Perhaps we will never know for certain, or perhaps the problem will be elucidated when more information becomes available. At present, the symbiotic hypothesis is certainly gaining favor, although it is probably not yet the most widely held view.

## PLANTS AND HUMAN WELFARE

While we may dispute how long plants have been on earth, we do know that man is a relative newcomer. If we were to measure geologic time on a 24-hour clock, man would have been on earth less than half a minute (about 1,040,000 years) (Figure 1.2). Yet in his short stay, man has done almost as much to change this earth as have the plants.

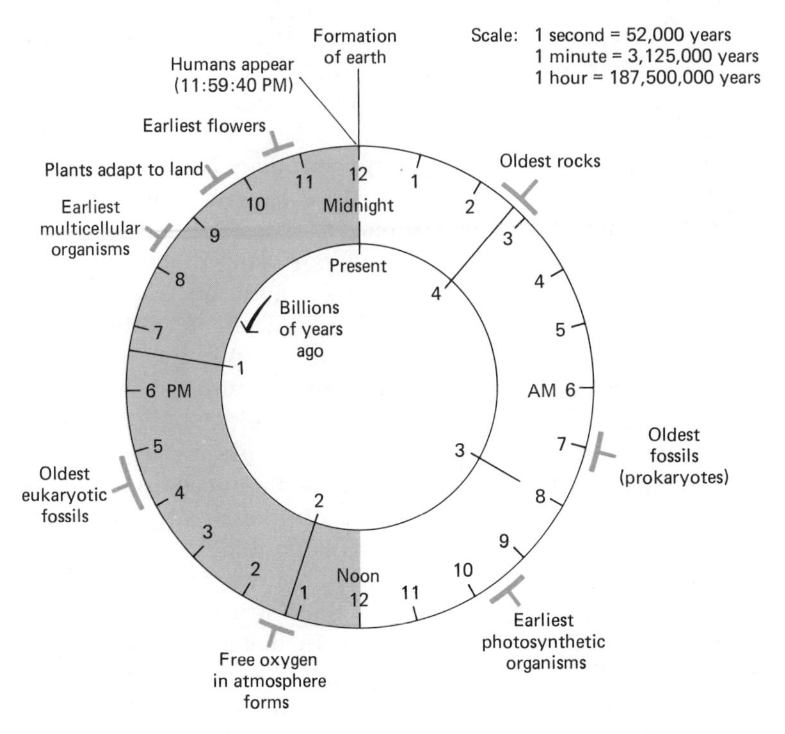


*Figure 1.2*
A clock of biological time. [Redrawn from P. H. Raven, et al., "Biology of Plants," figs. 1–7. Copyright © 1976 by Worth Publishers, Inc., New York].

We know little about the eating habits of earliest man. Many believe he was an opportunistic creature, eating either or both plants and animals as the occasion warranted. However, there is some evidence to indicate he was purely a meat eater. In any case, the primary source of food was probably the migratory herds of large mammals. Thus, earliest man was a nomadic hunter, following the herds as they wandered over the vast grasslands present during that time. Extensive glaciation created a cool climate which contributed to the development of such grassland, but as the glacier retreated and the climate grew warmer, these grasslands began to give way to forests, with many of the large mammals either becoming extinct or retreating far northward. But man adapted. With the migratory animals gone, he shifted his attention to smaller animals. These animals tended to be much less migratory. Thus, man too became less of a wanderer and began to supplement his diet with fish. With less of his energies spent in moving from place to place, he undoubtedly began more and more to test the suitability of the many and varied plants around him as possible food sources.

## Plant Domestication

Domestication of plants represented a major achievement. With it came the domestication of animals, the birth of agriculture, and in the opinion of most, the ultimate change from a more or less nomadic food-hunting and food-gathering society to a sedentary food-growing society. This event probably first occurred some 10,000 years ago in the near East, although there is evidence it arose independently in several different parts of the world.

How plant domestication arose is not known with certainty. James Michener, in his book *The Source*, relates one interesting legend. The wife of Ur (a pre-Israeli Caananite) noted that plants seemed to spring up where grains of seed had fallen, so she collected seed and planted it near her dwelling. Although her first attempts failed, she was later successful and her knowledge of seed planting and crop growing was given to all the people. This is only a fictional account, but it does serve to illustrate the genius theory for the origin of agriculture. It also illustrates a second theory: agriculture was invented by women. Man was supposedly a hunter, woman a wild-plant gatherer. It is thus reasonable to assume that the observant and reasonable woman would hit upon the idea of collecting seeds and planting them near her abode, thus saving her the effort of walking long distances and carrying back her finds.

Another widely held idea is that the origin of agriculture can be traced to the family or tribe dump heap. Primitive man, once he became more sedentary, had the same problem as modern man with regard to his waste products. Like modern man, he simply found a convenient place to dump them. Such rubbish heaps would be rich in organic matter and would also undoubtedly have contained discarded plant parts, including seeds. Under such conditions, the seeds would grow vigorously and produce a bountiful crop. This would have been noted, and being curious, members of the tribe would have attempted to discover how these plants came to be there. Eventually, they would have noted the plants growing from seeds, leading to the deliberate planting of such seeds.

Carl Sauer believes that agriculture arose among fisherman. As fish became a supplement to the diet, some peoples became more dependent upon fishing and less dependent upon hunting as their source of food. These people would have been the first to develop a sedentary living habit and, as a consequence of this living habit, would have been the first to domesticate plants. While most ethnobotanists believe that primitive peoples were at least partially nomadic until they discovered plant domestication, the theory of Sauer suggests that villages were formed first and agriculture developed as a consequence of this. Jan Jacobs proposes just such an idea. She contends that primitive people turned to communal living because of a variety of advantages it provided: more successful hunting, protection, shared labor, companionship, and others. As a result of communication, shared ideas, the necessity of providing food when hunting was bad, the very aspect of living in a more structured society, she believes that plant domestication was an obvious result.

## Plant Myths, Mysticism, Folktales

In ancient societies, and even in primitive societies existing today, certain plants were given magical or mystical properties, and religion was tied closely to natural events. Knowledge of the properties of plants was entrusted to only a select few, be they called priests, shamen, medicine men, or witch doctors. These men or women would have to be considered, in their own way, trained botanists, receiving their professional training from their predecessors. Even in our so-called modern society, particularly in the more rural or mountainous areas, certain local residents are much respected and often consulted because of their knowledge of "the herbs."

It is said that the arts, humanities, and sciences all have their origins in magic and the supernatural. In primitive societies, magic, religion, and medicine are often inseparable. Just as often, plants play a major role in all three activities. In any case, plants were often worshipped because of their general usefulness to man, their ability to cure disease, to cause death, or to create hallucinations or euphoria (a general feeling of well-being) in those partaking of them.

Ancient civilizations or societies often had one precept in common: the belief that man possessed a soul. This belief was quite natually transferred to other living things, and thus, plants too were thought to have souls or to be inhabited by godlike or manlike entities. Plants were recognized as possessing life and having the ability to transfer this life to the seed. Thus, a symbolic relationship was established between man's own fertility and that of plants. The earliest drawings known indicate that early civilizations established mother-goddess or earth-goddess fertility cults. The sexual features of the human female are generally shown greatly exaggerated (Figure 1.3).

Somewhat later, the sexual significance of the male was recognized and phallic worship established. For example, the god Min, from the dawn of Egyptian civilization, personified the generative force in nature as the bestower of procreative power. He represents the oldest god of fertility and sexual reproduction, and his plant was the lettuce (it has been suggested that the symbolism of lettuce as the plant of Min is that the wrinkled surface of the head of lettuce suggests the male scrotum). He personified the fertility of newly sown fields. Ritual public sexual intercourse between the king and queen apparently constituted an integral part of the festival of Min, and this supposedly ensured a good harvest. In other cultures, numerous couples would copulate in newly sown fields as a ritualistic or symbolic attempt to stimulate crop production. These activities should not

*Figure 1.3*
Ancient mother-goddess fertility figure. Note exaggerated breasts, abdomen, and vulva.

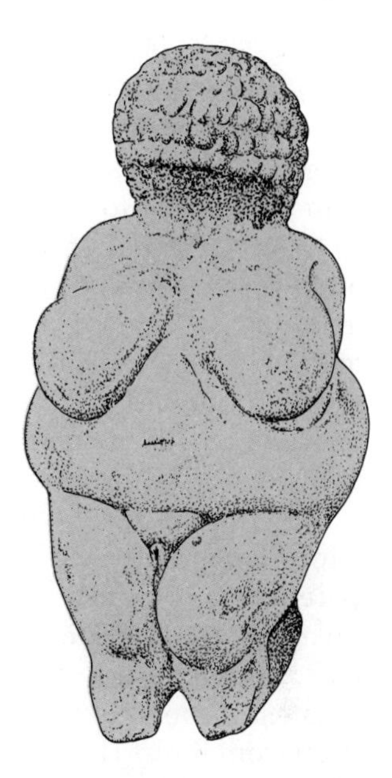

be considered as wild debauched sexual orgies; they were not. Sex was considered to be a sacred activity among those people. The ritualistic and symbolic spring fertility rites persisted through Greek and Roman times. Remnants of these activities exist today in the English celebration of May Day and the dancing around the Maypole (a phallic symbol).

The precept in many religions of some type of life after death can also be traced to ancient man's observation of plants. The seed, as mentioned earlier, was thought to contain the life force of the parent plant. When placed beneath the soil, it would bring forth the new generation. But if a seedling was dug out of the ground, the seed from which it arose appeared to have rotted away; the seed had "died." From these observations, arose the idea that man too must die before rebirth could occur. One theory concerning the origin of seed cultivation is also associated with seed planting and human death. It was a custom of certain primitive people to bury with an individual all the things he would need in the next world: tools, wealth, and (in some instances) servants and concubines. Food too was included in this burial. As burial was often fairly shallow and the food included many seeds of annual plants (seeds that would grow very well in the newly cleared soil of the burial site), a vigorous growth of food plants would soon be growing on the burial site. To the primitive mind, this indicated that the dead person's spirit was rewarding them for their good deed in providing food by returning grain from the body in an abundance far greater than that originally placed in the grave. It was thought, of course, that the dead body had some magical property by which it created the new plants. Thus, if a person did not conveniently die a natural death, it was necessary to sacrifice someone to provide for the people. Eventually, primitive societies learned that tilling the soil and planting the seed would accomplish the same purpose and a dead body was not really necessary. Thus, according to this theory, agriculture was born.

Even the names of plants can be traced to mythology, particularly Greek and Roman. Certainly, many of you have read Greek or Roman mythology and thus know many of the instances I allude to. For example, when Apollo killed Hyacinthus with his discus, a flower supposedly sprang up from the blood, and to this day, the flower is called a hyacinth. Narcissus was so entranced with his own beauty that he would sit for hours gazing at his own reflection in the water. As punishment for this behavior, the gods turned him into a flower which still bears his name. The beautiful orchid supposedly represents Orchis, the son of a nymph and a satyr. From such parents, he obviously inherited a tremendous passion, which got him into no end of trouble. Having drank a little too much wine at a festival of Bacchis, he attempted to rape a virgin priestess of Dionysis. This was the greatest crime one could attempt, and the enraged crowd literally tore him limb from limb. The father of Orchis was deeply saddened by the death of his son and prayed to the gods to put the boy back together again. But his crime had been so great, they refused. Orchis had been such a terrible nuisance in life that at least he should be beautiful in death, so they changed him into the flower that bears his name. You must beware though, because the flower was cursed to retain the temperament of Orchis. Should you eat its roots, you may be converted for a moment to a satyr state!

Many plants are associated with the Judeo-Christian religious tradition. The rod of Aaron may have been an almond branch brought out of Egypt. The burning bush of Moses was probably a thorn apple, *Acacia*, which bears a profuse growth of the red, parasitic mistletoe, *Loranthus acaciae.* The "manna from heaven" was undoubtedly of plant origin; some say the secretion from the tamarisk, others the blue-green alga *Nostoc.* The Nile turning to blood was probably due to a bloom of algae containing a red pigment. The crown of thorns, Judas tree, virgin's bower, and many other plants received their common names from their supposed associations with Christ, the Virgin Mary, and other New Testament persons and events.

These stories represent only a few of the literally thousands associating plants with religious events or with man, and these stories exist for all cultures, whether they be African, Far Eastern, Middle Eastern, European, South American, North American Indian, or other. Myths about plants and the relation of plants to the supernatural still exist today, even in the so-called civilized cultures. This can be seen quite clearly when reading a book published not, as you might imagine, several hundred years ago, but in 1973. Written by Peter Tompkins and Christopher Bird, *The Secret Life of Plants* contains many fanciful stories concerning plant behavior.

## Plants for Food or Drink

There are approximately 200,000 flowering plants known. Of the 3,000 or so are used by man, about 200 have been domesticated. Although a number of different plants may be of local significance and in fact may be indispensable for the people utilizing them, only 12 to 13 plants are of major, worldwide significance. Of these, 4 are in the grass family: wheat, rice, corn, and sugar cane. Not only are these used for food directly, but they may be made into flour, used for various industrial purposes including chemicals, cosmetics, and others, used for fuel, building materials, containers, malt beverages, livestock feed, and other purposes.

Another important group of plants are the legumes, sometimes called "poor man's meat" because of their relatively high protein content. Most important here are the common bean, soybean, and peanut. Hundreds of uses have been found for both the soybean and the peanut. In fact, over 300 uses for the peanut were discovered by just one man, George Washington Carver.

Root crops and other starchy foods include the white potato, introduced into Europe about 1570. It was at first not accepted because of its low productivity and its relation to the poisonous nightshade family. Ministers condemned it because of its "windie" property. But by the start of the sixteenth century, a rumor spread that the potato had aphrodisiac qualities, thus leading to an increase in its use. Perhaps no other plant has been so associated with calamities. As it was a cheap and readily available food to the poor, the birth rate in Europe began to increase, and in fact, some blame World War I on the potato because it allowed Europe to become overpopulated. By the early part of the nineteenth century, the potato had become the main food source in Ireland, and when the potato blight (a fungal disease) of 1845 and 1846 wiped out nearly the entire crop, famine followed. One and one-half million people died and another million emigrated, most to the United States. In the spring of 1970, five million pounds of potatoes were destroyed by fire in eastern Idaho. The fire was found to have been set deliberately in an attempt to raise prices.

Other starchy food plants include sweet potatoes, yams (these two are not the same: sweet potatoes belong to the genus *Ipomoea*, yams to the genus *Dioscorea*), manioc, taro, breadfruit, and bananas.

The palms constitute another important group of plants, with the coconut being called "man's most useful tree." The meat of the coconut (actually endosperm tissue) is dried to produce copra. Oil is extracted from the copra, and the remaining coconut cake is used primarily as a livestock feed. The fibrous husk of the coconut, the coir, is used in making rope, mats, rugs, and more. The shells are used for eating or drinking utensils, fuel, and bowls for hookahs, for making artistic novelties or grinding up as filler in plastics. The coconut milk (liquid endosperm tissue) is used as a drink or in plant tissue-culture work as a growth-promoting substance.

The leaves are used for thatching huts or for making baskets and hats, while the wood is used to some extent in construction and furniture making. Other palms include the date palm, African oil palm, rattan palm, wax palm, and sago palm.

Spice plants have been used by man for thousands of years, with cinnamon being mentioned in the Song of Solomon and the Book of Proverbs. The great explorations of the Dutch, Portuguese, and English in the sixteenth and seventeenth centuries were due to their attempts to establish a spice trade. Although according to Emboden, such activities were the result of reading the *Kama Sutra* and other similar Eastern books and then trying to find plants with aphrodisiac qualities.

The beverage plants constitute another important group to man. Plants from which nonalcoholic beverages are made include *Coffea arabica* (coffee), *Camellia sinensis* (tea), *Theobroma cacao* (cocoa), *Ilex vomitoria* (dahoon or Carolina tea), *Cola nitida* (cola), *Paullinia cupana* (guaraná), *Ilex guayusa* (guayusa), *Piper methysticum* (kava kava), *Catha edulis* (khat), *Ilex paraguariensis* (maté) *Ceanothas americanus* (New Jersey tea), *Smilax* spp. (sarsaparilla), *Mentha* spp. (mint tea), *Sassaffras albidum* (sassafras), and others.

Alcoholic beverages include malt beverages, wines, other fermented drinks, and distilled liqueurs, or cordials. Reference to beer is of great antiquity. In the Egyptian *Book of the Dead* (~ 4000 B.C.) reference is made to the intoxicant hek, made from grain. Beer has, at one time or another, been the drink of the elite or the drink of the peasants. Plants involved in the preparation of alcoholic beverages include various grains such as wheat, barley, rye, and millet, to which may be added hops (the seed pods from the plant *Humulus lupulus*), peppermint *(Mentha piperita)*, caramelized sugar (from sugar cane), honey, or other additives. Grapes and various other fruits are utilized in wine making, and both beers and wines undergo fermentation by yeast (a nongreen plant). Distilled spirits are made by the distillation of fermented mash of various plants. Brandy is a wine or fruit distillate, cognac is made from wine. Whiskies are cereal mash distillates: Scotch and Irish from barley, bourbon from not less than 51 percent corn, rye from not less than 51 percent rye, wheat from not less than 51 percent wheat. Gins include the added flavor of juniper berries. Rums are distillates of sugar cane juice. Vodka may be made from either wheat or potatoes. Akravit, a Scandinavian drink, may be made from either grains or potatoes. Tequila is made from agave.

The effect of alcoholic beverages on man has been known for a considerable period of time. Shakespeare perhaps states it best in *Macbeth:*

*Macduff* (addressing the tipsy porter): What three things does drink especially provoke?

*Porter:* Marry sir, nose-painting, sleep, and urine. Lechery, sir, it provokes and it unprovokes; it provokes the desire, but it takes away the performance: therefore, much drink may be said to be an equivocator with

> lechery; it makes him and it mars him; it sets him on, and it takes him off; it persuades him and disheartens him; makes him stand to, and not stand to: in conclusion; equivocates him in a sleep, and giving him the lie, leaves him.

## *Plants with Commercial Uses*

The commercial uses of plants or plant substances make an impressive list. One could write a book simply discussing those uses and the processes involved. We have already discussed the importance of plants in the food and beverage industry. Either directly or indirectly, wood is important in a variety of industries; lumber, construction, paper and paper products, furniture, and toys are some that come immediately to mind. Other products come from trees as well. Naval stores, such as turpentine and related products, come from certain pine trees, and these products, in turn, are used in paints and in varnishes, in producing wood tar from which gas can be made, and in other applications. In the chemical industry, wood can be used to make acetic acid, methyl alcohol, and a variety of other chemical compounds. Latex from the rubber tree forms the base of natural rubber, while other gums, such as chicle are used in chewing gum, and balata rubber is used in waterproofing. Plant fibers are important in the textile industry and in the production of cordage (rope, twine, etc.). Such fibers include cotton (from seed hairs of the cotton plant, *Gossypium*); hemp (from *Cannabis*); abaca or manila hemp (from *Musa textilis*, a relative of the commercial banana); kapok (from fruit hairs of a plant called *Ceiba pentandra;* used in life jackets and other flotation devices); ramie (from a plant called *Boehmeria*); sunn hemp (from *Crotalaria*); Mauritius hemp (from *Furcraea gigantea*); New Zealand hemp (from *Phormium tenax*); istle, sisal, and maguey (from various specise of *Agave*); and flax (from *Linum usitatissimum*). Charcoal is also a wood product, and its conversion into coke is important to the iron and steel industry. It is also used as an adsorbent, a reducing agent, a drawing material, and an ingredient in gunpowder.

A type of wood extract called tannin is important because of its ability to combine with protein forming nonputrescible compounds. When used on animal hides, it makes the hide decay resistant, strong, yet soft and pliable. Thus, it is used in the leather industry in an activity referred to as *tanning*. Tannins are also used in the manufacture of certain inks, where it reacts with iron salts forming colored compounds.

Fats, oils, and waxes extracted from plants have a wide variety of uses. They are used as fuels and as illuminants, in soaps, cooking oils, lubricants, and fertilizers, in dye manufacture, and in the rubber industry. Some have purgative properties and are thus used in medicine. Others are used in salad oils and oleomargarines, in varnishes and paints, in printers ink, in the manufacture of linoleum, in furniture waxes and polishes, to name a few products. A group of compounds called saponins have the property of reducing surface tension and thus are used as soaps. Several plants such as soapwort and soap berry are so called because they contain relatively high concentrations of saponin. These substances are also toxic to fish and are used in their capture in South America and elsewhere.

## Drug Plants

Drug plants have been classified by Emboden as hallucinogens, stimulants, inebriants, protoplasmic poisons (tobaccos and snuffs), and hypnotics. Such plants, or the active substance extracted from them, have been called *narcotic*, but this term is really a misnomer in that it is derived from the Greek word *narkotikos* which implies a state of lethargy, torpor, or sleep. Many plants have just the opposite effect. The term is now commonly used to include all those plants which, by virtue of their chemical properties, induce altered states of consciousness. In his book on drug plants, Emboden makes the following statement:

> No flora of the world is without plants that can provide a respite from reality. Given the tortures of the flesh and mind attendant on civilization, it is understandable that man turns to chemicals which are capable of altering states of consciousness. Whether these narcotics constitute a socially acceptable pastime or violate a social taboo is a question every civilization has had to answer, and the responses have been diverse. In his pharmacopoeia of 2737 B.C., the Chinese Emperor Shen Nung described marihuana as an important medicine: to the Hindus some centuries later it was "the heavenly guide," and to contemporary populations in the U.S. the same plant would seem to constitute a social threat. The physiological effect has not changed, quite obviously, but rather the judgements of societies.

Hallucinogens are plants which, in general, act upon the central nervous system to produce a dreamlike state in which there are changes in perception (of self, time, and space). Often these states are accompanied by an acute sensitivity to color and auditory sensations, and at times, one sensation is altered or displaced by another, or the senses become temporarily interchanged.

Many plants have been used as hallucinogens, some with dire results. Perhaps the most widely used is *Cannabis sativa* (marihuana, hashish, bhang, ganja, chares, pot, grass, hemp). The most prized plants for their hallucinogenic resins are those of Yarkland in Central Asia. Leaves from the female plant covered with a gummy resin which is mixed with milk or alcoholic beverages or made into candies such as *dwamsec*, have been smoked, snuffed, or eaten. The term hashish is said to come from the writings of Marco Polo. He recounts that a mountain fortress stood at 10,300 feet above sea level on the shortest route between the Caspian Sea and the Persian highlands. The master of this fortress was a Persian named Al-Hasan ibn-al-Sabbah, who regularly conducted raids to increase his domain, and during the Holy Wars, to murder the Christians of the Crusades. His followers, a ruthless band of cutthroats, were known as *ashishins.* In Marco Polo's words:

> He [Hasan] kept at his court a number of youths of the country, from twelve to twenty years of age, such as had a taste for soldiering. . . . Then he would introduce them to his garden, some four, or six, or ten at a time, having first made them drink a certain potion [hashish] which cast them in a deep sleep, and then causing them to be lifted and carried in. So when they awoke they found themselves in a garden. . . . When therefore they awoke, and found themselves in a place so charming they deemed that it was paradise in very truth. And the ladies and damsels dallied with them to their hearts' content. . . . So when the Old Man would have any prince slain, he would say to such a youth: Go thou and slay so and so; and when thou returneth my angels shall bear thee into paradise. And shouldst thou die, natheless even so I will send my angels to carry thee back into paradise.

It is said that from this band, we derive both the words *assassin* and *hashish.*

Another hallucinogen of great antiquity is soma, the stimulant, euphoriant, and hallucinogen in Aldous Huxley's *Brave New World.* It has been used as a narcotic since the time of India's earliest civilizations. It was well known from India to Persia, but no morphological descriptions can be found. Thus, the source of soma is lost in antiquity. This would appear to be a great loss, as soma was exhaulted as the giver of health, courage, long life, a sense of immortality, and almost every other virtue known to man. It was the liquor of the gods, and it was indicated that it was a stimulant, psychoactive agent, and even an aphrodisiac.

Some other plants having hallucinogenic agents include jimsonweed *(Datura),* used in Asia centuries ago to lure virgins into prostitution. Subsequently, they would employ extracts of the herb to rob their clients. As it was used as a stupefying agent, this represents one of the earlier "knockout drops." The opium poppy *(Papaver somniferum)* also has an ancient history. When one considers that the drug is addictive, that it generally leads to complete depersonalization and an early death, and that withdrawal is pure hell, its use in various societies is a mystery, particularly when other drugs may be more easily available, not addictive, less dangerous in side effects, and equally hallucinogenic. Sweet flag *(Acorus calamus)* rhizomes contain an oil which, when used in large amounts, is said to induce strong visual hallucinations. It was used for this purpose as well as for food, by North American Indians. In Central and South America, numerous drug plants exist, many of which are still being used by the natives. These include *sinicuichi* from *Heimia salicifolia;* peyote from several cacti, a composite, *Cacalia cordifolia,* and a crassulacean, *Dudleya caespitosa; mezcal* from *Agave; sophora* from *Sophora secundiflora; teonanacatl* from several mushrooms including *Psilocybe caerulescans* var. *mazatecorum, P. campanulatas* var. *sphinctrinus,* and *P. cubensis;* seeds of various varieties of morning glory *(Ipomoea violacae)* which contain amides of lysergic acid and thus give effects similar to those of LSD; and *Ayahuasca* from an admixture of *Banisteriopsis inebrians, B. rusbyana, B. quitensis,* and sometimes *Brunfelsia* spp. and *Psychotria viridis,* the latter, one of the few drug plants where considerable evidence seems to exist for a psychoerotic effect.

In Europe, deadly nightshade, mandrake root, and henbane were used in bacchanalian orgies to spike the wine. A mixture of these was also used in medieval witches' brew and introduced into the body via the mucous membranes of the vagina. This practice may account for the present-day association of a witch riding on her *anointed* broomstick.

Stimulants include many of the nonalcoholic beverages mentioned earlier. Tea, coffee, cola, dahoon, guaraná, maté, guayusa, and certain others contain caffeine, a mild stimulant to the cerebral cortex. Tea also contains theophylline, which causes dilation of the artery which supplies the heart. Cocoa contains a mild stimulant theobromine, while the plant *Erythroxylon coca* yields cocaine, which has an almost immediate action on higher levels of the brain, giving the user a sense of boundless energy and freedom from fatigue. Cocaine was, until the U.S. courts in 1904 ruled against its use, a basic ingredient, along with cola, of Coca-Cola. It would seem that Coca-Cola really did deserve its nickname Coke.

Inebriants are essentially those plants from which alcoholic beverages are made. It is a common misconception that alcohol is a stimulant. It is not! It acts as a primary and continuous depressant on the central nervous system. Although writers, comics, and others have drawn much humor from the aspects and antics of drunks, there is nothing humorous about the effects of excessive and long-term consumption of alcohol. It is an obvious and serious social problem, but in addition, its effect on the human body is to interfere with normal fat metabolism, especially in the liver, where the accumulation of fat causes it to become hard and swollen, a condition known as *cirrhosis.*

Tobaccos and snuffs, which owe their effect to nicotine, are classified as protoplasmic poisons. There is no question that nicotine is a very poisonous compound, and the proof relating smoking to lung cancer is vast and convincing, yet in still another quirk of human behavior, millions of people smoke, and that number is increasing daily.

The hypnotics are a very ill-defined class, as there are few, if any, plants whose only drug effect is to produce a sleep or sleeplike state. Such sleeplike conditions are generally referred to as somnolent, or soporific, states. Perhaps the best known drug plant that does produce such a state is the opium poppy, although many of the hallucinogenic plants will also have hypnotic effects.

It must be mentioned that nearly all drug plants have unpleasant, debilitating, and sometimes unpredictable side effects, and their physiological effects can be quite different in different people. Use of certain drugs, not even necessarily prolonged or excessive use, could lead to death. Thus, with *perhaps* the exception of marihuana and a few others, experimentation with drugs involves a great risk to future health and well-being.

## Plants and Medicine

The first medical uses of plants are lost in antiquity, but evidence exists that plants were widely used for such purposes long before written history. The first extensive compilation is in the Chinese *Materia Medica* of 2737 B.C., mentioned earlier. Plants were used by the ancients (and are still being used by some people today) for everything from growing hair to curing backaches and broken bones. Many of these uses were based upon quite fanciful or imagined curative properties, but some did have a real basis in medical fact, and plant drugs have an important place in modern medicine. Although the raw plant material is often not usable in modern medicine because of its adulterants, impurities, and other unknown or dangerous aspects, it has served as the basis for the extraction and purification of natural plant drugs or for the manufacture of synthetics having similar medicinal properties and fewer dangerous side effects.

Many plant products are used in medicines simply as sweeteners, coloring agents, suspending agents (emulsifiers), binding agents in pills, or agents adding a pleasant taste or smell to medication. Others, however, have a very specific medicinal or curative value. Some of the better known examples include digitalis, from the leaves of *Digitalis purpurea* (foxglove), which has a highly specific action on heart muscles; hesperidin, from the rind of unripe, green citrus fruits, indicated in the prevention or correction of capillary permeability in cerebrovascular or cardiovascular diseases, hypertension, treatment of hemorrhaging, nephritis, and habitual or threatened spontaneous abortion; atropine, from deadly nightshade and other plants, which is a parasympatholyte (blocks the action of the parasympathetic nerves); cocaine from *Erythroxylan coca,* which is used as a local anesthetic and cerebral stimulant; quinine from the bark of *Cinchoma* spp., used in treating malaria; mescaline from the cactus *Lophophora*, used as a tranquilizer and as a treatment of certain mental disorders; opium and its derivatives, used for its hypnotic, sedative, pain-relieving, and other qualities; reserpine, from *Rauwolfia serpentina*, used as a sedative and tranquilizer; antibiotics from various fungi, bacteria, and actinomycetes, used for treating various bacterial, fungal, or protozoan infections; and many, many others.

## General Significance of Plants

Many people are not totally aware of man's complete dependence upon plants. In addition to all the aspects mentioned previously, we must consider the importance or significance of plants in controlling future world population, as producers of oxygen, purifiers of waste water, providers of food and shelter for animals, and by their role in the water cycle, erosion control, and other natural phenomena. Certain of these aspects will be discussed in later chapters. Quite obviously, without plants, life as we know it, on earth could not exist.

## Nature of Plant Science

From its earliest origins in the supernatural to its virtual inseparability from medicine, from about 4000 B.C. to 1700 A.D., plant science evolved slowly into what we know it to be today. Important advances in knowledge were made when man changed his question from *what* to *why* and *how*. As man became more disinclined to attribute plant phenomena to the supernatural, he had to supply other answers to his questions. Thus evolved a methodology known quite generally as the *scientific method*.

All scientific knowledge comes from man's innate curiosity to know the unknown, and perhaps what separates the scientist from the nonscientist is that the former has more of a curiosity and asks more questions. Not only does he ask more questions, but he is more inclined to follow up those questions by actively searching for answers. In general, however, it is *observation* of natural events that leads to *questions* as to how or why the events occurred. A *hypothesis* is posed, which is really just an educated guess, to answer the question. A hypothesis is generally based on simple observation and intuition but may be based on additional information gathered from preliminary investigations. For this reason, some botanists would object to equating guess with hypothesis. In any case, a hypothesis is an *unproven* explanation and as such is not scientifically acceptable; it must be *tested*. It is this requirement for testability that places certain considerations outside the realm of science. Throughout man's existence on earth, he was observed events which he could not explain by the knowledge he possessed. Thus, the question has been posed by many different peoples in many different ways, "Is there a Supreme Being, a supernatural force which controls all events on earth?" As an individual, I can answer, "Yes, I believe there is a Supreme Being." But my statement is simply that, a *belief*. I cannot support it from observation, and thus, it does not even have the strength of a hypothesis. Nor can I ever give my belief scientific validity, as it is untestable. In fact, in my role as a scientist, I could not even ask such a question. But, for events or natural phenomena which are testable, a body of information, or *data*, can be assembled from which *inferences*, or *theories*, can be formulated. These inferences may or may not support the original hypothesis. For the inference to receive acceptance by the scientific community, the results must be *repeatable*, and they must be capable of being *duplicated* by other investigators. It is primarily in this way that a body of knowledge we call botany has been assembled.

Note that throughout this discussion, I have used the term or phrase *data, body of information, body of knowledge;* I have not used the term *fact.* The reason for this is that some would equate fact with something irrevocable and forever true. Actually, the term *fact* often refers to a transitory, or temporary, truth. For example, it is a fact that my hair is red. This does not mean that it will always be red or even that I will always have hair, and the way things are going, probably neither of these statements will be true. What it does mean is that this is the way it is now. So it is with botanical knowledge; what we know or think we know about plants is based upon information we believe to be true at present, and thus, our theories are really *best inferences* based on information currently available. The chemist G. N. Lewis puts it very succinctly:

> The Scientist is a practical man and his are practical aims. He does not seek the ultimate but the proximate. He does not speak of the last analysis but rather of the next approximation. His are not those beautiful structures so delicately designed that a single flaw may cause the collapse of the whole. The scientist builds slowly and with a gross but solid kind of masonry. If dissatisfied with any of his work, even if it be near the very foundations, he can replace that whole part without damage to the remainder. On the whole, he is satisfied with his work, for while science may never be wholly right it is certainly never wholly wrong; and it seems to be improving from decade to decade.

I have emphasized the importance of the scientific approach and carefully designed experiment in gathering botanical knowledge, but one should not overlook the importance of trial and error, the accidental acquisition of knowledge, or the genius insight. Important botanical knowledge has been obtained by all these means. In addition, the importance of instinct or insight on the part of the investigator, either in asking the right questions or in realizing the implications and importance of the information he has gathered must not be overlooked.

As you proceed through this book, you will encounter structures and processes that you would like defined. You will not find a list of definitions in this book. Should you have the irresistable urge for definitions, you can find them in the more encyclopedic general botany texts or in dictionaries of botanical names and terms. But definitions are static and thus bounded by the limitations we apply to them. For example, you could be asked to *define* a tree. All of you know what a tree is, and even without a formal course in botany, you know quite a lot about a tree; but can you give a definition of one? Not really. There are whole books written to describe characteristics of trees. But does this mean you know nothing about trees simply because you can not come up with a definition of one? Not at all. You have certain ideas about trees, and these ideas and inferences are the basis for forming *concepts.* Unlike definitions, concepts are dynamic. As you receive additional knowledge, you can enlarge, delete from, or otherwise modify your original concepts. As you encounter terms and processes in this book, attempt, on the basis of information given, to develop a concept of the term or process. As additional information becomes available to you, either from your instructor, from outside reading, or from a later discussion in this book, enlarge your concept accordingly. In this way, you will gain a conceptual knowledge of plants and hopefully a better *understanding*.

## SUMMARY

1. Life was formed apparently in the oceans under conditions of a reducing atmosphere and with energy from lightning, ultraviolet radiation from the sun, and radioactivity from the earth's crust.
2. Although we cannot state presently when life actually came into being, the first, very primitive protoorganisms probably lacked the ability to reproduce themselves.
3. With the evolution of a genetic apparatus, life evolved rapidly, and the first heterotrophic organisms were formed, utilizing the simple organic molecules in their environment both for food and as an energy source.
4. With the evolution of porphyrin-containing molecules, chlorophyll eventually was formed, and the organisms became autotrophic.
5. With the development of a photosystem in which oxygen was produced as a by-product, the atmosphere increased in oxygen.
6. Increased atmospheric oxygen allowed for development of the eukaryotic cell, the origin of multicellular plants, and the movement of plants onto the land.
7. The eukaryotic cells may have resulted through evolution of the existing prokaryotic cells or by serial symbiosis.
8. Man, throughout the time of his evolution, has been dependent entirely upon plants, both directly and indirectly.
9. The cultivation of plants and hence the advent of agriculture occurred possibly 10,000 years ago.
10. Humans depend upon plants for food and drink, certain drugs and medicines, shelter, fuel, light, clothing, and in industry, recreation, and in some way or another almost every other human endeavor.
11. Early knowledge and understanding of plants was based, in large part, on belief in mysticism and the supernatural.
12. Much of the current knowledge of plants comes from a common sense approach of observation, question, hypothesis, experimentation, and formulation of inferences based upon the information thus obtained.
13. Such information and inferences provide for a conceptual and causal knowledge concerning plant activities.

# Chapter 2

## INTERPRETATION OF BOTANICAL PHENOMENA

Primitive man could observe *what* was happening, but he had little concept of *how* or *why* it was happening. He tended to categorize all natural phenomena in relation to their effects upon him. These effects were considered to be either benevolent or malevolent. Thus, the sun, moon, wind, rain, waves, volcanoes, fire, plants, and animals seemed to have some mystical or supernatural power. Lacking the knowledge or technology to test his hypotheses, he attempted to explain *why* a natural phenomenon occurred without knowing *how* it occurred. Thus, all objects were given either human or godlike properties. The same emotions, needs, abilities, and motives that he perceived within himself or attributed to his gods, he attributed to these natural phenomena. This approach to account for natural events is termed *personification* or *deification.*

The use of personification, or deification, in explaining physical or chemical phenomena has passed or at least has taken more subtle forms. You would find little support for the idea that rain represents the tears of a god or falls in order to wet the land or that storms represent God venting his wrath on sinful man. Nor would you gain much support for the assertion that sodium combines with chlorine so that man will have something with which to salt his tomatoes.

But some authors would put us back to the dawn of man with their discussion of the mystical or supernatural powers of plants. Current pseudoscientific writings, such as Thompson and Bird, *The Secret Life of Plants* or Bolton, *The Secret Power of Plants,* are examples. Ripley's "Believe It or Not" often contains interesting examples of plant behavior, but its explanations are sometimes quite farfetched. My training, observations, and experiences tell me that more logical and reasonable explanations exist.

Some biologists even fall into the trap of personification occasionally. Consider the following statement: "While one thinks of a plant as a photosynthetic device, it is important to remember that it respires continuously, as an organism must, *in order to obtain energy for its activities*" (emphasis mine). It is unimportant at this point to be concerned about what photosynthesis or respiration is, or even *what* the statement is saying at all. Rather, it is *how* it is said that we are concerned with. I do not know what this statement conjures up in your mind, but to me, it implies that the plant somehow realizes its need for energy and thus respires to meet that need. I am certain the author did not mean to imply that. A plant respires because a particular sequence of physiological events occurs, and as a good biologist, the author was well aware of that. He was simply a little sloppy with his language.

This points up another rule for scientists: attempt to report investigations in a clear, precise, and concise manner. In other words, say what you mean, so that your information cannot give a false impression or be misinterpreted. Statements such as the one above (or the idea that a plant forms thorns in order to protect itself or that a leaf has chlorophyll so that it can photosynthesize), based upon the need or desire of a plant to reach a specific goal, are known as *teleological* statements and are scientifically unacceptable. The attribution of human characteristics to plants as an attempt to explain their behavior is equally unacceptable. Such explanations are known as *anthropomorphisms* or *personifications.*

## AN ALTERNATIVE VIEW

The attribution of supernatural or mystical properties to plants forms the basis for a philosophical viewpoint of life known as *vitalism.* This viewpoint holds that an organism is more than just the interaction of chemical molecules, that it is endowed with some supernatural or *vital force* which makes it a living being. The alternative view is that all aspects of life can be completely explained in purely physical or chemical terms. Such a viewpoint is termed *mechanism.* Most biologists subscribe to a modification of this mechanistic philosophy, that is, that life is composed of physical and chemical events, yet it is something more than just the sum total of those events.

How does a scientist approach the interpretation of natural phenomena? Take an example familiar to those who may have kept potatoes in a dark, moist basement for a period of time and then noted that they have sprouted, forming long, flaccid, yellowish, essentially leafless stems. Let us consider an experimental situation based upon this situation. Figure 2.1 shows the results in bean plants of several months growth. Plant A was grown in a greenhouse under normal light, plant B was grown in a green-

*Figure 2.1*
Differences in stems and leaves of bean seedlings that grew in different light intensities. *A.* 100% light; *B.* 50% light; C. No light. [Photos by F. H. Norris. From E. N. Transeau, et al., "Textbook of Botany," Figure 28. Copyright © 1953 by Harper & Row, New York].

house under shade or reduced light, and plant C was grown in a dark room. We can make a number of observations on each plant, but each one shows that plant C is the most different: the stem is flaccid, smaller in diameter, longer, nongreen, and bearing tiny, nongreen leaves. Our observations lead to a number of questions. Why are the stems smaller? Why are they more flaccid? Why is the plant nongreen? Why are the leaves reduced in size? Why is the stem more elongated? Each question may have a different answer. Let us deal only with the last question, the greater growth in length of the stem.

From our observations of the experimental results, let us attempt to draw an inference which would adequately explain stem elongation *in each plant.* At this point, some may be thinking, "Wait a minute, how can we draw any conclusions when all we know about this experiment is that one plant was in complete darkness and the others in light. Did plant C get more fertilizer, more water, better soil, or other favorable treatment?" These are excellent points. The plants were obtained from genetically similar seed, were planted in the same type soil, were equally fertilized, were watered just enough to keep the soil slightly moist in each pot; in fact, there were no appreciable differences in any factor other than light. Now how can we account for the differences in stem length?

Some might say, "As a green plant requires light to grow, the plant in the dark obviously grew longer because it was *searching* for light." Others might say, "It wasn't searching for light; it was simply trying to grow *towards* the light, just as a plant in a window does." But let us consider those two statements. Can a plant search; does it realize its *need* for light? If it is in a dark room, how does it know where the light is? These two statements are teleological. Plants simply do not have these powers of thought, so we can dismiss these assumptions.

Now some might reason that "The leaves of plant C are small; therefore, the energy went into making a longer stem rather than leaves." What is wrong with that conclusion? First, it is not based on the evidence available. We have no information concerning energy use. Second, the data actually disputes this idea. We may note that the leaves of plant B are actually larger than those of plant A, even though the stem of plant B is longer. At this point we may again recall the fact that the only thing really different was the amount of light or lack of it. Therefore, we might say, "Apparently darkness accelerates stem elongation," or the converse, "Apparently light inhibits stem elongation." Considering the first of these two statements, it would be difficult to say that plant B was subjected to greater darkness than plant A, as darkness is simply the *absence* of light. Plant B was subject to *less light* than plant A. Considering all of our observations, it would seem then that the *best inference* is that bright light inhibits the elongation of bean stems.

Having arrived at this inference only forms more questions in our minds: 1) If this is true, why is it that some stems grow more during the day than at night? 2) Why does light inhibit stem elongation? 3) Would we have gotten a different result if we had used another plant rather than a bean? 4) Would we have gotten the same results if we had grown the bean plants under a different temperature regimen? Obviously, each of these questions would require devising specific experiments to help arrive at the answers. Later in the text we may be able to answer some of these questions.

Having arrived at the inference that light inhibits stem elongation, what can you infer about the fact that a plant stem bends toward a more intense light source? You should now be able to give a reasonable inference for this phenomenon based on the principle just developed. Try it! Later in the text, you will be able to give a more detailed explanation.

## HEREDITY VERSUS ENVIRONMENT

The characteristics of the growth as affected by light illustrate the results of significant interaction between the *heredity*, or *genetics*, of the plant and the internal and external *environments* it is subjected to. This interaction produces all the plant characteristics we will be discussing throughout the rest of this book. Keep this foremost in mind when considering plant structures or processes.

How do these two important features interact? Let us look at two spring-wound toy cars as an analogy. If we assume these cars are *exactly* the same (the same build, the same top speed, etc.), given that they were started at the same time and ran over the same terrain, we would expect them to reach a specified point at exactly the same time. But if we run one of the two cars on an asphalt strip and the other through a shallow trough of water, we would expect the car on the asphalt to reach a specified point much sooner. Thus, both cars have a certain *potential.* This might be thought of as their heredity. The trough of water represents the effect of environment on that potential. A plant likewise has certain inherent potentials. It has the potential to produce cells with chlorophyll; it has the potential to develop into particular kinds of cells. Whether these potentials will be realized, or to what degree they will be realized, depends on both internal and external environmental effects.

## SUMMARY

1. In primitive times, all natural events were explained by some mystical or supernatural power.
2. Some attempts to give plants human attributes or godlike powers were made. Such anthropomorphism is not scientifically acceptable.

3. Some people attributed to plants the ability to perceive and thus to alter their behavior in order to reach certain needs or goals. Such teleology is not scientifically sound.
4. Most scientists employ a modified mechanistic approach to life and thus continue to experiment in a search for answers.
5. From facts come inferences and from facts and inferences are built concepts.
6. Heredity and environment are inseparable when plant characteristics are considered. They interact to produce all plant phenomena.

# Chapter 3

## PLANT CELLS AND TISSUES

## SOME REFLECTIONS

Before beginning our discussion of the specific aspects of cells, tissue and organ structures, developmental processes, and physiological activities, it might be well to reflect for a moment. In order to consider intelligently these various aspects of plants, it is obvious that certain terminology be utilized. In this and following chapters, you will encounter a number of terms that you might be quite unfamiliar with. Your first reaction might be, "Do I have to memorize all this?" The answer is a somewhat qualified no. But you should attempt to become familiar with the context in which each term is used and the general structures or processes to which each is related, and attempt to develop a concept of each term. The study of botany is much like the study of a foreign language. If you expect to be able to converse intelligently, you must know some vocabulary.

This reminds me of a story Mark Twain told in *A Tramp Abroad.* He was describing how they hitch up horses in Europe:

> The man stands up to the horses on each side of the thing that projects from the front end of the wagon, throws the gear on top of the horses, and passes the thing that goes forward through a ring, and hauls it aft, and passes the other thing through the other ring and hauls it aft on the other side of the horse, opposite to the first one, after crossing them and bringing the loose end back, and then buckles the other thing underneath the horse, and takes another thing and wraps it around the thing I spoke of before, and puts another thing over each horse's head, and puts the iron thing in his mouth, and brings the end of these things aft over his back, after buckling another one around under his neck, and hitching another thing on a thing that goes over his shoulder, and then takes the slack of the thing which I mentioned a while ago and fetches it aft and makes it fast to the thing that pulls the wagon, and hands the other things up to the driver.

Without the use of any technical terms, it is impossible from this description to know what is going on. Of course, if Twain had used the terms, tongue, trace, collar, singletree, doubletree, bridle, and bit, it would not have been at all humorous, and you probably still would not know how to harness a horse. In this day of the horseless carriage, I doubt if many of you are familiar with the parts of a harness. However, given a demonstration, diagram, or other factual information, you can then associate the terms with the process. I hope this will be accomplished by this text.

## HISTORICAL ASPECTS OF CELLS

All living matter is composed of cells, and these cells have been formed from preexisting cells. This sounds quite reasonable but it was well into the middle of the nineteenth century before biologists generally accepted these as valid theories.

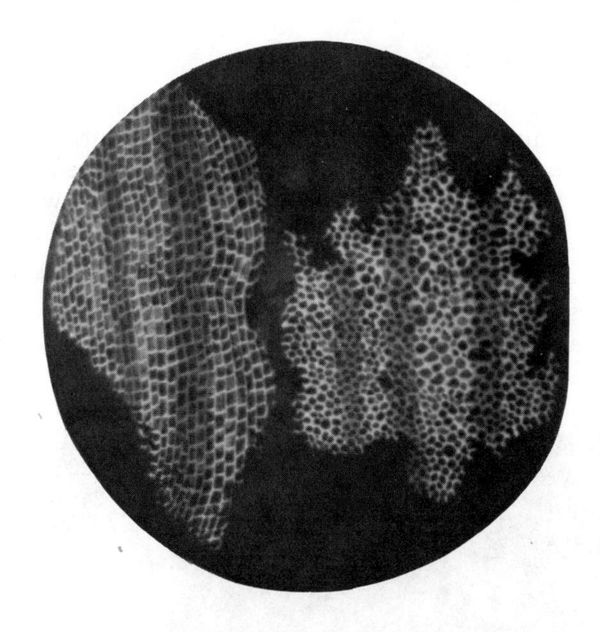

*Figure 3.1*
Robert Hooke's drawing of slices from a piece of cork, reproduced from his *Micrographia* (1665).

The nature of cells was certainly not known by the originator of the term, Robert Hooke; in fact, the name itself is a misnomer. Hooke was looking at a thin layer of cork from an oak tree (Figure 3.1). Cork is composed of the thickened walls of dead cells, so he was actually observing cell walls. The pattern of the walls, however, reminded him of the little rooms in which monks in monasteries lived, so he named the cells, from the Latin word *cella*, "little rooms or chambers." He had no idea of the significance of his discovery.

Although a number of people observed both plant and animal cells following Hooke's origination of the term, they were apparently impressed only by the great diversity of shapes, sizes, and forms. They saw no general significance in their observations. It was not until almost 200 years later (1838), after extensive observation of plant and animal cells, respectively, by two German biologists, Matthius J. Schleiden and Theodor Schwann, that a generalization was made: "We have seen that all organisms are composed of essentially like parts, namely, of cells." Another German, Rudolf Virchow, in 1855, extended the cell theory still further with his observation: "Where a cell exists there must have been a preexisting cell, just as an animal arises only from an animal, and a plant only from a plant." So from simple observation of fact came inductive generalizations of major importance.

## STUDY OF CELLS

The observation of cells was made possible by use of a magnifying lens. Simple lenses were known before the birth of Christ, and spectacles had been worn by the wealthy several hundred years before Hooke's observations. Thus, it is entirely likely that cells were observed long before the seventeenth century. But there is little doubt that the incorporation of the lens into an instrument called the microscope, by some accounts first invented by Jansen in 1590 (Galileo Galilei published a description of a microscope in 1610), was of major importance. Figure 3.2 illustrates one of

*Figure 3.2*
The Culpeper microscope (~1740).
[Courtesy of Carl Zeiss, Inc., New York.]

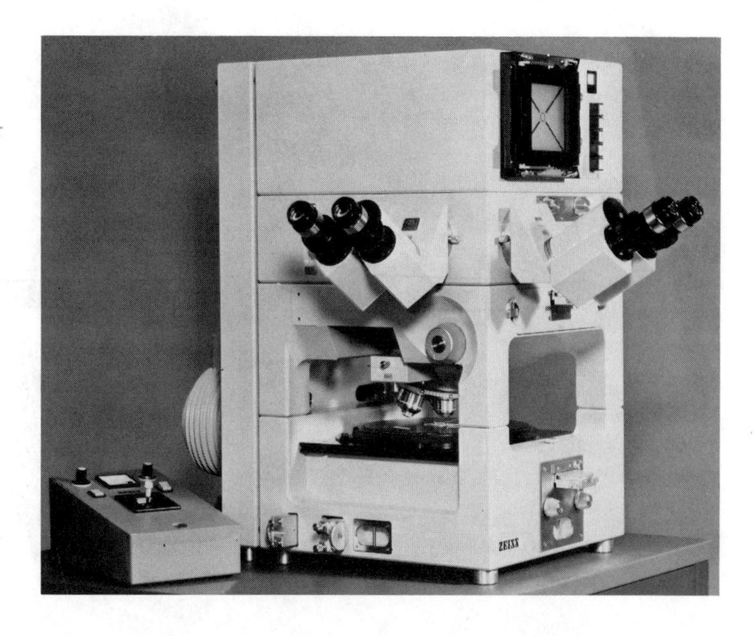

*Figure 3.3*
A modern light microscope. The Carl Zeiss Axiomat for transmitted light, and with 4 × 5 and 35 mm camera built in for photomicrography. [Courtesy Carl Zeiss, Inc., New York.]

these early microscopes. A technological achievement plus the new scientific curiousity of the Renaissance thus led to major advances in science.

Man's observation of cells was of course limited by the ability of the eye to perceive two closely placed points as separate points. This ability is known as *resolution.* A person of excellent eyesight can barely perceive points 0.1 to 0.2 millimeters apart. Such tiny distances are generally expressed in micrometers ($\mu$m), formerly microns, which is 1/1000 or $10^{-3}$ mm. Thus, the limit of human perception is approximately 100 to 200 $\mu$m. The microscope is simply a tool to expand or extend visual observations. The simple lens magnified the image so that structures approaching 50 $\mu$m in size could be seen. By use of two lenses, early microscopists could see objects as small as 1 $\mu$m. Modern light microscopes, using as many as 12 to 15 different lenses, are capable of magnifying objects, while maintaining resolution, approximately 1500 times (Figure 3.3). Above that magnification, the resolving power of the microscope rapidly falls off. The maximum theoretical resolving power is calculated as one-half the wavelength of the light being used; thus, with violet light having a wavelength of 400nm (nanometer or millimicron, m$\mu$ = 1/1000 $\mu$m), structures down to 0.2 $\mu$m can be resolved. (Wavelength was formerly measured in angstroms, A. There are 10 A in 1 nm.) Many modifications of the *brightfield* light microscope are now available, including *darkfield, polarized, phase, flourescent,* and *differential-interference contrast,* each with its own specific advantages and disadvantages.

De Broglie, a physicist, theorized, in 1924, that electrons emit energy in the form of waves. The length of these waves was calculated to be about

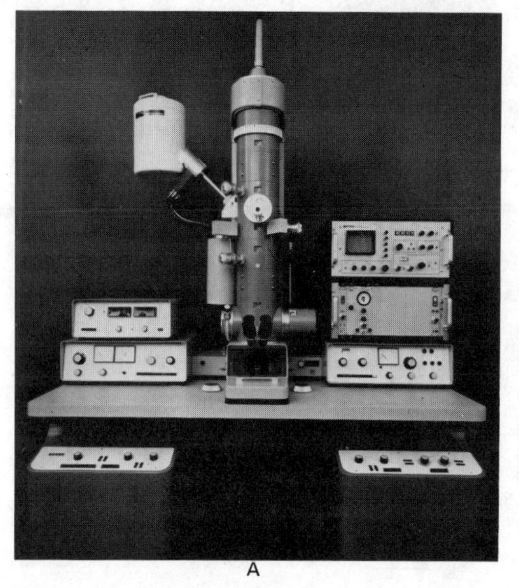


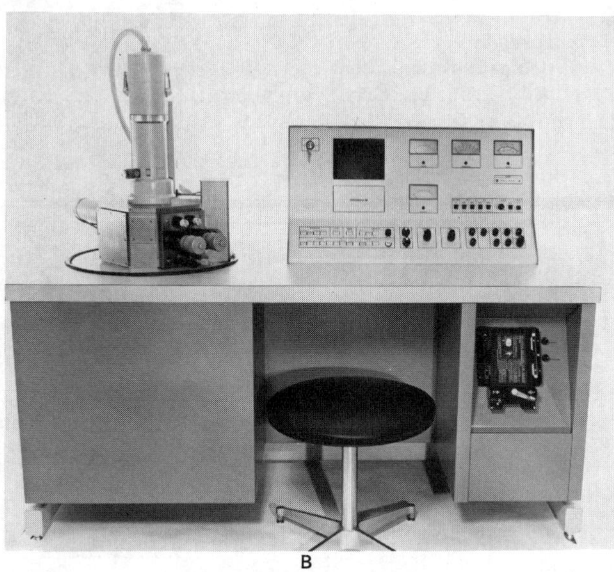


*Figure 3.4*
Electron microscopes. *A*. A high-resolution transmission microscope, 100X to 500,000X magnification. *B*. The Novascan 30 Scanning electron microscope. Magnification from 5X to 150,000X and 100 Å resolution. The former *(A)* provides for study of internal cellular details; the latter *(B)* for study of surface details. [Courtesy Carl Zeiss, Inc., New York.]

0.5 nm. He predicted that these electron waves, if substituted for light waves, could increase resolutions some 100 times. Thus, by use of an *electron* microscope, one can obtain magnifications of 200,000 to 300,000, while at the same time increasing resolution. Some electron microscopes pass the electron beam through the specimen, and thus, optical sections can be observed (Figure 3.4A). These are called *transmission* EM. In others, the electron beams are reflected from the surface of the specimen, and thus, only surface characteristics can be noted (Figure 3.4B). These are called *scanning* EM. In practice, somewhat lower magnifications than those theoretically possible are used because of better resolving power. Increased magnification is obtained by an enlarged photographic print.

Microscopes are not the only tools used by biologists to study cells. As many cellular structures have an index of refraction close to that of the water they are mounted in (i.e., bend the light rays to the same degree water does), many cellular details are difficult to see. The specimen can be treated by *dyes* or stains which differentially stain cellular compounds either to a different degree or in a different color, thus enhancing their visibility. In addition, the specimen can be mounted in material having a higher refractive index than water, or microscopic techniques may be utilized which increase contrast. Often, however, increased contrast is gained with the loss of resolution.

## PARTS OF CELLS

It is common practice to discuss cell structure in terms of a generalized cell. Such a practice is convenient but, like other generalizations, is not entirely valid. We must keep in mind that plant cells have many shapes and sizes and that not all plant cells contain all structures. Even the dimensions of the structures will vary from cell to cell. Our generalized cell then might be thought of as a composite of all plant cells and not necessarily representing any one of them (Figure 3.5).

### The Protoplast

A Bohemian microscopist, E. J. Purkinje, in 1838, gave the name *protoplasm* to the contents of cells. The derivation of the root *"plasm"* comes from the Greek *plasma* or *plasmatos*, which means anything formed or molded, while *plast* comes from *plastos*, which also means molded or formed. Thus, the terms *protoplast* and *protoplasm* would seem to mean the same thing. To most botanists, however, *protoplast* denotes the *living unit structure* of the cell, implying that it includes everything within the cell, both protoplasmic and nonprotoplasmic. *Protoplasm*, on the other hand, has come to refer to the *living substance* of the cell, thus excluding the nonliving inclusions.

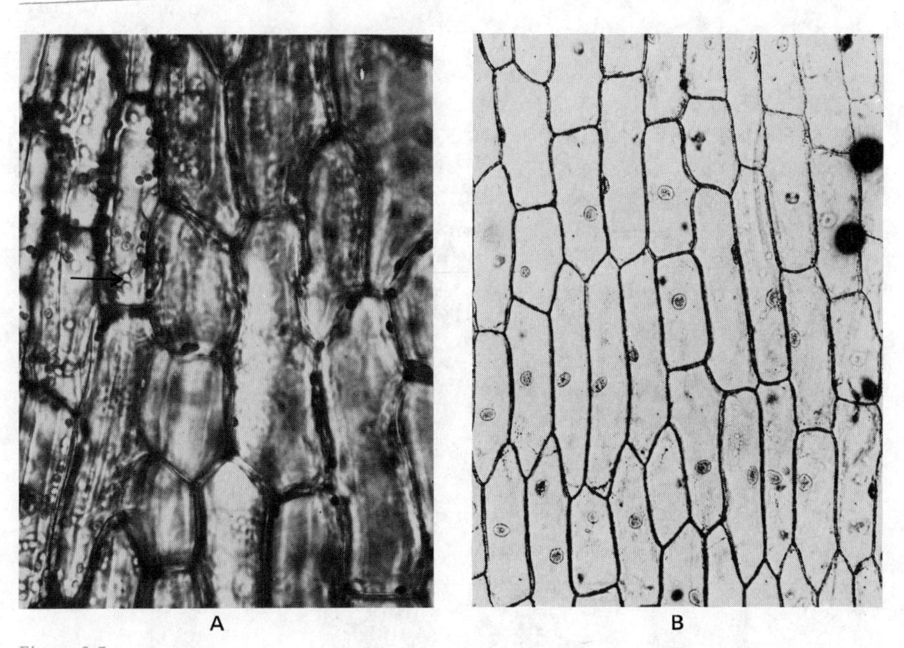


*Figure 3.5*
A. *Elodea* cell. The numerous disk-shaped structures are chloroplast. Note how they line the sides of the cell; the central portion of the cell is the vacuole. The location of the nucleus can be noted by the clustering of chloroplasts around it *(arrow)*. *B.* Onion epidermal cell. Note prominent nuclei.

The protoplast is bounded externally by the *plasma membrane* or *plasmalemma* (Figure 3.6 and Figure 3.7). The protoplast consists of various substances, membrane-bounded organelles, and ribosomes embedded in *cytoplasm* which surrounds a central *vacuole*. The membrane separating cytoplasm and vacuole is known as the *tonoplast*. Both the tonoplast and the plasmalemma are single membranes and appear to be similar in structure. The cellular organelles are generally bounded by two membranes.

The study of biological membranes began with the observation of Overton (1895) that on the basis of solubility the cell seemed to be covered essentially with lipid. Of course, due to the absence of cell wall, much of the work on membranes has been with animal cells. Langmuir had already demonstrated that fatty acid molecules at a water interface stand at right angles to the surface and thus form a monolayer. It was thought that this would likely be true for the plasmalemma as well. But in 1925, Gorter and Grendel, from their studies on red blood cells, showed that the extracted lipids occupied an area twice that for a single monolayer and postulated

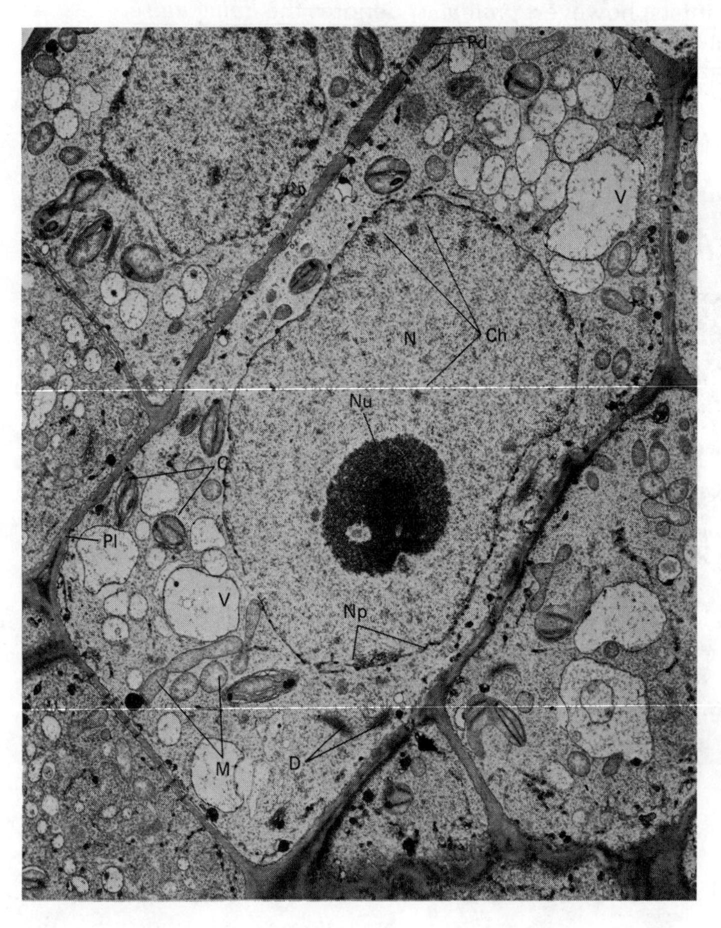


*Figure 3.6*
Young parenchyma cell of *Elodea* (X 9,600). Note the plasmolemma *(Pl)*, plasmodesmata extending through the membrane and wall *(Pd)*, chloroplasts *(C)*, mitochondria *(M)*, dictyosomes *(D)*, vacuole *(V)*, and the nucleus *(N)* with its nucleolus *(Nu)*, and chromatin *(Ch)*. Note the "pores" *(Np)* in the nuclear envelope. [Ledbetter and Porter, "Introduction to the Fine Structure of Plant Cells," figure 1-1. Copyright © 1970 by Springer-Verlag, New York, Inc].

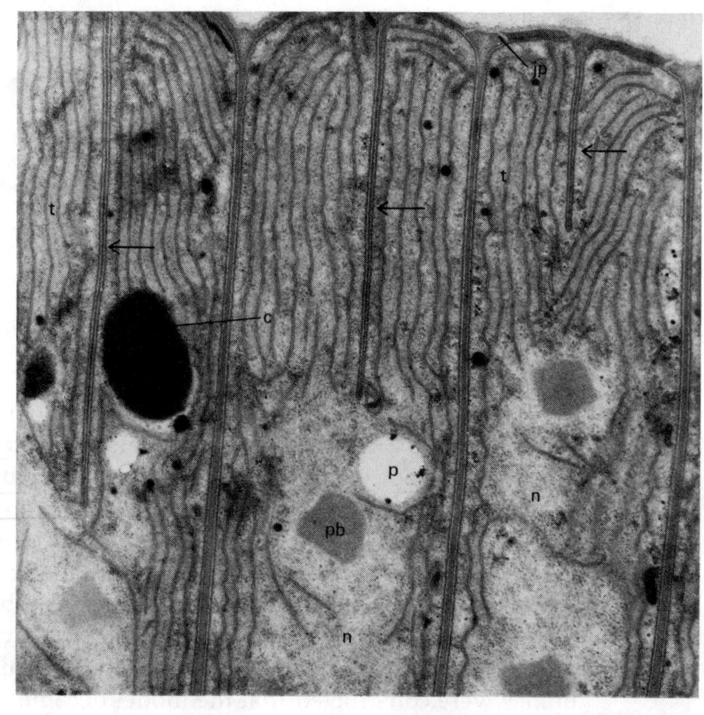


*Figure 3.7*
A prokaryotic cell. TEM micrograph of the blue-green algal cell *Starvia zimbabweënsis* (X 27,000). Note absence of discrete membrane-bound organelles and central vacuole prominent in the eukaryotic cell (Fig. 3.6). This longitudinal section shows photosynthetic thylakoids (*t*), cyanophycin granule (*c*), polyphosphate body (*p*), junctional pores (*jp*), and developing cross-walls (arrows). Scale = 0.5 mm. [Reproduced by permission of Norma J. Lang. Reprinted in *J. Phycol.* 13, 288–296 (1977).]

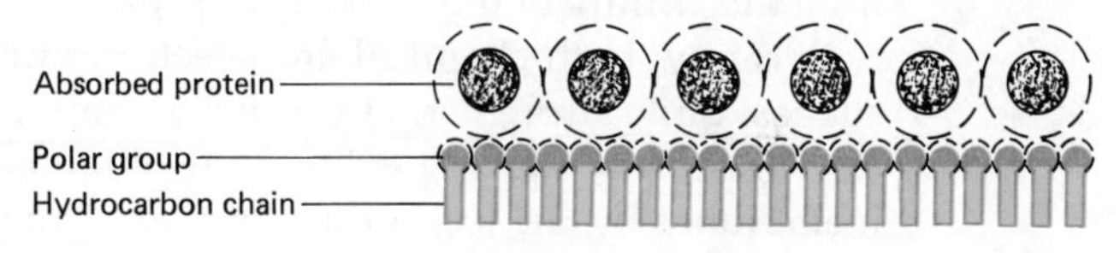


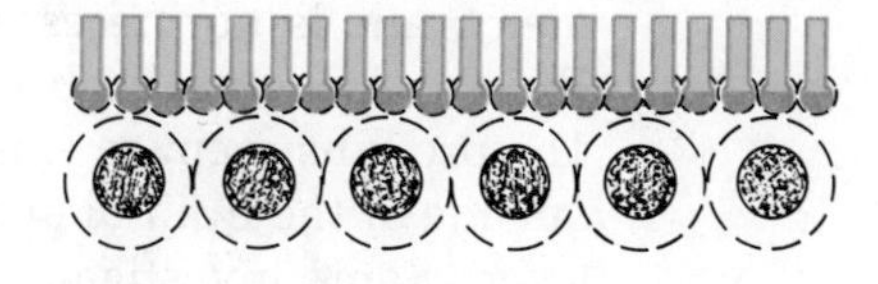


*Figure 3.8*
The unit membrane model proposed by Danielli and Davson (1935). It shows a bimolecular layer of lipid bounded by a monomolecular layer of globular proteins. [From J. F. Danielli and H. Davson, "A Contribution to the Theory of Permeability of Thin Films," *J. Cell. Comp. Physiol.* 5:495–500, fig. 5.3. Copyright © 1934 by Wistar Press, Philadelphia.]

that the membrane consisted of a double layer of lipid molecules, each with its hydrophobic end toward the other. Because the surface tension exhibited at the surface of a plasma membrane was much lower than that expected for a lipid alone, Davson and Danielli (1934) proposed that the lipid bilayer was coated with protein. They suggested the first molecular model of a *membrane*, similar to that of Figure 3.8.

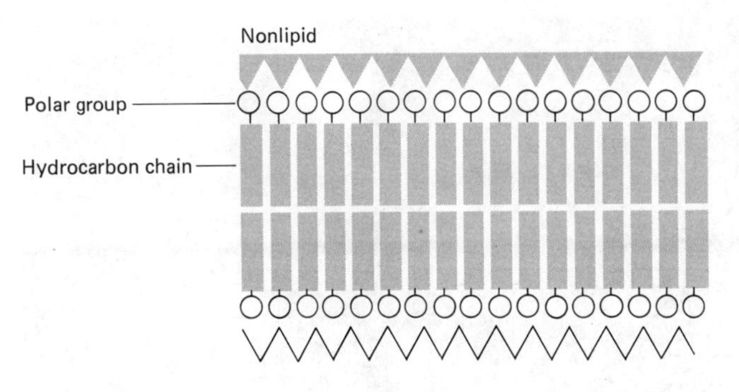


*Figure 3.9*
The unit membrane model proposed by Robertson (1959). The monomolecular layer of protein is extended rather than globular. [Reproduced with permission, from J. D. Robertson, and the Biochemical Society, London, © 1959.]

Robertson proposed a different model in 1959. From his study of the myelin sheath surrounding nerve fibers, he noted that there simply was not enough space for globular proteins and also that their chemical activity would be quite different from that of extended proteins. He thus proposed the following model, which he termed a *unit membrane*, similar to that of Figure 3.9. Robertson's presentations, both written and oral, were forceful, dynamic, and convincing, and he can probably be held as most responsible for the popularity of this so-called sandwich model of membrane structure. Many biologists were so convinced that this was the way all membranes were constructed that the model became accepted as the universal model for membrane structure. This occurred even though chemical evidence was unable to account for the protein; the evidence came purely from the affinity of the fixative for protein.

Recently, both chemical and electron micrographical evidence seem to suggest quite strongly that various membranes are quite different in structure. Especially striking evidence is obtained following preparation of the membranes by the techniques of freeze-fracturing and freeze-etching. By these techniques, the internal organization of the membrane can be examined. Most of the phospholipids of the membrane are in a bilayer, but the membrane is not necessarily only composed of a phospholipid bilayer. There is evidence for up to 50 to 60 different proteins in the membrane. The proteins seem to exist in two classes: peripheral or extrinsic (on the outside of the lipid bilayer), and intrinsic or integral (inside the lipid bilayer). Some investigators, such as Singer, Nicolson, and others, believe there is good evidence to support a fluid mosaic model, as illustrated in Figure 3.10. Some of the protein is thought to go all the way through the membrane, while some is not. It is also suggested that some proteins may be mobile within the lipid layer. The plasmalemma probably contains both structural and enzymatic proteins, while some intracellular membranes may have all of their proteins as enzymatic proteins. The plasmalemma seems to be continuous and without the pores suggested by Danielli (1954).

By 1975, Robertson seemed to be convinced that his unit membrane hypothesis was still correct but perhaps could not be accepted as a univer-

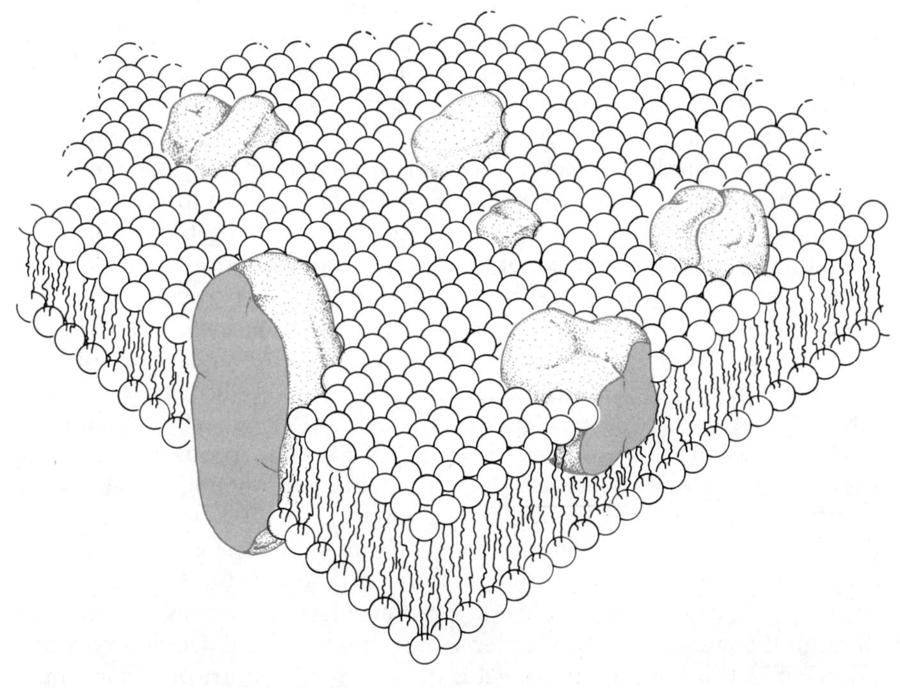

*Figure 3.10*
The fluid mosaic model of cell membranes proposed by Singer and Nicholson (1972). The protein layer is interspersed within the lipid bimolecular layer. [From S. J. Singer and G. L. Nicholson, "The Fluid Mosaic Model of the Structure of Cell Membranes," *Science* 175:720–731, 1972, fig. 3. Copyright © 1972 by the American Association for the Advancement of Science.]

sal model for membrane structure and in fact may only be true for the myelin sheath which he investigated (personal communication). The myelin sheath is a unique membrane. His summary of current knowledge on membrane structure, based on his observations and his interpretation of the observations of others, is as follows: 1) All membranes consist of lipid bilayers. 2) The polar heads of the lipids are always oriented toward the outside. 3) Some intramembrane inclusions may be protein but certainly not all, and those which are protein are probably not pure protein. 4) There are no proteins floating around as intramembrane particles. 5) The outside of a membrane is different from the inside. He postulates that this difference is due to the presence of mucopolysaccharides in the outer layer.

Singer, on the other hand, points out that thermodynamic requirements render highly unlikely the sandwich model proposed by Robertson. In addition, in membranes other than myelin, the lipids are in a fluid rather than crystalline state. Due to the asymmetry of the integral globular protein molecules, they could undergo translational diffusion in the plane of the membrane while maintaining their molecular orientiation. Thus, according to Singer and others, both the lipids and the proteins of plasma membranes are mobile to varying degrees. The asymmetry of membranes is due, in part, to oligosaccharides or mucopolysaccharides, but the integral proteins also contribute to the asymmetry of the membrane.

What difference does it make what the membrane is composed of or how it is structured? The membrane represents a boundary, whether it be the plasmalemma or a membrane of an organelle. Thus, an interface exists. The type of interface reactions that might occur will be based, in part, upon the nature of the surfaces of that membrane. But more importantly, if substances are going to move into or out of the cell or organelle, they obviously must pass through a membrane. Numerous experiments have shown that certain substances pass through membranes unimpeded, others do not pass through at all, while still others pass through at varying rates. Membranes are *differentially permeable*. Certain substances that pass freely into a cell may not pass freely out of the cell, an observation that supports Robertson's assertion that the inside and outside surfaces of a membrane are different. In addition, the nature of the permeability may differ for different membranes and for the same membrane under varying conditions. Thus, if we completely understood the structural and chemical nature of the membrane, we could predict what substances would pass through the membrane, and under what conditions this would or would not occur. A third important aspect is the enzymatic nature of membranes. Many extremely important reactions are associated with membranes.

Observations of membrane function have resulted in many incongruities. Very tiny ions may not pass through the membrane, while large molecules such as nucleic acids will pass through. Some investigators suggest that certain molecules move through the membrane via a carrier system, thus suggesting that the cell must expend energy to accomplish this activity. Others suggest an electrical potential exists across the membrane, thus accounting for the movement of charged ions. A proton gradient does exist for certain membranes, quite strong for membranes of the chloroplast and mitochondrion but much weaker for the plasmalemma. On this basis, an electrical potential would exist across the membrane, but the effect of this on ion mobility has not been clearly demonstrated.

## Organelles

The term *organelle* (little organ) refers to the membrane-bounded structures within the cytoplasm. Our discussion of these structures will be limited to eukaryotic cells for the obvious reason that such membrane-bounded structures do not exist in prokaryotic cells.

The most obvious intracellular structure in eukaryotic cells is the *nucleus*, which may be anywhere from 5 to 30 μm in diameter. As mentioned earlier, most mature plant cells are characterized by a central vacuole. Thus, the nucleus may often be found embedded in the peripheral cytoplasm and flattened against one of the walls. Within the nucleus, one or more small denser bodies, the *nucleoli* (*nucleolus*, sing.), which may range from 1 to 5 $\mu$m in diameter, can be noted. Embedded in the clear fluid portion (*nucleoplasm* or *karyolymph*) of the nucleus, as seen by electron micrography, is a densely staining fibrillar network, the *chromatin*. It is suggested that during nuclear division, the chromatin fibrils become the *chromosomes*, but this relationship has not been clearly demonstrated (Figure 3.6).

As mentioned in Chapter 1, it is well established that the nucleus stores and distributes genetic information. We might thus assume that the nucleus is necessary for normal cell activity. This may be true *for cells that normally have a nucleus*. Cells in which the nucleus has been experimentally removed will function normally for only a short time. But some cells do not have a nucleus at maturity. This is true for all prokaryotic cells, although discreet DNA-containing regions are present in such cells. The best example of a cell of a eukaryote that is enucleated is the mature red blood cell of humans. It lives for only about 30 days and does not carry out all the activities associated with a nucleated cell. In plants, the mature food-transporting cell, the sieve tube cell, carries on metabolic activity and yet does not appear to be nucleated. Recently, with special staining techniques, nuclei have been demonstrated in certain mature sieve tube cells (Figure 3.11).

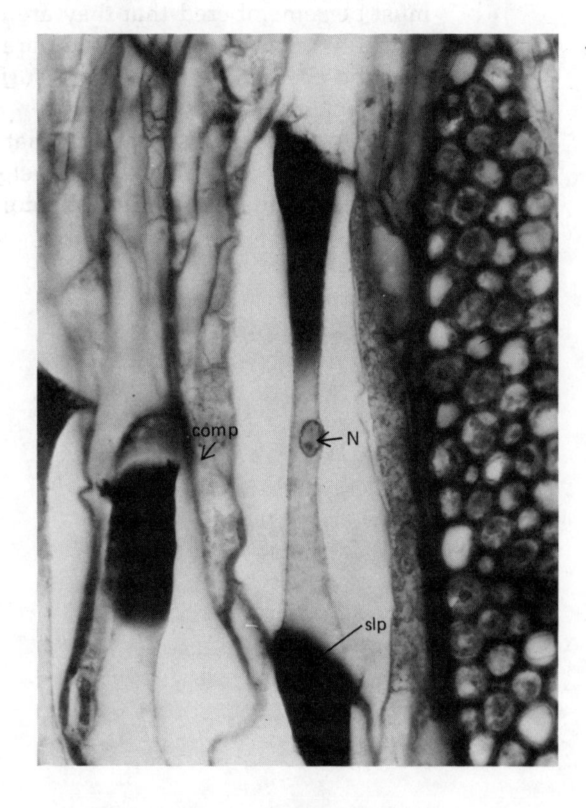


*Figure 3.11*
Mature sieve element of *Ulmus americana* (X 475). *N*, nucleus; *slp*, slime plug; *comp*, companion cell. [From R. F. Evert, et al., "Light Microscope Investigation of Sieve-Element Ontogeny and Structure in *Ulmus americana*," *Amer. J. Bot.* 56(9):999–1017, 1969.]

These, however, can in no way be considered normal nuclei, at least not in appearance. They do not have the dense regions normally associated with the nucleus, nor do they have the discrete form. The best way to describe such nuclei might be the term *diffuse.* It is not known whether such nuclei are capable of carrying out the usual activities associated with normal nuclei.

Regardless of whether the cell never has a nucleus or loses it at maturity, it is now clearly established that both DNA and RNA are found in certain cellular organelles. In the case of prokaryotic cells, the DNA and RNA seem to be associated with lamellar (layered-membrane) structures. Such material does have the capacity for carrying genetic information. In the case of eukaryotic cells, it has still not been clearly established whether the organelle DNA is completely autonomous or whether its activities are directly, or at least indirectly, mediated by the nucleus.

The nucleus is delimited by a double membrane referred to as the *nuclear envelope* (Figure 3.12). The envelope is described as having pores. Many investigators are not convinced that they are pores but rather only thin spots in the envelope. These thin spots are referred to as the poroid complex, as it is postulated that peripheral microtubules line the pore. The pore is covered by a thin membrane which has both fibrous and amorphous elements and may be filled by both protein and RNA. These characteristics of the poroid complex have not been clearly demonstrated, so it must be remembered that they are only hypothesized at this point. The functional characteristics of the pore vary, but it is thought that the pore must be charged, with the pore surface positive and the membrane face negative.

Some investigators postulate that vesicles (small, membrane-bounded sacs) pinched off the nuclear envelope may be involved in forming the Golgi apparatus. Other investigators note the seeming connection be-

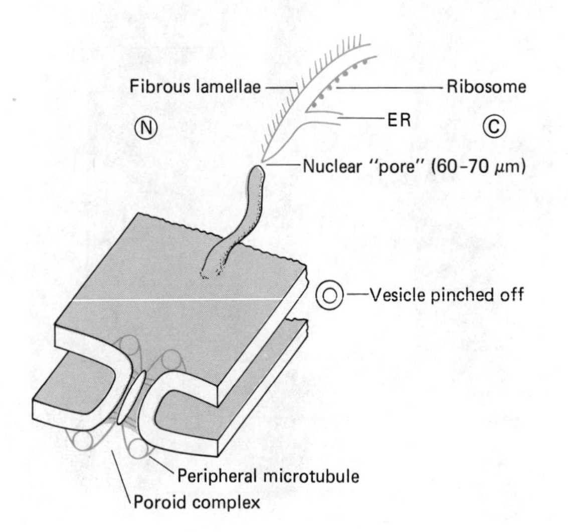


*Figure 3.12*
Sketch of nuclear envelope indicating both a surface and transectional view. A portion of the membrane may be continuous with the endoplasmic reticulum.

tween the endoplasmic reticulum (the double-membrane system traversing the cytoplasm), nuclear envelope, and plasmalemma and suggest that the endoplasmic reticulum and nuclear envelope may have been formed simply by an invagination of the plasmalemma (mentioned in Chapter 1 in regard to eukaryote cell formation). Chemically, however, these membranes are distinct, although the nuclear envelope is more similar to the endoplasmic reticulum than to the plasmalemma.

The *endoplasmic reticulum* (ER) consists of a system of two parallel membranes traversing the cytoplasm. The ER represents a closed system. At no point does it open to the cytoplasm. It can thus be thought of as an extremely flattened and elongated vesicle. Such vesicles are called *cisternae.* Some portions of the ER have ribosomes appressed to their outer surfaces, *rough endoplasmic reticulum* (RER), while other portions lack ribosomes, *smooth endoplasmic reticulum* (SER) (Figure 3.13). Although these two may be continuous with each other, evidence indicates that they are chemically different. The cytoplasmic connections between cells, the *plasmodesmata,* are extensions of the endoplasmic reticulum. There is considerable evidence indicating an association between RER and protein synthesis. In cells in which little protein synthesis occurs, such as certain leaf cells, the RER is poorly developed. In cells having considerable protein synthesis,

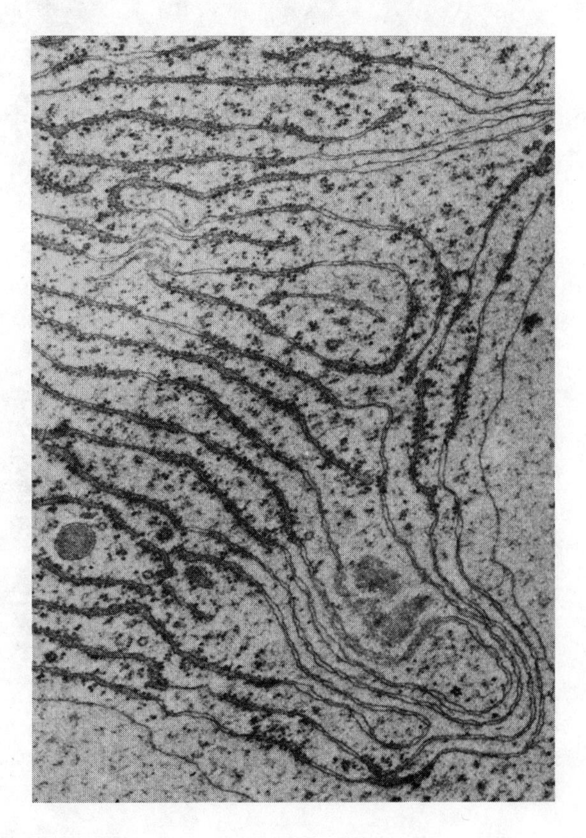

*Figure 3.13*
Endoplasmic reticulum in an apical cell of the alga *Chara.* Both rough and smooth endoplasmic reticulum are present and are continuous with each other (X 15,000). Note the ribosomes lining the RER. [Micrograph courtesy of Michael G. Barbour.]

the ER is well developed, with the majority being RER. The actual site of protein synthesis appears to be the ribosomes which line the surface of the RER. The endoplasmic reticulum may be associated with synthesis, secretion, or transport of carbohydrate and other substances to different parts of the cell or between cells.

The *plastids* are double-membrane organelles from 2 to 25 $\mu$m in diameter. Plastids develop from small structures called *proplastids*. These have little or no lamellae but may have an elaborate crystallinelike structure referred to as a *prolamellar body*. Mature plastids may be of three general types: 1) chloroplasts, in which chlorophyll pigments (green) predominate, with carotenoid pigments also generally present; 2) chromoplasts, in which carotenoid pigments (yellow, red, or brown) predominate and chlorophylls are entirely lacking (according to some botanists) or much reduced (according to others); and 3) leucoplasts, which contain no pigments (Figure 3.14). Leucoplasts are generally associated with storage and may

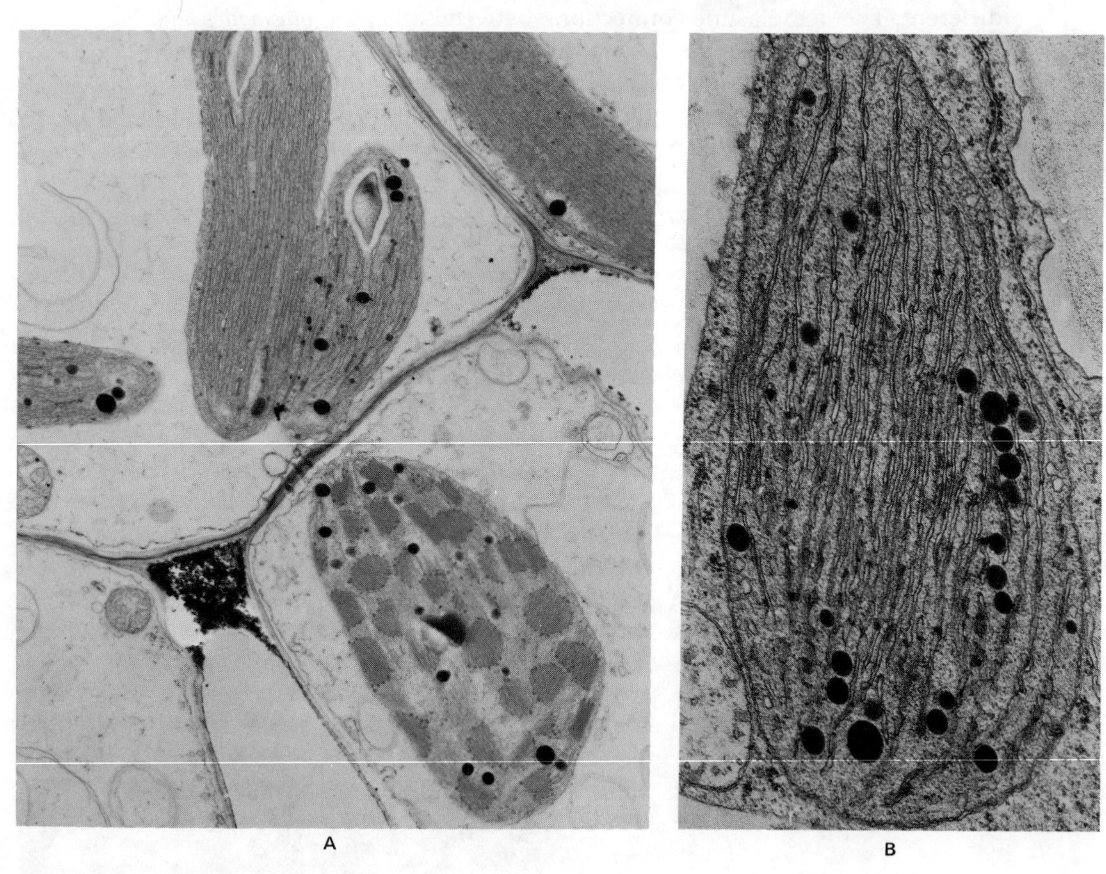


*Figure 3.14*
Plastids. *A.* Chloroplast. *B.* Chromoplast. [*A.* Courtesy of C. C. Black, Jr.; *B.* W. M. Harris and A. R. Spurr, "Chromoplasts of Tomato Fruits." I. Ultrastructure of Low-Pigment and High-Beta Mutants. Carotene Analysis. *Amer. J. Bot.* 56(4):369–379, 1969.]

be named on the basis of their storage product. If they store starch they may be called *amyloplasts*; if oil, *elioplasts* (or *elaioplasts*); if protein and if in certain grains associated with the aleurone layer, *aleuroneplasts*. One type may be converted to another, and it has recently been shown that this is reversible; for example, chloroplast → chromoplast → chloroplast.

The chloroplast membrane is thought to be the most complex of all membranes. As it is a double membrane, it is appropriate to refer to it as the *chloroplast envelope*. All chloroplast membranes seem to be prokaryotic in nature, particularly the inner membrane. That is, they are similar to those of the blue-green algae or bacteria. This has led to the hypothesis, stated in Chapter 1, that chloroplasts were actually prokaryotic cells that became incorporated into eukaryotic cells and, through time, became a part of the cell. The inner membrane is quite different from the outer one, as is the inner face of a membrane different from the outer face. At various points, an infolding of the inner membrane occurs forming an elaborate system of internal chloroplast membranes termed *lamellae*. At intervals, the lamellae form helical stacks called *grana* (Figure 3.15). The lamellae between the stacks are termed *stroma lamellae*. Figure 3.16 indicates a more recent model for chloroplast lamellar arrangement. The *matrix* or *stroma* which the lamellae traverse, has a granular appearance and seems to contain ribosomelike particles. The term *thylakoid* was coined to refer to these lamellae and to each member of a grana stack. It was thought that the internal spaces between thylakoids were not interconnected. Even though the grana lamellae seem to be continuous in some regions with the stroma lamellae, there is some evidence that these two membranes are not the same.

The location of the pigment is somewhat open to question. Some investigators believe that it is found entirely within the lamellae. Others sug-

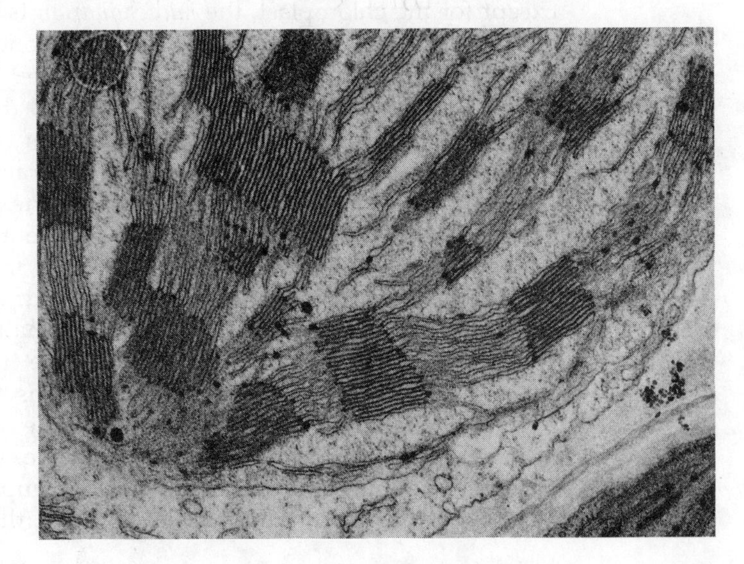

*Figure 3.15*
Granal compartments and frets. [Courtesy C. C. Black, Jr.]

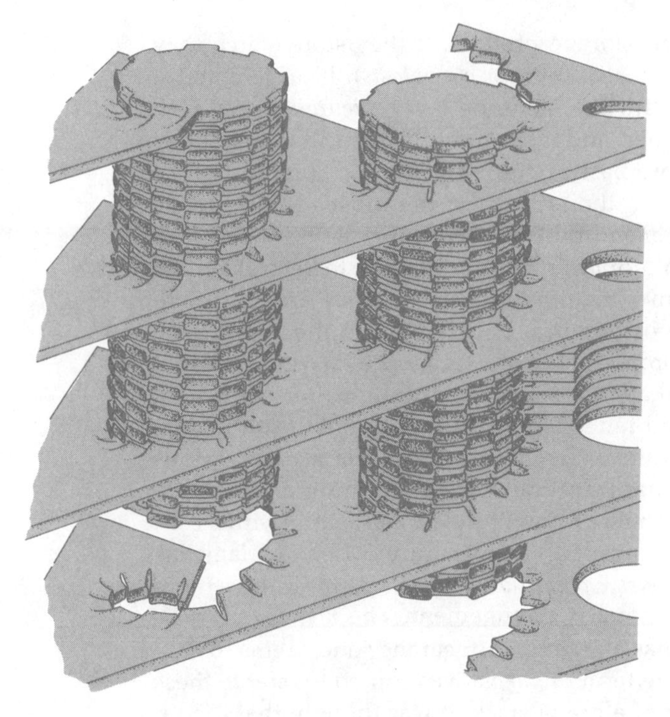

*Figure 3.16*
Model for higher plant chloroplast lamellar arrangement. Each compartment of the granum can contact a number of different stroma lamellae, which, in turn can connect with other grana stacks. [Redrawn from D. J. Paolillo, "The Three-Dimensional Arrangement of Intergranal Lamellae in Chloroplasts," *J. Cell Sci.* 6, 1970, fig. 1. Copyright © 1970 by *Journal of Cell Science.*]

gest that most is contained within the partition regions of the grana, with some occurring within the stroma lamellae.

It has been recognized for some years that the chloroplast is the sole site for photosynthesis in eukaryotic cells. The relationship of the various regions will be discussed later when the process of photosynthesis is considered.

Except for the chloroplast, the *mitochondrion* is perhaps the most complex organelle. Like the chloroplasts, the mitochondria are enclosed by two membranes comprising a *mitochondrial envelope.* The inner membrane is greatly infolded forming a closed system called the *cristae*, which extend into the inner space, the *mitochondrial matrix*, of the mitochondrial body. A second space exists between the two membranes of the envelope and extends out into the cristae. The outer membrane, intermembranal space, inner membrane, and matrix all contain different chemical components. For example, in some mitochondria the outer membrane was found to contain 55 percent protein and 45 percent lipid, the inner membrane 75 percent protein and 25 percent lipid, while the matrix contained 60 percent water and 40 percent protein. In addition, the matrix also contains the major cations $K^+$ and $Mg^{2+}$, and the major anions $ATP^{4-}$, $ADP^{3-}$, and $Pi^{2-}$. The protein in the membranes seem to be all enzyme protein; thus, the higher the amount of protein in the membrane, the more enzymatically active it is. The inner membrane is generally impermeable to polar ions, while the outer membrane is quite leaky. Thus, the two membranes comprising the mitochondrial envelope are quite different, from a physical, chemical, and enzymatic point of view.

The function of mitochondria, which are generally small rods from 0.5 to 2 μm in diameter and from 2 to 10 μm long (Figure 3.6), is known to be respiration and the subsequent production of energy in the form of ATP.

*Dictyosomes* appear to be the plant equivalent of animal Golgi bodies. There are normally many dictyosomes in a plant cell, and collectively, they have been referred to as the *Golgi apparatus.* The dictyosomes are normally about the same size as certain of the smaller mitochondria (0.5 × 2 μm) and are comprised of a series of stacked discs, *cisternae,* with each member of the stack bounded by two membranes. The margins of each cisternae form a coarse net. Vesicular structures are formed at the outermost margins of the net.

The dictyosomes seem to be involved in the synthesis of enzymes, particularly the precursors of cell wall material. The peripheral dictyosome vesicles (Fig. 3.17) become pinched off, and during cell division, migrate

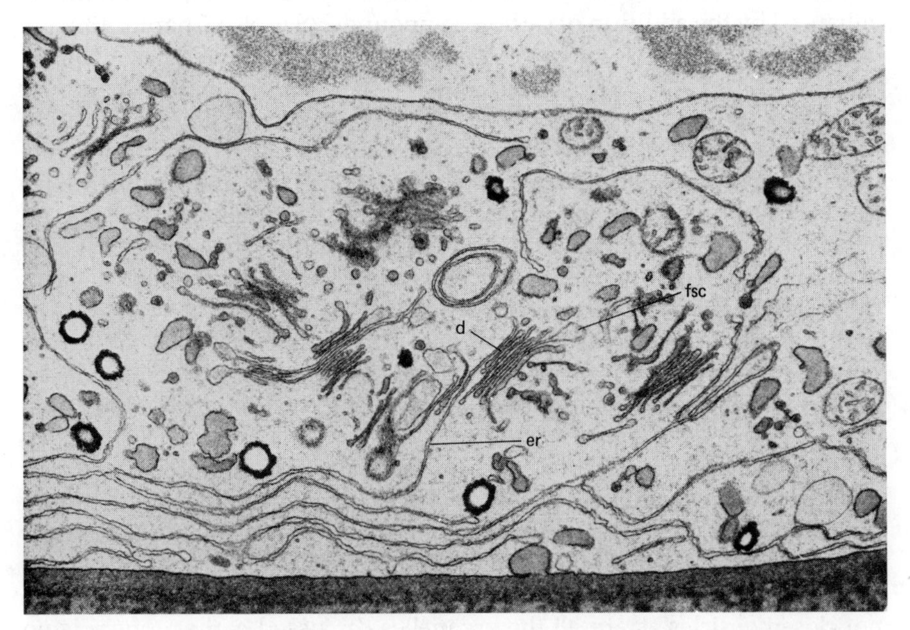


*Figure 3.17*
Electron micrograph of mantle cell from a *Zea mays* seedling showing dictyosomes with attached secretory vesicles, free secretory vesicles, endoplasmic reticulum, and other cytoplasmic components. Dictysomes are confined to the central portions of the cytoplasm where they appear to be organized into small groups of closely associated dictyosomes. Each group of dictyosomes is partially separated from other groups by segments of endoplastic reticulum. But, all dictyosomes of the cell are always at the same developmental state and, therefore, must be functionally, if not directly, interconnected (X 10,000). *d,* dictysome; *fsc,* free secretory vesicles; *er,* endoplasmic reticulum. [From H. H. Mollenhauer et al., "Endoplasmic Reticulum—Golgi Apparatus Associations in Maize Root Tips," *Mikroskopie* 31:257–272, 1975.]

to the equatorial plane of the spindle, where they fuse with spindle fibers and vesicles from the endoplasmic reticulum, forming the cell plate. Dictyosome vesicles have also been noted to migrate to the plasmalemma and fuse with it. The significance of this phenomenon is not well understood.

*Microbodies* are widely distributed in plant cells and are associated with the endoplasmic recticulum. Two distinct types of microbodies are generally recognized, peroxisomes and glyoxysomes. Both are essentially spherical bodies 0.2 to 1.5 $\mu$m in diameter, bounded by a single membrane.

The glyoxysomes seem to have an important function in the metabolism of fatty acids. Fatty acyl-CoA derivatives are converted in the glyoxysomes to acetyl-CoA, which is then, in turn, converted to succinate. The succinate can be transferred from the glyoxysome to the mitochondria, where it is respired. (Refer back to this discussion after you have read Chapter 13.)

Compounds from the photosynthetic carbon cycle of the chloroplast can be transferred to the peroxisome, where they are converted to carboxylic acids and finally to amino acids. The process in the peroxisome may also be reversible and thus may be a mechanism by which photosynthesis is enhanced by the addition of carbon compounds. (See also Chapter 13.)

## *Nonmembrane-Bounded Components of the Protoplast*

Ribosomes are tiny (20 nm), electron-dense particles. As mentioned earlier, they are associated with the RER (Figure 3.13) but may also be found free in the cytoplasm, in the plastids, and in the mitochondria. There is no evidence that these particles are membrane bounded, thus their exclusion from the section on organelles. Ribosomes may occur in groups or clusters called *polyribosomes* or *polysomes*.

Ribosomes are exceedingly rich in RNA and, as mentioned earlier, are important in protein synthesis. It is at the ribosomes that amino acids are sequenced and peptide linkages formed, producing the polypeptide chains.

*Microtubules* and *microfilaments* are recognized as components of virtually every eukaryotic cell. Microtubules, which are about 24 to 28 nm in diameter, have been described as comprising an open membranous system in the cytoplasm. This refers to the fact that the microtubules are in direct communication with the ground substance of the cytoplasm and are not closed vesicles as are all the organelles discussed earlier. There is no evidence, however, that they are membranes. The wall of a microtubule is apparently characterized by a basic protein subunit called *tubulin* (Figure 3.18). Microfilaments, on the other hand, may contain the protein *actin*, one of the contractile proteins associated with muscle fibers in animals.

Microtubules may constitute the spindle in dividing plant cells and may in fact control either or both the position and orientation of the spindle. Microtubules may play a role in nuclear migration in algae and fungi but may not be involved in nuclear migration in higher plant cells. There is also good evidence that microtubules, at least in some cells, are the ele-

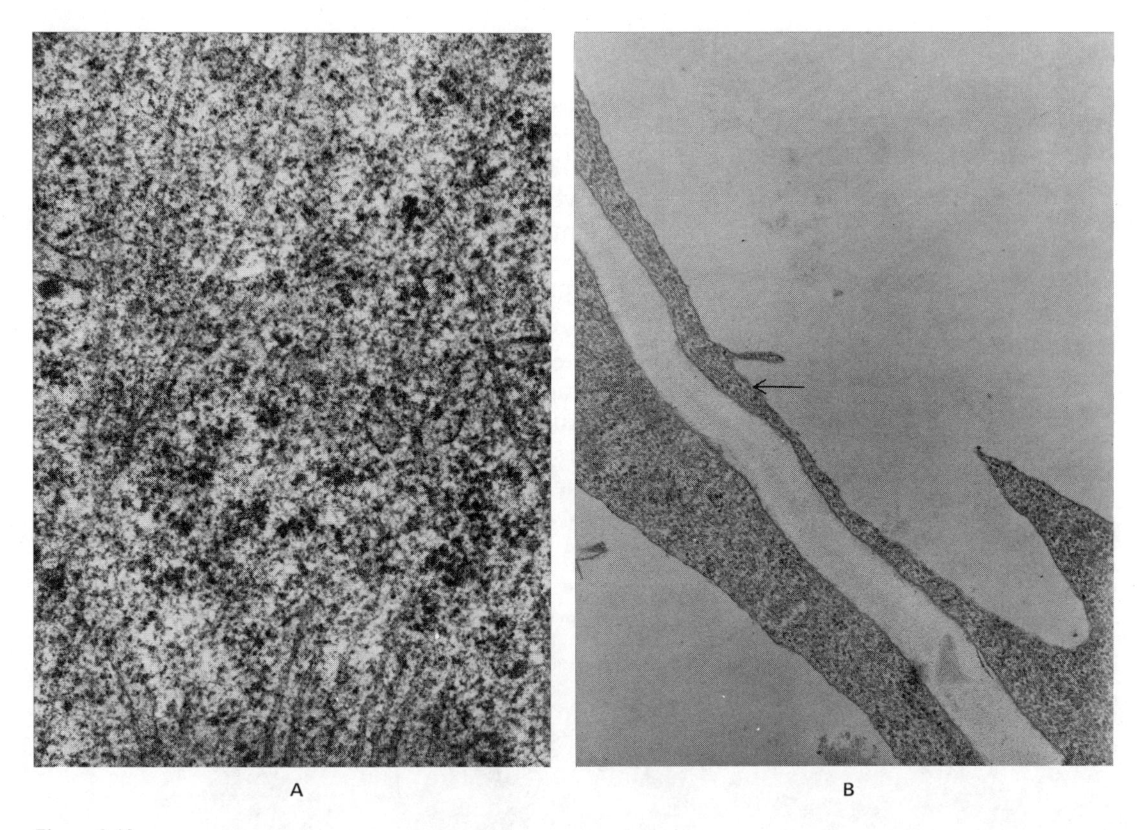


*Figure 3.18*
Microtubules. *A.* Longitudinal view. *B.* Cross-sectional view *(arrow).* [Courtesy of James S. Marks.]

ments that control the orientation of cellulose during cell wall formation. They may also control cell shape, perhaps by acting directly as structural, cytoskeletal elements. Microfilaments have been held to be responsible for such phenomena as amoeboid movement, cytoplasmic streaming (cyclosis), and nuclear migration. These microfilaments are very small, measuring only 5 to 8 nm in diameter. They usually occur in parallel bundles or as a loose network just beneath the plasmalemma.

Some investigators have suggested that mocrotubules and microfilaments might function cooperatively, with microtubules providing the framework upon which microfilaments could bind and generate force.

Nonvacuolar crystals have been observed in the cytoplasm, nucleus, chloroplast, mitochondria, and microbodies. Rather than being inorganic in nature, such as those found in the vacuole, these crystals appear to be proteinaceous. The individual components may be quite small, but crystalline bodies up to 10 $\mu$m in diameter are found. Their function is unknown, but as they do seem to be composed of protein, they may represent some type of protein storage component.

## Vacuoles

A nonprotoplasmic part of the protoplast is the *vacuole* (Figure 3.19). Vacuoles are separated from the protoplasm by a single membrane called the *tonoplast*, which seems to be quite similar to the plasmalemma. The tonoplast is differentially permeable, and while some materials can move passively through the membrane, others must be moved from the cytoplasm into the vacuole at the expense of cellular energy.

In most young plant cells, the vacuoles are numerous but small. As the cell becomes mature, these small vacuoles coalesce, so that most mature plant cells have a single, large, central vacuole.

The vacuoles contain an aqueous substance termed *cell sap*. This cell sap is composed of water highly diluted with a wide variety of different water soluble substances. These soluble substances include inorganic ions or salts, gases, organic acids, sugars, water soluble proteins, alkaloids, and pigments. Not all components of a vacuole are necessarily soluble. Cal-

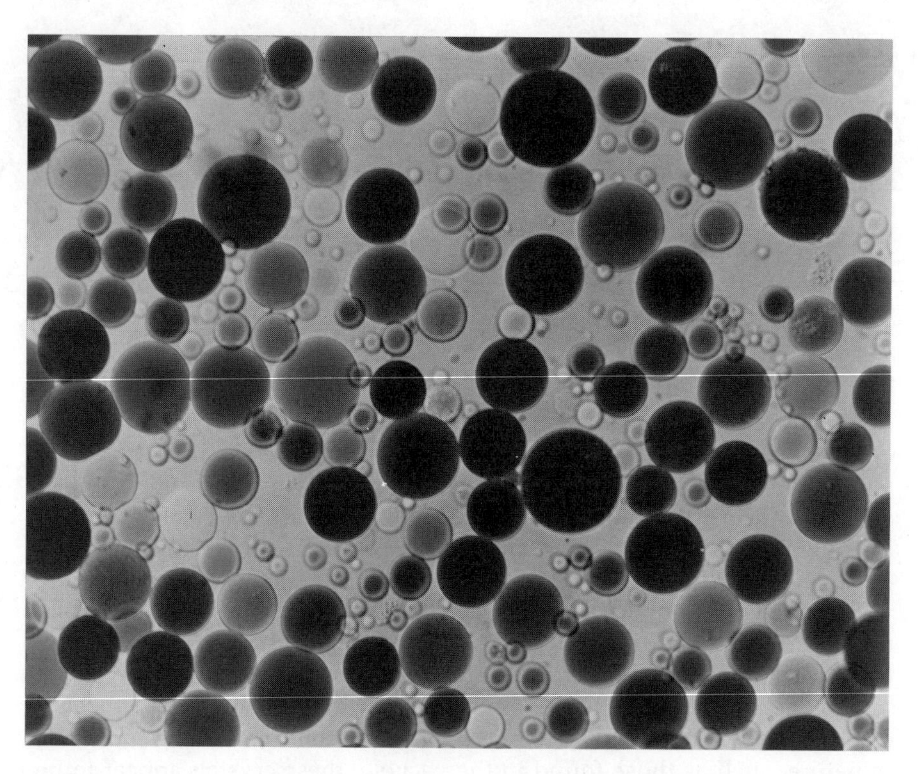

*Figure 3.19*
Intact vacuoles isolated from plant cells. [From C. J. Wagner et al., "Large Scale Isolation of Intact Vacuoles from Protoplasts of Mature Plant Tissues," *Science* 190:1299, 1975, fig. 2a. Copyright © 1975 by the American Association for the Advancement of Science.]

cium may combine with oxalic acid in the vacuole forming insoluble crystals of calcium oxalate. Crystals may be in a variety of forms depending upon chemical composition and conditions under which they formed. It is the presence of long, needlelike crystals in the vacuoles of leaf cells of dumb cane and rhubarb that makes the eating of these raw leaves so dangerous. Literally thousands of these tiny, needlelike crystals become embedded in the tongue and throat causing considerable pain and, occasionally, swelling of the throat tissue to the degree that strangulation occurs.

It is the water soluble pigments, belonging to a class of pigments called the *flavonoids*, that are responsible for flower color in many flowering plants. The *anthocyanins* are most commonly seen in red, purple, and blue flowers. The pH of the vacuole has a strong controlling influence on the color of most anthocyanins. Most are red under acid conditions, become purplish as the pH approaches 7.0, and finally, blue under basic or alkaline conditions. The *flavonols* and *flavones* are closely related to the anthocyanins and also contribute to flower color. These normally produce yellowish to ivory or cream coloring.

## Cell Walls

The protoplast secretes and is thus surrounded by a *cell wall*. Cell walls are frequently thought of as hard, rigid objects, but they may be both elastic and plastic, depending in part upon 1) age of the cell, 2) presence of primary wall only or primary and secondary walls, or 3) internal environmental factors such as hormone concentration and other chemical aspects.

All cell walls of multicellular plants seem to consist of at least two layers: the intercellular substances called the *middle lamella*, and the *primary wall*. Some cells lay down additional layers on the primary walls. These are termed the *secondary wall*, the *tertiary wall*, and so on.

The primary walls of most plant cells are composed of cellulose fibers, interwoven and impregnated with pectin. Cellulose is a polysaccharide, formed from repeating molecules of glucose attached end to end. (If you are not sure what a polysaccharide is or what glucose is, you might well at this point read Appendix I: A Few Basic Chemical Concepts.) The cellulose molecules are united forming *microfibrils*, each of which may contain as many as 2000 cellulose molecules (Figure 3.20). The microfibrils wind around each other forming *fibrils*, which in turn wind around each other forming *macrofibrils*. Thus, the macrofibril might be thought of as being constructed very much like a steel cable or piece of rope. Each macrofibril contains about 500,000 cellulose molecules, may be as much as 0.5 $\mu$m in diameter, and reach 4 $\mu$m in length. The cellulose macrofibril is, in fact, as strong as a steel cable of equivalent size.

Cellulose and pectin are not the only substances found in the primary

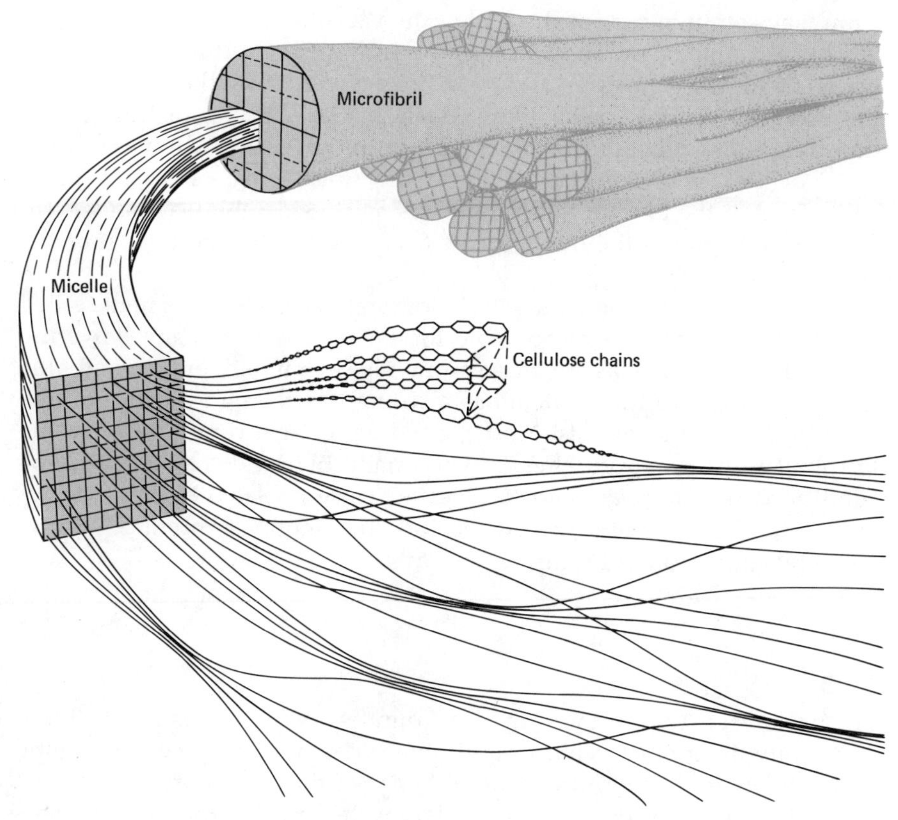


*Figure 3.20*
Cell walls are composed of long interwoven and interconnected strands, the microfibrils. The microfibril is composed of micellar strands, which, in turn, are made of many long-chain cellulose molecules.

cell wall. Other carbohydrates such as hemicellulose, xylans, mannins, lipids, and even proteins are constituents of the cell wall.

In many cells, additional layers of cellulose are laid down on the primary wall. These cellulose fibrils, rather than being interwoven (called *intusseception*), are deposited in layers (termed *apposition*). The secondary walls typically lack pectin and other polysaccharides commonly present in the primary wall, with the exception of *lignin*. The cellulose, and lignin when present, tend to make the secondary wall much more rigid than the primary wall. Additionally, the fibrils in each cellulose layer are at right angles to those of the next layer (much like the construction of plywood), adding to the rigidity. Proteins are absent in secondary walls, but *suberin*, certain other fats, and tannins may be present in addition to the previously mentioned lignin. The increase of lignin in the wall, called *lignification*, is extremely important in the formation in certain wood cells such as vessel members, tracheids, and fibers, as we will see in Chapter 7. Lignification typically begins in the middle lamella, spreads to the primary wall and finally to the secondary wall. Thus, lignification progresses from the outside toward the inside. But the deposition of *cutin*, a waxy material, proceeds in the opposite direction. In the process of *cutinization*, cutin is secreted towards the outside to be deposited as a surface layer called the *cuticle*.

Unlike the plasmalemma, the cell wall, unless heavily cutinized, lignified, or suberized, is not only completely permeable to water, but it is completely permeable to any substance dissolved in water. Some individuals refer to the wall as porous, but this is not true in the sense of holes going straight through the wall from inside to outside. However, water freely moves through the spaces created by the fibrillar network of the wall. Although not having pores, the primary wall does characteristically have thin areas called *primary pit fields*. The cytoplasmic connections between cells, the plasmodesmata discussed earlier, are commonly aggregated in the primary pit fields. When the secondary wall is laid down, it is not deposited over the primary pit fields of the primary wall. Consequently, a depression, or *pit* is formed. A pit in one cell wall usually occurs opposite one in an adjacent cell wall. The middle lamella and the two primary walls form what is called the *pit membrane*, with the two opposite pits plus its membrane constituting a *pit pair*. Two commonly occurring types of pits are *simple* and *bordered*. In bordered pits, the secondary wall arches over the *pit cavity*; in simple pits, there is no such overarching (Figure 3.21).

The cell wall and its middle lamella is, as discussed in Chapter 1, an important structure to man. Wood pulp for paper, lumber for construction,

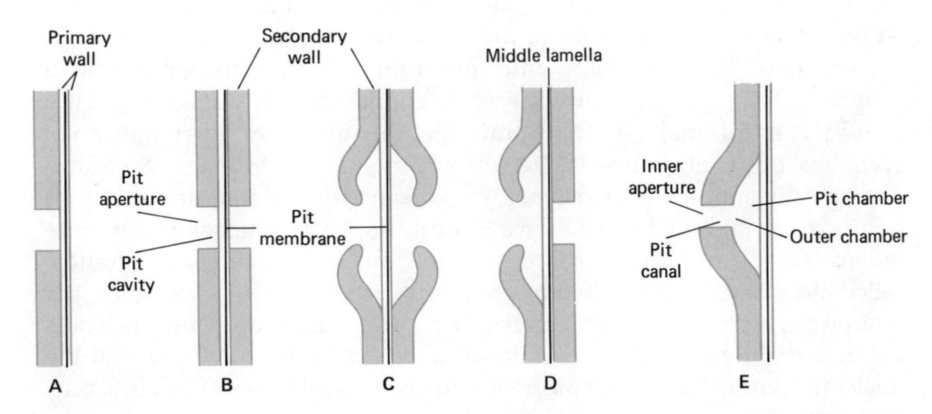


*Figure 3.21*
Structure of pits. *A*. Simple pit; *B*. Simple pit-pair; *C*. Bordered pit-pair; *D*. Half-bordered pit-pair; *E*. Bordered pit. [Redrawn from A. Fahn, *Plant Anatomy*, 2d ed. Copyright © 1975, Pergamon Press, Inc].

rayon, cellophane, explosives, fibers such as cotton, flax, and linen, and chemicals are only a few of the products associated with the cellulose wall. The pectin materials of the middle lamella constitute the jelling agent used in the preparation of jellies.

Although the presence of a cellulose cell wall is considered to be a characteristic of plant cells, and animal cells are said not to have cell walls, some plants, such as the slime molds and members of one euglenoid group of algae have no cell walls in the vegetative state. Also, another group of algae, called the diatoms, have cell walls composed entirely of silicon. We might mention, however, that in a newer system of classification, which we will discuss in Chapter 5, these organisms are not considered to be true plants.

## CELLULAR AND NUCLEAR DIVISION

### *Mitosis*

The modern view of mitosis considers the process as taking place in four stages. These stages should not be thought of as discrete steps because they are a continuum. That is, the stages follow one another in continuous fashion. The major stages are called *prophase*, *metaphase*, *anaphase*, and *telophase.* The period of time between mitotic divisions is called *interphase.* (The major phases of mitosis are shown in Figure 3.22).

During interphase, the nucleus has a typically granular appearance. Electron microscopy reveals the basic structure as an array of quite long, slender, tangled fibers forming a reticulum (network) in the nucleus. This material has been termed *chromatin.* The chromatin is of a nucleoprotein nature, consisting of *DNA* and a simple protein, usually one called *histone.* At the onset of nuclear division, the chromatin resolves into visible double-stranded chromosomes at which point the cells are considered to be in prophase. The strands are closely associated, particularly during very early prophase, and it may be at this point that the infrequent phenomenon of exchange of genetic material between homologous chromosomes (morphologically similar chromosomes, one maternal, one paternal in origin) occurs, an event called *somatic crossing over.* As the chromosomes continue to shorten and thicken by coiling, it can be noted that the two strands, called *chromatids,* are attached to each other at a single specific point, the *centromere.* As prophase progresses, the nuclear envelope and nucleolus begin to disintegrate. Structures called *centrioles* become evident near the nuclei in some cells. During prophase, the two membranes of a pair of centrioles migrate to opposite poles of the cell and apparently organize the formation of *spindle* fibers. Nonmotile cells lack centrioles, but spindle fibers form just the same. The spindle fibers become attached to the centromere. As the *spindle* forms, the chromosomes are thus brought to lie in a

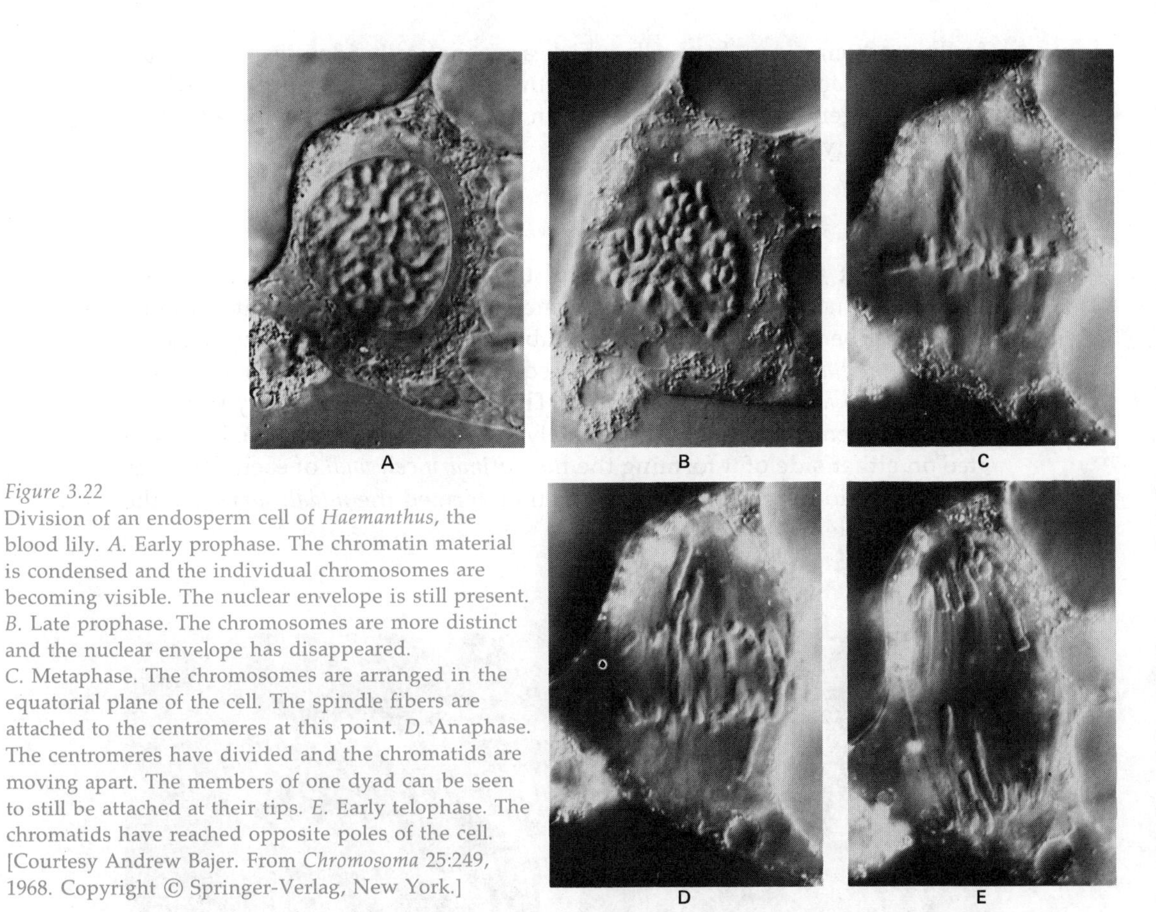


*Figure 3.22*
Division of an endosperm cell of *Haemanthus*, the blood lily. *A.* Early prophase. The chromatin material is condensed and the individual chromosomes are becoming visible. The nuclear envelope is still present. *B.* Late prophase. The chromosomes are more distinct and the nuclear envelope has disappeared. *C.* Metaphase. The chromosomes are arranged in the equatorial plane of the cell. The spindle fibers are attached to the centromeres at this point. *D.* Anaphase. The centromeres have divided and the chromatids are moving apart. The members of one dyad can be seen to still be attached at their tips. *E.* Early telophase. The chromatids have reached opposite poles of the cell. [Courtesy Andrew Bajer. From *Chromosoma* 25:249, 1968. Copyright © Springer-Verlag, New York.]

more or less linear order on the equatorial plane of the spindle. The formation of the spindle is considered to initiate metaphase. Metaphase passes into anaphase when the centromere divides and the daughter chromosomes, now single-stranded, begin to move toward opposite poles, presumably by contraction of the spindle fibers. By this process, two genetically identical sets of chromosomes are produced, barring somatic crossing over or mutation.

In the final stage of mitosis, telophase, there is a regrouping of the chromosomes into a nuclear structure. The nuclear envelope and nucleolus are reconstituted, and the chromosomes uncoil and elongate to give the appearance of an interphase nucleus.

During late telophase, the following interphase, or in very early prophase of the next division series, the DNA of the single-stranded chromosome, or *monad*, will replicate forming the double-stranded chromosome, or *dyad*, observed during prophase. The consequence of mitosis is theoretically producing daughter cells which are genetically identical to the parent cell. In fact, many events can occur to produce daughter cells which will not be genetically similar. Thus, in mitosis, the *potential* for introducing variability exists.

## Cytokinesis

Usually, but not always, division of the cytoplasm *(cytokinesis)* follows mitosis so that two new cells are formed. This process is initiated by the fusing of fibers, vesicles, and other substances, forming a structure called the *phagmoplast* in the region of the equatorial plane. This results in the formation of a *cell plate* (Figure 3.23). The cell plate is essentially pectin in nature. It eventually extends completely across the cell. Cellulose is deposited on either side of it forming the new *primary cell wall* of each daughter cell. The region of the old cell plate is then termed the *middle lamella* of the two new sister cells.

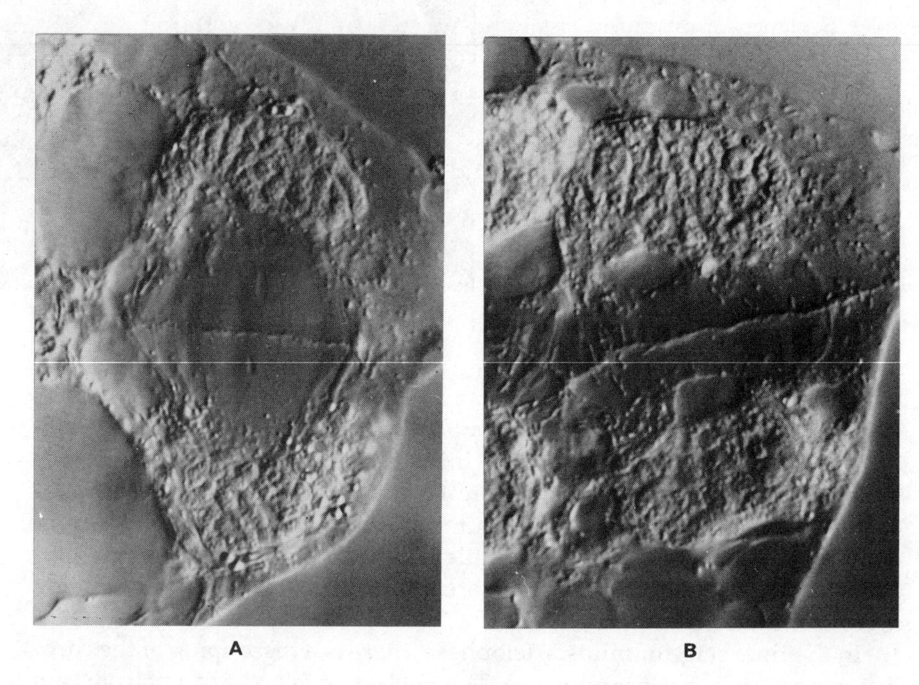


*Figure 3.23*
*A.* Late telophase. Mitosis is completed and the chromosomes are becoming diffuse. Cytokinesis is being initiated through formation of the cell plate across the equatorial plane of the spindle. *B.* The cell plate has reached the margin of the cell. The primary wall will be laid down on either side of the cell plate, resulting in the formation of two new cells. [Courtesy Andrew Bajer. From *Chromosoma* 25:249, 1968. Copyright © Springer-Verlag, New York.]

## Meiosis

In 1883, van Beneden, working with the round worm *Ascaris*, called to attention the fact that the gametes contained only *half as many chromosomes as the somatic cells.* The process by which the chromosome number was reduced was later termed *meiosis.* He noted that the normal chromosome number was restored at fertilization. The chromosome number in the somatic cells would later be known as the *diploid* number, while that of the gametes would be called the *haploid* number. But van Beneden was unaware, or at least did not suggest, that the chromosomes were related to the hereditary material, and he falsely assumed that all the paternal chromosomes became segregated into polar bodies, and thus, the egg cell contained only maternal chromosomes. It was not until 1901 that a young assistant professor at the University of Pennsylvania, Thomas Montgomery, observed that *maternal chromosomes paired only with paternal chromosomes during meiosis.* Sutton, in 1903, while reviewing Mendel's work, noted a clear parallel between the meiotic behavior of pairs of chromosomes and the behavior of pairs of Mendelian traits. He referred to the maternal and paternal pairs observed by Montgomery as *homologous* members, which he defined as "those that correspond in size." Sutton also introduced the idea that the characters mentioned by Mendel (see Chapter 6) are actually carried on the chromosome. As the number of characters for an organism clearly exceeds the number of chromosomes, it was conceivable that the chromosome may be divisible into smaller entities. These entities would later (1909) be called *genes* by Johannsen. He further noted that during meiosis, the homologous members separate with one entering one daughter nucleus, the other entering the second daughter nucleus. This, he stated, was the cytological basis for Mendel's principle of segregation. As the fate of one pair of homologous chromosomes is completely independent of the fate of any other pair, the cytological basis for Mendel's principle of independent assortment was established. But Sutton had several misconceptions concerning the process of meiosis. We will discuss it in terms of modern evidence.

As a rule, meiotic divisions follow a rather standard scheme, although variations are known. It is convenient, as in mitosis, to discuss meiosis as occurring in steps or stages. The active division occurs in four stages: *prophase, metaphase, anaphase,* and *telophase,* with the first of these being subdivided into an additional five stages. The entire process requires two successive series of divisions, generally referred to as meiosis I and meiosis II. (The major phases of meiosis are shown in Figure 3.24.)

During the initial stages of meiosis, the reticulum of the nucleus (chromatin) begins to resolve into long, slender threads. These threads, the chromosomes, continue to shorten and thicken throughout prophase I apparently by coiling. It is possible that the chromosomes become double-stranded during the initial stages (i.e., early prophase) by *replication* of the

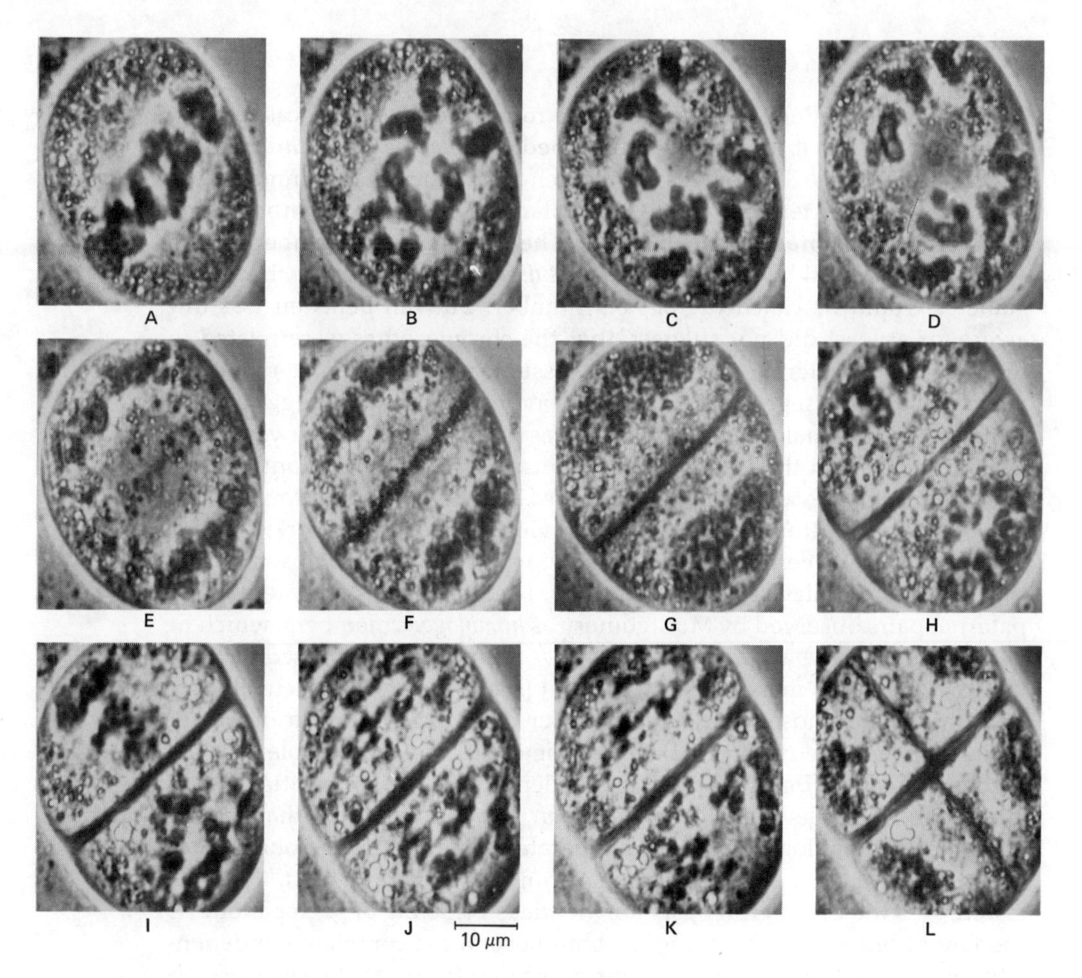


*Figure 3.24*
Meiosis. Living microsporocyte of *Tradescantia paludosa.* Micrographs taken out of 16 mm film. Phase contrast (time in minutes): A: 0 min; B: 18; C: 25; D: 34; E: 40; F: 50; G: 105; H: 200; I: 235; J: 255; K: 270; L: 310. *A.* Metaphase I showing 6 bivalents in side view. *B.* to *D.* Different stages of Anaphase I with homologous chromosomes moving towards the poles. *E.* and *F.* Telophase I with cell plate formation. *G.* Prophase II. *H.* Nuclear envelope breakage in both haploid cells. *I.* Metaphase II. The spindles have slightly different orientation. *J.* and *K.* Anaphase II with synchronous poleward movements of the chromatids. *L.* Telophase II and formation of 4 haploid microspores. [Courtesy of Dr. Ann-Marie Lambert, Strasbourg, France.]

chromosome material. During prophase the homologous chromosomes seem to attract each other and enter into a very close association (called *synapsis*), becoming tightly coiled around one another. At this stage each chromosome consists of two sister *chromatids*, held together at some point along their length by a *centromere.* The closely bound group of two homologous chromosomes or four chromatids is known as a *bivalent* or *tetrad.* It is during this tight pairing that exchanges of genetic material can occur between the homologous chromatids, thus leading to the partial linkage observed by Bateson and others, which will be described later. Such exchange of genetic information is called *crossing over* or *recombination*, and will also be discussed later in more detail.

During the later stages of prophase I, the homologous chromosomes act as though they repulse each other, begin to uncoil and separate along their length, except at specific regions where an actual physical crossing over appears to have occurred. These crossing areas, called *chiasmata* (*chiasma*, sing.), can be seen as *x*-shaped attachments between homologous chromatids. They seem to be the only things holding the bivalents together until metaphase. It is also during the later stages of prophase I that the nuclear membrane and nucleolus disappear, the *spindle* forms, the bivalents attach to the spindle fibers by their centromeres and begin to migrate toward the equator of the spindle.

Metaphase I is characterized by the presence of only single terminal attachments between homologous chromosomes, created by the chiasmata moving toward the end of each chromosome *(terminalization).* These remaining chiasmata prevent the separation of homologous chromosomes, and thus, they come to lie side by side at the equator of the spindle (Figure 3.25).

In anaphase I, the two homologous chromosomes are pulled apart, apparently by contraction of the spindle fibers, and the chiasmata slip off the ends. The homologous chromosomes thus separate from each other, with each poleward-moving chromosome bound at only one point, the centromere. Each pair of sister chromatids compose a *dyad.*

Telophase I varies considerably between organisms. In some plants, it is eliminated altogether. In organisms exhibiting a telophase, once the

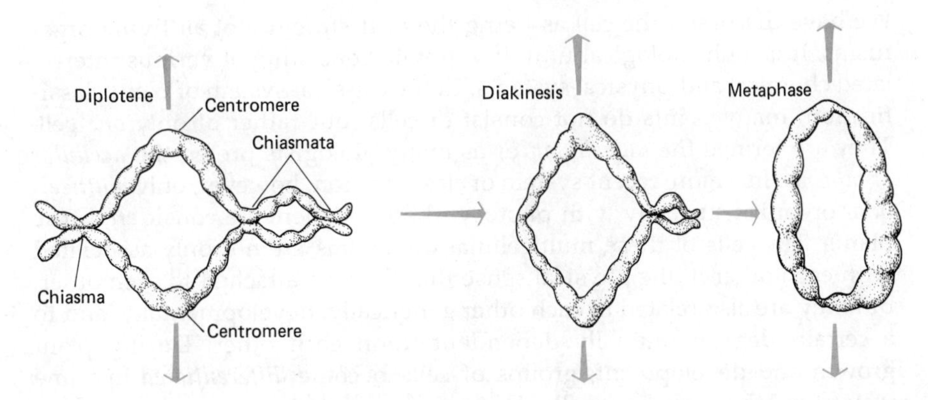


*Figure 3.25*
Terminalization of chiasmata in a bivalent during meiosis. The three chiasmata become two during terminalization, with these slipping off the ends of the chromosomes as the chromosomes are pulled apart in early anaphase.

dyads reach the spindle pole, a nuclear membrane is formed around them and the chromosomes pass into a typically very short period of inactivity termed *interphase I* or *interkinesis.* Cell division *(cytokinesis)* may occur at this time or may be postponed.

Prophase II is relatively short. It is characterized by the dissolution of the nuclear membrane, if formed, and the shortening and thickening of the dyads, which normally become somewhat long and slender during the preceding interkinesis. The formation of a second spindle occurs normally at a right angle to the first. The attachment of the centromeres to the spindle fibers and migration of the dyads toward the equator of the spindle also characterize this phase.

In metaphase II, the dyads are positioned across the equatorial plane. Upon contraction of the spindle fibers, the centromeres split, and each daughter chromosome *(monad)* separates from its sister and moves toward opposite poles. This initiates anaphase II. Telophase II and cytokinesis rapidly follow, producing four haploid cells from each diploid cell that entered meiosis. In plants, these products are generally *spores,* with gametes being formed by *mitosis.*

The term cell division has been used very loosely by some people to refer to both mitosis and cytokinesis. Do not think of it that way! Cell division is just that, the division of the cell, in another word, cytokinesis. The terms mitosis and meiosis should be restricted to divisions of the nucleus or, in a word, *karyokinesis.* Such restricted concepts must be utilized as cytokinesis may not immediately follow karyokinesis, and conversely, cytokinesis may occur some time after karyokinesis has occurred.

## PLANT TISSUES

We have discussed the cell as being the unit structure of all living organisms. It is a physiological unit, if you will, consisting of various interrelated chemical and physical systems. In the classical system of plant classification, many plants do not consist of cells, but rather of only *one* cell. They are termed the *unicellular,* or as many biologists prefer, the *acellular organisms.* In a more recent system of classification, however, only *multicellular* organisms usually with photosynthetic pigment are considered true plants. The cells of these multicellular organisms are not only associated with each other in the physical sense that they are attached to each other, but they are also related to each other genetically, developmentally, and to a certain degree, mutually dependent upon each other. During plant growth and development, groups of cells become *differentiated* in some way from other groups of cells. These differentiated groups or regions are called *tissues.*

Some biologists describe a tissue as "a group of similar cells performing a similar or related function." But function is a rather dubious criterion to hang your hat on. Certain similar cells may not be performing the same function at the same time. A particular tissue may perform differing functions at different times, and one tissue may perform the same function as another distinct tissue. Therefore, we should try to develop other criteria for the naming or identifying a tissue.

In attempting to arrive at these criteria, it might be well to consider some plant tissues and their characteristics. The cells and tissues of a plant are generally derived during the development of the embryo from the zygote. But plants, unlike anmals, have the property of open growth. Thus, certain embryonic tissue zones, the *meristems*, continue to produce cells throughout the life of the plant. This is called *primary growth*, and the resultant tissues, *primary tissues*. In many plants, certain of these primary tissues retain the ability to produce new cells. Because of their position parallel with the sides of stem and root, they are called *lateral meristems* or *cambia*. This growth from a lateral meristem is termed *secondary growth* and the resultant tissues, *secondary tissues*. Thus, one criterion for naming tissues might be *mode of origin*.

Another general way to classify plant tissues is based primarily on the nature of *wall thickening*. *Parenchyma cells* are characteristically isodiametric living cells with relatively *thin primary walls*, although secondary walls are not uncommon. The walls are essentially *evenly thickened* all the way around. Such cells compose *parenchyma tissue*. *Collenchyma cells* are also characteristically living cells but are elongated rather than isodiametric. The most common type has an *uneven thickening* of their primary walls, with the walls being thicker in the corners or angles of the cell. Such cells comprise *collenchyma tissue*. *Sclerenchyma cells* have thick, secondary, or even tertiary, often lignified walls. They are characteristically nonliving at physiological maturity. There are two common types of sclerenchyma cells: those irregular in shape, called *sclereids;* and those elongated, termed *fibers*. If such cells exist in groups, they may be termed *sclerenchyma tissue*. A type of parenchyma is termed *chlorenchyma* because of the presence of chloroplasts in those cells. Thus, color or presence of chloroplasts may also be a criterion for naming a tissue.

Specific plant tissues may illustrate other criteria. *Hypodermis* refers to a tissue located below the skin, or dermis; thus, you would expect to find this tissue lying immediately below the epidermis. *Mesophyll* has the roots *meso* ("middle") and *phyll* ("leaf") thus, you would expect this tissue to occupy the middle region of a leaf. *Epidermis* has the roots *epi* ("upon") and *dermos* ("skin"), so this would be the outermost layer of tissue, although not all outer tissues are necessarily epidermis. In this case, origin is also an important consideration. But in these examples, location is of primary concern.

In addition to the above considerations, *structure, shape, size, arrangement*, and physiological activity or *function* are also useful criteria for recognizing or characterizing tissues. So what is a tissue? We could perhaps describe it best by saying that it is *a group of cells that are similar in some manner*.

## SUMMARY

1. Cells are the unit structures of all living organisms.
2. Plant cells are bounded by a normally relatively rigid cell wall secreted by the protoplast, although under certain conditions the wall may be either plastic or elastic.
3. Protoplast is the term applied to everything within the cell and is composed of protoplasm and various inclusions.
4. Protoplasm is the living substance of the cell and can be conveniently considered as composed of cytoplasm and nucleus.
5. The protoplast is bounded by a membrane, the plasmalemma, and contains various membrane-bounded organelles and nonmembrane-bounded structures and inclusions, both organic and inorganic.
6. The plasmalemma and other cellular membranes consist of a lipid bilayer which may or may not include protein or other organic substances.
7. Membranes are important for their differential permeability and their enzymatic functions.
8. The nucleus seems to control cell division and mediate many of the cellular activities via transcription of genetic information from the DNA of chromosomes and nucleolus.
9. The endoplasmic reticulum may or may not be continuous with the outer membrane of the nuclear envelope. It may bear ribosomes or it may be smooth, and it may function in both intra- and intercellular chemical communication.
10. Chloroplasts contain, along with other pigments, chlorophyll and are associated with the process of photosynthesis. Chromoplasts are associated with the synthesis and retention of carotenoid pigments. Leucoplasts are colorless and generally associated with food storage.
11. Mitochondria are the principal sites of the respiration of organic molecules yielding energy.
12. Dictyosomes are groups of flat, disk-shaped bodies, that bud off vesicles. They may be associated with plasmalemma formation.
13. Microbodies are associated with various metabolic reactions, while microtubules apparently play a role in karyokinesis, cytokinesis, cell wall formation, and other activities.
14. Ribosomes are normally associated with the endoplasmic reticulum, although they may be found in certain organelles or free in the cytoplasm, where their role is associated with that of protein synthesis.
15. Vacuoles are surrounded by a membrane called the tonoplast and contain an aqueous solution of material called the cell sap.
16. Karyokinesis, or nuclear division, is of two types: mitosis and meiosis.
17. Mitosis consists of a series of phases whereby the genetic material is equally redistributed to two daughter cells.
18. Meiosis consists of essentially two series of phases whereby the genet-

ic material to the daughter cells is reduced by one-half, with each daughter cell receiving a single set of chromosomes.

19. Karyokinesis is generally followed by cytokinesis, the division of the cell. The cytoplasm is divided by means of a cell plate, with new cell walls being secreted on either side of the plate.
20. Cells are associated normally into tissues.
21. A tissue is a group of cells that are similar in some respect: origin, location, size, shape, wall thickening, presence of chloroplasts, genetic makeup, function, or a combination of these.

# Chapter 4

## PRIMARY ORGANS OF VASCULAR PLANTS —AN OVERVIEW

Most of you are familiar to some degree with the flowering plants. If you have not observed the growth and development of a plant, I suggest that you obtain a small pot of soil, plant several bean seeds in it, and watch the changes that occur from the planting of a seed to the harvesting of seeds. This period of development, from seed to seed, includes part of two generations. Like humans, the bean plant begins its existence as a single-celled *zygote*. The young plant, or embryo, within the seed developed from the zygote.

If we remove the outer covering of the seed, or *seed coat*, we will note the embryo, consisting of two relatively large *cotyledons* connected by a very short stalk to the top of the *hypocotyl* (Figure 4.1). As there are two cotyledons, the bean belongs to a group of flowering plants called the Dicotyledonae (*dicots* for short). Above the point of attachment of the cotyledons is the *shoot primordium*, which may consist of little more than a *meristem* (a region capable of active cell division) or, as in the bean, an *embryonic shoot* or *plumule* ("little feather" in Latin). At the base of the hypocotyl is the *embryonic root*. Some botanists like to refer to that portion of the embryo below the attachment of the cotyledons as a hypocotyl-root axis, with a root primordium covered by a root cap at its lower end. The old term *radicle* for embryonic root is not held by modern plant anatomists to be particularly appropriate so is not generally used.

Two or three days after planting, the embryonic root will have elongated, forming the *primary root*, which breaks through the seed coat and penetrates the soil. From the primary root, *lateral* or *secondary roots* develop (Figure 4.2). Next, rapid elongation of the hypocotyl lifts the hypocotyl and finally the plumule above the soil surface.

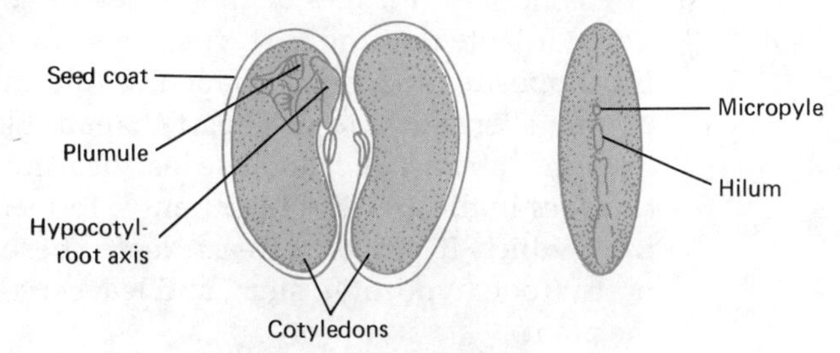


*Figure 4.1*
Seed of the garden bean, *Phaseolus vulgaris*. Seed shown open and from external edge.

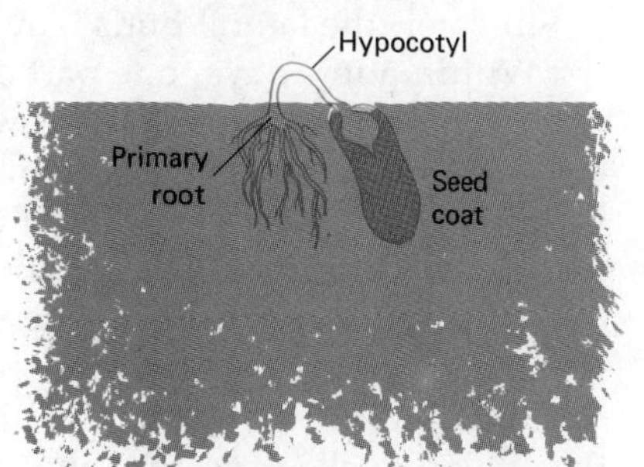


*Figure 4.2*
Germinating seed of *Phaseolus*.

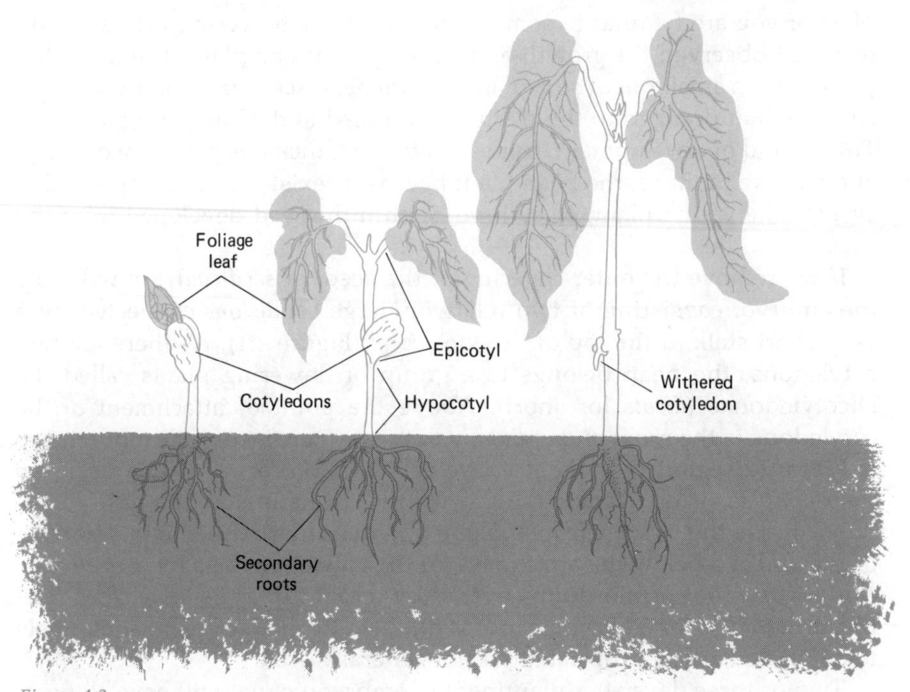


*Figure 4.3*
The seedling of *Phaseolus.* The cotyledons become green and photosynthesis occurs within them. After development of the leaves, the cotyledons wither and eventually drop off.

As the cotyledons increase in size, the seed coat falls away and a *stem* (the shoot) with *leaves* begins developing from the plumule (Figure 4.3). You will note that the first true leaves developing from the stem are simple, opposite, and heart shaped, while those closer to the top are compound, alternate, and not heart shaped. Elongation of the stem continues from the *terminal bud,* and the leaf-bearing branches begin to develop *lateral buds* in the axil (the upper angle formed by a twig or leaf and the stem from which it grows) of each leaf. These four morphologically distinct parts, (root, hypocotyl, stem, and leaves) constitute the *vegetative organs* of the plant.

Still later, the lateral buds may produce either lateral branches or *flowers.* Within a few days, one part of the flower, the *pistil,* will begin to enlarge to form a pod or *fruit.* It is within the fruit that *seeds* develop. Flowers, fruits, and seeds have been referred to as the *reproductive organs* of the plant, but of course the latter two organs actually occur as the result of reproduction, in most instances (Figure 4.4).

For comparative purposes you might also like to plant some corn grains. If you attempt to peel off the outer covering as you did the bean seed, you will find it to be very difficult. This is because the corn grain is a fruit with the ovulary wall *(pericarp)* and ovule wall *(seed coat)* tightly fused (Figure 4.5). If you soak the grains overnight, this layer will be more easily re-

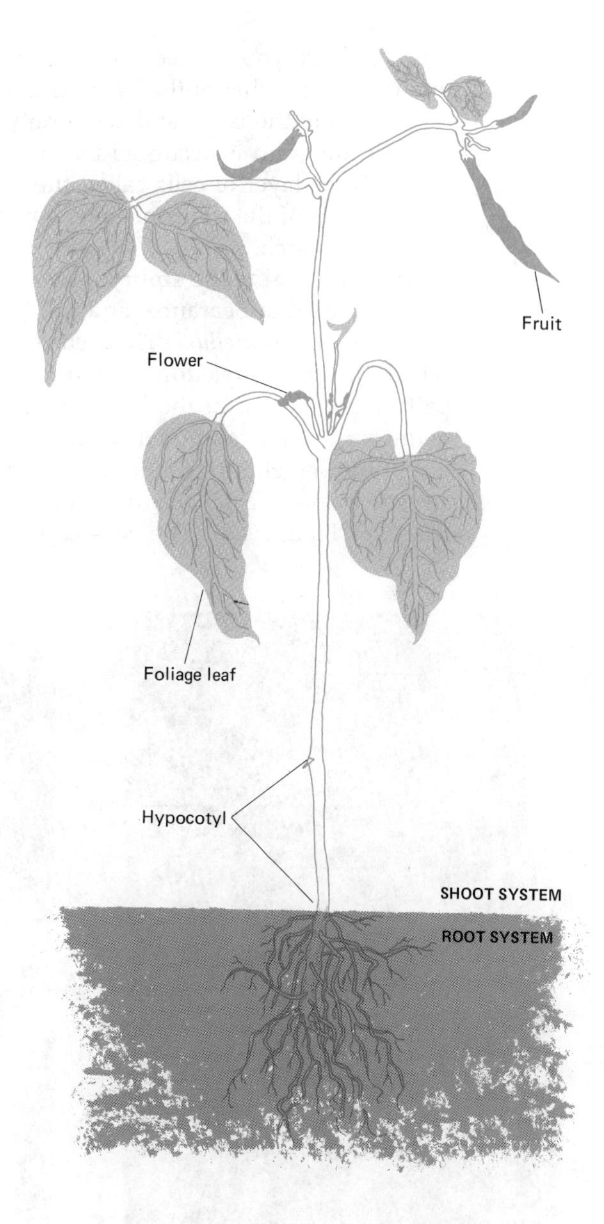


*Figure 4.4*
The mature plant of *Phaseolus*. Flowers are borne from axillary buds. Eggs are produced within the ovule and, following pollination and subsequent fertilization, an embryo will be formed within the ovule. The integuments of the ovule become modified forming the seed coat, with the modified ovule and its enclosed embryo being called the seed. The ovulary also undergoes modification, forming the fruit.

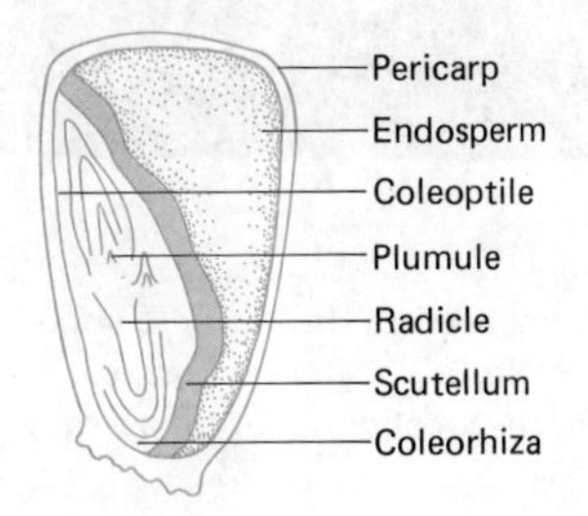


*Figure 4.5*
The corn grain. This grain is actually a one-seeded fruit, with the seed coat and fruit wall fused together.

moved. Once you succeed in removing the coat, you will note that the embryo, unlike that of the bean, is a relatively small, shield-shaped structure found at the base, and seemingly on only one side, of the grain. The bulk of the grain is occupied by the *endosperm*, composed of 1) an outermost, single layer of cells called the *aleurone layer* and 2) a starchy endosperm. Cells of the aleurone layer contain fats and proteins, pigments, but little or no starch.

As you examine the embryo, you will note that the structure giving it a shield-shaped appearance, and the largest part of the embryo, is a single cotyledon, the *scutellum.* Thus, corn is representative of a group of plants called the Monocotyledonae or simply *monocots.* This cotyledon, unlike the pair of cotyledons in the bean embryo, remains within the seed during germination. The main axis of the embryo *appears* to be a lateral appendage of the scutellum. It consists of a *shoot apex* surrounded by a sheathing structure, the *coleoptile*, and a *root apex* at the opposite end, covered by a *root cap* (Figure 4.6). The root cap is surrounded by another sheathing

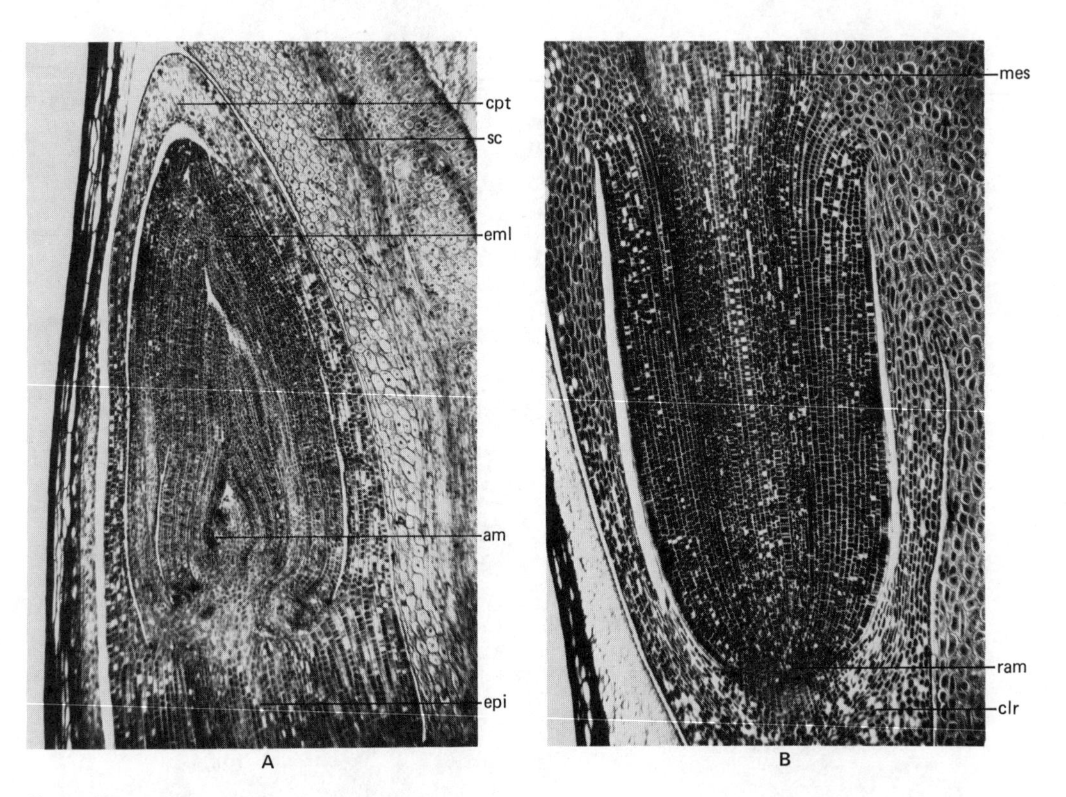


*Figure 4.6*
*A.* Longitudinal section through the shoot of a corn embryo. *B.* Longitudinal section through the root of a corn embryo. *Sc*, scutellum; *cpt*, coleoptile; *eml*, embryonic leaves; *am*, apical meristem; *epi*, epicotyl; *mes*, mesocotyl; *ram*, root apical meristem; *clr*, coleorhiza.

structure, the *coleorhiza*. A short internode or *mesocotyl* develops between the point of attachment of the scutellum *(cotyledonary node)* and the point of attachment of the coleoptile *(coleoptile node)*. Not all monocot embryos have a mesocotyl. Most botanists consider a hypocotyl to be lacking in corn. But if one is present in a monocot seed, it would be located between the cotyledonary node and the root primordium, and would be extremely short. It does not elongate during germination of the seed.

As you observe the germinating seed, you will note that the coleorhiza is the first structure to pierce the seed coat. The embryonic root later elongates the breaks through the coleorhiza. From the primary root, secondary or lateral roots develop. The coleoptile elongates and pierces through the seed coat and is the first structure to appear above ground. This is followed by elongation of the embryonic shoot which breaks through the coleoptile (Figure 4.7). Unlike the bean, the primary roots of corn are

*Figure 4.7*
The germinating corn embryo.

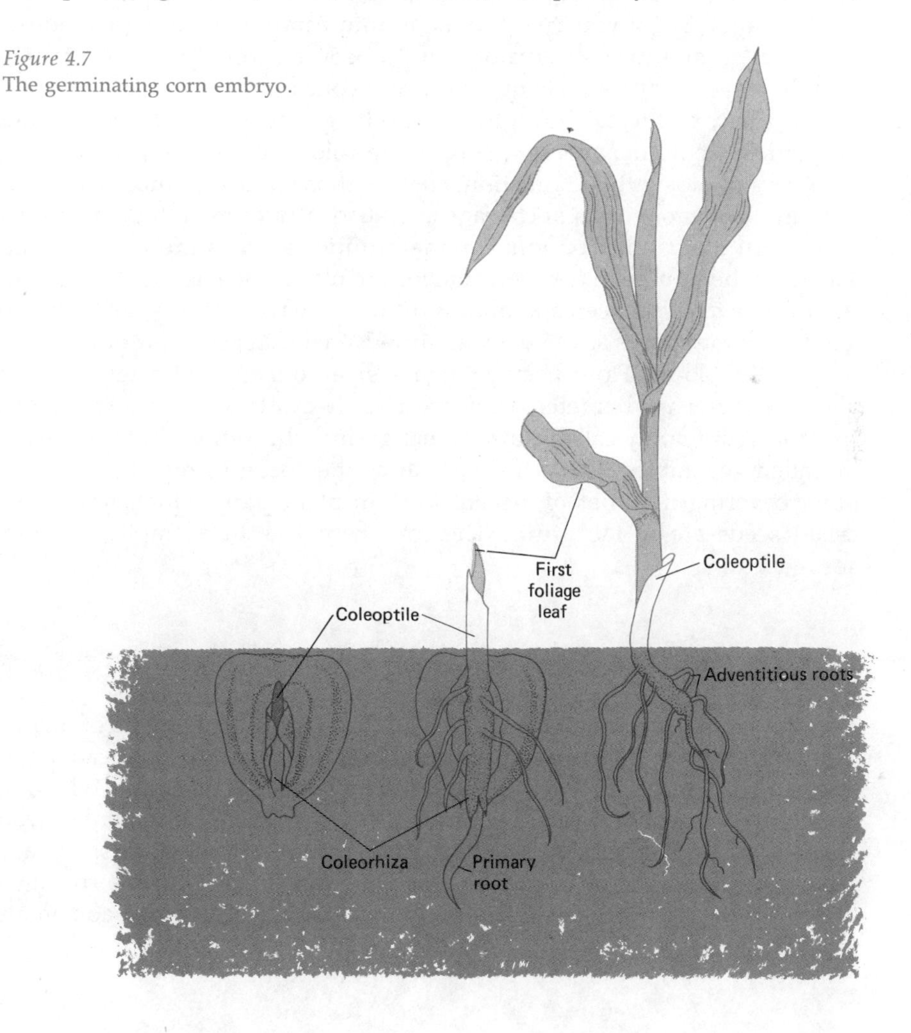

*ephemeral* (short lived). Thus, the root system of mature corn plants and many other monocots are *adventitious* (not originating from the base of the hypocotyl or from other roots). Adventitious roots are borne by both the mesocotyl and the stem. Certain of these become the aboveground *prop roots*, while others form the underground *fibrous root system* of the mature plant.

You had probably already formed certain concepts regarding plant organs before reading this chapter. You may think of a root as the underground, colorless, anchorage and absorption organ of the plant; the stem as the aboveground elongated, cylindrical, supportive structure; and the leaves as green, broad and flat, lateral appendages of the stem. You may think of flowers as showy, brightly colored structures, fruit as something like the apple or orange, and seeds as structures covered by a hard outer layer. These concepts may serve to help you identify the organs of a number of plants. But if you take a walk around campus, visit a greenhouse, botanical garden, or conservatory, and observe carefully the plants you find there, you may begin to question your concepts.

Roots may be found aboveground and be green such as the aerial roots of orchids. Roots may serve a supportive role, such as the prop roots of corn, or a photosynthetic function, such as those of the orchids. They may not have any more of an anchorage and absorption function than another plant part. Stems may be found underground, such as the potato tuber. They may be nongreen and help anchor the plant, such as the rhizomes of *Iris*, or they may be green and photosynthetic. Leaves are not always broad and flat. Some spines are actually modified leaves. Leaves may be colorless or brightly colored. Flowers may range in size from approximately 1 mm to almost a meter in diameter. They are not always brightly colored. Some fruits are commonly called nuts, others grain, while still others may even be called vegetables. Some fruits, such as the coconut, may have a hard outer covering like that of a seed. Certain plant parts which have been called seeds are in fact fruits. The corn kernal is an example of a one-seeded fruit.

## EDIBLE PLANT PARTS

Man at one time or another has eaten all of the principal organs of plants. You might find it interesting to make a list of plants you have eaten and attempt to determine whether you were eating stem, hypocotyl, root, leaves, flowers, fruit, or seeds. For example, when you eat a cucumber, you are eating a fruit; a potato, a stem; a carrot, a hypocotyl-root axis; a pea, a seed; green beans, a fruit; cabbage, leaves (until you get to the so-called "heart" which is the stem); sweet potatoes, a root. You may add many others to your list.

## USUAL FEATURES OF SEED PLANTS

It is sometimes taken for granted that all seed plants have the six principal organs in some form or another. This is not true. Certain seed plants may lack one or more of these organs. A tiny plant called *Wolffia* (water meal) has neither stem, root, nor leaves and bears one of the smallest flowers known in the plant kingdom. A close relative, *Lemna* (duck-weed), has a single root attached to a flattened, undifferentiated (no stems or leaves) plant body. Other plants, such as the carrot or common dandelion, have conspicuous roots and leaves but very inconspicuous stems. Some cacti have stems and roots but no leaves. The Spanish moss (not really a moss but a member of the pineapple family of flowering plants) of Florida and other Southern States has stem and leaves but no roots. A tropical parasitic plant *Rafflesia* has only a short stem, and neither roots nor leaves.

In plants other than flowering plants, various organs are lacking. The pine lacks flowers and fruits, but bears seeds and has stems, roots and leaves. Certain of the so-called lower vascular plants lack flowers, fruit, seeds, roots, and leaves, having only a stem.

## A SMALL CONTROVERSY

This section is directed primarily to those students who may continue studies in the field of botany. It is intended to stimulate some thought and introduce some aspects that will be covered in greater detail in later chapters on organ morphology and anatomy. Obviously, until you have studied those chapters, some of this discussion will be difficult to understand, so you may wish to come back and read this through again.

There are some differences of opinion concerning the nature of the hypocotyl and the cotyledon. Some botanists consider the hypocotyl to be rootlike (hypocotyl-root axis), while others consider it to be stemlike (hypocotyl-shoot axis). Still others consider it, as I do, to be a separate organ. If we consider for a moment, the characteristics of the hypocotyl, we find that in some plants, it is morphologically indistinguishable from the stem. In other plants, it is indistinguishable from the root. But it can also be quite morphologically distinct from both stem and root, in size, in position, and even in color. This is particularly true in the seedling. Differences exist among plants concerning the anatomical aspects of the hypocotyl as well. In some plants, its anatomy may be essentially similar to that of stems; in others, to that of roots. But in some plants, the hypocotyl may contain tissues not found in either stem or root, and its tissue *organization* may also be distinctive. Table 4.1 summarizes some of the similarities and differences between hypocotyl, stem, and root.

*Table 4.1*
**COMPARISONS BETWEEN HYPOCOTYL, EPICOTYL (STEM) AND ROOT**

| HYPOCOTYL | EPICOTYL | ROOT |
|---|---|---|
| Origin from intercalary meristem | Origin from apical meristem | Origin from apical meristem |
| Lateral appendages exogenous in origin (derived from outer tissues), and not derived from apical meristem | Lateral appendages exogenous in origin, derived from apical meristem | Lateral appendages endogenous in origin (derived from inner tissues), not derived from apical meristem |
| Well-differentiated in embryo | Not completely differentiated in embryo | Fairly well differentiated in embryo |
| Grows for only a short time | May grow over a long period of time | May grow over a long period of time |
| Lacks epidermal hairs | May have epidermal hairs | May have epidermal hairs |
| Cortex relatively thick | Cortex relatively thin | Cortex relatively thick |
| Pith small | Pith larger | Pith lacking (in many dicots) |
| Hypodermis not well differentiated | Hypodermis well differentiated | Hypodermis well differentiated |
| Endodermis not very distinct | Endodermis indistinct | Endodermis prominent |
| Usually has four vascular bundles | Usually has four or more vascular bundles | Vascular tissue not in bundles |
| Cork cambium differentiates in same or deeper tissue than in stem or root | | |

You are probably thinking, "What difference does it make how we consider the hypocotyl?" To the morphogeneticist, plant breeder, and others concerned with the practical application of genetic expression, the structural and ontogenic (pattern of development) affinity of different structures is of utmost importance.

What about the cotyledons? These are commonly known as the *seed leaves*, which suggests that they are leaves or modified leaves. This idea gained prominence as early as 1790 when the great German philosopher-biologist-writer Goethe introduced his concept of homology (similarity of origin but not necessarily of function) in his essay *Metamorphosis in Plants.* Goethe stated that there was no real difference between cotyledons, foliate leaves, bracts, and flower parts. All are leaves, as they are all lateral outgrowths of the shoot axis. One of the foremost American plant anatomists has stated, "The structural relation between hypocotyl and cotyledons is comparable to that between stem and leaves. Thus, the beginning of shoot organization is found in the hypocotyl-cotyledon system in which the hypocotyl is the first stem unit and the cotyledons the first leaves." Note that this concept entails the acceptance of a stemlike nature for the hypocotyl, a notion I find unacceptable. Let us consider just one simple point of comparison: origin. Foliage leaves, bracts, tendrils, spines, flower parts all have their origin from cells of the apical meristem. This is not so for cotyledons. Thus, simply on the basis of origin, I would not consider cotyledons to be leaves. "But," you say, "they are green." So are some stems and roots. "But," you continue, "they are broad and flat and vascularized like leaves." So are some stems. In addition, the vascularization of cotyledons and foliage leaves may be different in the same plant. "But at least they are lateral in position like leaves." Not always, they may be terminal in position. For these and other reasons, I see little purpose in retaining the term *seed leaves* for cotyledons. Foliar leaves may also be already formed in some seeds, why not call them seed leaves too? Rather, I would simply call them cotyledons and let it go at that. It is interesting to note that a foremost botanist has postulated recently that flowers are not modified leaves either. At present his ideas are not widely accepted. Sometimes new ideas find very slow exceptance in the scientific community, particularly if such ideas are at variance with those of the experts in that field.

## SUMMARY

1. The principal organs of flowering plants include stems, roots, leaves, flowers, fruits, and seeds.
2. The first three are termed vegetative organs, the latter three reproductive organs.
3. Certain flowering plants may lack one or more of these organs.
4. Organs may be modified in various ways so that function, shape, and location may not provide a very reliable way for distinguishing one organ from another.
5. The hypocotyl may be considered as a vegetative organ separate from that of root and stem.
6. The consideration of cotyledons as seed leaves may be debated.

# Chapter 5

## PLANT TAXONOMY AND BOTANICAL NOMENCLATURE

When primitive man determined that certain plants were edible and others inedible, he was assuming one of the roles of a plant *taxonomist*, because taxonomy, in part, is concerned with the grouping of plants on the basis of certain similarities. Any system of classification presupposes the adoption of names for the items being classified. Thus, *nomenclature* (a system of names) is also an integral part of taxonomy.

The difficulties generated by the use of local common names for plants is well illustrated in Biblical references. The lily is probably the most famous plant of the Bible, being referred to in both Old and New Testaments. But Biblical scholars believe the term *lily* refers to five or six different plants, including *Anemone*, *Nymphaea*, *Iris*, *Cyclamen*, *Anthemis*, and *Lilium*, the true lily. Obviously, a better system than common names is needed. Each plant should be given a name that would identify that plant in every part of the world. This is exactly what plant taxonomists have done.

## THE GREEK AND ROMAN AGE

Modern plant taxonomy had its beginnings with the philosophers of ancient Greece. Theophrastus (370–285 B.C.) is considered by many scholars to be the father of modern botany (Figure 5.1). In the manner of ancient

*Figure 5.1*
Theophrastus. [Courtesy of Hunt Botanical Library, Pittsburgh, Pa.]

Greek philosophers, his extant writings on plants mix sound scientific inquiry with philosophical speculation, but little better was to come along for 18 centuries. Theophrastus brought together the names of nearly 500 plants and, because his books were widely read, gave those names general recognition. He classified plants into four groups: herbs, undershrubs, shrubs, and trees.

The Romans were inclined to be more practically oriented. The applied fields of agriculture and pharmacy flourished, yet little of real botanical substance resulted during the period of Roman culture. There are some scientists that go so far as to state that Rome did little or nothing for pure science. What was done was accomplished by transplanted Greeks or descendants of Greeks, such as Galen. Be that as it may, there were two books written that had a great influence on botanical studies. True to the Roman spirit, the plants listed therein were considered to be either food plants or of medicinal value. The books are Pliny's *Historia Natralis* (77 A.D.) and Dioscorides' *Materia Medica* (written about the same time). Many of the vernacular names used by Dioscorides are still being used today. His was the first known book to contain illustrations of plants, all laboriously copied by hand. The influence these books had on botanical studies was, however, largely negative. They were held in such high esteem by European scholars that it was considered heresy to dispute them. Individual initiative and inquiry was thus stymied. For almost 1500 years, little was added to botanical knowledge.

## THE AGE OF HERBALS

The European awakening of botanical investigative spirit began about 1500 with the interest, naturally enough, directed toward plants having medicinal value. Because of this overwhelming interest in herbs (nonwoody plants), the next 200 years has become known as the Age of Herbals. Publications containing beautiful woodcuts of plants (some of them admittedly quite fanciful because of the bias of the artist) characterize this period (Figure 5.2). Many of these writers were still strongly influenced by the *Materia Medica* and the beliefs and superstitions of the Middle Ages. With this background in mind, it is not difficult to conceive of certain of these investigators having the belief that *all* plants must have medicinal properties if they could but discover them. This led to a curious development known as the *Doctrine of Signatures.* This doctrine was probably stated most clearly by the botanist Robert Turner in 1664. He wrote, "God hath imprinted upon the plants, herbs, and flowers, as it were in hieroglyphicks, the very signature of their virtues." For example, it was believed that the adder's-tongue fern would cure the bite of an adder; the maidenhair fern would prevent baldness; the lungwort, diseases of the lung; the liverwort, diseases of the liver; the bloodwort, diseases of the

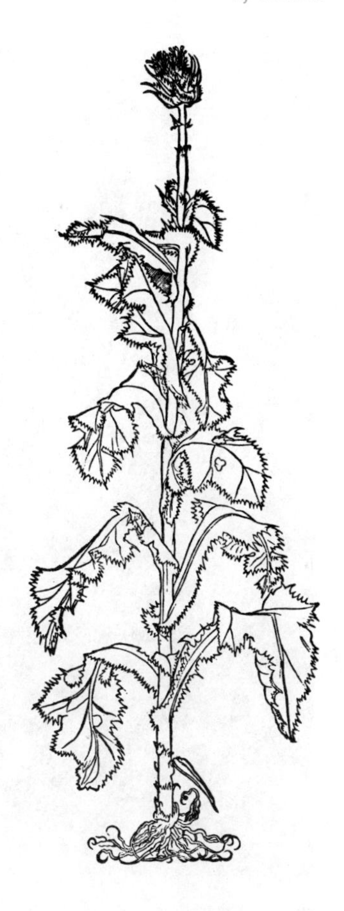

*Figure 5.2*
A plate from Otto Brunfel's *Herbarus Vivae Eicones*, the first volume of which appeared in 1530 with the third and final volume issued posthumously in 1536. [Courtesy of Hunt Botanical Library, Pittsburgh, Pa.]

blood; the walnut, as it had the appearance of the brain, headaches or diseases of the brain. Some horrible tasting concoctions were thus forced down "for the good of the body." Like some folk medicine today, the worse it tasted, the more it would help.

Botany eventually began to acquire the role of an independent science and withdrew from its dependency upon medicine. Efforts were made to classify plants according to their *natural* relationships. This type of system was associated largely with the so-called French school of botanists of the latter part of the eighteenth and early nineteenth centuries. The term *natural* in this case did not imply evolutionary. Rather, they looked at *all* visible characters of the plant and grouped then on the basis of their *overall* similarity. This was a radical departure from previous systems, which had been based on just a few characters.

Although the work just mentioned came somewhat later, the Age of Herbals essentially ended with a system of classification by the Swedish biologist Carolus Linnaeus (Figure 5.3). Linnaeus was an unusual individual. He held both the post of professor of medicine *and* professor of botany

*Figure 5.3*
Linnaeus. [Courtesy of Hunt Botanical Library, Pittsburgh, Pa.]

at the same time in the University of Uppsala. The most important aspect of Linnaeus' work was his ability to recognize the value in earlier works, value of which even their authors were possibly unaware, and fashion them into a serviceable whole. Most scholars consider the greatest significance of his work to be the development of a precise system of nomenclature. In 1753, Linnaeus published a two volume work titled *Species Plantarum* ("The Kinds of Plants"). See Figure 5.4. This work leaned heavily on the past works of many botanists. In it he described every known plant, giving each a *generic* name, an idea which did not originate with him. A number of descriptive names followed the generic name, as was also the custom at that time. The resulting *polynomial* (*poly*, "many"; *nomius*, "names") was considered by Linnaeus to be the proper name for the species. But he made one very important deviation from the established customs for naming plants. In the margin of the page, opposite each polynomial, he entered a single word which he called the *trivial* name for the species. For example, catnip was designated *Nepeta floribus interrupte spicatus pedunculatis* (i.e., "Nepeta with flowers in an interrupted peduncu-

*Figure 5.4*
Sample page from Linnaeus' *Species Plantarum.* [Courtesy of Hunt Botanical Library, Pittsburgh, Pa.]

1464 DIOECIA OCTANDRIA.

Populus alba. *Dod. pempt.* 835.
*Habitat in* Europa *temperatiori.* ♄.
*Flores omnino ut in P. tremula.*

tremula. 2. POPULUS foliis ſubrotundis dentato-angulatis utrinque glabris. *Hort. cliff.* 460. *Fl. ſvec.* 819; 909. *Roy. lugdb.* 82. *Hall. helv.* 156. *Gmel. ſib.* 1. *p.* 151. *Dalib. pariſ.* 301.
Populus tremula. *Baub. pin.* 429.
Populus lybica. *Dod. pempt.* 826.
*Habitat in* Europæ *frigidioribus.* ♄.
Folia *involuta baſi interne glandula duplici.*

nigra. 3. POPULUS foliis deltoidibus acuminatis ſerratis. *Hort. cliff.* 460. *Fl. ſvec.* 821; 911. *Mat. med.* 462. *Roy. lugdb.* 82. *Gron. virg.* 194. *Hall. helv.* 156. *Gmel. ſib.* 1. *p.* 151. *Dalib. pariſ.* 301.
Populus nigra. *Baub. pin.* 429. *Dod. pempt.* 836.
*Habitat in* Europa *temperatiore.* ♄.
Folia *involuta abſque glandulis ad baſin, hinc diverſa a præcedentis.*

balſamifera. 4. POPULUS foliis ſubcordatis denticulatis. *Hort. cliff.* 460. *Roy. lugdb.* 82.
Populus foliis cordatis crenatis baſi nudis, petiolis teretibus. *Wach. ultr.* 294.
Populus nigra, folio maximo, gemmis balſamum odoratiſſimum fundentibus. *Cateſb. car.* 1. *p.* 34. *t.* 34. *Duham. arb.* 2. *p.* 178. *t.* 38. *f.* 6.
Populus foliis ovatis acutis ſerratis. *Gmel. ſib.* 1. *p.* 152. *t.* 33.
*Habitat in* America *ſeptentrionali.* ♄.

heterophylla. 5. POPULUS foliis cordatis primoribus villoſis.
Populus foliis cordatis obſolete ſerratis: infimis ſerratis glanduloſis, petiolis lateraliter utrinque planis. *Wach. ultr.* 294?
Populus magna virginiana, foliis ampliſſimis, ramis nervoſis quaſi quadrangulis. *Duham. arb.* 2. *p.* 178. *t.* 39. *f.* 9.
Populus magna, foliis amplis: aliis cordiformibus; aliis ſubrotundis: primoribus tomentoſis. *Gron. virg.* 194. 157.
*Habitat in* Virginia. ♄.

RHO-

DIOECIA ENNEANDRIA. 1465

RHODIOLA.

1. RHODIOLA. *Fl. lapp.* 378. *Fl. ſvec.* 831; 912. *Mat. med.* 477. *Hort. cliff.* 470. *Roy. lugdb.* 457. Roſea.
Rhodia radix. *Baub. pin.* 286. *Cluſ. hiſt.* 2. *p.* 65. *Raj. hiſt.* 690.
Telephium luteum minus, radice roſam redolente. *Moriſ. hiſt.* 3. *p.* 468. *ſ.* 12. *t.* 10. *f.* 8.
*Habitat in Alpibus* Lapponiæ, Auſtriæ, Helvetiæ, Britanniæ. ♃.

*ENNEANDRIA.*

MERCURIALIS.

1. MERCURIALIS caule ſimpliciſſimo, foliis ſcabris. *Hort. cliff.* 461. *Fl. ſvec.* 823; 913. *Roy. lugdb.* 203. *Dalib. pariſ.* 302. perennis.
♀. Mercurialis montana teſticulata. *Baub. pin.* 122.
CynoCrambe mas. *Cam. epit.* 999.
♂. Mercurialis montana ſpicata. *Baub. pin.* 122.
Cynocrambe femina. *Cam. epit.* 999.
*Habitat in* Europæ *nemoribus.* ♃.

2. MERCURIALIS caule brachiato, foliis glabriuſculis, floribus verticillatis: femineo maſculiſque. *Linn. dec.* 1. *p.* 15. *t.* 8. *. ambigua.
*Habitat in* Hiſpania.
*Simillima M. annuæ.*

3. MERCURIALIS caule brachiato, foliis glabris, floribus ſpicatis. *Hort. cliff.* 461. *Hort. upſ.* 298. *Mat. med.* 464. *Roy. lugdb.* 263. *Dalib. pariſ.* 302. annua.
Mercurialis teſticulata ſ. mas. *Baub. pin.* 121. ♀.
Mercurialis mas. *Dod. pempt.* 658. ♀.
♂. Mercurialis ſpicata ſ. femina. *Dod. pempt.* 658.
*Habitat in* Europæ *temperatæ umbroſis.* ☉.

4. MERCURIALIS caule ſubfruticoſo, foliis tomentoſis. *Hort. upſ.* 461. *Roy. lugdb.* 203. *Sauv. monſp.* 128. *Gouan. monſp.* 507. tomentoſa.
Phyllon teſticulatum. *Baub. pin.* 122. ♀.
Phyllum marificum. *Cluſ. hiſt.* 2. *p.* 48. ♀.
♂. Phyllon ſpicatum. *Baub. pin.* 122.
Phyllon feminificum. *Cluſ. hiſt.* 2. *p.* 48.
*Habitat in* G. Narbonenſi, Hiſpania. ♄.

Zzzzz 5 HY-

late spike''), and in the margin, he placed the trivial name, *cataria* (''of cats''). Although he probably did not intend it so, users of the books found it obvious and convenient to combine the generic and trivial name, a custom Linnaeus also adapted in later publications. Thus was established the current system of *binomial nomenclature* by which every particular kind of plant is given two names, a generic name and a specific name. Together, these two names establish the *species.* As Latin was the language of scholars at the time of Linnaeus, plant names were given in Latin, a practice which persists today.

To classify plants under his new system, he devised what is known as the sexual system of classification. Dr. William T. Stearn, in a recent article in *Taxon* (1976), recounts this system in an entertaining fashion:

> Linnaeus' adolescent imagination sought human parallels with what he observed as the sexuality of plants. Comparing the stamens to husbands, the carpels to wives, the disc to a marriage bed, and the ray florets to harlots or concubines [*Note:* Here Stearn speaks of a type of flower similar to an aster or daisy], he found in the sexual arrangements of plants an entertaining array of situations, such as *Polygamia superflua, Polygamia necessaria,* etc. Gradually his so-called ''sexual system'' of classification took shape; basically it was a method of arranging genera into major groups (classes), such as *Monandria, Diandra, Triandria,* etc., according to the number of their stamens, and then into subgroups (orders), such as *Monogynia, Digynia, Trigynia,* etc., according to the number of stigmas; although arithmetical, it was made more interesting by Linnaeus' anthropomorphic metaphorical names and explanations.

## THE MODERN AGE

Several important taxonomic works followed those of Linnaeus. But the event which had the most impact, and is still being debated today, occurred July 1, 1858. For the numerous publications, debates, and sermons it has since elicited, it had a rather inauspicious beginning. A joint paper by C. Darwin and A. R. Wallace, "On the Tendency of Species to Form Varieties; and on the Perpetuation of Varieties and Species by Natural Means of Selection," was read before the Linnaean Society of London (Figure 5.5). The paper elicited little excitement among the assembled scientists, and Darwin himself was not even present. But the effect of Darwin's book *On the Origin of Species by Means of Natural Selection, or the Preservation of Favoured Races in the Struggle For Life*, published little more than a year later in 1859, was immediate, astounding, and controversial. Until that time all botanists believed, or at least felt compelled to express belief, in the doctrine of *constancy of species.* Each organism was specifically formed by the Creator and existed in exactly the form in which He had created it. Darwin's work (understandably more widely known by its shortened name *Origin of Species*) presented so many examples and arguments in such a straightforward, logical, unemotional way, that little doubt could remain concerning the mutability and evolution of species.

Once the theory of evolution became established, it was necessary for taxonomists to devise systems that would attempt to account for the evolutionary or *phylogenetic* relationships of plants. As little information was available concerning the simpler plants, many systems of classification

*Figure 5.5*
Charles Darwin. [Courtesy of Hunt Botanical Library, Pittsburgh, Pa.]

considered only the seed plants; some only the flowering plants. Of those systems which considered all organisms, one must again go back to Linnaeus for the classical or traditional system. Linnaeus recognized matter as being classified into three groups which he termed *kingdoms:* animal, mineral, and plant. The term *kingdom* for the major groups of organisms is retained, with the traditional system grouping organisms into either Plantae or Animalia.

Obviously, if we are to consider grouping different kinds of *plants*, it is germane to ask, "Just what is a plant?" You probably have a pretty good notion what a plant is, and hopefully, after completing this text, you will have formed a conceptual idea of a plant. But even then you, and I, would have great difficulty providing a simple definition of the word *plant.* Before the advent of the microscope, it seemed a simple matter to distinguish plants from animals: plants were green and nonmobile, while animals were nongreen and mobile. It is obvious that a horse is an animal, and oak tree a plant.

As more information was accumulated concerning the chemistry, reproduction, and structure of organisms, particularly the microscopic or near-microscopic forms, the former sharp distinctions between plant and animal became hazy. Additional criteria were added to plant characteristics: presence of a cellulose wall; production of spores by meiosis, gametes by mitosis; lack of organ systems such as those concerned with irritability or excretion; storage of food in the form of starch; indeterminate growth; absence of centrioles; different mechanism of cytokinesis; and others. Animals, on the other hand, lack a cellulose wall, produce gametes by meiosis, do not produce spores, have organ systems, store food in the form of glycogen, have determinate growth, have centrioles, and so on.

Even with all the additional criteria mentioned above, certain organisms seem to defy pigeonholing. For example, some organisms have chlorophyll and starch, and exhibit many of the other characteristics of plants mentioned above, but may have their cell wall composed of silica rather than cellulose; or they may lack entirely a cell wall. Other organisms may exhibit many plant characteristics but either lack chlorophyll, or possess motility, or form gametes by meiosis, or experience determinate growth, or have centrioles associated with their nuclei, or do not store starch, etc. Although basically plantlike, such organisms exhibit one or more animallike characteristics. Other organisms, while basically animallike, may lack organ systems, be nonmotile, have indeterminate growth, etc.

How do we resolve this quandary? Obviously, another kingdom is needed. All those organisms that are either motile, lack chlorophyll, or are one celled we will place in the kingdom *Protista.* But wait a minute. Some of these organisms are prokaryotic, others eukaryotic. Some are unicellular but have chlorophyll and are nonmotile. How do we resolve these problems? Let's create more kingdoms. The prokaryotic organisms can all be

*Table 5.1*
KINGDOM CLASSIFICATIONS

| SYSTEM 1 | SYSTEM 2 | SYSTEM 3 | SYSTEM 4 |
|---|---|---|---|
| Traditional, see Altman and Dittmer (1972) | Dodson (1971) | Curtis (1968) | Stanier et al. (1970) |
| **PLANTAE** | **MONERA (MYCHOTA)** | **PROTISTA** | **PROTISTA** |
| Bacteria | Bacteria | Bacteria | Bacteria |
| Blue-green algae | Blue-green algae | Blue-green algae | Blue-green algae |
| Green algae | **PLANTAE** | Protozoa | Protozoa |
| Chrysophytes | Green algae | Slime molds | Green algae |
| Brown algae | Chrysophytes | **PLANTAE** | Chrysophytes |
| Red algae | Brown algae | Green algae | Brown algae |
| Slime molds | Red algae | Chrysophytes | Red algae |
| Fungi | Slime molds | Brown algae | Slime molds |
| Bryophytes | Fungi | Red algae | Fungi |
| Tracheophytes | Bryophytes | Fungi | **PLANTAE** |
| **ANIMALIA** | Tracheophytes | Bryophytes | Bryophytes |
| Protozoa | **ANIMALIA** | Tracheophytes | Tracheophytes |
| Multicellular animals | Protozoa | **ANIMALIA** | **ANIMALIA** |
| | Multicellular animals | Multicellular animals | Multicellular animals |

placed together in the kingdom *Monera;* the eukaryotic but not truely multicellular organisms can be left in the kingdom *Protista*. The fungi, being so different from everything else, can have their own kingdom *Fungi.* Green plants not possessing characteristics of the above can constitute the kingdom *Plantae*, while multicellular animals constitute the *Animalia*. Now we have solved the problem of what is plant and what is animal, right? No, we have just created a new set of problems. In 1975, Margulis modified the five kingdom system, not by creating new kingdoms, but by shifting around certain groups of organisms into other kingdoms. Some of the various kingdom classification systems which have been formulated are summarized in Table 5.1. Have her modifications solved the problem of plant classification? Undoubtedly not, for plant classification will always be in a state of flux due to increased information and knowledge. Certain species will be so unique as to defy classification. Does this mean we should create a new kingdom for such organisms? Obviously it makes no sense to do so, as such severe splitting will only add to the confusion. When we get right down to it, what difference does it make what we label an organism? The characteristics of the organism remain the same no matter what man-made scheme is employed to classify it. It has the same morphology, anatomy, genetic characteristics, physiology, growth and de-

| SYSTEM 5 | SYSTEM 6 | SYSTEM 7 |
|---|---|---|
| Copeland (1956) | Whittaker (1969) | Whittaker (modified by Margulis, 1974a, 1971a, and in this chapter) |
| **MONERA** | **MONERA** | **MONERA** |
| Bacteria | Bacteria | Bacteria |
| Blue-green algae | Blue-green algae | Blue-green algae |
| **PROTOCTISTA** | **PROTISTA** | **PROTISTA** |
| Protozoa | Protozoa | Protozoa |
| Green algae | Chrysophytes | Green algae |
| Chrysophytes | **PLANTAE** | Chrysophytes |
| Brown algae | Green algae | Brown algae |
| Red algae | Brown algae | Red algae |
| Slime molds | Red algae | Slime molds |
| Fungi | Bryophytes | Flagellated fungi |
| **PLANTAE** | Tracheophytes | **FUNGI** |
| Bryophytes | **FUNGI** | Amastigote fungi |
| Tracheophytes | Slime molds | **PLANTAE** |
| **ANIMALIA** | Fungi | Bryophytes |
| Multicellular animals | **ANIMALIA** | Tracheophytes |
| | Multicellular animals | **ANIMALIA** |
| | | Multicellular animals |

velopment, response to environment no matter which group we place it in, and is that not what we are *really* trying to learn?

Admittedly, we have approached this whole idea of plant classification with tongue in cheek. The various classifiers are trying to group the organisms on the basis of what seems, to them at least, to be natural affinities or evolutionary relationships. If we can determine along what lines a plant evolved, we will have greater knowledge concerning the nature of that plant. The major criterion for a classification system is that it be workable and make sense to the individual using it.

The fact that changes are being made in the way a particular organism, or group of organisms, may be classified reflects the new tools and techniques available to the modern plant taxonomist. No longer are morphology and anatomy the only criteria available. Plants produce thousands of chemicals, each being formed via specific biochemical pathways. If particular plants have a great number of certain types of compounds in common, it is reasoned that they are probably closely related. The biochemical pathways necessary to form those compounds probably did not arise independently many different times. Thus, phenolic compounds can be analyzed via a technique called two dimensional paper chromatography. Alkaloids, terpenes, flavones, fatty acids, and various types of pigments can be analyzed by chromatography or other techniques. Enzymes and other proteinaceous compounds may also be important in the study of plant relationships. These can be analyzed in several ways. One way is an immunological analysis, involving methods similar to that of blood typing or serum analysis. Amino acids can also be used, but probably will not yield much systematic data. Amino acids and certain other compounds are what might be called *primary* compounds. All plants require them. Thus, we would expect that all plants would have essentially the same amino acids, although not necessarily in the same proportions. It is the so-called *secondary* compounds that chemical taxonomists are most interested in. Those that do not seem essential in metabolism are perhaps by-products of certain metabolic reactions.

Another promising method is the mathematical approach of the numerical taxonomists. They utilize as many different plant characters as possible and note whether the various plants being compared exhibit those characters. This so-called presence-absence data is set up in a matrix. From it an index of similarity can be calculated. It is assumed that the numerically closer two or more indices of similarity are, the more closely the organisms are related.

Regardless of what system is used, how one assembles the data, or how those data are interpreted, the groups (or *taxons*) within the system, in general, remain the same. The kingdoms are generally divided into *subkingdoms*, the subkingdoms into *divisions*, the divisions into *classes*, the classes into *orders*, the orders into *families*, and the families into *genera*. Each of these major taxa, in turn, may be further subdivided. By convention, specific endings are generally used to denote the rank of each group. For the major group, these would be

Division: -phyta
  Class: -ae
    Order: -ales
      Family: -aceae

Let us look at how the common red maple would be classified according to the most commonly used system:

Kingdom: Plantae
  Subkingdom: Embryophyta
    Division: Pterophyta
      Class: Angiospermae
        Subclass: Dicotyledoneae
          Order: Sapindales
            Family: Aceraceae
              Genus: *Acer*
                Species: *Acer rubrum*

In the Margulis system, the same plant would be classified in the following manner:

Kingdom: Plantae
  Subkingdom: Euchlorophyta
    Branch: Metaphyton
      Grade: Tracheophyton
        Division: Angiospermophyta
          Class: Dicotyledoneae
            Order: Sapindales
              Family: Aceraceae
                Genus: *Acer*
                  Species: *Acer rubrum*

Having spent a few pages describing taxonomy and taxonomic systems, let us now consider in the next few chapters how the various plants came to possess their particular characteristics.

## SUMMARY

1. Plant taxonomy is concerned with all aspects involved in naming, classifying, and determining the phylogenetic relationships of plants.
2. Modern taxonomy can be thought of as having its beginnings around 300 B.C.
3. Botany and medicine were two closely allied sciences well into the eighteenth century, with the period from approximately 1500 to 1700 being known as the Age of Herbals.
4. An important doctrine of many herbalists was the Doctrine of Signatures.
5. Linnaeus was an important influence in the development of our modern system of binomial nomenclature.
6. An important advancement to the understanding of phylogenetic relationships was Darwin's *Origin of Species.*
7. Modern taxonomists are involved in numerous areas, including studies in cytogenetics, biochemistry, anatomy and morphology, statistics and computer technology, and others.

# Chapter 6

## INHERITANCE AND VARIATION IN PLANTS

It is not known with certainty when, or even if, primitive man had any notion of sex and sexual reproduction in plants. Anthropologists have found that primitive tribes today have little knowledge of the sexual activities of plants. The knowledge of "good trees" (trees that bear fruit) and "bad trees" (trees that do not bear fruit) is well known, but obviously, plants do not copulate, thus, the consequence of fertilization is completely lost to these peoples.

However, in Babylon and Syria the unisexual nature of the date palm was recognized in 5000 B.C. (Figure 6.1), and Theophrastes (third to fourth century B.C.) compared the fertilization (pollination) of plants to that of fish (i.e., both were external). But no widespread understanding of the relationship between pollination and fertilization existed. After the work of Camerarius, in the latter part of the seventeenth century, scientific circles began to become interested in demonstrating the existence of sex in higher plants. The issue soon became a hotly debated one, and prizes were offered (by the Academies of Science of St. Petersburg in 1759, Prussia in 1819, and Haarlem or Holland in 1830) for proof of sexuality in plants, the possibility of obtaining plant hybrids, and data concerning the characteristics of such hybrids. Linnaeus won the first of these, Naudin the last.

*Figure 6.1*
Alabaster relief of *Winged Genius Pollinating the Sacred Tree.* [From Palace of Assurnasirpal II (Assyrian). Purchased by Andover-Newton Theological School, Charles Amos Cummings Fund. Courtesy of Museum of Fine Arts, Boston.]

The process of inheritance in plants was likewise difficult to understand. It was not too difficult to imagine that as a single-celled organism divides, the characters of the original cell may somehow be assigned to each of the new cells. But how and in what form are these characters transferred? What about the multicellular plant—or animal for that matter? They are composed of many different kinds of cells, organs, and so on. How can these characteristics be transferred to offspring? If each cell of an organism contains all the characters of the parent organism, does this mean that species are therefore constant and unchanging, or can characteristics of a species change over time? In other words, do species evolve? Can different individuals of the same species show variation? Can individual offspring of the same parents show variation? If so, what are the probable causes of such variation? Can specific characters as well as the potentiality for variation be transmitted from parent to offspring? Is sexual reproduction really important in plants? These are only a few of the many questions that could be asked when one considers the development of structure and form in plants. We will attempt to provide at least partial answers for some of these questions.

## CLASSICAL THEORY OF DIRECT INHERITANCE OF CHARACTERS

### *Hippocrates' Ideas (ca. 400 B.C.)*

Hippocrates believed that the reproductive material came from all parts of the individual's body. Thus, the characters for each part were directly handed down to the progeny. He supported his hypothesis by reference to a race of people called Macrocephali. It was customary among these peoples to shape the head of each newborn child by gently squeezing the head by hand until it assumed an elongated shape. Round heads were considered quite socially unacceptable. Hippocrates states that after several generations the elongated head became a natural characteristic; in other words, children were born with elongated heads. As he could not possibly have been an eyewitness to the change occurring over several generations, Hippocrates was obviously relying on hearsay to support his theory. But he questioned, if such characters as hair color and eye color could be inherited, then why not elongated heads?

### *Aristotle's Views (ca. 350 B.C.)*

Aristotle questioned the Hippocratic view and pointed out the difficulties, and in fact the impossibilities, that it presented. He pointed out that certain characters such as voice, mannerisms, nails, and hair are inherited, yet they could not *in his estimation* contribute to reproductive material as they were either intangible aspects or concerned with dead tissues. He also mentioned that some things, such as gray hair, may not even be present at the time of reproduction. In addition, Aristotle found that some children would resemble a grandparent more closely than they would resemble either parent. The dilemma was confounded even more when the Hippocratic theory was applied to plants. How, questioned Aristotle, could a particular plant part (such as what we now call a cotyledon) be inherited when it wasn't even present at the time of reproduction? How also could each of two parents contribute something from all of their parts, and yet their offspring end up with not two but only one of each?

Aristotle, in attempting to take into account the difficulties presented by the Hippocratic view, argued that reproductive material, instead of being derived from all parts of the individual, should be regarded as composed of nutrient substances which, while on their way to the various parts, were diverted to the reproductive pathway. These would thus differ from one another depending upon which part of the organism they were destined for. He also favored the idea that the two sexes did not contribute equally to the progeny. The reproductive material all came from the female, while the male contributed only something to define the general form the embryo would assume.

## *Darwin's Theory (1868)*

Darwin, unaware of Mendel's work two years earlier, suggested that all cells and tissues of organisms threw off minute granules, both during development and when the individual reached maturity. He suggested that these granules circulated through the organism, were multiplied, and passed to the reproductive cells. Thus, each reproductive cell contained a multitude of components thrown off from each individual part of the organism. When transmitted to the progeny via the reproductive cells, these granules were responsible for the development of cells and tissues corresponding to those of the parent from which they originated. Darwin assumed that some granules might be transmitted in a dormant state for several generations before revealing some ancestral character. He remarked that his theory resembled that of Hippocrates but differed in specifying that granules were responsible for heredity transmission.

Darwin's theories were based, in part, on his work with normal (asymmetrical flower) and peloric (radially symmetrical flower) *Antirrhinum majus* (snapdragon). His results were as follows:

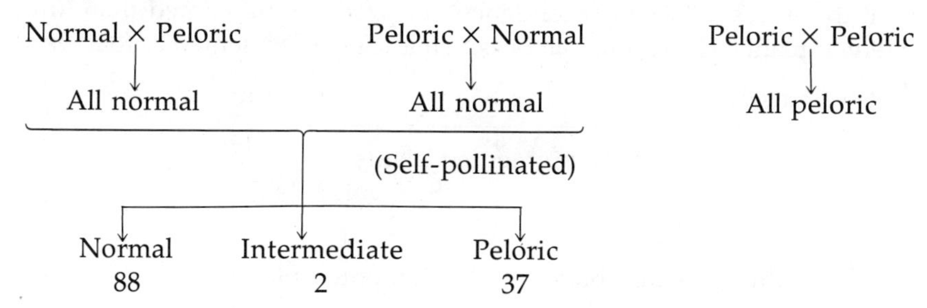

These results would later be interpreted by Bateson and Saunders (1902) as essentially following the Mendelian inheritance pattern.

Various experiments were performed during the eighteenth and nineteenth centuries in regard to crossing different varieties of both plants and animals. Although these studies failed to support the classical theory, they did not lead to the development of an alternate hypothesis either. Thus, the classical theory of Hippocrates, even though lacking confirmation, went virtually unchallenged for 23 centuries, except for Mendel's work. Weismann (1883) was the first to challenge effectively the idea that characters are transmitted directly from parent to offspring when he proposed his theory of the continuity of the germ-plasm. Weismann removed the tails of new born mice for 22 successive generations and found that the tails of the final generation still grew to normal length. As the tailless mice could not have transmitted the trait for tails directly, it must have been carried in the germ-plasm from generation to generation. Somewhat later (1891), he concluded that germ-plasm was solely nuclear substance and that this nuclear substance was the sole bearer of hereditary tendencies.

## PLANT HYBRIDIZATION AND THE DEVELOPMENT OF MENDELIAN MECHANISMS

### *Knight (1799)*

Knight conducted experiments on plant hybridization using the edible pea because of its short generation time, self-fertilizing habit, and easily recognizable characters. His primary intent was to obtain information that would lead to new and improved varieties of fruits and vegetables.

Knight crossed two varieties of peas which differed in color. One was unpigmented, with white flowers, white seed coats and green stems, while the other had purple flowers, grey seed coats and purplish stems. He found that when an unpigmented plant was pollinated by a pigmented one, only pigmented progeny were obtained. But the progeny, when self-pollinated, produced some pigmented and some unpigmented offspring. This was true even when peas from the same pod were sown. But Knight failed to record the number and kinds of seeds, so he did not discover the underlying mechanisms later cited by Mendel. He was only able to state that there was a "stronger tendency" to produce pigmented than unpigmented plants. His results can be schematically represented as follows:

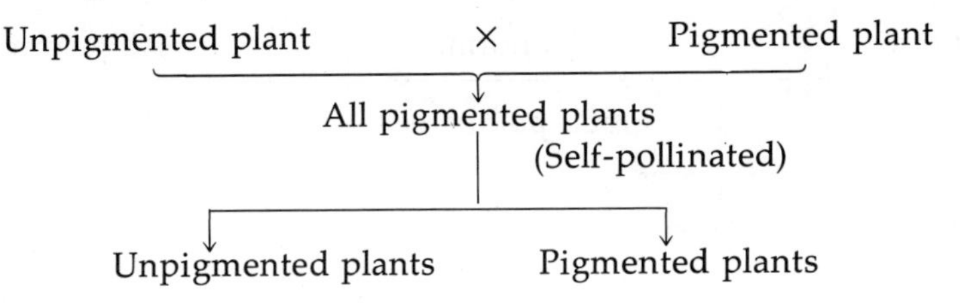

## *Goss (1824)*

Goss, also working with the garden pea, made discoveries similar to those of Knight. He removed the stamens ("male" reproductive structure) from a green-seeded pea plant, pollinated it from a yellow-seeded variety, and was surprised to find that all the progeny had yellow seeds like those of the "male" parent. He sowed these seeds the following year, allowed the plants to self-pollinate, and was equally surprised to find, in the second generation, some plants with all green seeds, some with all yellow seeds, and a great many with both green and yellow seeds in the same pod. He sowed these second generation seeds the following year and, upon allowing them to self-pollinate, found that only the green-seeded plants bred true, that is, only the green seeds produced all green progeny. The yellow seeds produced some plants having all yellow seeds, but more often yielded pods with both green and yellow seeds. But Goss, like Knight, did not count the number and kinds of seeds,or if he did he failed to discover the mechanism pointed out 42 years later by Mendel. Goss's results can be represented as follows:

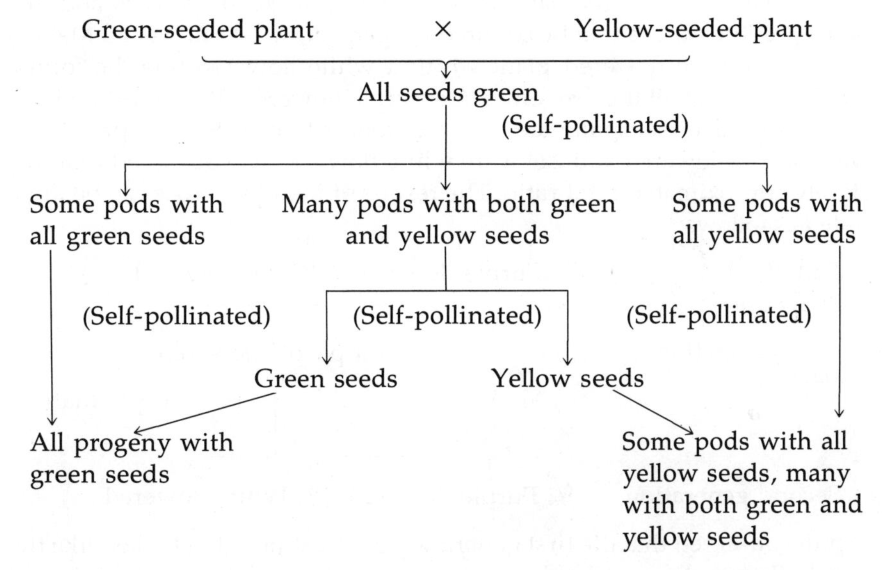


## *Mendel (1866)*

Mendel, like Knight and Goss, worked with varieties of the garden pea. Mendel's reasons for selecting the pea can be best stated in his own words, as translated by E. R. Sherwood:

> From the start, special attention was given to the *Leguminosae* because of their particular floral structure. Experiments with several members of this family led to the conclusion that the genus *Pisum* had the qualifications demanded to a sufficient degree. Some quite distinct forms of this genus possess constant traits that are easily and reliably distinguishable, and yield

perfectly fertile hybrid offspring from reciprocal crosses. Furthermore, interference by foreign pollen cannot easily occur, since the fertilizing organs are closely surrounded by the keel, and the anthers burst within the bud; thus the stigma is covered with pollen even before the flower opens. This fact is of particular importance. The ease with which this plant can be cultivated in open ground and in pots, as well as its relatively short growth period, are further advantages worth mentioning. Artificial fertilization is somewhat cumbersome, but it nearly always succeeds.

There is no indication in Mendel's paper that he was familiar with the work of either Knight or Goss, but he was familiar with the experiments of other early workers in plant hybridization, namely, Kölreuter, Gärtner, Hebert, Lecoq, and Wichura.

Mendel selected seven traits for experimentation: 1) difference in shape of the ripe seeds (round or angular), 2) difference in coloration of seed albumen (yellow or green), 3) difference in coloration of the seed coat (white or grey brown), 4) difference in shape of ripe pod (smooth or constricted), 5) difference in color of the unripe pod (green or yellow), 6) difference in position of flowers (axillary or terminal), and 7) difference in stem length (long or short). The principal difference between Mendel's work and that of his predecessors is that he counted progeny of each kind. When Mendel crossed a purple-flowered plant whith a white-flowered one, he found, like Knight, that all the progeny were purple flowered. When these plants were allowed to self-pollinate, Mendel found that in 929 offspring, 705 were purple flowered and 224 were white flowered. He observed that this closely approximated a 3:1 ratio. The results of Mendel's cross can be illustrated as follows:

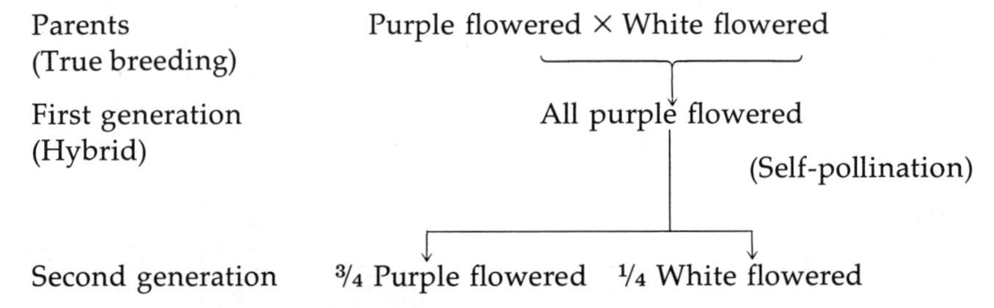


Mendel observed that the first generation plant, although it looked like the purple-flowered parent, could not be identical to it, as the purple-flowered parent, when selfed (self-pollinated) produced all purple-flowered offspring, but the first generation purple-flowered plant when selfed produced some plants with white flowers. He called the trait that passed into the hybrid association unchanged, the *dominating*, and the trait which became latent in the hybrid, the *recessive*.

The simplest way Mendel could account for his observations was to postulate that each hybrid contained two pieces of information for any single character. He assumed the union of parental traits through union of the "germinal cell" (female) and "pollen cell" (male). He stated further "that of the seeds formed by the hybrids with one pair of differing traits, one-half again develop the hybrid form while the other one-half yield plants that remain constant and receive the dominating and the recessive character in equal shares." To simplify representation of this information, Mendel substituted letters. He designated the dominating trait *A* and the recessive *a*. The cross could now be illustrated as follows:

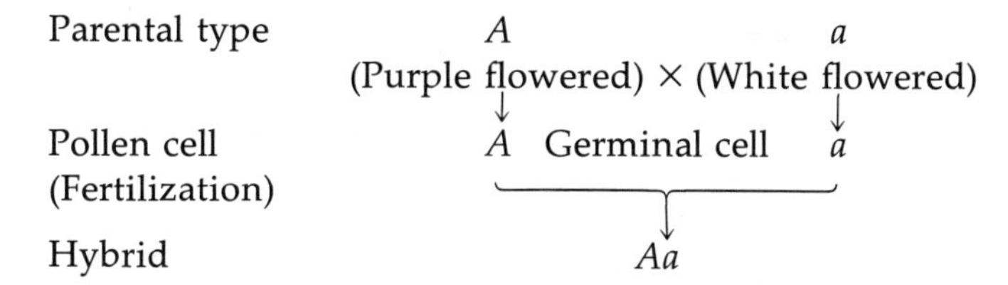


Mendel assumed "that pea hybrids form germinal and pollen cells that in their composition correspond in equal numbers to all the constant forms resulting from the combination of traits united through fertilization." As the hybrid was produced from constant forms *A* and *a*, it must then produce pollen cells and germinal cells of the forms *A* and *a*. These two forms will, on the average, participate equally in fertilization, and as four individuals are produced, each form must manifest itself twice. It is purely a matter of chance which pollen cell combines with each germinal cell, but according to the laws of probability, every pollen cell *A* and *a* will unite equally often with every germinal cell *A* and *a*; thus:

| | | | | |
|---|---|---|---|---|
| Pollen cells | *A* | *A* | *a* | *a* |
| | ↓ | ↘ | ↙ | ↓ |
| Germinal cells | *A* | *A* | *a* | *a* |

Mendel visualized the results of this fertilization by writing the designations as fractions, with pollen cells above the line and germinal cells below the line, as follows:

$$A/A + A/a + a/A + a/a$$

As the germinal and pollen cells were alike in the first and fourth terms, Mendel assumed that the products of their association must be constant, in other words *A* and *a*. Mendel recognized the fact that $A/a$ and $a/A$ were actually the same association, *Aa*. Therefore, he represented the hybrid progeny as follows:

$$A/A + A/a + a/A + a/a = A + 2Aa + a$$

Thus, of the three purple-flowered plants produced by the hybrid, one was of the parental type while two were hybrid.

If Mendel's hypothesis is correct, then a cross between the hybrid and a true-breeding white-flowered plant should give equal numbers of purple-flowered and white-flowered progeny, as follows:

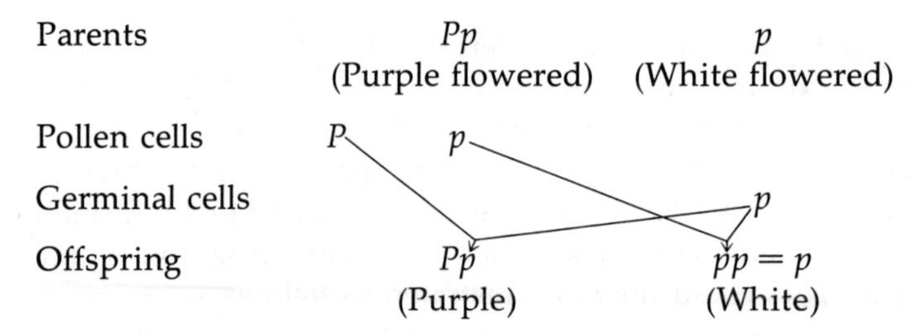


Mendel found that this was so. He obtained 84 purple-flowered plants and 81 white-flowered plants; in very close agreement with the predicted 1:1 ratio. This type of cross, where the hybrid is crossed to one of the parent types is called a *back cross*. This special case where the hybrid is crossed to a recessive parent is called a *test cross*.

Mendel also repeated, among other types of crosses, the work of Goss. Among 8023 second generation seeds, Mendel found 6022 yellow seeds and 2001 green seeds, again close to the expected 3:1 ratio. He also confirmed Goss's third generation observations that green peas bred true but yellow often did not. Of 519 such yellow seeds, he found that 166 bred true, in other words produced only yellow seed, while 353 did not. The latter group produced both yellow and green seeds in the ratio of 3:1 as in the previous generation. Mendel observed that the ratio 166:353 was close to 1:2, and assumed this to imply that the second generation ratio of 3 yellow to 1 green was again really a ratio of 1 true breeding yellow to 2 impure yellow to 1 true breeding green, the 1:2:1 ratio suggested earlier.

To observe the result of considering two traits simultaneously, Mendel crossed a true-breeding strain having both round and yellow seeds with one having angular and green seeds. The hybrid was found to be round and yellow. Selfing of the hybrids produced the following:

315 round, yellow seeds
101 angular, yellow seeds
108 round, green seeds
32 angular, green seeds

He noted that these offspring of the hybrid appeared in nine different forms in the ratio of 1:1:1:1:2:2:2:2:4. Mendel used the symbol *A* for round shape, *a* for angular shape, *B* for yellow albumen, and *b* for green albumen. He recognized that the progeny ratio he observed for the dihybrid cross was, as he called it, a "combination series" in which the traits *A* and *a*, *B* and *b* are combined term by term. If one considered round and angular as a monohybrid cross the progeny ratio would be 1:2:1 as observed earlier. Yellow seed and green seed considered as a single cross would also produce a 1:2:1 ratio. Thus, in considering the two series simultaneously, there is simply a combining of the two 1:2:1 ratios in the following manner:

$$(1 + 2 + 1)(1 + 2 + 1) = 1 + 2 + 1 + 2 + 4 + 2 + 1 + 2 + 1$$

corresponding to the results Mendel actually observed from his cross.

Mendel further observed that when three traits were considered, the results were again predictable: a combination series of three 1:2:1 ratios combined with each other. He therefore postulated the following generalization: $3^n$ is the number of terms in a combination series, $4^n$ the number of individuals belonging to the series, and $2^n$ the number of combinations that remain constant, that is, breed true; where *n* designates the number of character differences in the two parental plants. In our monohybrid cross, we are considering only a single character difference, thus $3^1$ is the number of terms (1:2:1), three terms; $4^1$ is the number of individuals (1 + 2 + 1), four individuals; and $2^1$ is the number breeding true (*A* and *a*). For our dihybrid cross, two character differences are being considered, thus $3^2$ or 9 terms, as we noted above, $4^2$ or 16 individuals, $2^2$ or 4 that breed true.

## MODERN INTERPRETATION OF MENDEL'S RESULTS

It might be best to digress for a moment and move ahead in time so this information might be presented in a more modern context. Mendel was concerned primarily with the offspring of the hybrids, and although he recognized that the hybrid acquired its form from a union of two differing parental traits, he either did not recognize or chose to ignore the fact that constant or true-breeding plants would also acquire their form by the union of two parental traits. It was later understood that the characters discussed by Mendel could each be represented by two factors, and these factors could be differing as in the hybrid or they could be similar as in the parental or constant types. Bateson and Saunders (1902) referred to the alternate factors associated with the character as *alleles.* They also stated that when both genetic factors are similar (*AA* or *aa;* not simply *A* or *a* as used by Mendel) the organism is *homozygous* for that character (developed from a homozygote). When the two genetic factors are different *(Aa)* the organism is *heterozygous* for that character (developed from a *heterozygote*). They introduced the symbols *P* for parental generation, $F_1$ for first filial generation, $F_2$ for second filial generation, and so on. Later, the outward expression of these factors, purple flowers or white flowers, round seeds or angular seeds, would be called by Johannsen (1903) the *phenotype*, and the 3 purple-flowered: 1 white-flowered ratio observed by Mendel would be called a *phenotypic ratio.* The genetic factors responsible for the phenotype would be called the *genotype*, i.e., *AA*, *Aa* or *aa*, and the 1*AA*:2*Aa*:1*aa* ratio of Mendel would be called a *genotypic ratio*.

We might now look at Mendel's monohybrid cross between purple-flowered and white-flowered plants in the following way. As suggested by Mendel, each gamete of the purple-flowered plant would receive a single *A* while each gamete of the white-flowered plant would receive a single *a.* These traits would combine to produce the hybrid *Aa.* When the hybrids are selfed, each male gamete would receive *either A* or *a*, while each female gamete would likewise receive *A* or *a*. These traits would then recombine *in all possible combinations* to form the $F_2$. Thus, plants of the $F_2$ would appear in the phenotypic ratio of 3 purple-flowered to 1 white-flowered as purple is dominant to white; or a genotypic ratio of 1*AA* to 2*Aa* to 1*aa*, as illustrated below:

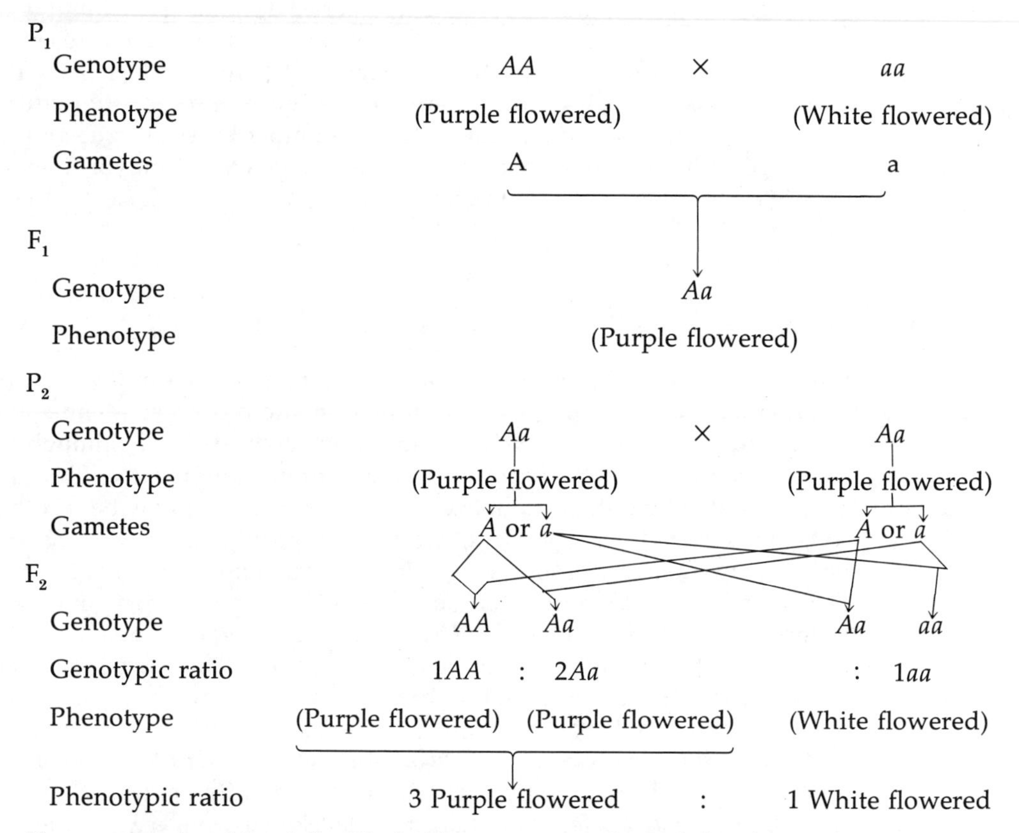


If we reexamine Mendel's dihybrid cross consisting of the characters round, angular, yellow, and green; the round, yellow parent would be signified as *AABB*, as both round and yellow are dominant, the angular, green parent would be signified as *aabb*. The gametes of the round, yellow plant would all contain a trait for round *and* a trait for yellow and thus would be denoted *AB*, while the gametes of the angular, green plant would all contain a trait for angular and green and thus would be denoted *ab*. Upon combining, the hybrid would thus be AaBb as noted by Mendel. Upon selfing the $F_1$, each plant would now produce four different types of gametes, following the observation of Mendel that "the behavior of each pair of differing traits in a hybrid association is *independent* of all other differences in the two parental strains." Thus, gametes having *AB*, *Ab*, *aB*, and *ab* would be formed, and they would be formed in equal numbers.

The $F_2$ produced by union of these gametes would consist of nine different genotypes in the 1:1:1:1:2:2:2:2:4 ratio specified by Mendel, following his observation that "constant traits occurring in different forms of a plant kindred can . . . enter into all the associations possible within the rules of combination." As it is somewhat difficult to determine all the possible combinations when one is dealing with several different traits, Punnett devised a scheme whereby one set of gametes are set as a horizontal row and the other set of gametes as a vertical file and all gametes in the row are combined with all gametes in the file. This scheme is known as a *Punnett Square*. It can also be noted that the phenotypic ratio in the $F_2$ is 9 round, yellow: 3 round, green: 3 angular, yellow : 1 angular, green, a ratio not mentioned by Mendel. The cross can be illustrated as follows:

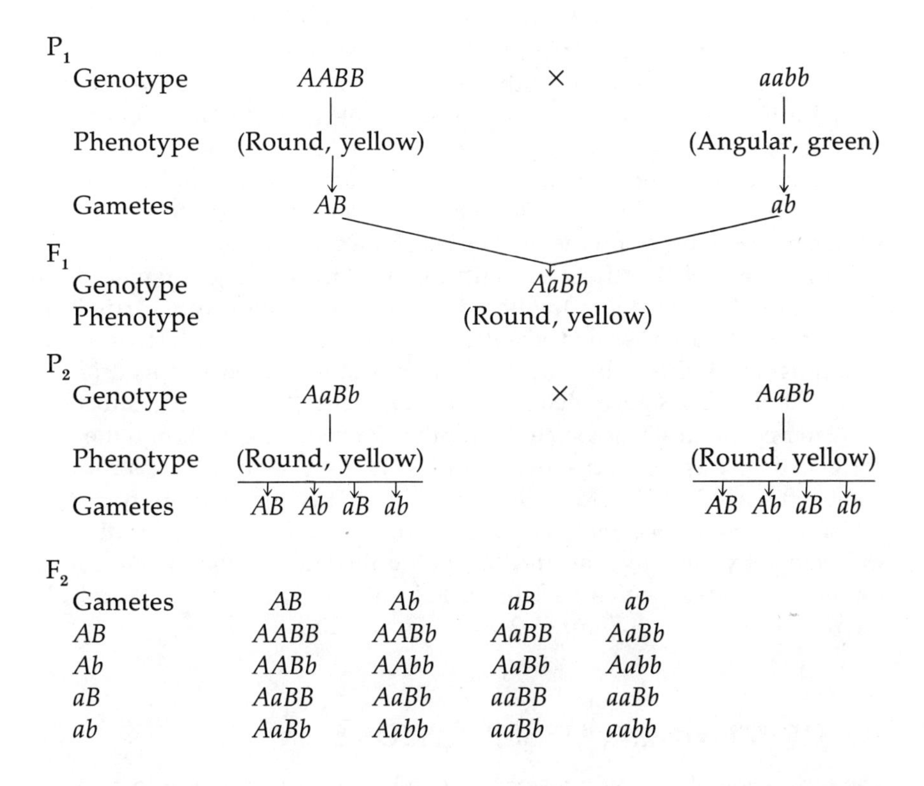


Genotypic ratio:

1*AABB* : 2*AABb* : 2*AaBB* : 4*AaBb* : 1*AAbb* : 2*Aabb* : 1*aaBB* : 2*aaBb* : 1*aabb*

Phenotype:

Round yellow, Round yellow, Round yellow, Round yellow, Round green, Round green, Angular yellow, Angular yellow, Angular green

9 Round yellow : 3 Round green : 3 Angular yellow : 1 Angular green

## THE ROLE OF THE NUCLEUS

The basis for the inheritance pattern observed by Mendel was unknown to him, although he did recognize that the genetic factors were somehow carried in the "germinal cells" and "pollen cells," now referred to more specifically as egg and sperm.

The German botanist Strasburger studied the process of fertilization in seed plants, and was particularly interested in the orchid, *Orchis latifolia.* He noted that only the sperm *nucleus* was supplied to the egg, and publishing the results of his studies in 1884, he stated that in general 1) the process of fertilization is based on the copulation of the sperm nucleus with the egg nucleus, 2) the cytoplasm is not involved in this process, and 3) the sperm and egg nuclei are genuine cell nuclei. He concluded, "I believe in this work to have definitively furnished the proof that fertilization is based only on the union of the cell nuclei and that from the previous sexual differentiation only a sperm nucleus needs to be introduced from the paternal organism into the egg and indeed often alone is introduced. Inasmuch as the child thus inherits the characteristics of the father only through conveyance by the cell nucleus; the specific characters of the organism must be based on the properties of the cell nucleus."

Carl von Nageli, another German and one of the leading botanists of the last century, theorized that the hereditary characteristics were carried in a substance which he termed *idioplasm.* A few years later, Hertwig (1885) would associate this substance with the nucleus and term it *nuclein.* We now call this substance deoxyribonucleic acid (DNA). This substance was later cytologically associated with the *chromosomes.* Although the term chromosome was not introduced until 1888, by Waldeyer, the structures had been noted by Schneider 15 years earlier in his observations of nuclear division in flatworm eggs. The most definitive studies of nuclear division were made by Fleming, and in 1879, he published a plate of figures illustrating the process which he later called mitosis (Figure 6.2). It might be interesting to compare Figure 6.2 with Figure 3.23.

## SUMMARY OF MENDEL'S OBSERVATIONS

The separation of homologous genes (genes that correspond in location on homologous chromosomes) in the formation of gametes was the situation actually implied by Mendel's *law of segregation.* Also, as the result of meiosis, which nonhomologous chromosomes become associated in a single gamete is purely a matter of chance. If one assumes a single gene to be responsible for a single character and further assumes these genes are all on different chromosomes, we then have the situation implied by Mendel's, *law of independent assortment,* formulated as the result of observations of di- and trihybrid crosses. This is not a true law in that a number of genes

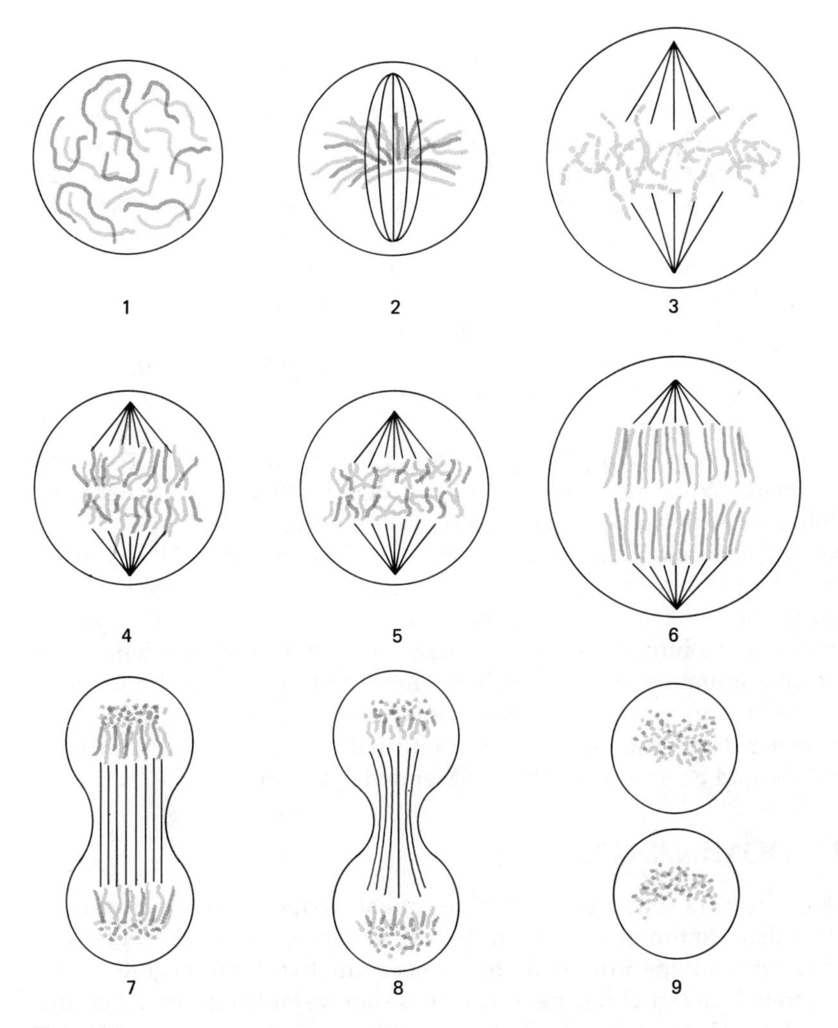


*Figure 6.2*
Selected stages of mitosis in the salamander as figured by Walter Fleming in 1879. [W. Fleming, "Contribution to the Knowledge of the Cell and its Life Appearance," *Arkiv. Für Mikrokopische Anatomie* 16:302–406, 1879.]

will be carried on the same chromosome and thus must be inherited as a unit; they cannot "independently assort," and secondly, many characters of plants are not the result of a single gene but the expression of several genes. We shall later discuss examples of both these instances. A third observation of Mendel was incorporated in his *law of dominance*. Stated in modern terms, this indicates that the expressivity of the recessive gene will be completely repressed in the hybrid. Again, this is not a true biological law, as many contrasting traits do not show a simple dominance.

## REDISCOVERY OF MENDELIAN INHERITANCE

Mendel's work made no impression on contemporary thought and in fact was almost entirely overlooked for 34 years. It was rediscovered independently by DeVries (1900), Correns (1900), and Tschermak (1900). Each had made experiments like those of Mendel and had obtained comparable results. DeVries had worked with a number of different flowering plants, Correns worked with *Zea mays* (corn), and both Correns and Tschermak worked with the garden pea. The characters of the pea studied included those used by Knight and Goss, and unknown to them, also by Mendel. Thus, when Mendel's work was finally rediscovered, his results had already been confirmed independently by several investigators. Bateson and Saunders (1902) provided further evidence for the wide applicability of *Mendel's Laws* (so named by Correns) in their work with both plants and animals. Working independently, Cuénot (1902) established that Mendelian inheritance applied equally well to animals.

The fundamental difference between Mendel's theories and the classical theory of Hippocrates is that in the classical theory inheritance is direct, and therefore, the extent of transmission of a particular character to the offspring may be influenced by the character itself. In the Mendelian theory the determinant of a character is not modified in any way by its presence in an organism; thus, the inheritance of a character is unaffected by the character itself. The inheritance of mutilations, as referred to by Hippocrates, would be impossible by the Mendelian theory.

## INCOMPLETE DOMINANCE

Bateson, Saunders, and Punnet (1905) established one of the first instances of incomplete dominance, or blending inheritance, where the heterozygote has a phenotype intermediate between the two homozygotes. They demonstrated this in chickens, but it was soon established in other animals and in plants as well. Certain varieties of *Mirabilis jalapa* (four-o'clock) demonstrate such a pattern of inheritance. If a true-breeding red-flowered plant is crossed to a true-breeding white-flowered plant, the $F_1$ will all be pink flowered. If the $F_1$ plants are selfed, their progeny will be red, pink, and white, in the approximate ratio of 1:2:1. This is clearly just a simple modification of the 3:1 phenotypic ratio, as follows:

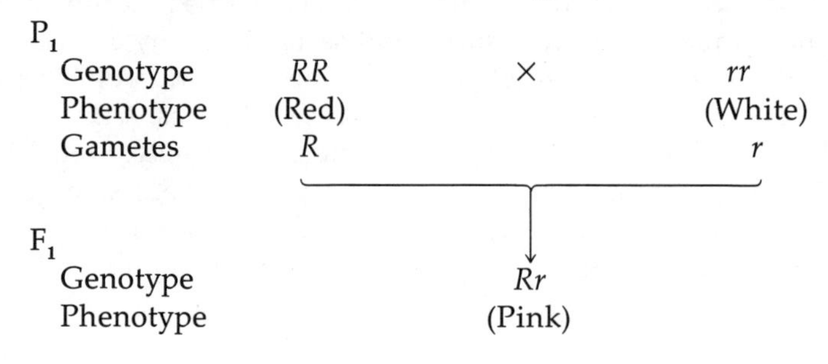

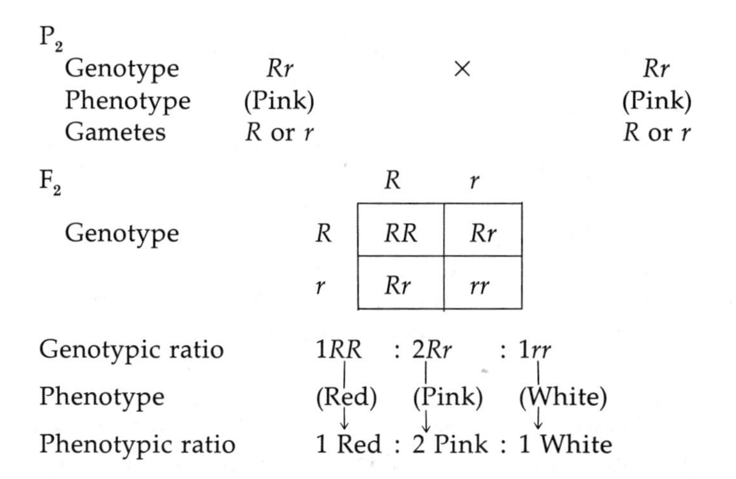


## PARTIAL OR INCOMPLETE LINKAGE

Bateson, Saunders, and Punnet (1905) examined flower color and pollen shape in *Lathyrus odoratus* (sweet pea). When they classified 427 $F_2$ progeny, they obtained the following results:

296 (69.3%purple flowers, long pollen
19 (4.5%)—purple flowers, round pollen
27 (6.3%)—red flowers, long pollen
85 (19.9%)—red flowers, round pollen

Although the 315 purple to 112 red, and 323 long to 104 round was not a perfect 3:1 ratio, later work did establish that complete dominance existed (with purple dominant to red, and long dominant to round) and a single gene was associated with each character. However, in considering both traits simultaneously, the results certainly did not fit the expected 9:3:3:1 ratio for independent segregation. The investigators assumed that purple and long were somehow *coupled* to each other.

Sutton (1903), as mentioned earlier, had pointed out in his formulation of the chromosome theory of Mendelian heredity that the number of distinct characters in an organism must exceed the number of chromosomes. He stated that each factor was only a part of a chromosome and if chromosomes retain their individuality, then all factors represented by one chromosome must be inherited as a unit. But it was quickly discovered that in both garden peas and snapdragons, the number of different character sets exceeded the known number of chromosomes.

DeVries (1903) offered a solution to this dilemma. He accepted Sutton's view that many factors must coexist on each chromosome, but he suggested that an exchange of material took place between the maternal and paternal *homologous chromosomes*, when they were closely associated during sex cell formation. He argued that if two like factors lay opposite each other, a simple exchange could occur. In any given instance it was simply a matter of chance whether such an exchange did in fact occur. He argued further that a factor could only be exchanged for one representing the corresponding hereditary character, otherwise each chromosome would not retain the entire set of factors. If such an exchange did occur, then some chromosomes would contain both maternal and paternal units. As the two chromosomes of a homologous pair pass into different sex cells as the result of reduction division, this redistribution of factors between homologous chromosome pairs would lead to the formation of gametes having all possible combinations of the parental character differences, just as Mendelian theory required.

These theories of DeVries gained little support at the time. Evidence for such an exchange was lacking and the idea was thought to be contrary to the concept of permanence and individuality of chromosomes. Thus, when Bateson, Saunders, and Punnett discovered partial linkage between two different characters in their sweet pea studies, they did not accept the chromosome theory as a possible explanation. But in 1909, Janssens reported seeing point fusions of two of the four strands of synapsing homologous chromosomes. He termed these *chiasmata*, and postulated that through breakage at the fusion point, followed by reunion, exactly equal chromatid regions could be exchanged (Figure 3.25). When Morgan (1911) in his fruit fly work, turned up similar examples of partial linkage, he had no alternative but to postulate a DeVriesian exchange and agreement with the theories of Janssens. The results of Bateson, Saunders, and Punnet were explainable in the same manner.

## CROSSING OVER

As stated earlier, genes show linkage because they are on the same chromosome. Genetic variability can be introduced by a mutual break and reciprocal exchange between two homologous chromatids. Such an event is termed *crossing over*.

Much of the evidence for crossing over has been obtained from studies of the fruit fly *Drosophila*. However, it would be more convenient and more reliable if one could examine *all* the end products of meiosis. This can be done through studies of a red bread mold *Neurospora*. Physiologically different haploid nuclei (+ and −) fuse forming a diploid zygote, and the zygote nucleus then undergoes meiosis forming four haploid nuclei enclosed within an elongated cell called the *ascus*. These nuclei are contained in a row so that one can easily determine which nuclei are the result of the first meiotic division and which are products of the second meiotic division, as the nuclei remain in the order in which they are produced (Figure 6.3). Each nucleus then undergoes a mitotic division resulting in a total of eight nuclei within each ascus, with each meiotic product being represented

*Figure 6.3*
Nuclear pairing between vegetative plus and minus hyphae of the ascomycete *Neurospora*. Note final linear arrangement of the eight ascospores.

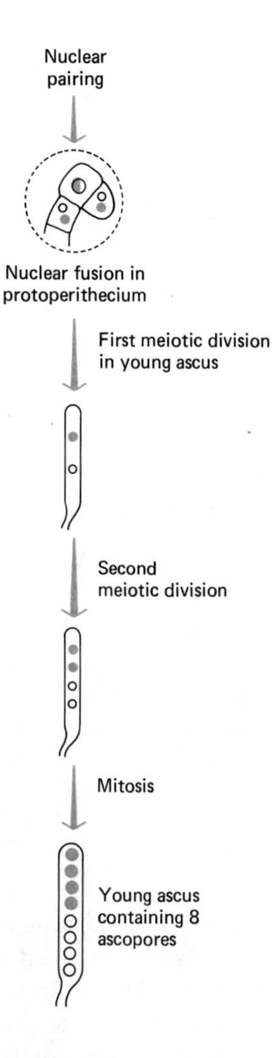


twice. The nuclei become surrounded by a small amount of cytoplasm and a cell wall develops to form ascospores. The ascospores can then be dissected out one by one, their position in the ascus recorded, and upon culturing of each ascospore in separate test tubes, with appropriate tests, the genotype of each can be determined.

One genetic trait of *Neurospora* can result in the production of either a colored or a colorless mycelium (the vegetative structure of *Neurospora*). These could be symbolized as *col* (colored) and *noncol* (colorless). With no crossing over, the products could be diagrammed as in Figure 6.4. You can note that without crossing over, there is a 1:1 distribution of ascospores within an ascus, with physiological mating type + linked with *colored mycelium* and physiological mating type − linked with *colorless mycelium.*

Numerous experiments have shown that when crossing over occurs, it occurs between two of the homologous chromatids at the four-strand

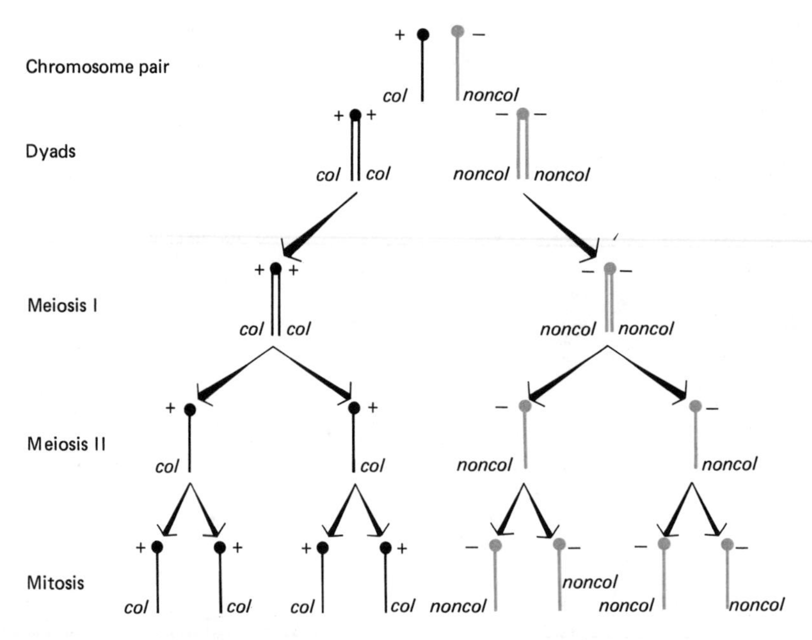


*Figure 6.4*
Types of ascospores produced within an ascus of *Neurospora* with no crossing over. Gene *col* determines colored mycelium; *noncol* determines colorless mycelium; the gene + determines one mating type; − determines the other mating type. As + is linked to *col*, − to *noncol*, they will segregate in this fashion in the ascospores.

stage. Actual results of the above cross indicate that only two of the four strands are involved and that a single crossover occurs. But such a crossover could occur in one of four possible ways, as illustrated in Figure 6.5. Several aspects can be noted from Figure 6.5. Study of the products will indicate, by the position of the ascospore in the ascus and the nature of its product, which of the above possible crossovers did, in fact, occur. Secondly, you can see that in this example, assuming that crossover occurred in every ascus, one-half of the progeny will be parental, that is physiological mating type + will be linked with *colored* and physiological mating type − will be linked with *colorless*, while one-half will be new combinations, physiological mating type + linked with *colorless* and physiological mating type − linked with *colored*. These new linkage arrangements are called *recombinants*.

Crossovers may occur several times within the same homologous pair of chromosomes, and the same or different strands may participate. In a double crossover, for example, the genetic consequences can be quite variable depending upon which strands are involved.

The consequences of crossing over are several. The most obvious is that it introduces variability. An important aspect to the geneticist, however, is the indication that genes are in a definite serial order on the chromosome

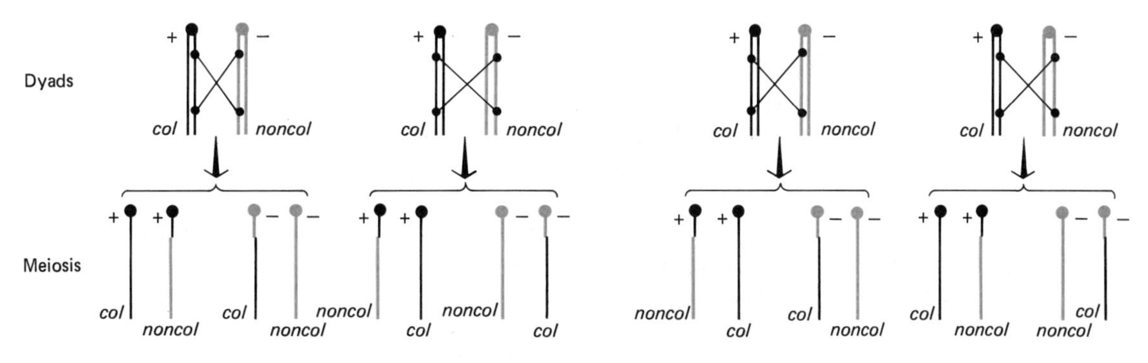


*Figure 6.5*
Types of ascospores produced within an ascus of *Neurospora* with two strand crossing over. New linkage arrangements are now produced between + and −, *col* and *noncol*.

with each gene having a point on the chromosome at which it begins and another point at which it ends. This region is called the *locus* of the gene. The frequency of crossing over appears to be closely related to the physical distance between genes. Thus, the study of crossover patterns can serve both to establish the order of the genes on a chromosome and the relative distance between genes. Such an analysis is referred to as *gene mapping*.

## MULTIPLE ALLELES

Up to now we have been considering the possibility of only two alleles at any one locus. But Cuénot (1904) found in his studies of different strains of mice that the number of homologous genes was not limited to two. It has now been determined that numerous organisms, both plants and animals, have systems of *multiple alleles*, whereby, in some instances, hundreds of different combinations of gene pairs may exist.

One type of multiple allelic system found in plants is that for self-sterility. Although pollination can occur in such plants, if the pollen grain contains a self-sterility factor the same as that carried by the female plant, the pollen tube will not grow and thus fertilization cannot occur. The tobacco plant *Nicotiana* has the following self-sterility alleles: $S^1$, $S^2$, $S^3$ and $S^4$. Self-sterility can be demonstrated in this plant by the following examples. If we assume the pollen-producing plant to have the genotype $S^1S^2$ in all cases, then the pollen will carry either $S^1$ or $S^2$. The result of pollination by such pollen of an $S^1S^2$, $S^2S^3$ or $S^3S^4$ egg-producing plants would be as follows:

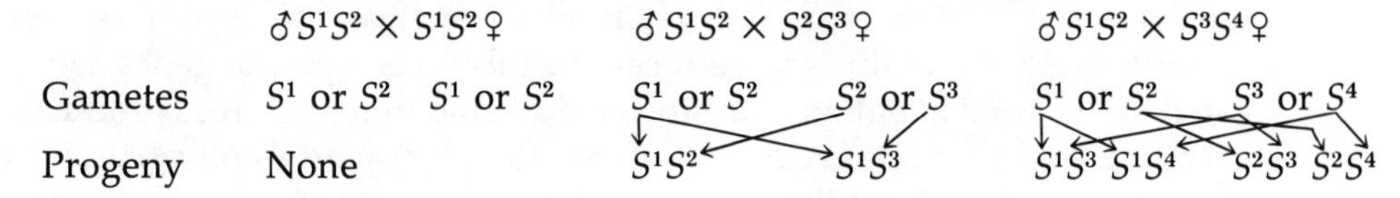


This type of self-sterility system has been established with as many as 200 or more alleles, as found by Bateman (1947) for red clover.

## GENE INTERACTION

Just as allelic genes can interact in various ways to determine the traits of an organism, so can nonallelic genes. In some instances, two gene pairs can "complement" each other in their effect on the same characteristic. In some instances, each genotypic combination will result in a specific phenotype; in other instances, one gene pair may hide the effect of the other. This latter type of interaction is called *epistasis.* Bateson and Punnet in their studies of 1905, 1906, and 1908 discovered an excellent illustration of gene interaction in sweet peas. They crossed two white-flowered plants and were very surprised to find that the $F_1$ were all purple flowered. When $F_1$ plants were crossed the $F_2$ was produced in the ratio of nine purple to seven white. These results can be diagrammed as follows:

$P_1$ *AAbb* (White) × *aaBB* (White)

$F_1$ AaBb (Purple)

$P_2$ *AaBb* × *AaBb*

| Gametes | *AB* | *Ab* | *aB* | *ab* |
|---|---|---|---|---|
| *AB* | *AABB* | *AABb* | *AaBB* | *AaBb* |
| *Ab* | *AABb* | *AAbb* | *AaBb* | *Aabb* |
| *aB* | *AaBB* | *AaBb* | *aaBB* | *aaBb* |
| *ab* | *AaBb* | *Aabb* | *aaBb* | *aabb* |

1 *AABB* : 2 *AABb* : 2 *AaBB* : 4 *AaBb* = 9 Purple
1 *AAbb* : 2 *Aabb* : 1 *aaBB* : 2 *aaBb* : 1 *aabb* = 7 White

Purple thus arose as the complementary effect of dominant alleles at two different gene pairs. But when either homozygous recessive was present, the effect of the dominant was masked. Thus, the homozygous recessives were epistatic to the effects of the other gene.

## MULTIPLE FACTOR INHERITANCE

Nilsson-Ehle reported in 1909 that there were three individual gene pairs involved in the determination of grain color in wheat. Three of these were dominant *(ABC)* and three were recessive *(abc).* Each pair of genes segregated in the usual Mendelian fashion. Thus, a monohybrid cross produced a ratio of 3 red : 1 white; a dihybrid cross, 15 red : 1 white; and a trihybrid cross, 63 red : 1 white. But they found that some of the red grains appeared more red than others (Figure 6.6). They could place the grains into seven classes, according to degree of redness: red, dark, medium dark, medium, medium light, light, and nonred or white. The cross can be diagrammed as follows:

$P_1$ *AABBCC* (Red) × *aabbcc* (White)

$F_1$ *AaBbCc* (Medium)

$P_2$ *AaBbCc* (Medium) × *AaBbCc* (Medium)

| | Gametes | *ABC* | *ABc* | *AbC* | *Abc* | *aBC* | *aBc* | *abC* | *abc* |
|---|---|---|---|---|---|---|---|---|---|
| | *ABC* | *AABBCC* | *AABBCc* | *AABbCC* | *AABbCc* | *AaBBCC* | *AaBBCc* | *AaBbCC* | *AaBbCc* |
| | *ABc* | *AABBCc* | *AABBcc* | *AABbCc* | *AABbcc* | *AaBBCc* | *AaBBcc* | *AaBbCc* | *AaBbcc* |
| | *AbC* | *AABbCC* | *AABbCc* | *AAbbCC* | *AAbbCc* | *AaBbCC* | *AaBbCc* | *AabbCC* | *AabbCc* |
| $F_2$ | *Abc* | *AABbCc* | *AABbcc* | *AAbbCc* | *AAbbcc* | *AaBbCc* | *AaBbcc* | *AabbCc* | *Aabbcc* |
| | *aBC* | *AaBBCC* | *AaBBCc* | *AaBbCC* | *AaBbCc* | *aaBBCC* | *aaBBCc* | *aaBbCC* | *aaBbCc* |
| | *aBc* | *AaBBCc* | *AaBBcc* | *AaBbCc* | *AaBbcc* | *aaBBCc* | *aaBBcc* | *aaBbCc* | *aaBbcc* |
| | *abC* | *AaBbCC* | *AaBbCc* | *AabbCC* | *AabbCc* | *aaBbCC* | *aaBbCc* | *aabbCC* | *aabbCc* |
| | *abc* | *AaBbCc* | *AaBbcc* | *AabbCc* | *Aabbcc* | *aaBbCc* | *aaBbcc* | *aabbCc* | *aabbcc* |

Progeny:

| | | | | | | |
|---|---|---|---|---|---|---|
| 1 *AABBCC* | : 2 *AABBCc* | : 1 *AABBcc* | : 2 *AABbcc* | : 1 *AAbbcc* | : 2 *Aabbcc* | : 1 *aabbcc* |
| | : 2 *AABbCC* | : 1 *AAbbCC* | : 2 *AAbbCc* | : 1 *aaBBcc* | : 2 *aaBbcc* | |
| | : 2 *AaBBCC* | : 1 *aaBBCC* | : 2 *AaBBcc* | : 1 *aabbCC* | : 2 *aabbCc* | |
| | | : 4 *AABbCc* | : 8 *AaBbCc* | : 4 *AaBbcc* | | |
| | | : 4 *AaBBCc* | : 2 *AabbCC* | : 4 *AabbCc* | | |
| | | : 4 *AaBbCC* | : 2 *aaBBCc* | : 4 *aaBbCc* | | |
| | | | : 2 *aaBbCC* | | | |
| 1 Red | : 6 Dark | : 15 Medium dark | : 20 Medium | : 15 Medium light | : 6 Light | : 1 White |

In this example, each dominant gene has an equal and additive effect, and thus, this character for grain color is a quantitative one. Although there are obvious steps in the above distribution, as one increases the number of gene pairs involved, the distribution will approach a normal curve. As there are many genes for one quantitative character, these are called *multiple factors.* But the effect of many multiple factors are so small that their

Parents: AABBCC X aabbcc
Red White

| | | |
|---|---|---|
| $F_1$: | AaBbCc<br>Medium red | |
| $F_2$: | Genotype | Phenotype |
| 1 | AABBCC | Red |
| 6 | AABBCc<br>AABbCC<br>AaBBCC | Dark |
| 15 | AABBcc<br>AAbbCC<br>aaBBCC<br>AABbCc<br>AaBBCc<br>AaBbCC | Medium red |
| 20 | AABbcc<br>AAbbCc<br>AaBBcc<br>AaBbCc<br>AabbCC<br>aaBBCc<br>aaBbCC | Medium |
| 15 | AAbbcc<br>aaBBcc<br>aabbCC<br>AaBbcc<br>AabbCc<br>aaBbCc | Medium light |
| 6 | Aabbcc<br>aaBbcc<br>aabbCc | Light |
| 1 | aabbcc | White |

*Figure 6.6*
The genetic control of color in wheat kernals. This is an example of multiple-factor inheritance.

individual quantitative effect is difficult to measure, thus the term multiple factors has been replaced by the term *polygene*. Polygenes are genes which produce only a small effect on a particular character but can supplement each other to produce observable quantitative changes.

## NON-MENDELIAN INHERITANCE

Thus far, we have been concerned only with the genetic material of the nucleus. However, various cellular organelles contain their own endogenous species of DNA and RNA; thus, it would be logical to assume that this might also act as genetic material. Evidence for this might be demonstrated by studies of *Mirabilis*, the four-o'clock. Plants were found which were variegated, that is, had white sectors and green sectors. Flowers born from the white sectors would, of course, bear the genes for white, those from the green sectors, the normal genes. If a normal green plant was pollinated with flowers from a white branch, the $F_1$ were all green. But if flowers from a white branch were pollinated by a normal green plant, the $F_1$ were all white and of course eventually died. This behavior does not fit Mendelian patterns at all. If one assumes white to be recessive and green dominant, then it should make no difference in the reciprocal cross; the $F_1$ should always be green. Obviously, nuclear genes must not control this inheritance pattern. Examination of the results indicate that the $F_1$ always assume the phenotype of the maternal parent. If we consider the nature of the gametes, we see that the egg contains considerable cytoplasm whereas the sperm contains little if any. Thus, it seems appropriate to examine a cytoplasmic factor to account for the results. Examination of the white sectors indicated that the chloroplasts were abnormal with no chlorophyll, thus accounting for the lack of green color. If normal chloroplasts were inherited through the cytoplasm of the egg cell, then the offspring would be normal green, but if abnormal chloroplasts were inherited, the offspring would be colorless. Occasionally some plastids could be transmitted through the sperm, but this was usually negligible. Such an inheritance pattern is termed *maternal inheritance* and is probably the most common instance of *non-Mendelian* or *extrachromosomal inheritance.* These terms are often used interchangeably.

A similar pattern is seen in streptomycin resistance in *Chlamydomonas. Chlamydomonas* has two mating types, designated *mt*$^+$ and *mt*$^-$. A strain of *Chlamydomonas* was found that was resistant to streptomycin *(sm-r);* others are streptomycin sensitive *(sm-s).* Figure 6.7 illustrates the results obtained from various crosses. As you can see, there is a uniparental inheritance pattern similar to that of maternal inheritance. Only the *mt*$^+$ parent can transmit streptomycin resistance. Although these results are obviously based upon extrachromosomal inheritance, the mechanism is quite unknown. Fusion in *Chlamydomonas* is *syngamous,* in other words, the two

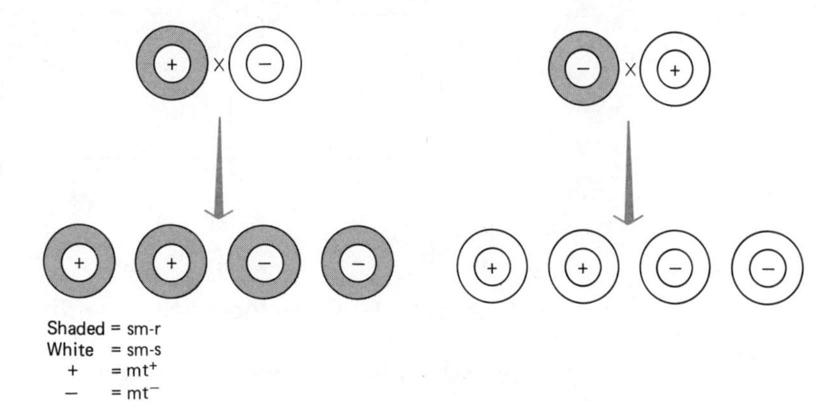


*Figure 6.7*
When a streptomycin resistant (sm-r) strain of *Chlamydomonas* is mated with a streptomycin sensitive (sm-s), all progeny are resistant. However, only mating type + (mt$^+$) cells transmit streptomycin resistance.

mating cells look alike, and thus, both contain essentially the same amount of cytoplasm.

Another example of a uniparental inheritance pattern can be seen in the fungus *Neurospora.* A very slow growing strain called *poky* was discovered. Reciprocal crosses of poky and wild type gave the following results:

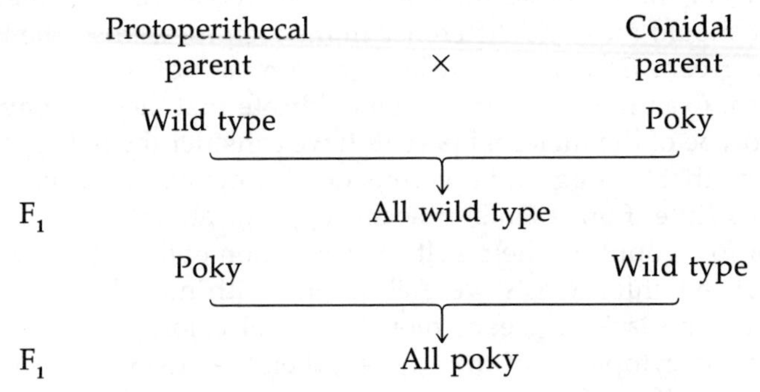


Investigation of the causes for the slow growth in poky showed it to be due to the deficiency of an enzyme associated with aerobic respiration. The enzymes of aerobic respiration are known to be associated with the mitochondria. Thus, these results provide indirect evidence for mitochondrial DNA as genetic material.

But some extrachromosomal inheritance shows the influence of nuclear genes as well. In corn, a cytoplasmic factor induces pollen sterility. A single cross between two different strains of corn will, of course, produce a hybrid $F_1$. If a second hybrid $F_1$ is produced by crossing two other strains, the two $F_1$'s can then be crossed to produce a double hybrid. This is usual technique for producing hybrid seed corn. When pollen sterile strains were crossed with pollen fertile strains, the following results were obtained:

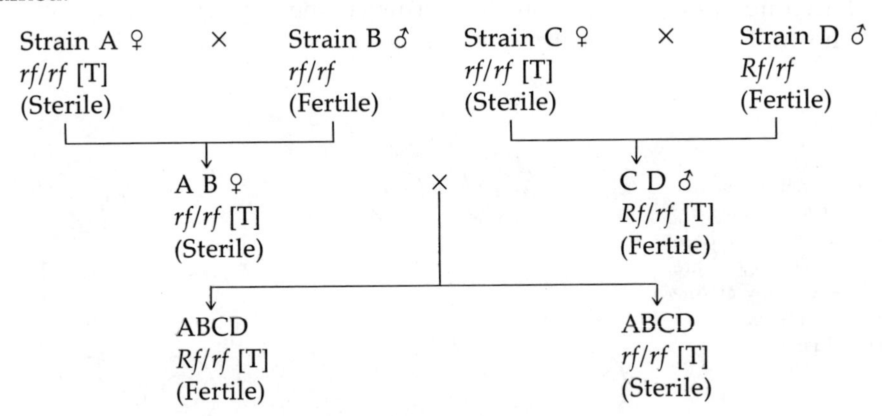


*key:* T = Texas cytoplasm — induces pollen sterility
*Rf* = dominant nuclear gene — restores fertility
*rf* = recessive nuclear gene — has no effect on fertility

Pollen sterility in this instance can be clearly shown to be a maternally inherited factor. But in the presence of the dominant allele Rf of a nuclear gene, pollen sterility will not occur. Thus, we have the interaction of nuclear and extrachromosomal factors. Another example of interaction can be noted in Iojap corn (corn with green and white striped leaves). In this instance, a homozygous recessive nuclear gene can cause mutation of the chloroplast DNA resulting in loss of chlorophyll production.

Many other examples of non-Mendelian inheritance are noted in plants and represent an important source of variability. At this time, however, the only organelles which have been rather conclusively involved in such inheritance are the chloroplasts and mitochondria.

## OTHER ASPECTS OF VARIABILITY

Still other sources of variability exist in plants, in addition to these already mentioned. One of the more important is change in chromosome number. *Euploidy* concerns whole sets of chromosomes and includes haploid (N), diploid (2N), triploid (3N), tetraploid (4N), and so on. A multiple of the *N* number of chromosomes is called *polyploidy*. Polyploidy in plants is quite widespread in nature and may occur either by doubling of the chromosome sets of an individual (autopolyploidy) or by doubling of the chromosome sets due to hybridization. In the latter instance, the hybrid may be sterile due to lack of homology between the two chromosome sets, but fertile hybrids may also result. Polyploidy may be of significant evolutionary importance, as the increase in genetic material might result in increased mutation or variation.

Another change in chromosome number can involve the addition or deletion of individual chromosomes. The term *aneuploidy* is used to refer to this situation. One of the most comprehensive studies of such a situation in plants was made by Blakeslee and Belling in 1924, on the jimsonweed, *Datura stramonium*. They found that the plant has 12 pairs of chromosomes as its diploid number, but through experimental breeding they produced 12 aneuploids, each having different chromosomes present in triplicate (a trisomic). Each was recognized by the appearance of its capsule (Figure 6.8).

## ADDITIONAL INSIGHT

This section is intended for those of you who may wish to explore more specific aspects of variation. It was my intention to discuss how variation could occur, and I think our discussion of Mendelian inheritance and variations of it should have suggested several possibilities to you. We have limited ourselves to the whole organism or at best cellular aspects, and

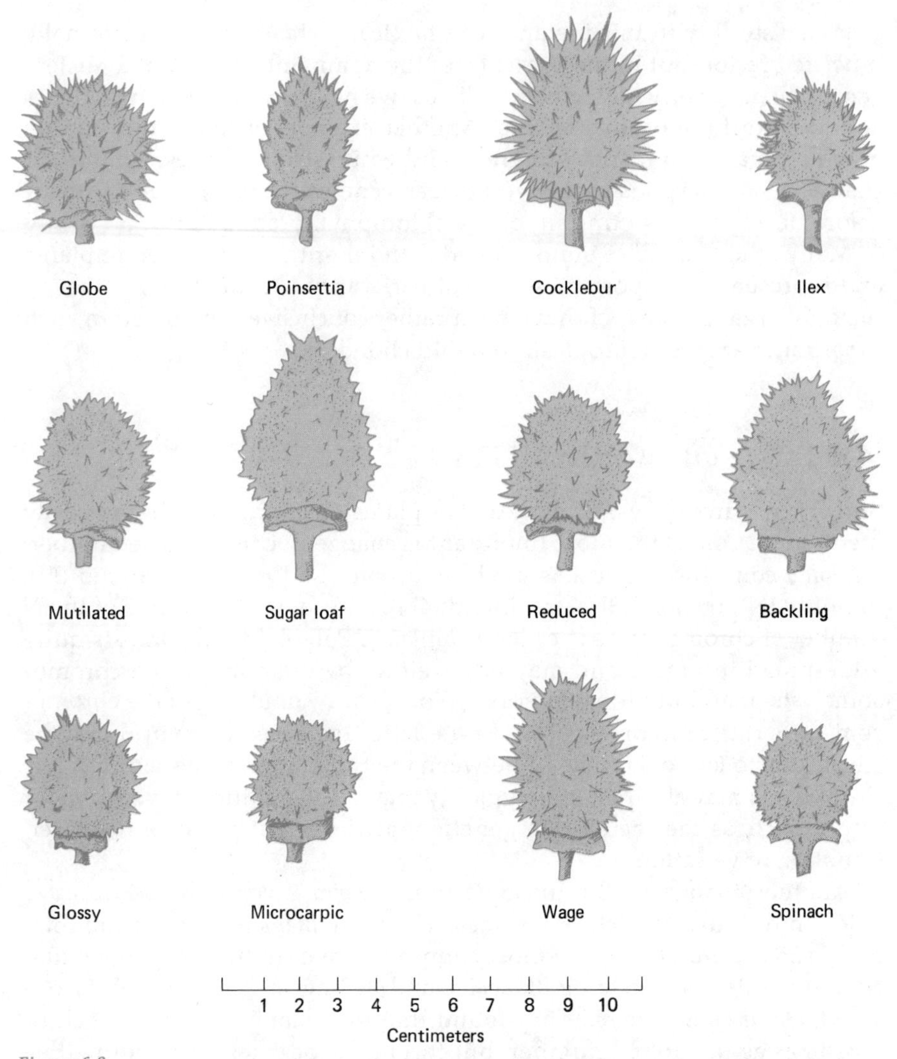


*Figure 6.8*
Seed capsule of the normal diploid *Datura stramonium* and those of the 12 kinds of primary trisomics. [From A. F. Blakeslee, "Variations in *Datura* Due to Changes in Chromosome Number," *Amer. Nat.* 56:16–31, 1924.]

even if we went no further, you would have the basis for drawing a number of basic inferences regarding plant development. But you have little insight into the underlying mechanisms involved. These must be investigated at the molecular level and such discussion obviously will involve some knowledge of basic chemistry. If you feel you do not have sufficient background for the discussion that follows, you may wish to study the supplemental material in Appendix I.

Before proceeding further with this discussion, it might be interesting to consider the remarks of Paul J. Kramer in the prefatory chapter to volume 24 of the *Annual Review of Plant Physiology:*

> Molecular biology created a better understanding of how genetic material is replicated, of how enzymes are produced, and of how they are turned on and off. However, the dogmas of molecular genetics really have contributed very little toward a better understanding of morphogenesis or the physiology of whole plants. Our increasing knowledge of plant processes results more from improvements in instrumentation and in experimental techniques than from development of new concepts.

I will not comment on his remarks. Certainly, many botanists would agree with him, and I am equally certain that many would disagree. Yet I must mention that little of the information presented in this section on biochemical aspects was obtained from work with vascular plants. But let us proceed.

In order for orderly development to occur in organisms and for their production of offspring in turn, the organisms must possess properties based upon information that is: 1) *preserved,* 2) capable of being *replicated,* and 3) *transmissible.* Such information is contained in the genetic material, and this ability at the molecular level depends upon the presence and arrangement of essentially one type of giant molecule: *nucleic acids.*

Experimentation with viruses shows many of them to be composed of a strand of *RNA (ribonucleic acid)* surrounded by a protein coat. The protein alone cannot produce either more protein or RNA, but the RNA alone (mediated by a host cell) can produce hundreds of viruses identical in both RNA and protein to that of the infecting virus. Other viruses contain *DNA (deoxyribonucleic acid)* surrounded by a protein coat, and it too is capable of producing progeny. If we use the ability to replicate as a criterion for genetic material, then both RNA and DNA can be considered genetic material.

## Nucleic Acid Structure

As nucleic acids seem to be of utmost importance, it might be well to consider their molecular structure. We first consider their *primary structure,* the organization of the component parts into a strand.

Nucleic acids are composed of nucleotides. Each nucleotide consists of a *phosphate* ($PO_4$) group, a *pentose,* (five-carbon sugar), and an *organic base.* The organic base is related to benzene and is either a *pyrimidine* (nitrogen replaces a CH at the 1 and 3 carbon positions on the benzene ring) or a *purine* (a pyrimidine plus a side ring called an *imidazole*).

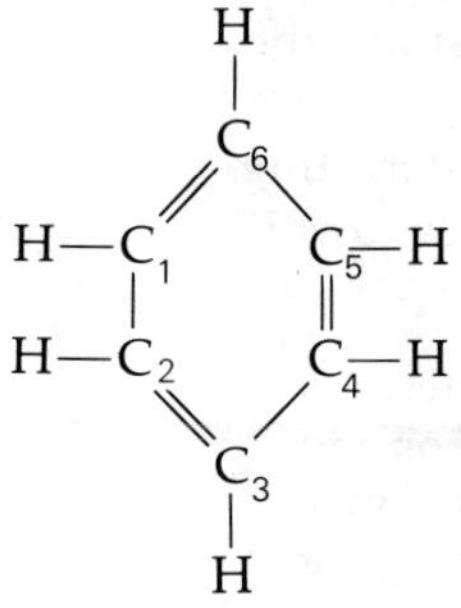

BENZENE

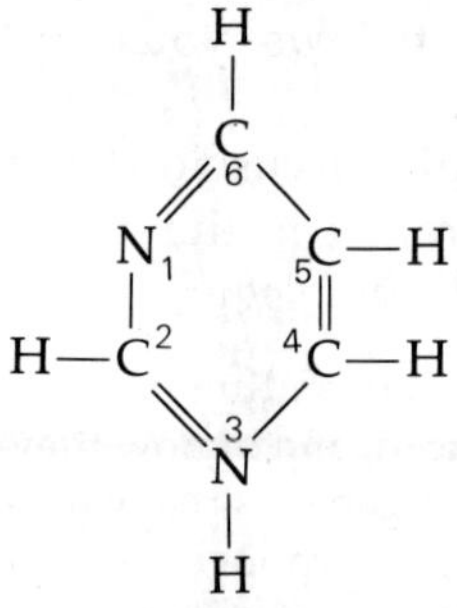

PYRIMIDINE

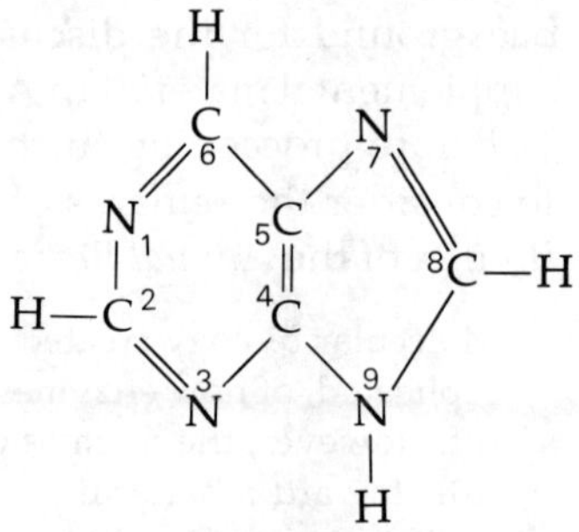

PURINE

The most common pyrimidines in RNA are *cytosine* and *uracil,* while *cytosine* and *thymine* are usually found in DNA. The most common purines in RNA and DNA are *adenine* and *guanine.*

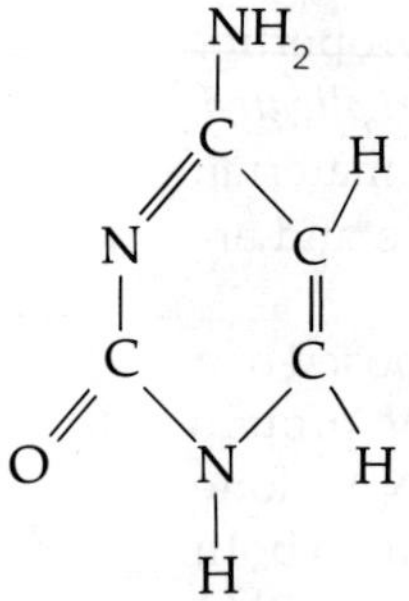

CYTOSINE
(6-amino-2-oxypyrimidine)

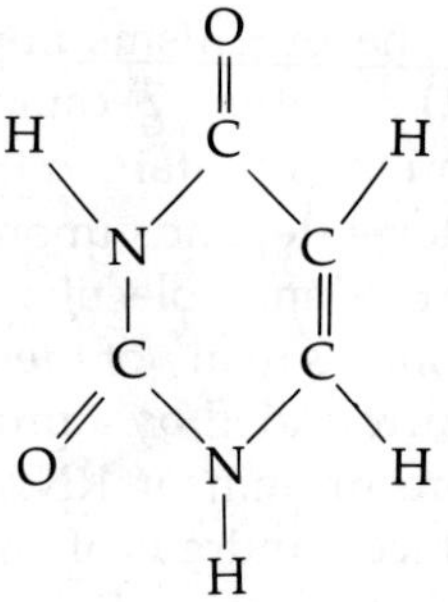

URACIL
(2,6-oxypyrimidine)

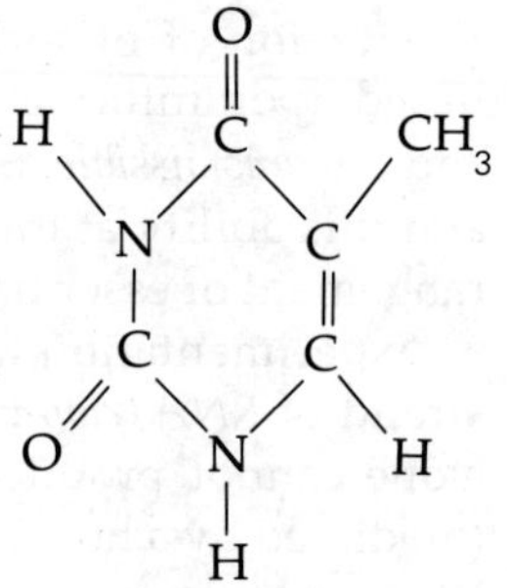

THYMINE
(2,6-oxy-5-methylpyrimidine)

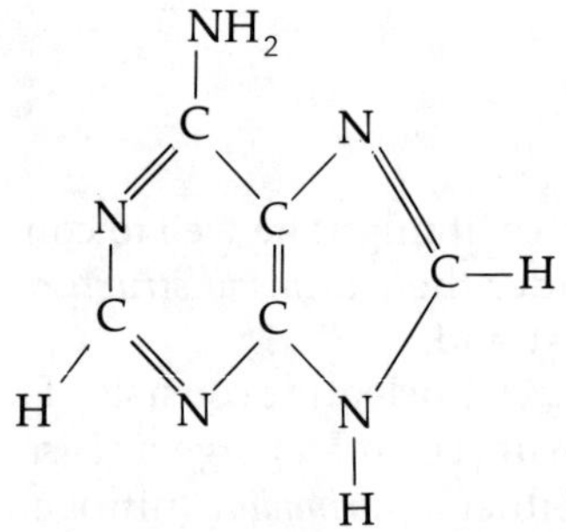

ADENINE
(6-aminopurine)

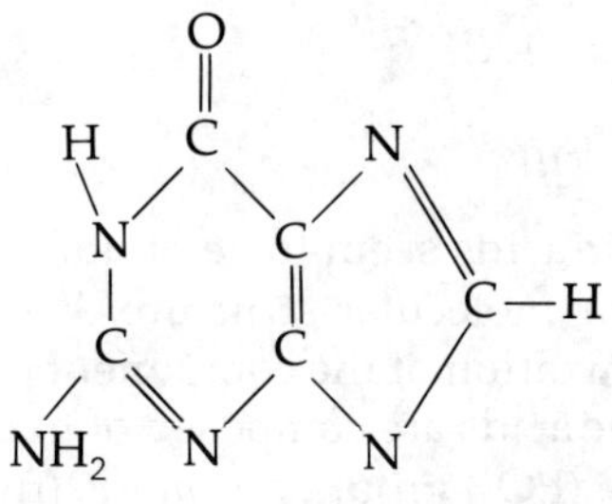

GUANINE
(2-amino-6-oxypurine)

Although these are the most commonly occurring bases, other bases are found in certain nucleic acids.

The most common pentose sugars are *ribose* and *deoxyribose.*

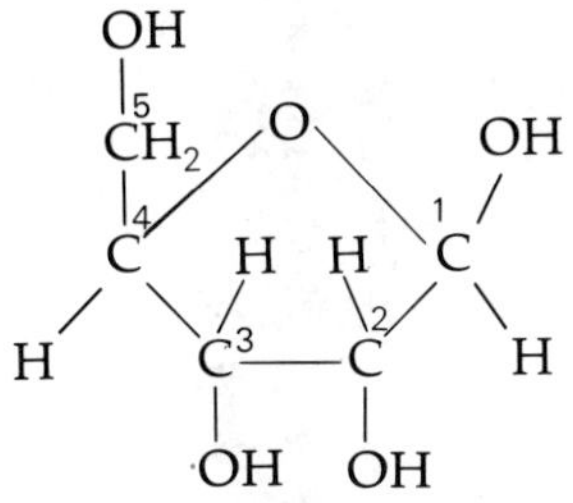

RIBOSE

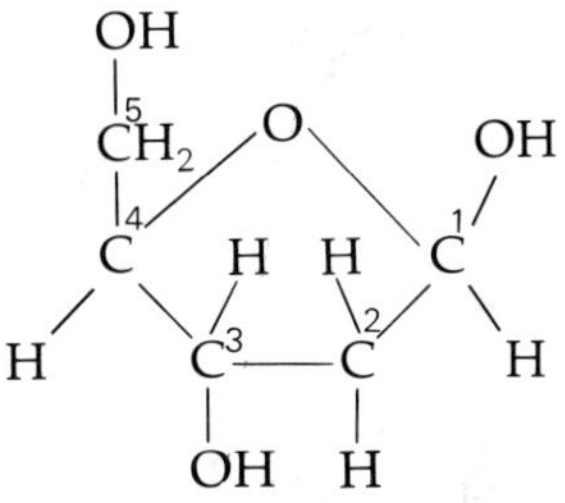

2-DEOXYRIBOSE

A purine or pyrimidine base linked with a pentose sugar forms a *nucleoside.* The usual nucleosides in DNA are *deoxycytidine* (dC), *deoxythymidine* (dT), *deoxyadenosine* (dA), and *deoxyguanosine* (dG). In RNA, the usual nucleosides are *cytidine* (C), *uridine* (U), adenosine (A), *and* guanosine (G).

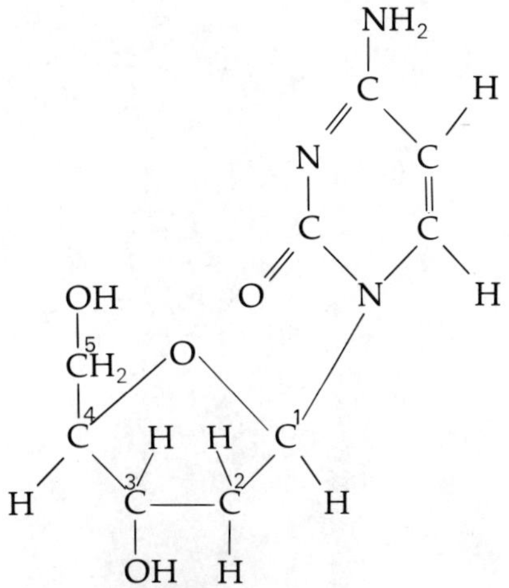

DEOXYCYTIDINE

As you will note by the above example, the pyrimidine or purine links to the sugar at the number one carbon, and all nucleosides are formed in the same fashion.

A nucleoside plus a phosphate group forms a nucleotide. The nucleotides of DNA are *deoxycytidylic acid,* (dCA), *deoxythymidylic acid* (dTA), *deoxyadenylic acid* (dAA), and *deoxyguanylic acid* (dGA). The nucleotides of RNA are *cytidylic acid* (CA), *uridylic acid* (UA), *adenylic acid* (AA), and *guanylic acid* (GA).

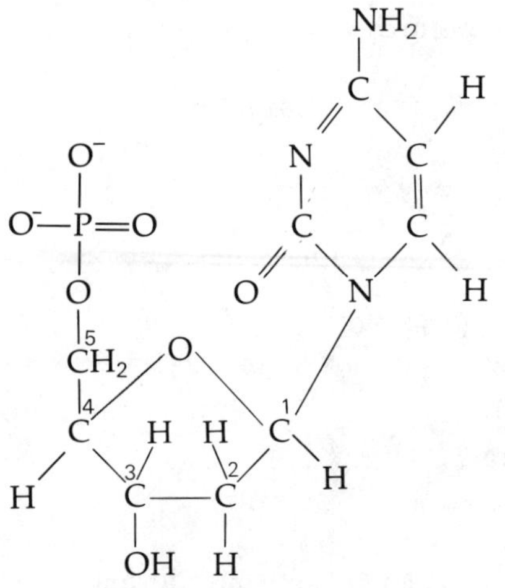

DEOXYCYTIDYLIC ACID
(Deoxycytidine 5-monophosphate)

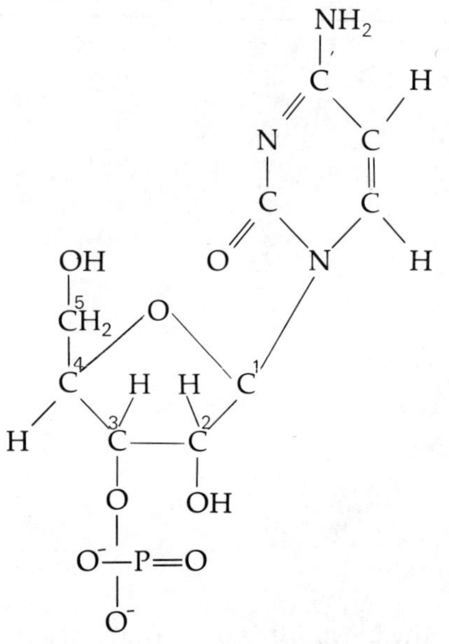

CYTIDYLIC ACID
(Cytidine 3-monophosphate)

The phosphate group may join to either the 3 or 5 carbon of the sugar. The nucleotides may join forming a chain. Thus, DNA is a chain in which deoxyribonucleotides are the links; it is a *polydeoxyribonucleotide*.

```
        O⁻
        |
   O⁻—P=O
        |
        O                      B
        |                     /
       CH₂    O              /
        |   /     \         /
        C    H   H    C
     /   \   /     \  /  \
   H      C——C          H
          |      |
          O     H
          |
    O⁻—P=O
          |
          O                     B
          |                    /
         CH₂   O             /
          |   /     \        /
          C    H   H    C
       /   \   /     \  /  \
     H      C——C          H
            /      \
          O       H
          |
     O⁻—P=O
          |
          O⁻
```

POLYDEOXYRIBONUCLEOTIDE

B is a pyrimidine or purine base (usually cytosine, thymine, adenine, or guanine).

Because of its physical and chemical properties, a polynucleotide strand can interact with itself or with other polynucleotide strands. It can wind back on itself (rarely) or wind around another strand forming a double coil (Figure 6.9). This coiling represents the *secondary structure* of the nucleic acid.

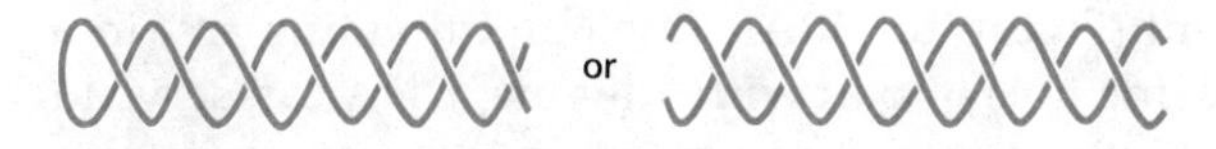


*Figure 6.9*
A polynucleotide strand may wind upon itself *(left figure)* or around a second strand *(right figure)* forming a double coil.

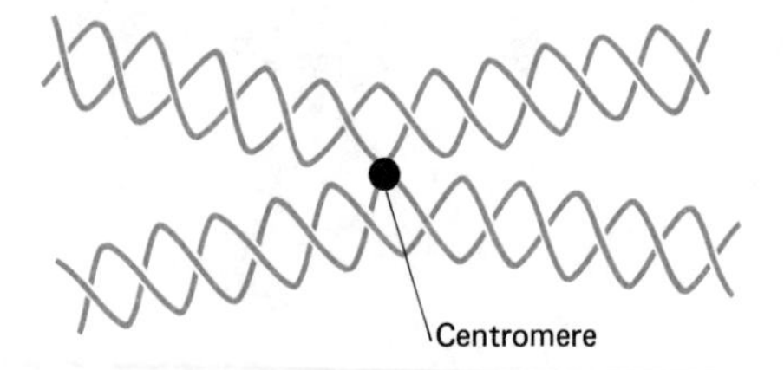


*Figure 6.10*
Model of a dyad composed of DNA double helices.

Normally, in forming the secondary structure of DNA, adenine in one strand is hydrogen bonded to thymine in the other strand, while cytosine of one strand is linked to guanine in the other. According to the Watson-Crick model, all the sugars of one strand face in one direction while those of the second strand face in the opposite direction. As these two strands coil around each other, they form a double helix as postulated by Watson and Crick. As the usual base-pairing in the double helix is that stated above, if we know the base sequence of one strand, we know it for the opposite strand, thus the strands are *complementary*.

Genetic material in bacteria and phages (viruses) seem to consist simply of a single or double strand of DNA or RNA. In eukaryotic cells, however, the nucleic acids are generally found in combination with basic proteins, but this is not always so. Such combinations are called nucleoproteins, and form the *chromosomes* of cells' nuclei. During certain stages of nuclear division such chromosomes may consist of at least two DNA double helices, held together at some point by a centromere (Figure 6.10), while at all other times they consist of a single double helix. Thus, the chromosome consists of a linear array of nucleotides. Successive portions of the chromosome are different, due to the organic bases present and their sequence; thus, we might expect different portions of the chromosome to contain different genetic information. The smallest unit of genetic material that contains information used by the organism is called a *gene*.

## *Replication and Transcription*

As genetic material in daughter cells is identical with that of the parents, barring mutation, the nucleic acids must be able to reproduce themselves. This process is called *replication* and may occur in one of three ways. Most often, however, it occurs semiconservatively whereby the DNA or RNA duplex divides to produce two duplexes, each composed of a parental strand and its complement (Figure 6.11). As uracil and thymine are, in a sense, equivalent, both being able to pair with adenine, DNA can serve in the replication of RNA or vice versa. The synthesis that proceeds from DNA to RNA is normally referred to as *transcription;* from RNA to DNA, *reverse transcription.*

The genetic information contained in genetic material is utilized in the formation of specific proteins. The synthesis of proteins occurs in the cyto-

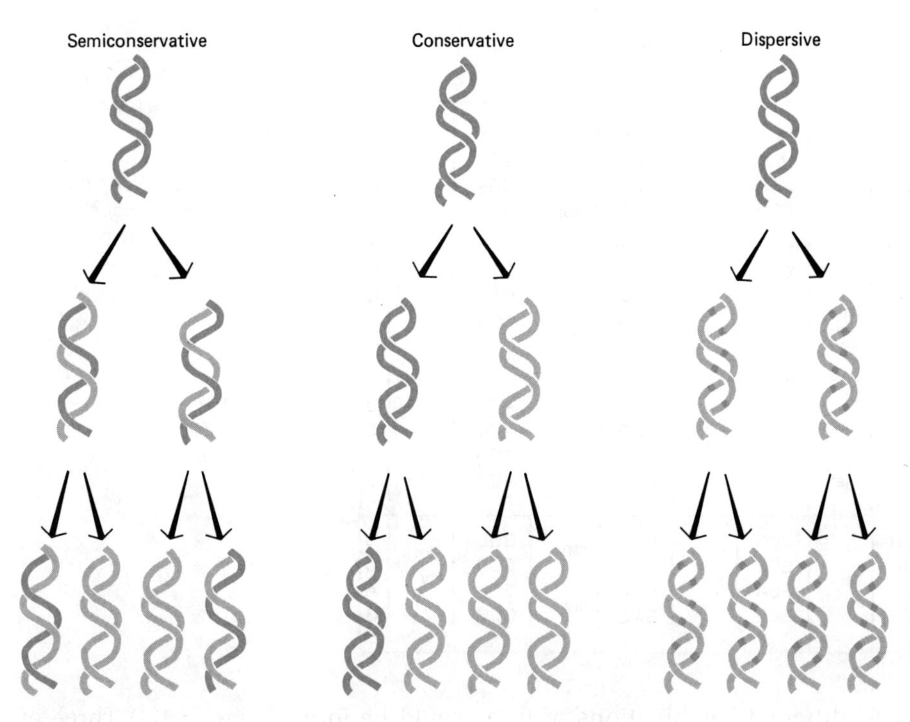


*Figure 6.11*
Three modes of replication of DNA molecules: *A.* semiconservative; *B.* conservative; *C.* dispersive.

plasm and is associated with the ribosomes through the activity of several different RNAs. In eukaryotic cells *messenger RNA* (mRNA) and *transfer RNA* (tRNA) are transcribed from chromosomal DNA, while *ribosomal RNA* (rRNA) is transcribed from a portion of the DNA in the nucleolus. This synthesis is initiated and terminated by a specific type of enzyme called *RNA polymerase*.

## *Translation*

You may recall that *proteins* are composed of *amino acids*, with the amino acids joined by *peptide linkages*. The question then, is how does this genetic information contained in the nucleus bring about the formation of many different kinds of proteins, each protein having specific amino acids in a specific sequence? After much patient and painstaking work, primarily with viral or bacterial systems, investigators postulated the following mechanisms. The sequence of nucleotides in the nucleic acid determines the sequence of amino acids in the protein. It was found that a sequence of three nucleotides codes for one amino acid. These units of three in the nucleic acids are referred to as *codons*. As there are four different nucleotides,

| First letter | Second letter: U | Second letter: C | Second letter: A | Second letter: G | Third letter |
|---|---|---|---|---|---|
| U | UUU phe | UCU ser | UAU tyr | UGU cys | U |
| U | UUC phe | UCC ser | UAC tyr | UGC cys | C |
| U | UUA leu | UCA ser | UAA stop | UGA stop | A |
| U | UUG leu | UCG ser | UAG stop | UGG trp | G |
| C | CUU leu | CCU pro | CAU his | CGU arg | U |
| C | CUC leu | CCC pro | CAC his | CGC arg | C |
| C | CUA leu | CCA pro | CAA gln | CGA arg | A |
| C | CUG leu | CCG pro | CAG gln | CGG arg | G |
| A | AUU ile | ACU thr | AAU asn | AGU ser | U |
| A | AUC ile | ACC thr | AAC asn | AGC ser | C |
| A | AUA ile | ACA thr | AAA lys | AGA arg | A |
| A | AUG met | ACG thr | AAG lys | AGG arg | G |
| G | GUU val | GCU ala | GAU asp | GGU gly | U |
| G | GUC val | GCC ala | GAC asp | GGC gly | C |
| G | GUA val | GCA ala | GAA glu | GGA gly | A |
| G | GUG val | GCG ala | GAG glu | GGG gly | G |

*Figure 6.12*
The genetic code consists of 64 triplet combinations (codons) and their corresponding amino acids. Since there are only 20 amino acids, different triplet combinations code for the same amino acid. Only 61 codons specify specific amino acids. The other three are known as nonsense codons or stop signals, because it is thought that they cause the polypeptide chains to terminate. The code shown here is as it would appear in mRNA.

64 different combinations of three could be formed (Fig. 6.12)) Three of these codons do not appear to code for any amino acid and are thought to terminate a given protein. The remaining 61 codons code for the 20 or so amino acids; obviously, several different codons must code for the same amino acid. The chromosomal DNA can produce a number of different RNAs. Thus, we might think of the chromosome as a large train yard with many trains, the RNA as a single, very long train, and the codons as specific boxcars making up the train. The RNAs synthesized in the nucleus are transferred to the cytoplasm with the rRNA becoming associated with the ribosome and apparently acting in bringing the mRNA and tRNA into position for establishing the peptide linkages between amino acids. Messenger RNA contains the codons and may code for one or several proteins. It becomes associated with the smaller of two subunits of the rRNA molecule. The tRNA contains complementary nucleotides to the mRNA, and thus, the 3 nucleotide units of tRNA are called *anticodons.* The function of tRNA is to transfer amino acids from the cytoplasmic amino acid pool to the site of protein synthesis in the ribosome. Which amino acid a tRNA picks up depends upon the particular anticodon it bears. Protein synthesis might be diagrammetically illustrated as in Figure 6.13.

Thus, we have a system for translating genetic information in the nucleus into specific proteins of the cell. One of the more important classes of proteins are *enzymes.* These proteins control all the chemical activities of the cell, and it is these activities, in turn, that are responsible for the very nature of the organism.

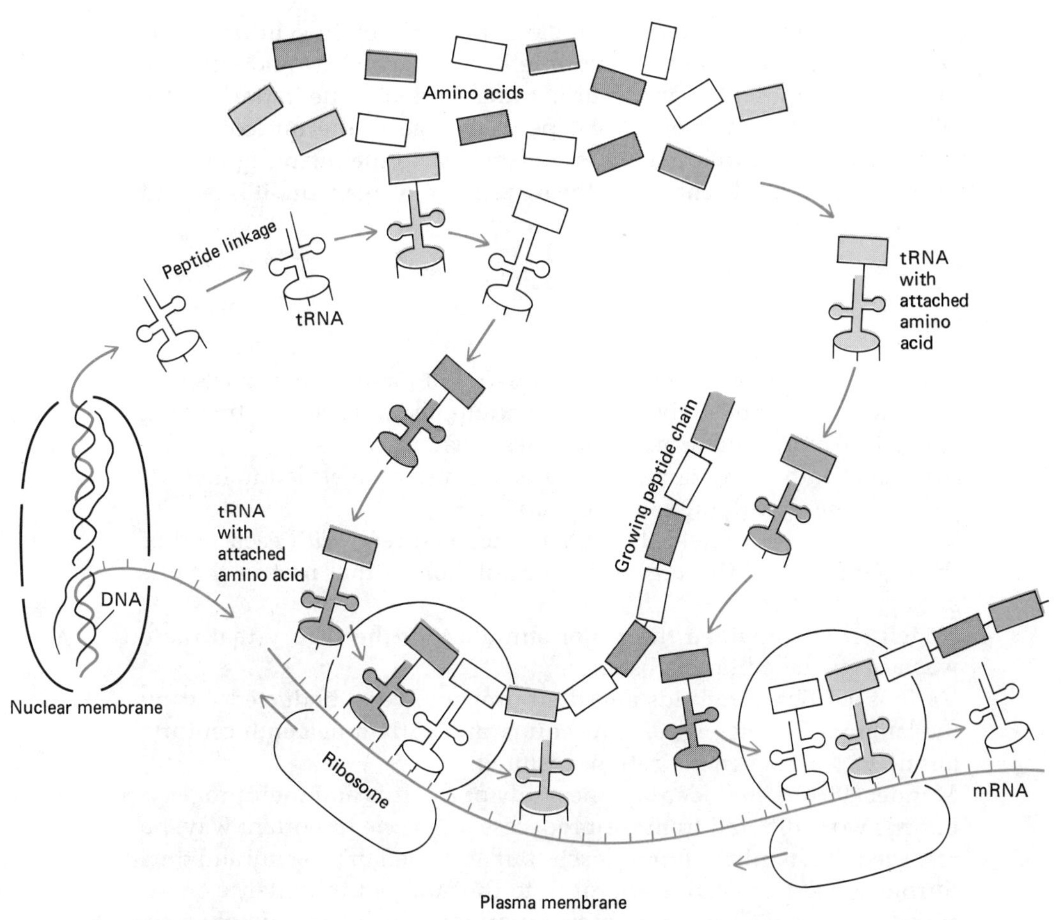


*Figure 6.13*
Protein synthesis. As many as 64 different kinds of tRNA molecules are transcribed from chromosomal DNA. The tRNAs attach to specific amino acids and each carries an anticodon that fits an mRNA codon for that particular amino acid. Protein synthesis is initiated when an mRNA strand attaches to a subunit of the ribosome and a matching tRNA (with its attached amino acid) attaches momentarily to the codon of the mRNA. As the ribosome moves along the mRNA strand, the amino acids are released from the tRNA, link together, and the tRNA, now minus its amino acid, is released from the mRNA strand.

In considering these mechanisms, it is well to keep in mind two points: 1) we have not excluded the possibility that proteins may be synthesized by some other mechanism, and 2) most of the evidence for these events have come from experimentation with viruses, bacteria, fungi, or animal cells, not with higher plants. Although it has been widely generalized that the mechanisms in higher plant cells are similar, this does not necessarily mean that it is a fact.

With such a complex yet systematic process for protein synthesis, it is logical to assume that many events might occur which would alter the scheme. Some of these changes occur in transcription, some in translation, but all occur as the result of some type of change in the nucleotide sequence or makeup. If this change is transmitted to the offspring and results in some observable change in the nature of the organism, it is termed a *mutation*.

## SUMMARY

*1.* Long ago man noted certain characteristics of plants and animals that he believed to be desirable ones and attempted by selective breeding to produce more organisms having these characteristics.
*2.* Interest in improving domesticated plants and animals led to investigation of the underlying causes of variation.
*3.* The classical theory held that inheritance is direct, with each part of the organism contributing to the formulation of that part in the offspring.
*4.* Nägeli (1883) provided the major stimulus for the theory that there was a *single* hereditary substance.
*5.* Various studies of hybrids and their progeny were conducted during the latter part of the eighteenth century and early nineteenth century, but no important generalizations resulted.
*6.* Mendel (1866) also undertook the study of hybrids and their progeny, but his work differed from his predecessors in one important way: he recorded the number of times each trait appeared in progeny and thus introduced a quantitative approach to the study of inheritance.
*7.* Mendel's studies resulted in three important principles: 1) dominance, 2) segregation, and 3) independent assortment.
*8.* The nucleus is of major importance in inheritance and contains structures termed chromosomes, with the hereditary units being carried on the chromosome.
*9.* These hereditary units are called genes, with alternate genes associated with the same region of the chromosome termed alleles.
*10.* Morphologically similar maternal and paternal chromosomes exist; they are called homologs.
*11.* During sexual reproduction, the homologous chromosomes become separated from each other and independently assorted into different cells by the process of meiosis.
*12.* As the result of meiosis, the chromosome number of the daughter cells will be one half that of the dividing cell, whereas in the alternate process of nuclear division, mitosis, the number of chromosomes in the daughter cell will be identical to that of the dividing cell.
*13.* An exception to Mendel's law of dominance is shown in a pattern of

inheritance called incomplete dominance, whereby the $F_1$ is intermediate in appearance between the dominant and recessive parents.

14. An exception to Mendel's law of independent assortment is established by the occurrence of many genes on the same chromosome. Thus, they will not independently assort but will be inherited as a unit. Such genes are referred to as linkage groups.
15. Linkage is generally not complete, as portions of homologous chromatids can be exchanged, thus creating new combinations. This is called crossing over, and study of cross-over results can establish both the position of genes on a chromosome and their relative distance from each other.
16. The physical position of a gene is called its locus, and many different alleles can exist for the same locus. Such a system whereby different alleles affect the same trait is termed multiple alleles.
17. Nonallelic genes may interact in various ways to produce effects different from those of the individual gene. Such interactions may be such that different gene pairs complement each other, are epistatic to each other, or result in the production of similar and additive effect. The latter instance is referred to as multiple factor or polygenetic inheritance.
18. At the molecular level, the basis for the above patterns of inheritance is in the nature of the DNA in the chromosomes and nucleolus.
19. DNA can replicate itself and can also, by transcription, produce RNA.
20. Several different types of RNA are found in cells, including messenger RNA, transfer RNA and ribosomal RNA, and these are involved in the synthesis of protein.
21. The sequence and kinds of amino acids in a particular protein are determined by sequences of three nucleotides, called codons, in the nucleic acids.
22. One way of envisioning a gene is to describe it as a region of the chromosome coding for a single specific polypeptide chain. As polypeptides are the basic part of an enzyme, we might think of one gene coding for one enzyme. Thus, the primary effect of the gene could be related to the specific activity of that enzyme coded for by the gene.
23. Genetic material may originate outside of the nucleus in specific cellular organelles: the chloroplast and the mitochondrion. Thus, these organelles can produce their own inheritance patterns. Such patterns are termed non-Mendelian or extrachromosomal inheritance.
24. There may be interaction between chromosomal and extrachromosomal genes.
25. Variability in plants can be associated with both chromosomal and nonchromosomal factors. Thus, sexual reproduction is not the only nor necessarily the major source of variability. Important new characteristics might be introduced through mutation, while environmental effect can also produce variation.

# Chapter 7

## THE NONVASCULAR PLANTS – A SURVEY

As stated in Chapter 5, there are various systems for the classification of plants. We will mix a little of the old with the new in our considerations of plant groups. From a practical view, it is convenient to consider those plants without a vascular system together, as these organisms lack true stems, roots, and leaves. On the other hand, in order to consider adequately the vascular plants, we must have some knowledge of the morphology and anatomy of the vegetative organs. From the modern view, the systematics of Margulis, although not yet widely adopted, do have some basic advantages and therefore will be the system followed in this text. From the traditional view, we will be discussing a number of groups of organisms not now considered to be true plants.

## VIRUSES

It may seem strange to begin our discussion of plants with the viruses. Not only are they not plants, but there is considerable doubt that they should even be considered living organisms. However, because of their importance in scientific research, their implication in the genetic mechanisms of bacteria, and more importantly their great importance as agents of disease in both plants and animals, we shall consider them briefly.

Electron microscopy has shown many viruses to be icosahedral (20 sided), although other shapes also exist. The virus consists of an outer shell of protein composed of identical subunits, and an inner core of single- or double-stranded DNA or RNA (Figure 7.1). Unlike all living organisms viruses do not contain both DNA and RNA. As it lacks ribosomes, the virus is unable to make its own protein. Thus, it depends on its host cell for replication.

The mode of replication of a virus depends upon the nature of its nucleic acid constitution. The protein of their shell is host specific, thus determining what type of cell a virus will attach to. After attachment, the virus enters the host cell, leaving its protein covering outside in some instances. This is true of those viruses (bacteriophages) that attack bacteria, but in some, the protein is shed within the cell, while in others, enzymes of the host cell digest the protein. By whatever means the protein coat is lost, once it has occurred, one of two things happens: 1) The virus nucleic acid directs the host cell to produce viral enzymes which synthesize viral nucleic acids and structural proteins which are then assembled into new viral particles. This results in the production of thousands of viral particles, the lysis of the host cell, and the escape of the particles, each capable of infecting another host cell. 2) The viral DNA may be inserted into the host chromosome. In this state, multiplication of the viral particles does not occur and the virus is referred to as a *prophage*. Phages that can exist in this form are called *temperate phages*. The phage DNA can, however, "escape" from the host chromosome, usually carrying a bit of

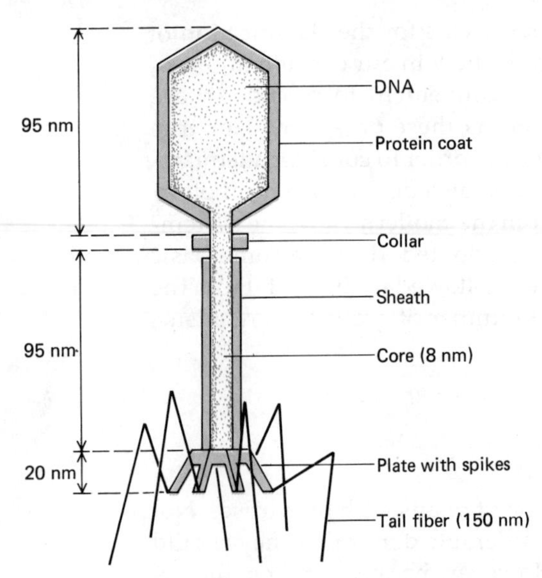


*Figure 7.1*
Tobacco mosaic virus (300 × 18 nm).

the chromosome with it, and once this happens, multiplication of new viral particles begins, resulting, as before, in lysis of the host cell. The implications of viral infection of bacteria cells are several, not least of which is the effect such infection has on the metabolism of the host cell. The bacterial *Clostridium,* which we will mention later, form toxins which are responsible for producing botulism and gas gangrene. It was thought two different species of *Clostridium* were involved, one producing botulism, the other gas gangrene. It has been shown recently that only virus-infected bacteria produce the toxins and that the same species of *Clostridium* produced both botulism and gas gangrene; it is simply infected with different viruses.

## MONERA

The Monera have also been called the prokaryotes because all members of this kingdom have prokaryotic cells. The nature of the prokaryotic cell and its possible evolution has been discussed previously in Chapter 1.

To reiterate, no prokaryote is truly multicellular in the sense there is a division of labor between cells. Their members are either unicellular, colonial, filamentous, or mycelial, with the latter three types having their cells connected together only because the cells failed to separate following cell division or because they are within a common mucilaginous sheath or capsule. There are no plasmodesmata between adjacent cells.

Members of this kingdom lack an organized nucleus, cellular organelles, and sexual reproduction (some, however, do have mechanisms that lead to genetic recombination). Their storage product is never starch. Nearly all have a cell wall which is characterized by incorporation of polypeptides as links between the polysaccharide units and by the presence of muramic acid and diaminopimelic acid. These features of the wall also serve to distinguish prokaryotes from eukaryotes.

## *Bacteria*

The bacteria are the smallest organisms known, ranging from 0.1 μm to 2 μm in diameter. Most are less than 10 μm long, although a few may be as long as 500 μm. They may be placed into 13 different divisions (phyla).

Bacteria exist in nearly all habitats, from the poles to hot springs, from lighted areas to the darkest depths of the oceans. As discussed earlier, they are the most ancient group of organisms, with extant types being very similar to those postulated to exist over 3.3 billion years ago. Some are *obligate anaerobes* (cannot live in the presence of oxygen), some are *microaerobes* (can live in the presence of minute amounts of oxygen but do better without it), while still others are *aerobes* (can live only in the presence of oxygen).

Bacteria have variable modes of nutrition. Most are *heterotrophic* (cannot make their own food), and most heterotrophs are *saprobes* (organisms that obtain their nourishment from dead organic matter). Other heterotrophs may be either *parasitic* or *symbiotic.* Parasitic bacteria live either in or on another organism and may have little effect on their host or they may injure that host *(pathogenic).* Symbiotic bacteria, on the other hand, usually benefit their host. Examples of bacterial symbiosis include those living in the digestive tract of animals, including man; the cellulose-digesting bacteria in the rumen of cows; the nitrogen-fixing bacteria in the roots of legumes; and many others. A few groups of bacteria are autotrophic (can make their own food). The photoautotrophic, or photosynthetic, bacteria utilize light as an energy source and, under anaerobic conditions, produce sugar. Chemoautotrophic bacteria utilize energy gained by the oxidation of certain inorganic molecules to reduce carbon dioxide to carbohydrate, and thus require oxygen.

Although the prokaryotic cell is far less complex than those of eukaryotes, you can tell from the previous discussion that bacteria represent a diverse group of organisms. This diversity can also be seen in their morphology and cellular organization. Although bacteria may aggregate in various ways to form the diverse growth habits observed, the individual cells are one of three morphological types: the spherical forms are called *cocci;* the straight, rod-shaped forms *bacilli;* and the long, curved to spiral forms, *spirilli.*

Some bacteria are motile by means of long (3–12 μm), slender (10–20 nm) flagella. As these flagella are so small in diameter, they ordinarily cannot be seen under a light microscope unless a special staining technique is used. Studies seem to indicate that the bacterial flagella is composed entirely of a single type of protein called *flagellin.* Motile bacteria may have a single polar flagellum, a single flagellum at each pole, two flagella at each pole, or flagella inserted all around the cell.

Another characteristic of some of the bacilliform bacteria is the formation of thick-walled, highly resistant structures called *endospores.* As the name implies, these structures are formed within the parent cell, being released upon the death of that cell. They germinate producing a single bacterium. Thus, these structures are not considered to be reproductive bodies. They are produced by a single organism which dies; they give rise to a single organism. Hence, there is no increase in the population and therefore no reproduction. Endospores are however extremely important structures. Bacteria are normally quite sensitive to desiccation, heat, cold, ultraviolet light, and chemicals of various kinds. The endospores however are quite resistant. This becomes important if the bacteria produce disease or toxin (poison).

The chief mode of reproduction in bacteria is simply by an invagination of the plasma membrane and cell wall forming two new cells, a process known as *fission.* Accompanying this infolding is the replication of the cell's DNA and subsequent distribution of the daughter molecules to the two new cells. Modes of vegetative reproduction are illustrated in Figure 7.2. As mentioned earlier, bacteria do not exhibit true sexual reproduction, but there is a fragmentary or partial exchange of genetic material in some. This process has been called *parasexuality* or *conjugation* and represents a method of genetic recombination. Another method of bacterial recombination involves the transfer of genetic material from one bacterium to another by a virus (bacteriophage). This mechanism is termed *transduction.* Another mechanism involves the transfer of small pieces of DNA from a heat-killed bacterium into a living bacterium, where it becomes incorporated into that cell's DNA. This process, called *transformation*, was demonstrated in 1944 and established the first clear evidence that DNA was the genetic substance of cells.

The effects of bacteria on man and his affairs are numerous and varied. Many of man's diseases are caused by bacteria. Species of *Clostridium* may cause tetanus, botulism, or gangrene. Members of the genus *Streptococcus* are associated with rheumatic fever, scarlet fever, strep throat, and other infections. *Staphylococcus* may cause food poisoning or skin infection. *Corynebacterium diptheriae* is the causative agent of diptheria. *Diplococcus pneumoniae* causes most bacterial pneumonia. Typhoid and paratyphoid fever are caused by members of the genus *Salmonella.* Tuberculosis is associated with the bacterium *Mycobacterium tuberculosis.* Undulent fever is caused by a species of *Brucella.* Bacterial dysentery is caused by *Shigella dysenteriae. Neisseria gonorrhoeae* is the causative agent of gonorrhea, while the spirochete, *Treponema pallidum* causes syphilis. Typhus is caused by *Rickettsia prowazekii,* while a related organism causes Rocky Mountain spotted fever.

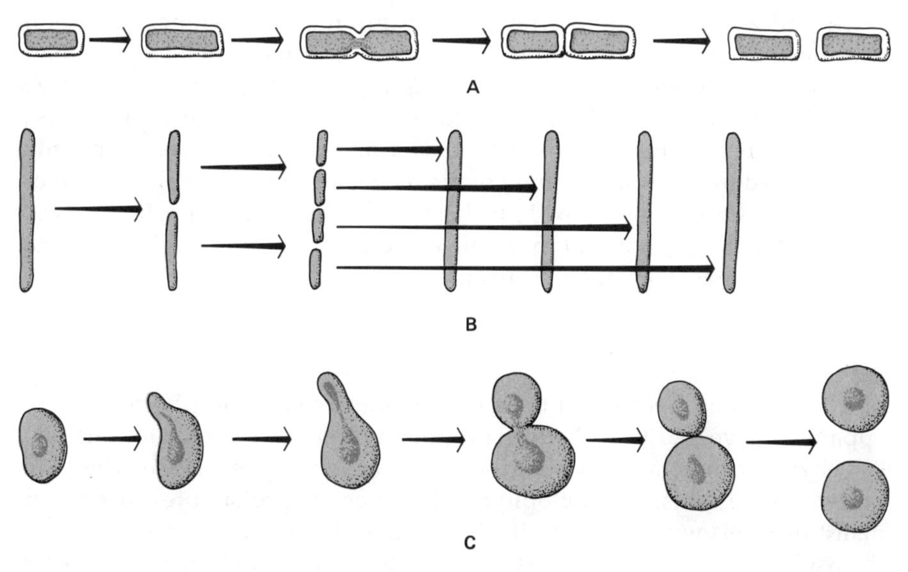


*Figure 7.2*
Types of asexual reproduction in bacteria. *A.* Fission in *Escherichia coli. B.* Fragmentation in an Actinomycetales. *C.* Budding in a Hyphomicrobiales.

Plant diseases caused by bacteria are also a major problem, particularly to farmers, gardeners, nurserymen, florists, and others. Some of these diseases are commonly referred to as blights, galls (not to be confused with insect galls), wilts, or soft rot.

Bacteria, however, are indispensible organisms; without them (and the fungi we will be discussing later), we would be up to our necks in our own rubbish and waste. Bacteria are among our major decomposers, breaking down organic matter into soluble compounds. It has been said, perhaps not facetiously, that for every known naturally occurring (as opposed to synthetic or man made) organic compound, there is a bacterium that can decompose it. We shall be discussing later the importance of bacteria in soil ecology, particularly in the nitrogen cycle and sulfur cycle. The making of many cheeses depend upon bacteria, and certain bacteria are important to the chemical industry. For example, acetone can be produced through the activities of a species of *Clostridium*.

It is hard to imagine, but that tiny little microscopic organism may have been responsible for the political geography of the world being what it is today. Acetone is a major ingredient in the manufacture of cordite, an explosive. During World War I, England could not get acetone from its usual source. A lecturer at the University of Manchester had developed a method whereby through bacterial fermentation by a species of *Clostridium*, acetone was formed as a by-product. He took the process to the then First Lord of the Admiralty, Winston Churchill, convinced him it would work, and was assigned by Churchill to begin making acetone. They soon became good friends, and through Churchill, the man came to know and become friendly with Churchill's successor, Lord Balfour. As an active Zionist, the man persuaded Balfour to establish a homeland for homeless Jews, and in 1917 the famous Balfour Declaration was made. Eventually this resulted in the establishment of the new state of Israel with the former lecturer of the University of Manchester, Chaim Weizmann, as its first President. Perhaps, without that little bacterium, England might have lost the war and Israel may not have been established.

## *Blue-Greens*

The blue-greens, because of their morphology, color, and photosynthetic apparatus have traditionally been grouped with the algae and, for this reason, have been commonly referred to as the blue-green algae. They are placed in the division Cyanophyta. However, as prokaryotes, they bear many affinities with the bacteria. Although cellulose has been reported in the wall of blue-greens, this has not been confirmed, and the wall appears to be morphologically and chemically similar to that of bacteria. Most blue-greens have an outer mucilagenous layer, or sheath, similar to that of bacteria. As in the bacteria, blue-greens do not have starch as their storage product, but rather a substance similar to glycogen and called *cyanophycean starch* by some investigators.

Like bacteria, blue-greens exist in a wide variety of habitats: hot springs, ice and snow, desert soils, and many others. They are extremely resistant to desiccation, due perhaps in part to their gelatinous sheath and in part to the nature of their protoplasm.

Some blue-greens are unicellular, some are colonial, while others are filamentous. The filaments may be unbranched or branched, straight or coiled, uniform in width or tapered.

Blue-greens differ from bacteria in several respects, not least of which is the presence of photosystem II in their photosynthetic mechanisms, resulting in oxygen evolution by the photosynthesizing cell. Their pigments include chlorophyll *a* and $\beta$-carotene in common with every other photosynthetic plant except the photosynthetic bacteria (a major argument of some investigators for the relation of blue-greens to the eukaryotic green plants and not the bacteria). In addition, they have a variety of pigments called xanthophylls, and two belonging to a class called phycobilins. The latter two pigments are called phycoerythrin (a red pigment) and phycocyanin (a blue pigment). These phycobilin pigments are different from all others in that they are complexed with proteins.

As might be imagined, the blue-greens are often not blue-green in color, and depending upon the concentration of the various pigments present, may be blue-black, blue, olive-green, yellow-green, a pale golden-

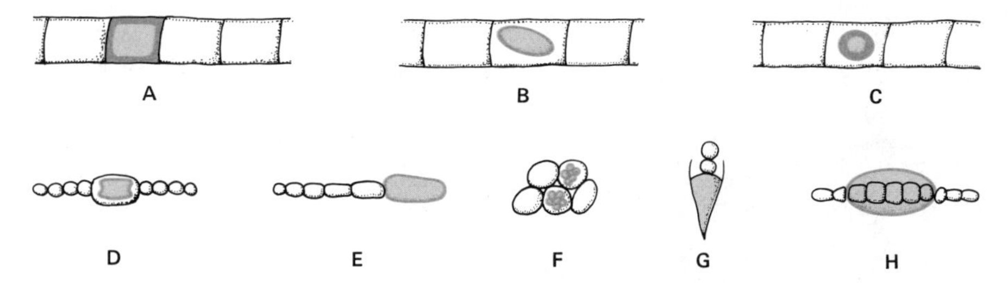


*Figure 7.3*
Types of asexual spores in blue-greens. *A.* Akinete. *B.* Aplanospore. *C.* Hypnospore. *D.* Heterocyst. *E.* Gonidium. *F.* Endospores. *G.* Exospores. *H.* Hormospore.

yellow, gray-green, emerald green, yellow-brown, brown, purplish, or even red. The same species of blue-green, if grown under different light regimes, may show significant variations in pigment composition, and thus in color. This property has given rise to the theory of *complementary chromatic adaption*, or the *Gaidukov phenomenon*. It is thought that the color of the pigment is complementary to the color quality of light waves reaching the cell. In a vertical water column, the longest wave lengths are absorbed most rapidly; the shorter wave lengths penetrate the deepest. According to the theory of complementary chromatic adaptation, this explains why the algae found at the greatest depths in the ocean are red in color. Only the blue end of the light spectrum will penetrate that deeply, and maximum absorption of that light could only be possible with a red pigment. In most instances, however, the blue pigment phycocyanin predominates in the blue-green algae, and it has been shown to be a primary light-absorbing pigment in the photosynthetic process, with the energy being transferred directly to chlorophyll *a*.

A second difference from bacteria is the degree of differentiation shown by blue-green algal cells. We have previously mentioned branching of some filaments, the tapering of others. Certain vegetative cells may differentiate into asexual reproductive bodies termed *aplanospores* or *akinetes* (Figure 7.3). Algal taxonomists recognize as many as six different types of spores in the blue-greens. Certain vegetative cells, either within *(intercalary)* or at the ends *(terminal)* of filaments may metamorphose to form a different type of specialized cell. These cells are called *heterocysts*, and are formed by the laying down of a second wall layer internal to the original plasma membrane, leaving a pore at either end where the heterocyst adjoins other cells of the filament. Thus, an intercellular connection is formed between the heterocyst and adjacent vegetative cells. During maturity of the heterocyst, these pores become plugged with mucilage, and under light microscopy these plugs appear as bright nodules. The contents of the heterocyst lose their granular appearance, with these morphological

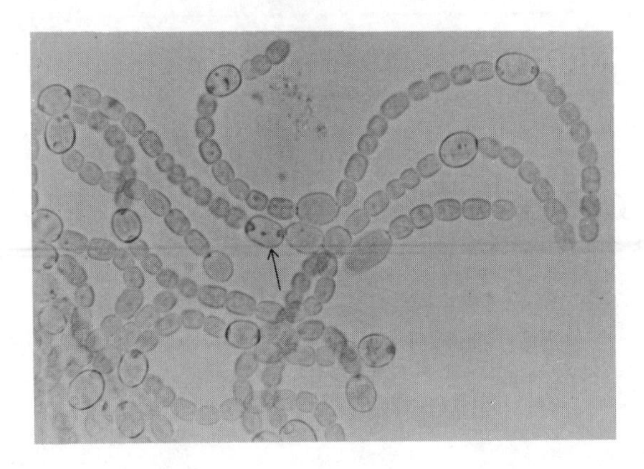

*Figure 7.4*
Heterocysts *(arrow)* in the blue-green alga *Anabaena azollae.* [Courtesy of Norma J. Lang.]

changes being accompanied by changes in the chemical composition of the protoplasm and, most importantly, the loss of photosynthetic ability (Figure 7.4).

An important property of some blue-greens is their ability to fix atmospheric nitrogen, that is, to convert $N_2$ to $NO_3$ (nitrate). Nitrate is the form of nitrogen which can be utilized by higher plants. It has been known for many years that the ability of blue-greens to fix nitrogen was associated only with those genera possessing heterocysts; thus, the heterocyst appeared to be implicated in some way with nitrogen-fixing ability. Recently, it became possible to isolate a suspension of heterocysts from the vegetative cells and show that the nitrogen-fixing ability lies solely in the heterocyst. In addition, the enzyme *nitrogenase,* which catalyzed the reaction, was isolated from the heterocyst suspension but was not found in the vegetative cells. But the problem has been complicated somewhat by the observation that under certain environmental conditions, a nonheterocystic genus was also capable of nitrogen fixation.

The nitrogen-fixing ability of blue-greens has practical implications. In Southeast Asia, rice has been grown on the same land for years without the addition of fertilizer because of the presence of blue-green algae in the rice paddies. Recently, strains of blue-greens having high nitrogen-fixing capacity have been developed, and are "seeded" into the rice paddies, resulting in as much as 20 percent increase in rice production.

Some investigators have noted the germination of heterocysts; thus, they can, at least under certain conditions, act as asexual spores. Heterocysts have been implicated by other investigators in promoting spore formation in adjacent vegetative cells. Heterocysts are also involved, perhaps incidentally, in vegetative reproduction by fragmentation of the filament.

Although blue-greens are primarily photosynthetic, some are reported to be facultatively heterotrophic. That is, in the light they are photosynthetic, but if placed in the dark, they become heterotrophic. At least one has been reported to be solely heterotrophic.

Blue-greens lack organs of locomotion, but some do possess the ability to move. This is especially evident in an order called the Hormogonales (Oscillatoriales). The movement consists of a gliding backward and forward (a movement similar to that of the gliding bacteria) and a rotational movement of the forward end of the trichome. Halfen and Castenholz (1971) attribute the movement to a band of continuous fibrils located in the cell wall and organized in a helix, supporting the explanation advanced by Jarosch in 1964.

Reproduction in the blue-greens is largely vegetative, primarily by fission, although fragmentation commonly and successfully occurs. Asexual reproduction occurs, as mentioned earlier, by the production of several different types of nonmotile spores. True sexual reproduction is unknown, although a pseudosexual process similar to that of certain bacteria has been reported in the blue-green *Anacystis nidulans.*

The blue-greens are an important group of organisms, and as most of their qualities are detrimental, perhaps notorious would be a better word than important. As photosynthetic organisms, they are involved in oxygenation of the water, and they serve as food for certain aquatic organisms. Some terrestrial forms are early colonizers and thus important in the process of soil building, while others may be important in reducing soil erosion and improving the water-holding capacity of the soil. Their importance in nitrogen fixation has been mentioned previously. Certain terrestrial species of *Nostoc* have been reportedly used as food by man in parts of China ("Fat Choy") and South America ("Yuyuko"), but such use is probably not very extensive. Blue-greens occur commonly as symbionts in green algae, certain diatoms, some higher plant cells, coenocytic fungi, and protozoa. They are frequently the algal component of lichens.

The deleterious effects are many. Under certain conditions blue-green algae may undergo rapid reproduction to produce massive growths. As the cells age, structures called gas vacuoles develop in their protoplasts, adding buoyancy, and the cells break loose and float to the surface producing *blooms.* These blooms are not only unsightly, but as the cells die, the oxygen-requiring agents of decay (bacteria and fungi) may cause severe oxygen depletion of the waters, thus resulting in the death of zooplankton, fish, or other aquatic animals. In water supply reservoirs, blue-green algae may constitute a major problem. They may impart objectional tastes and odors to the water and clog filters. In addition, they may render an area unfit for swimming, boating, fishing, or other recreational activities. Some blue-greens produce potent toxins which may kill fish, birds, or domestic animals. Blue-green algae intoxications in humans may be both chronic and acute, but rarely lead to death, although death through blue-green intoxication has been documented. Some blue-green algae produce an extrametabolite which stimulates the reproductive rate in certain zooplankton while inhibiting it in others. The significance of this is not yet known.

## PROTISTS

Of the protists, we shall consider only those groups that have traditionally been treated as plants. These include the green algae, charophytes, chrysophytes, xanthophytes, diatoms, brown algae, red algae, dinoflagellates, euglenoids, slime molds, and flagellated fungi.

Protists are eukaryotic organisms. They divide mitotically, with the presence of centrioles in many cells, generally by an inward pinching of the plasma membrane, similar to that of animal cells. However, some protists do not have centrioles and during cytokinesis form a cell plate as do higher plant cells. Protists may or may not have a cell wall, photosynthesize, store starch, possess motility by means of flagella, or exhibit multicellularity. In no case however, is true tissue formed, nor does the organism pass through embryonic stages during its development. The zygote (fertilized egg) develops directly into the mature organism. As implied by this statement, sexual reproduction does occur.

The term *algae* has been used for centuries to refer to the photosynthetic members of this group, and although abandoned as a formal term in modern systematics, common usage of the term is not likely to be discarded. This very diverse group has been divided into 15 different divisions (phyla). We shall be considering nine of those. Nearly all members are photosynthetic, but heterotrophic forms are known. The plant body may be unicellular, colonial, a simple or branched filament, or truely multicellular. The reproductive structures are generally single cells not surrounded by a sterile jacket of cells such as those of higher plants. Most divisions, with the exception of the red algae, have members with either flagellated vegetative cells, flagellated reproductive cells, or both. The nature of their pigmentation and storage products is quite variable, and although most groups possess cellulose as the primary cell wall component, this characteristic is also variable.

### *The Green Algae*

The green algae are an extremely diverse group. The division name Chlorophyta, if literally translated, means "green plants." This name is widely used although some botanists feel it is not a very good name as there are many groups of green plants which do not belong to the chlorophyta, yet *are* green plants (*chloro*, "green"; *phyte*, "plant"). Thus, they suggest the division name Chlorophycophyta ("green algal plants") as more appropriate.

The green algae are found in both marine and fresh water (with perhaps more being found in fresh water), on moist soil, rock surfaces, tree trunks, and in snow. They are symbionts in lichens, the coelenterate animal called Hydra, and many protozoans. Some even grow and reproduce while suspended in droplets of water in the atmosphere. The division is a large one, estimated by some algologists to contain as many as 20,000 species, with the desmids (a group in the order Conjugales, an order that Margulis elevates to the division rank Gamophyta) alone having 11,000 species.

There is considerable morphological variation within the group, including motile and nonmotile unicells, motile and nonmotile colonies, pseudofilaments, branched and unbranched simple filaments, coenocytic filaments (composed of cells having more than one nucleus), siphonaceous filaments (without cross walls), and pseudoparenchymatous (false parenchyma) aggregates or sheets of cells.

Most green algae are characterized by the presence of cell walls, although some have no wall at all; instead the outer membrane becomes somewhat rigid, forming a *pellicle*. The wall typically consists of three layers; an inner cellulosic layer, a middle pectic layer, and an outer mucilaginous layer. However, the cellulosic, pectic, or outer mucilaginous layer may be lacking in some. In siphonaceous forms, the cellulose may be replaced by callose (another carbohydrate); in others, the walls may be slightly impregnated with silica or may contain deposits of chitin (a compound composed of *N*-acetyl-2-glucose amine), calcium, or various iron salts. Pickett-Heaps questions the presence of pectic compounds in the middle layer of certain chlorophytes and would prefer to call that layer the *reticulate layer*. The absence of an outer mucilaginous layer in members of the order Cladophorales and certain others accounts for the high degree of epiphytism (growth of other organisms upon them) upon those taxa.

The most distinctive characteristics of the green algae are their pigments, which are contained in variously shaped organelles called chloroplasts. Chlorophyll *a* predominates; thus, most are grass green in color, at least during their vegetative stages. Chlorophyll *b* is also present, being found in no other group of algae, with the exception of the euglenoids and charophytes, but being a feature of higher green plants. This feature, and others, has led many investigators to postulate that the euglenophytes and charophytes must be closely related to the green algae, and that higher plants probably evolved from some chlorophycean ancestor. In addition to the chlorophylls, $\alpha$-, $\beta$-, and $\epsilon$-carotenes plus five or six xanthophylls may also be found in either or both the vegetative and reproductive cells. Some investigators have postulated that the chloroplasts originated from blue-green algal cells incorporated into colorless eukaryotic cells as endosymbionts. Bodies called *pyrenoids* are present in the chloroplasts of many green algae, and consist of a proteinaceous core surrounded by layers of starch. Pyrenoids are involved in starch synthesis, starch storage, or perhaps both.

Motility in the green algae is associated with organelles called flagella, and these have the typical 9 + 2 arrangement of microtubules. The flagella are typically two per cell, inserted either apically or subapically, and are similar in length and structure. There are, however, a few exceptions in regard to number, length, and structure. Some of these motile cells possess a red pigment body called an *eyespot*, or *stigma*, a body which may or may not be associated with light perception.

Reproduction is highly variable in the green algae. Fragmentation and cell division are common vegetative means, while a wide variety of asexual spores are formed, both motile *(zoospores)* and nonmotile. Sexual reproduction is likewise quite variable, but three general types emerge: *isogamy*, in which the gametes are similar both in size and motility but differ physiologically (a characteristic generally designated as + and −); *anisogamy*, in which the gametes are dissimilar in size but similar in their motility (or more rarely, lack of it); and *oogamy*, in which a smaller motile gamete, the *antherozoid* or *sperm*, unites with a larger nonmotile gamete, the *egg* (Figure 7.5). We shall be discussing sexual reproduction and the life histories of plants in much more detail in Chapter 16.

One of the simplest, and perhaps most closely related to the ancestral green algal type, is the organism called *Chlamydomonas*. It is small (6–25 μm), ovoid, ellipsoid, or spherical, and is motile by means of two equal length flagella of the *whiplash* (smooth) type. The cell is bounded by a cellulose wall, and the protoplast contains a single nucleus and a single, large, cup-shaped, parietal (near the wall) chloroplast containing one or more

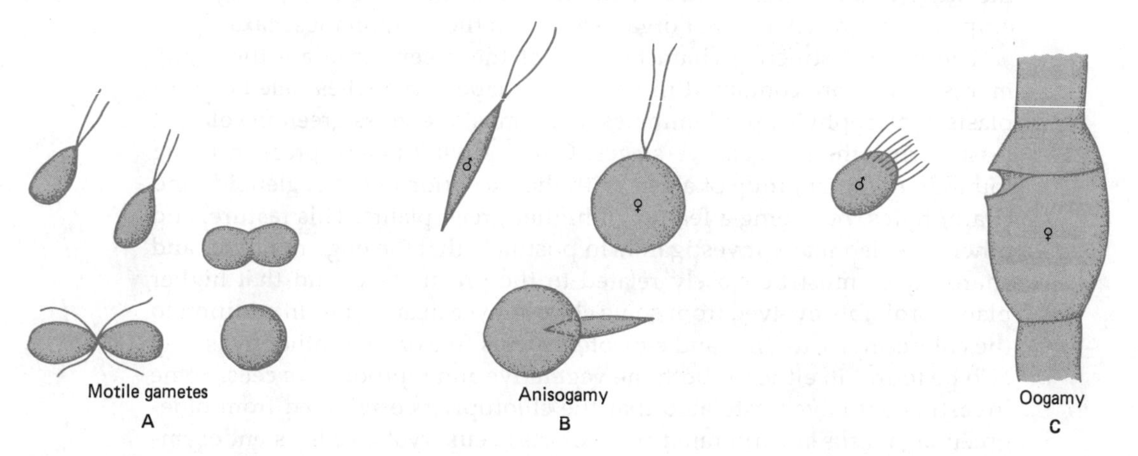


*Figure 7.5*
Type of sexual reproduction based upon morphology of gametes. *A*. Isogamy—the fusion of morphologically similar gametes. *B*. Anisogamy—the fusion of morphologically dissimilar gametes, both motile (or rarely both nonmotile). *C*. Oogamy—the fusion of morphologically dissimilar gametes, one large and nonmotile, one smaller and motile.

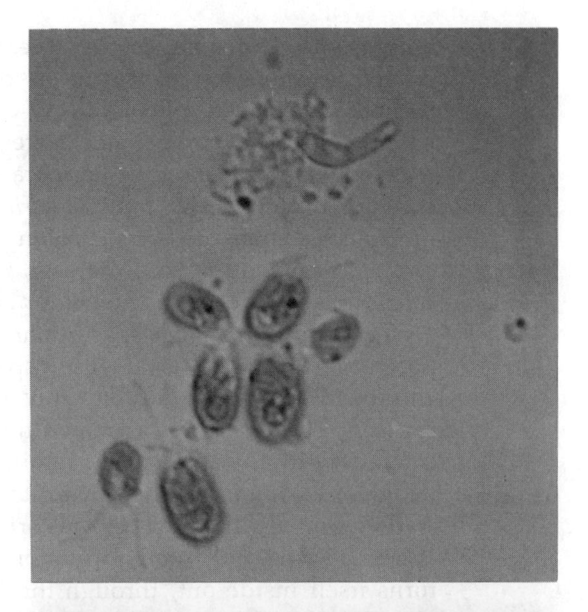

*Figure 7.6*
*Chlamydomonas* sp. An example of a unicellular, motile green alga. Note equal-length flagella. Both are of the whiplash type. [From Kodachrome by David R. Williams.]

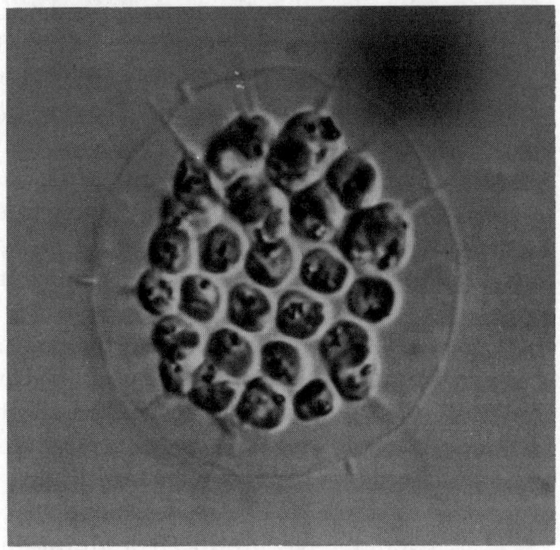

*Figure 7.7*
Member of the Volvocine line. *Gonium*. Members of this line are colonial flagellates, with the individual cells being replicas of, or similar to *Chlamydomonas*. [From Kodachrome by Robert G. Trumbull, III.]

pyrenoids and an eyespot (Figure 7.6). Although normally motile, cells of *Chlamydomonas* have been noted to clump; their walls become somewhat gelatinous, and they lose their flagella. A *Chlamydomonas*-like cell is postulated as the ancestral type for many green algae.

A colonial alga is *Gonium* (Figure 7.7), a small colony of 4, 8, 16 or 32 cells, held together in a gelatinous material with each cell flagellated and each capable of producing a new colony. In *Pandorina*, consisting of 16 or 32 cells, each cell of the colony divides several times until the characteristic number of cells for the colony is reached. A miniature colony is called a *daughter colony*, and as it consists of a fixed number of cells, it is also termed a *coenobium*. When all cells of all daughter colonies within the old parent colony have developed flagella, there is dissolution of the parent matrix, like the opening of Pandora's box (hence the generic name), and the young colonies swim out. *Eudorina* is larger, being composed of 32, 64, or 128 cells. Some cells of the colony are smaller, grouped at the anterior end, and function only in motility, while the others are capable of reproduction as well. Here is the beginning of specialization of function.

The largest and most spectacular colonial green alga is *Volvox*. Composed of as many as 60,000 cells, it spins through the water due to the synchronized beat of the flagella. Only a few of the cells are capable of reproduction. These cells, by repeated divisions, form characteristic daughter colonies, with the anterior ends pointed inward towards each other. Each daughter colony is attached by protoplasmic threads to neighboring cells. At the point at which the cell was originally attached to the parent colony, is a small hole. Following formation of the daughter colony, it suddenly turns itself inside out, through the hole, like pushing a finger out of a glove, so that the anterior ends of the cells now face outward. The cells develop flagella and are now ready for independent existence .

An example of a nonmotile colonial green alga is *Tetraspora*. The vegetative cells of this colony are very *Chlamydomonas*-like in character with the exception that they are usually nonmotile. Under certain conditions, however, they may develop flagella. A curious feature of this organism is the presence of two long, extremely fine, flagellalike structures extending from the anterior end of the cell. These are termed *pseudocilia*, as they are incapable of movement, nor do they have the typical 9 + 2 arrangement of microfibrils.

The organism *Palmodictyon* is in some biologists' minds illustrative of how the filamentous growth habit may have come about. Cells of this colony divide in a single plane, with the daughter cells then separating from each other. Thus, a row of cells invested by a common gelatinous sheath is formed. This arrangement is called by some algologists a *pseudofilament*. If the daughter cells did not happen to separate following cell division, they would have produced a simple unbranched filament similar to that of *Ulothrix*. If, in addition, cell division in a different plane occurred, a branched filament such as *Stigeoclonium*, *Coleochaete* or *Fritschella* might result. *Coleochaete* is an interesting organism because it is the one in which meiosis in the algae was first demonstrated. *Fritschella* and reportedly *Ulothrix* as well are of special importance because cytokinesis in these cells is by formation of a cell plate rather than by furrowing as in most green algae. All of the foregoing organisms have cellular characteristics similar to that of *Chlamydomonas*, although some have isogamous, some anisogamous, and others oogamous sexual reproduction.

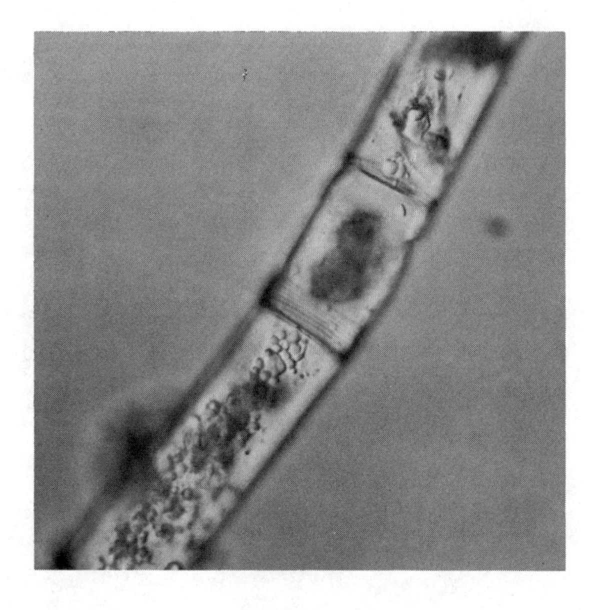

*Figure 7.8*
*Oedogonium*. Note the reticulate chloroplasts and apical caps. The apical caps are the result of successive unequal cell divisions. [From Kodachrome by Robert G. Trumbull, III.]

In *Cladophora*, we find a branched alga with distinctly different characteristics. The cells are multinucleate, and the chloroplast either forms a parietal network or there may be many disklike chloroplasts. There are many pyrenoids in each cell. The walls are very thick, often formed in layers (lamellated), composed of callose, and without a mucilagenous sheath. Thus, older filaments of *Cladophora* are often heavily epiphytized by diatoms and other algae. Reproduction in *Cladophora* can also be unusual. One species produces its spores by mitosis, its gametes by meiosis (as in animals)—just the reverse of what occurs in most plants. Occasionally, some gametes may fail to fuse with other gametes, but give rise to a new plant anyway. This production of a new organism from a gamete rather than a zygote is called *parthenogenesis*.

Representative of another group of algae with netlike chloroplasts is *Oedogonium* (Figure 7.8). Although this organism is unbranched, other members of the order are branched. All rise from a modified basal cell called a *holdfast*. *Oedogonium* is unusual in that growth of the filament is *intercalary* (occurs between the base and apex) rather than *apical* (from the apex).

A large order of green algae is one called the Chlorococcales. Members of this group are either one celled or form colonies of rather definite shape. This group is unusual in that its members are unable to multiply by cell division in the vegetative state. Some cells are uninucleate others multinucleate, and chloroplast structure is quite variable. Pyrenoids may be present, or they may be lacking. Some members of this order are illustrated in Figure 7.9.

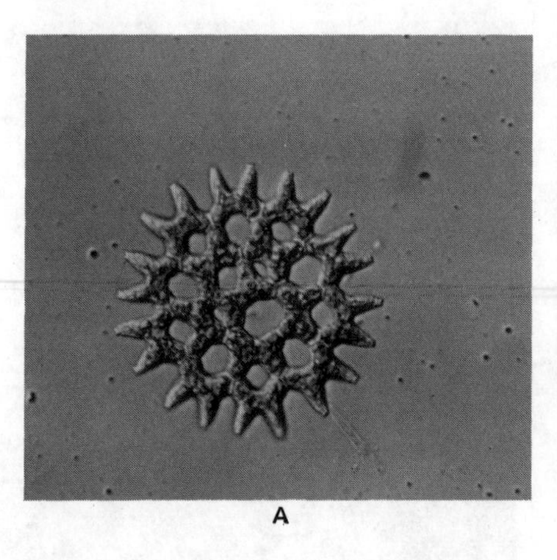


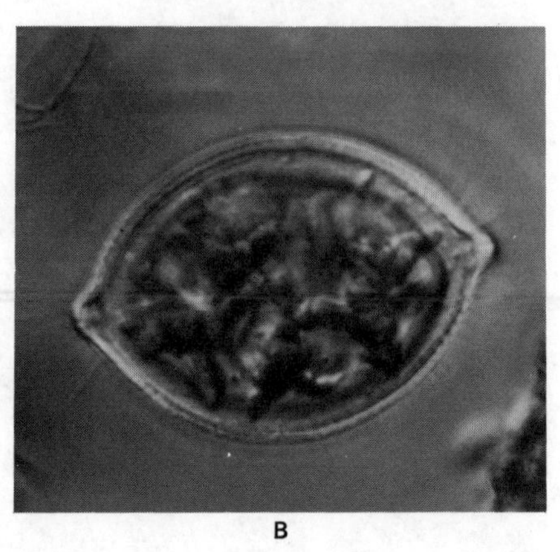


*Figure 7.9*
Members of the Chlorococcales. Members of the order do not have the ability to undergo cell division. *A. Pediastrum. B. Oocystis.* [From Kodachromes by Robert Trumbull, III.]

Some green algae are not only multinucleate but have no crosswalls as well. Such a condition is called *coenocytic* or *siphonaceous.* The alga is essentially one long continuous cell, although partial septa may be present. Most members of this group are marine. Margulis places this group in a separate division, the Siphonophyta. This group includes the largest known green alga, *Codium magnum,* which may reach 8 to 10 meters in length.

Another interesting group of green algae reproduce sexually via a mechanism termed *conjugation.* Members of this group may be either filamentous or unicellular and include some of the most beautiful of the green algae. Motile cells are completely lacking in this group. The chloroplasts may either be parietal or axial (down the middle of the cell). Figure 7.10 shows some members of this group.

As in the blue-greens, the effects of green algae on man and his activities are largely detrimental. They represent a major problem in water supply reservoirs, where they may cause taste and odor problems in the water. The primary odor-causing members of the green algae include *Hydrodictyon, Dictyosphaerium, Actinastrum, Chlorella, Cladophora, Chlamydomonas, Pandorina, Eudorina* and *Volvox.* Some are problems in the water treatment plant itself, covering the walls of galleries, aerating, settling, and clear-water basins. Most of these are the filamentous forms *Cladophora, Stigeoclonium,* and *Oedogonium,* although the desmid *Cosmarium* might also be included in this group. *Chlorella* can be a major problem in greenhouses where it clogs the filters, and in the greenhouses, laboratories, and other areas where distilled water or culture media is kept in glass containers. *Scenedesmus, Ankistrodesmus,* or both often develop in the water.

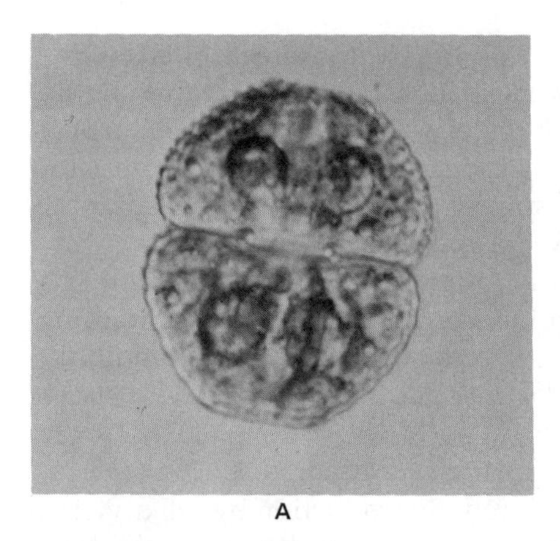


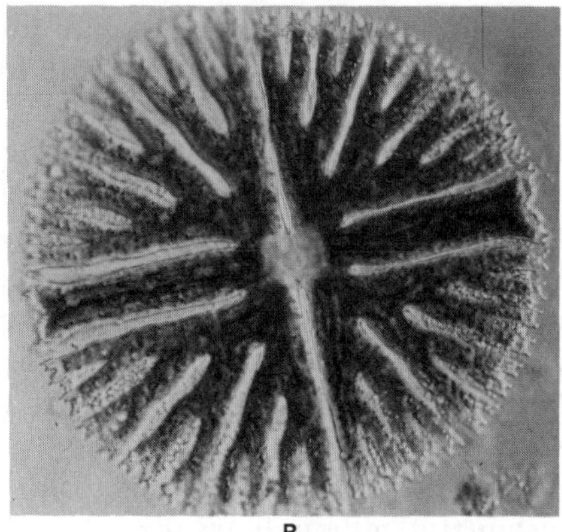


*Figure 7.10*
Members of the Conjugales. They are unique in their form of sexual reproduction and the absence of the flagellated cells. *A. Cosmarium. B. Micrasterias.* [From Kodachromes by Robert G. Trumbull, III.]

*Chlorella* will sometimes even develop in commercial bottled water, and the organism along with others, is often a nuisance on outdoor drinking fountains. In aquaria, the green algae *Ankistrodesmus, Scenedesmus, Cladophora* or *Rhizoclonium* may be problems; clogging the gills of fish and shading out rooted aquatics thus causing their death. In recreational areas, the growth of odor-causing green algae and other forms create a nuisance situation and certainly not a very aesthetically pleasing one.

Green algae do have some beneficial aspects. Some are important members of the aquatic food web, particularly in fresh water, and are important oxygenators of such waters. Some, such as *Chlorella*, are widely used in scientific research, and this organism has even been utilized experimentally as a source of human food. In the future, this alga may be of major importance to man. The marine alga *Codium*, and the freshwater alga *Rhizoclonium* have both been used as a vermifuge (agent for expulsion of worms or other parasites from the intestinal tract). It has been reported that the ancient Polynesians used a filamentous green, perhaps *Spirogyra*, as a poultice for abrasions and for inflammation of the eye. Some green algae are being grown in sewage lagoons because of their capacity to remove nitrates, phosphates, and certain heavy metal ions from the water.

## *Chrysophytes*

This highly diversified group, of the division Chrysophyta, is found in both fresh water and salt water. It contains many members having characteristics not considered plantlike, such as the presence of amoeboid movement, heterotrophic nutrition, presence of centrosomes, pigment spots, and contractile vacuoles. Growth forms similar to those exhibited by the green algae can be found. Motile forms, if not amoeboid, are motile by means of two equal or unequal length flagella. These flagella are morphologically dissimilar, with one being of the smooth or whiplash type, the other being of the tinsel type. Many species in this group are planktonic (free floating), cold-water forms.

The one characteristic common to all photosynthetic members of this group is the presence of one or two golden-brown chloroplasts containing chlorophylls *a* and *c*, β-carotene, and several xanthophylls. Pyrenoidlike bodies may be associated with the plastids, but these never have a starch sheath.

Photosynthates are normally stored as the carbohydrate *leucosin (chrysolaminarin)* although *oils* may also be stored. Reproduction by cell division is common, and zoospores or amoeboid spores may be formed. A common reproductive body is a *statospore*, or *cyst*. These are siliceous and often ornamented in various ways. Sexual reproduction is apparently rare, and to date, only isogamy has been reported. An example of a chrysophyte is shown in Figure 7.11.

Major taste and odor producers in water supplies include some chrysophytes: *Mallomonas*, *Synura*, *Uroglena*, *Uroglenopsis* and *Dinobryon*. They have little, if any, economic importance, although it is now thought that planktonic chrysophytes may be a major food-producing organism in the ocean. One marine group, the coccolithophores, have an outer membrane with structures called *coccoliths* embedded in it (Figure 7.12). These coccoliths are of calcium carbonate and the organisms form a very minor part of

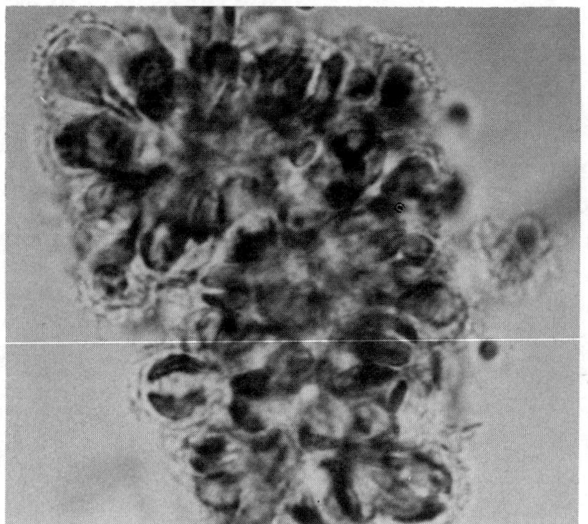

*Figure 7.11*
Member of the Chrysophyceae. *Synura*. [From Kodachrome by Robert G. Trumbull, III.]

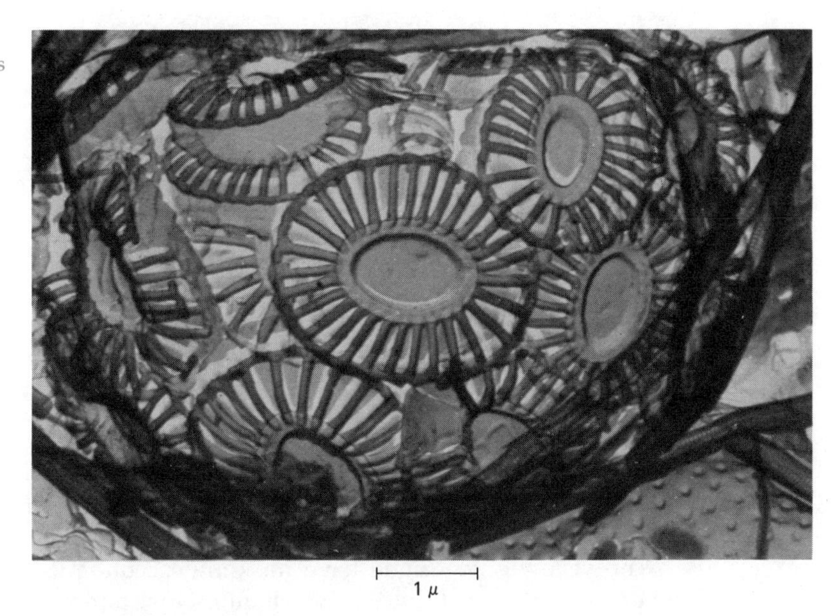


*Figure 7.12*
*Coccolithus huxleyi.* This belongs to a marine group having calcium carbonate tests. [Photograph courtesy of Harold G. Marshall.]

chalk deposits such as the White Cliffs of Dover. These chalk cliffs are primarily deposits of the exoskeleton of an animal called *Globigerina*, a foraminiferan.

## Xanthophytes

The Xanthophyta, or yellow-green algae, are thought by some investigators to be closely related to the chrysophytes, and in fact are placed in the same division by some botanists. The yellow-green color of the chloroplast has been shown to be due to an excess of xanthophylls. In addition, chlorophylls *a* and *e*, and β-carotene are also present. Chlorophyll *c* has also been reported, but the evidence, in my view, is not very conclusive. Pyrenoidlike bodies may be present, but these are not associated with a starch sheath. They may be associated with the plastid or may be free from them. Storage products are generally in the form of leucosin (chrysolaminarin). The cell wall is composed of cellulose and pectin, and in some it is impregnated with silica. In several genera the cell wall dissociates at the midregion rather than at the cross wall, resulting in the formation of an H-shaped section (Figure 7.13). Some algologists, such as Geitler, consider the flagellation as the most distinctive feature of the group. The flagella are always unequal in length, although in some only slightly so, and different in morphology, with one being the tinsel, one the whiplash type.

Most xanthophytes are found in fresh water, and like the chrysophytes are typically cold-water forms. Reproduction is generally by zoospores or

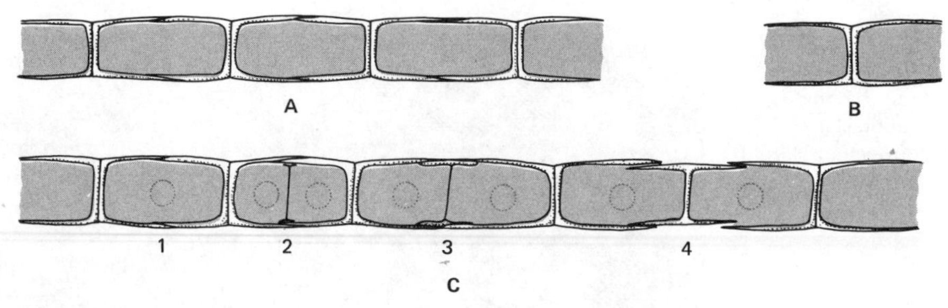


*Figure 7.13*
*Tribonema. A.* Vegetative filament. *B.* H-piece. *C.* Stages of cell division illustrating H-piece formation.

by nonmotile spores, although sexual reproduction is known for the group. Fragmentation may occur commonly in certain filamentous forms. Sexual reproduction is isogamous or anisogamous, although oogamy is known for the genus *Vaucheria*.

The yellow-green algae have no known economic importance. Most, when present, are found in small numbers, although the filamentous forms may produce extensive growths. Certain terrestrial species of *Vaucheria* may be significant for their soil-holding ability. Several are taste and odor producers in water supply reservoirs.

## Diatoms

The division Bacillariophyta, the diatoms, grouped by some botanists with the chrysophytes and xanthophytes, are the most numerous, best known (in terms of familiarity, not necessarily taxonomically), and most important economically of the three groups. They are found almost everywhere: in or on soil, in caves, on tree trunks, moist rocks or cliffs, in spray or surf zone of beaches, on tidal flats, in the atmosphere (one species reported), or among *Sphagnum* and other mosses. Many diatoms are epiphytic, being attached by a jellylike secretion to certain filamentous algae or higher plants, some are endophytic, being found within other plants. Most diatoms however are aquatic, being found in fresh, brackish, salt water. They occur from pole to pole and at all longitudes. Some are truly planktonic (free floating), some are benthic (attached to the bottom), while some spend a part of their lives in each habitat. Some diatoms form a brown-colored, slippery coating on the rocks of streams, as many fishermen have found out, sometimes too late to avoid an unscheduled bath. Marine diatoms are found in the open ocean as well as in the littoral (intertidal) areas, while freshwater diatoms are found in lakes, reservoirs, ponds, rivers, streams, springs, bogs, marshes, in fact almost everywhere a little moisture is present. Certain diatoms represent some of the few living species to be found in acid mine waters. Many diatoms seem to be cosmopoli-

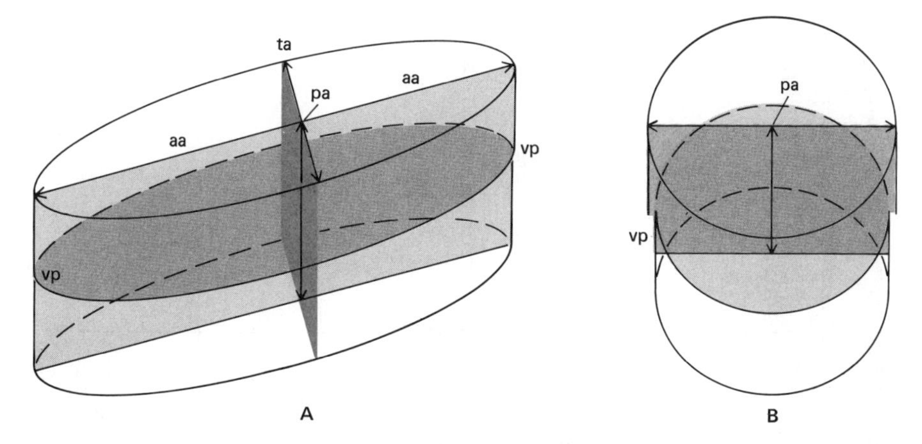


*Figure 7.14*
Symmetry in diatoms. *A.* A pennate form. *B.* A centric form. *VP,* valvar plane; *pa,* pervalvar axis; *aa,* apical axis; *ta,* transapical axis. The transapical plane dotted, valvar plane cross-hatched, apical plane diagonal lines.

tan, with certain species being found in many parts of the world in many different habitats. Others seem to be restricted to specific locations or specific habitats or both.

Diatoms are unicellular (most), colonial, or filamentous. The individual genera are usually divided into the major groups on the basis of their symmetry. The *centric* diatoms are placed in the class Centrobacillariophyceae, the order Centrales, and have their symmetry about a point. The *pennate* diatoms are placed in the class Pennatobacillariophyceae, the order Pennales, and have their symmetry about a line (Figure 7.14). Recent studies have indicated that this is probably not a natural system of classification. For our purposes, however, we will retain the common terms centrics and pennates for the two groups.

Probably the most characteristic feature of the diatoms is the silicified cell wall, with the degree of silicification varying from almost 95 percent or more to one diatom which is very weakly silicified. There is no cellulose in the wall, and the silica is impregnated with pectic compounds. Most diatoms do have an abundance of silica in their walls. These walls, or *frustules,* are highly structured and ornamented, a feature which makes the diatom perhaps the most beautiful of microscopic structures. Identification of diatoms is based almost solely on these wall characteristics.

The wall of all diatoms is composed of two overlapping halves, the *valves.* The diatom frustule has been likened to a petri dish or shoe box, being composed of two valves, with the slightly larger, coverlike portion, called the *epitheca,* overlapping the smaller bottom portion, or *hypotheca.* The two valves are held together by the *girdle band,* a band of pectin impregnated with silica (Figure 7.15). Additional bands may be inserted be-

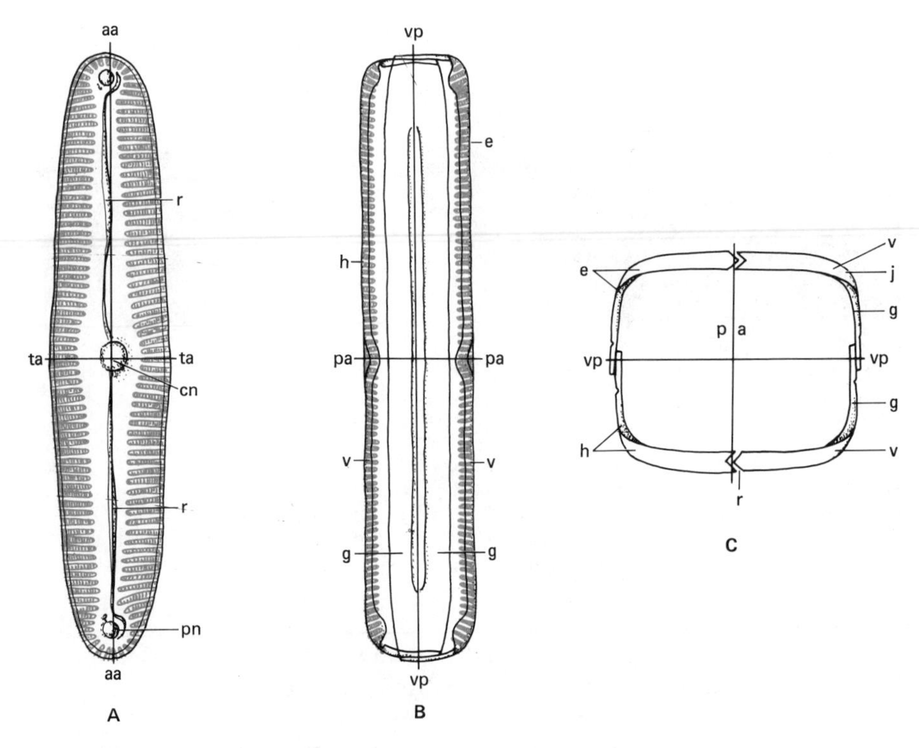


*Figure 7.15*
Structure of the diatom frustule (wall). *Pinnularia viridis* (after Pfitzer). *A*. Valve view. *B*. Girdle view. *C*. Transection. *AA*, apical axis; *cn*, central nodule; *e*, epitheca; *g*, girdle; *h*, hypotheca; *j*, valve-jacket; *pa*, pervalvar axis; *pn*, polar nodule; *r*, raphe; *ta*, transapical axis; *v*, valve; *vp*, valvar plane.

tween the junction of the two walls. When viewing a diatom, one will be either looking at the valve surface (valve view) or at the girdle area (girdle view). In certain diatoms, the valve is bent in such a fashion that both valve and girdle are seen at the same time.

The most obvious features of the protoplast are the brownish plastids. In centrics, these are tiny, numerous, discoid structures, while in pennates they are usually two very large structures. Pigments include chlorophylls *a* and *c*, β-carotene, and a number of xanthophylls, one called fucoxanthin being largely responsible for the brown color. Pyrenoidlike bodies may be present, but starch is never found as a storage product; rather, the diatoms, like the chrysophytes and xanthophytes, store oils and leucosin (chrysolaminarin). A single large nucleus is usually observed in the center of the pennate cell, suspended in a large central vacuole by cytoplasmic bridges leading to the peripheral cytoplasm.

Normally, the details of the frustule are difficult to observe in living diatoms unless the gelatinous covering surrounding the wall and the inner protoplast is destroyed. This can be done in several ways. The diatom can simply be allowed to sit in water until normal bacterial decomposition occurs. However, fungal growth may also occur during this time, with the fungal filaments growing around, over, and into the diatom frustule, making observation difficult if not impossible. An alternative is to boil the diatoms in concentrated nitric, sulphuric, or hydrochloric acid, or a combination of these. Noxious fumes are produced during this process; thus, it must be carried out under a well-ventilated hood sometimes an inconvenience. Cold concentrated acid can be used, allowing the diatom-acid mixture to set for some seven to ten days. This obviously is time consuming. The simplest procedure I have found is to add an amount of 30 percent hydrogen peroxide equal to that of the sample (50 ml or less), then add a pinch of potassium dichromate crystals. The mixture does not have to be heated, nor does it need to be placed in a hood, because the only gas produced during the oxidation of the organic material is oxygen.The process is generally completely finished within a few minutes, as denoted by the change in dichromate from its purple oxidized form to its yellow reduced form. The yellow dichromate is removed by adding distilled water to the sample, allowing it to set four to six hours, then decanting much of the water. Repeat this procedure until the yellow color has disappeared from the samples. A few drops of the cleaned diatoms are then placed, along with distilled water, on the surface of a cover glass, and the suspension allowed to air dry. The cover glass is then mounted onto a glass slide with a highly refractive medium such as Hyrax (refractive index 1.65).

When diatom frustules prepared in the above fashion are observed with a quality light microscope, a number of ornamentations can be noted. The valves of diatoms contain a series of regularly spaced lines called *striae* (Figure 7.16). The ability to resolve striae of certain diatoms is used to de-

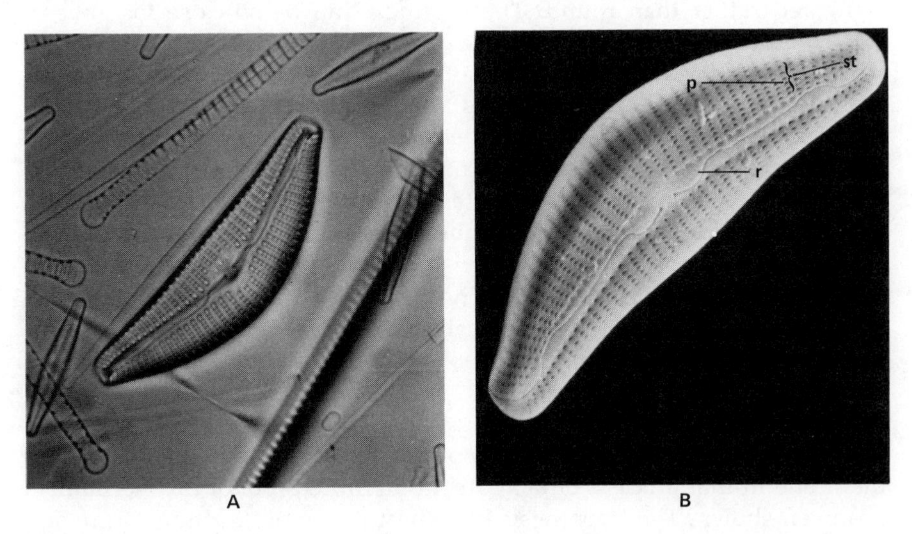


*Figure 7.16*
*Cymbella cistula*. *A*. Light micrograph (× 1,000). *B*. Scanning electron micrograph (× 2,000). *ST*, striae; *p*, punctae; *r*, raphe.

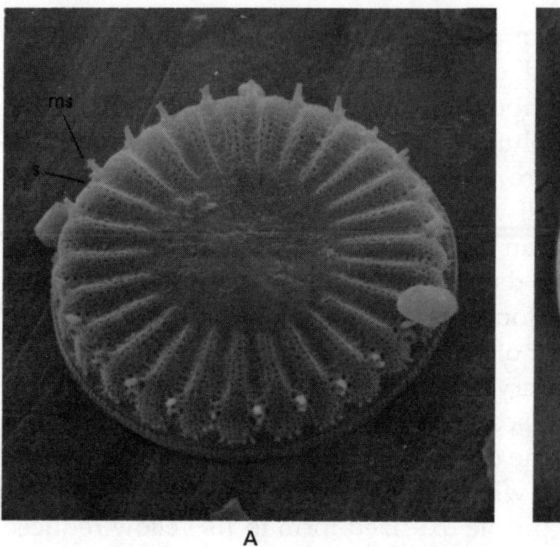


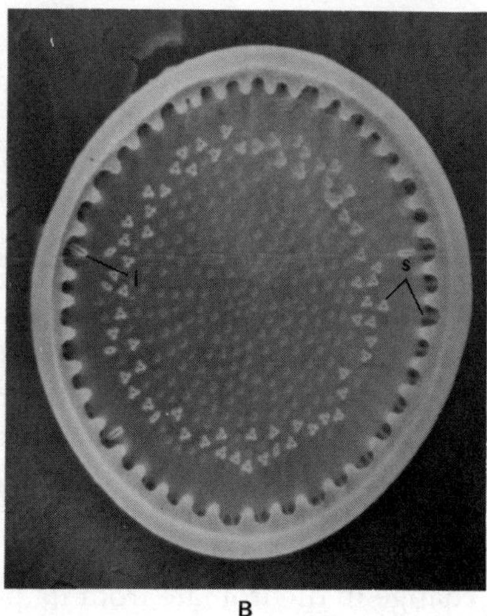


*Figure 7.17*
The centric diatom *Cyclotella meneghiniana* showing marginal spines *(ms)*, strutted processes *(s)*, labiate processes *(l)*. *A.* External view of valve. *B.* Internal view. [From R. L. Lowe, "Comparative Ultrastructure of the Valves of some *Cyclotella* species (Bacillariophyceae)," *J. Phycol.* 11:415–424, 1975.]

termine the quality of microscope lenses. In the specimen illustrated in Figure 7.16A, the striae are clearly seen to be composed of dotlike structures. These are termed *punctae.* But when this same organism is observed under the SEM, the structures are seen not as dots, but rather openings or holes in the cell wall (Figure 7.16B). In this organism, the holes are slightly elongated rather than round. These puncta can be so close together in some diatoms that the striae look like continuous lines under a light microscope. Sometimes these pores are not open but are covered with a membrane. In other instances, the pores are depressed chambers in the wall, with small pores in the basement membrane of the chamber. In addition, the wall may bear internal or external spines or various other processes (Figure 7.17). Internal septa or plates are present in some diatoms, and these may divide the cell into chambers.

In many pennate diatoms, the striae do not traverse the entire cell but end somewhat short of the center. The resulting clear area is called the axial area. In some diatoms a channel called the *raphe* traverses this area (Figure 7.16). In others the raphe is absent, with the resulting clear area being called the pseudoraphe.

For one hundred years or so, it has been observed that only those diatoms with a raphe are capable of motility. This would seem to implicate the raphe in that process, but to date, no definite proof of the mechanism has been shown. The hypothesis I like best is one put forth by Drum and Hopkins in 1966. They reported the secretion of a substance through the raphe, with its subsequent hydration and adhesion to the substrate. Other, unhydrated secretion flowing against this adhesion created a force pushing the diatom forward. Thus, the diatom would be moving in a direction opposite that of the flow of secreted substance (Figure 7.18).

Reproduction in diatoms is principally by nuclear and cell division. Normally, the protoplast divides right down the middle. As the daughter protoplasts enlarge, each uses the old parent wall as the epitheca, forming a new hypotheca. Consequently, one of the daughter cells will be slightly smaller than the other. In most instances, the walls have a compensating ability, or the protoplast may be partially or completely extruded from its

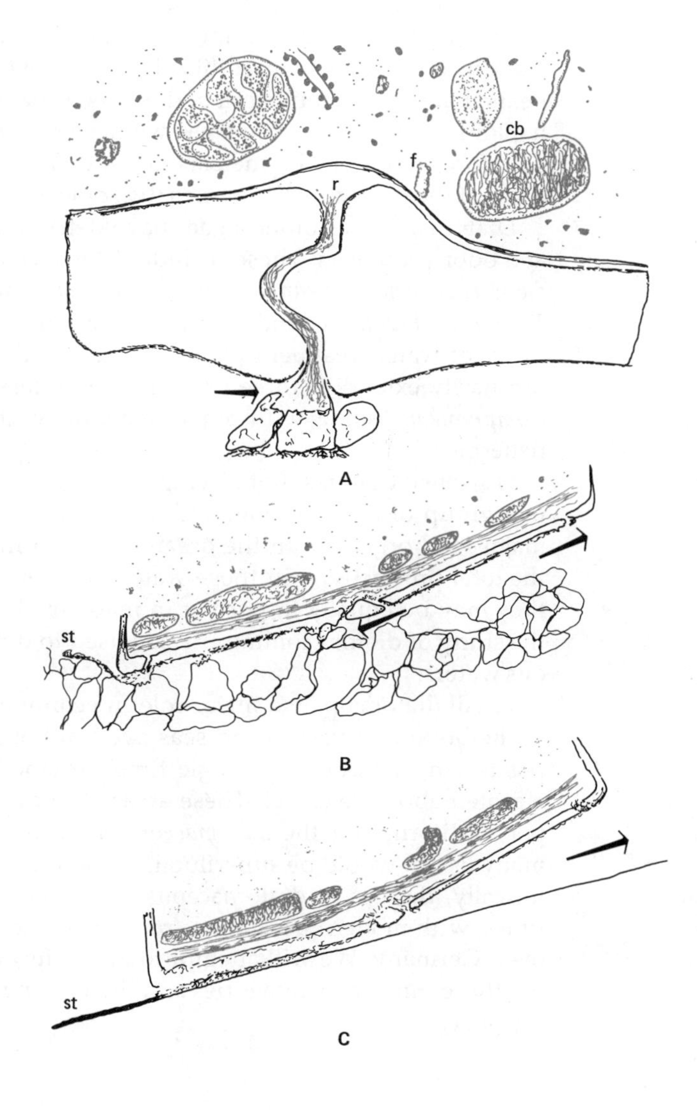


*Figure 7.18*
Schematic diagrams (not drawn to scale) of diatom raphe systems during locomotion. *A.* Transverse cross-section of a raphe fissure and adjacent cytoplasm of a diatom moving over a particulate substratum. The locomotor adhesion seal *(arrow)*, firmly attached to particles, is being pushed along raphe by streaming directed against that seal. The fibrillar bundle *(f)* lies next to the raphe *(r)*; longitudinal and transverse sections of crystalloid bodies *(cb)* are also shown. *B.* Longitudinal section through two raphe systems of ventral valve of a naviculoid diatom moving over particulate substratum. Locomotor secretion material is being deposited by the posterior raphe system on all particles over which the diatom moves; if particles are close together, a continuous trail *(st)* results, if not, a small clump *(arrows)* or pad of material is left on each isolated contact-adhesion site. *C.* Longitudinal section as in *B*, but the diatom is pictured as when moving over a smooth surface; locomotor secretion material is deposited in a continuous trail *(st)* as the diatom moves forward. [Drum and Hopkins, "Diatom Locomotion: An Explanation." *Protoplasma*, Bd. 62, 1966, Heft 1. Copyright © 1966, by Springer-Verlag, New York, Inc.]

frustule and form a completely new frustule. In either instance, daughter cells the same size as the parent cell will result. However, it has been demonstrated clearly that some diatoms do not have this ability; thus, with each successive generation, some of the progeny become progressively smaller. Obviously, the diatom does not literally divide itself away, but rather, at some point meiosis and gamete formation occur, resulting in formation of a zygote. These zygotes are very unusual, because after their formation they increase markedly in size, thus are called *auxospores* (from the Greek *auxo*, "increase"; *spora*, "spore"). Within the auxospore, new cell walls are laid down, producing an organism roughly similar to a normal vegetative cell. This is the *postauxospore* cell, and within it, the normal vegetative cell is formed. Thus, the diatom is returned to its larger size. The mechanisms of sexual reproduction in diatoms are quite variable and still not completely known for many taxa. It is not within the scope of this book to discuss this variability. However, we might mention that it appears that the vegetative cell of diatoms is a 2N or diploid cell, rather than haploid as are most of the algae we have discussed thus far. The gametes therefore are formed by meiosis rather than mitosis. Figure 7.19 illustrates a few of the nearly 10,000 species (by some accounts) of diatoms.

Diatoms, just as other algae, have detrimental aspects. Some are taste and odor producers. These include *Asterionella, Cyclotella, Diatoma, Meridion, Tabellaria, Synedra, Fragilaria, Melosira,* and *Stephanodiscus. Synedra, Tabellaria, Fragilaria,* and *Asterionella* are notorious for clogging the sand filters of water treatment plants, at times reducing filter runs that would ordinarily extend to three or four days to less than an hour. The diatom *Gomphonema*, along with other algae, often clog the nets of commercial fishermen.

However, diatoms, both living and dead, probably are the most important group of algae to man and aquatic organisms as well. They are the major food organism in the ocean, contributing as much as 85 percent of the total productivity in those waters. Current research seems to indicate they may be equally important in many freshwater habitats. In addition, the study of diatom communities is used to determine the quality of various waters.

Fossil diatoms play a major role in economy. Diatom frustules settling to the bottom of vast inland seas over millions of years formed thick deposits which through geologic time became compacted and eventually elevated above sea level. These areas are now mined in places like Lompoc, California for the *diatomaceous earth*, or *diatomite*. Diatomite has so many uses it would be superfluous to mention and try to list them all. Historically, the use of diatomaceous earth goes back to Greek and Roman times, with modern use dating from 1860 when it was first mined in Hanover, Germany. We will attempt to give a few examples of its use, grouping these into major categories and then listing some examples under each category.

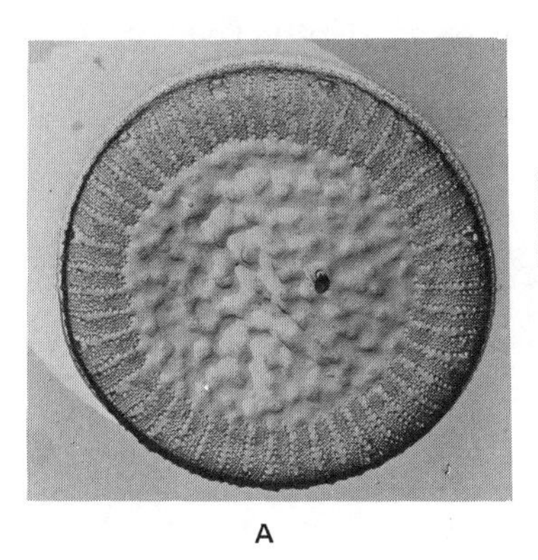
A

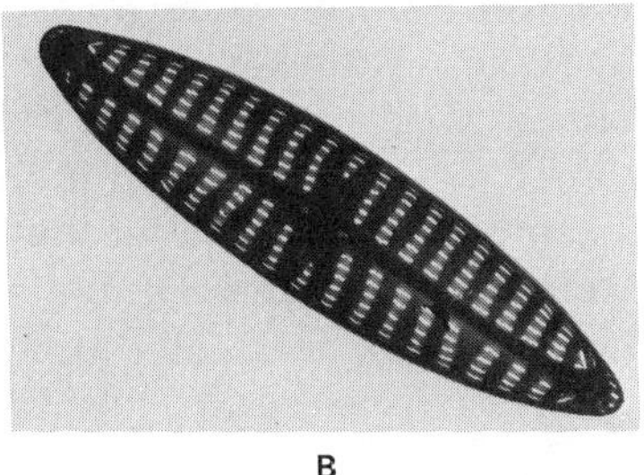
B

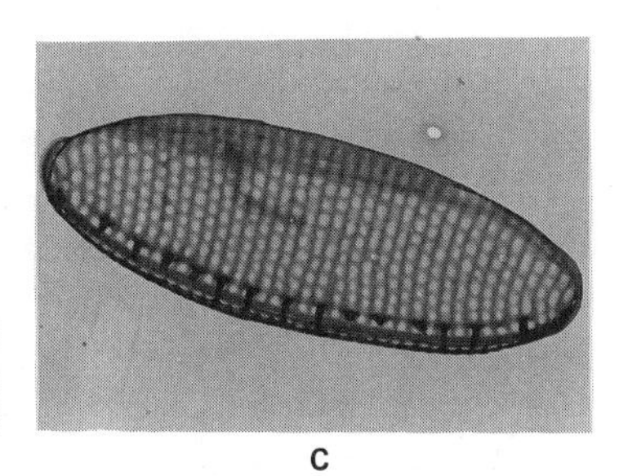
C

*Figure 7.19*
Some examples of freshwater diatom genera. *A. Cyclotella*, TEM carbon replica (× 7000). *B. Navicula*, TEM (× 6000). *C. Nitzschia*, TEM (× 6000). [Micrographs by Paul Hargraves.]

*Filtrant:* for pharmaceuticals, beverages, chemicals, dyes, soaps, perfumes, syrups, and oils, characterized by rapid filtration, long filter run, and color removal;

*Polish:* silverware, automobile, furniture, ceramics, lenses, fingernails, teeth (no longer used in toothpaste in the United States because of high abrasive action), cannot be surpassed as polish as it does not leave visible surface scratches;

*Insulation:* homes, factories, offices, refrigerators, cars, pipes, high temperature ovens, boilers, steam pipes, furnaces, used more than any other type of insulation, will not shrink at temperatures of 1600°F;

*Absorbant:* nitroglycerine, corrosive liquids, catalytic agents, disinfectants, fungicides, will absorb 150 to 300 percent of its own weight in liquid, inert and will not combine with compounds absorbed in it;

*Filler:* paper pulp, calcimine, rubber, sealing wax, erasers, linoleum, phonograph records, fertilizers, insecticides, modeling clay, prevents afterglow in safety matches, base for paints and varnishes (flat finish);

*Construction:* limited use as mined because of friableness but mixed with other materials to make roofing tile, stucco, artificial stone, concrete, and soil pipe;

*Miscellaneous:* mix for porcelains and glazes, fireproofing for safes and vaults, separating agent for molds, blackboard chalk, coating for coffins, nuclear power plants (unknown use).

## Brown Algae

The brown algae are almost entirely marine, being characterized by some authors as found primarily in cool, shallow waters. But, this characterization is not entirely appropriate for the group because some of the largest members are in temperate waters, intertidal to 30 meters in depth. The brown alga *Sargassum* is an important member of some tropical and subtropical waters. Some members of this division, the Phaeophyta, because of their thick gelatinous covering, are able to withstand considerable dessication (the kelps).

The protoplasts contain a single nucleus which may have centrosomes associated with it. The one to several plastids contain chlorophyll *a* and *c*, B-carotene, and several xanthophylls including the predominant one, fucoxanthin. Storage products consist of *laminarin* (a mixture of polysaccharides), the sugar alcohol *mannitol*, and fat droplets. The protoplast is bounded by a primary wall of cellulose and a middle lamella composed of *algin*, or *alginic acid.*

The evolution of the brown algae is difficult to determine. Although postulated to have arisen from a motile unicell, neither unicellular nor colonial forms are found. The simplest type of plant body is a branched filament, with many of the brown algae having considerable complexity of structure, with rootlike, stemlike, and leaflike organs, and growth to as much as 50 m in length. Many have specific regions of cell division, much like the meristems of higher plants.

Only the reproductive cells exhibit motility, these by two morphologically dissimilar, laterally inserted flagella. Reproduction may be vegetative by fragmentation, asexual by motile or nonmotile spores, or sexual by isogamous, anisogamous, or oogamous reproduction. Representative members are shown in Figure 7.20.

The brown algae have many uses by man, but as many of those are similar to the uses of the red algae as well, we will defer this discussion until after our consideration of the red algae.

## Red Algae

The red algae, or Rhodophyta, while having more freshwater representatives than the browns, are nontheless predominantly marine. They are typified by some authors as being found in warmer and deeper waters. But, reds predominate from the subantarctic to tropical waters. Some members produce calcium carbonate, and thus are important in reef building processes.

Vegetative cells of red algae may be either uninucleate or multinucleate and lack centrosomes. The plastids may be single and massive or numerous and discoid and may contain pyrenoidlike bodies. Pigments include chlorophylls *a* and *d*, $\beta$-carotene, several xanthophylls, and the phycobilin pigments *phycoerythrin* and *phycocyanin.* The storage product is a complex carbohydrate called *floridean starch.* The cellulose cell walls are often covered by a slimy layer, and in some members, by a copious gelatinous substance as well. Associated with the wall, in certain species, are hydrocolloids known as *agar-agar, carrageenin,* and *gelan.*

*Figure 7.20*
Some representative brown algae. *A. Alaria. B. Colpomenia* spp. *C.* Cutleria. *D. Sargassum. E. Duruillea;* the attached fronds wash back and forth in the surge of a rocky cliff at Otago Peninsula, South Island, New Zealand. *F. Macrocystis;* an underwater view showing the basal differentiation of the fronds. *G.* Harvesting of a *Macrocystis* canopy by the Stauffer Chemical Kelp Company. [*A.* from R. G. Buggeln; *B., C.,* and *D.* by James Norris; *E.* by Katherine Muzik; *F.* and *G.* by David Coon.]

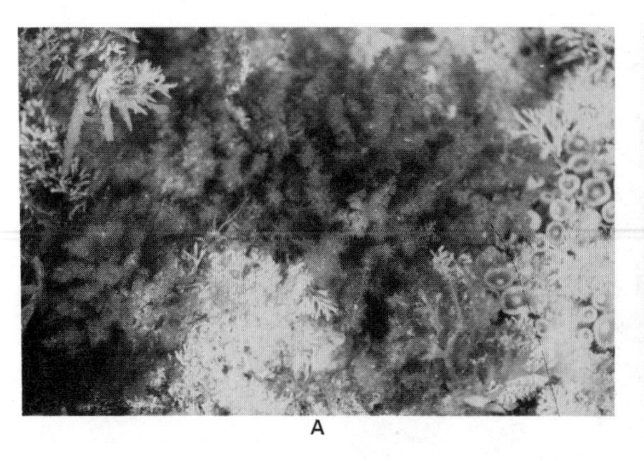

*Figure 7.21*
Some representative red algae. *A. Laurencia. B. Eucheuma. Euchemia* is now the most important alga commerce. It is the major source of the hydrocolloid, carraghenin.

Although unicellular and colonial red algae are known, the vast majority are filamentous with one being sheetlike in appearance. They usually range from a few centimeters to a meter or slightly more in length, never reaching either the size or internal complexity of the large brown kelps. Growth is normally through divisions of an *apical cell* or apical cell row. Flagellated cells are not known for the red algae and reproduction is by asexual nonmotile or amoeboid spores or by oogamous sexual reproduction. Isogamy and anisogamy are rare. The life history of certain red algae, which we shall discuss in Chapter 16, is quite complex. Two entirely different morphological phases may be produced, and historically, this has led to considerable confusion concerning the taxonomy of the group. Early taxonomists sometimes identified the two phases as two different plants. A freshwater red, *Batrachospermum* is macroscopic, with tufts of filaments projecting from the axial filament, giving it a somewhat beaded appearance. The freshwater red *Chantransia* is a microscopic, branched filament, looking somewhat similar in form to the green alga *Cladophora*. But when the life history of *Batrachospermum* was studied carefully, the organism was found to produce spores which, upon germination, produced *Chantransia*, which, in turn, produced spores which gave rise to *Batrachospermum*. We now call the microscopic plant the *Chantransia* stage of *Batrachospermum*. Figure 7.21 illustrates some representative red algae.

The importance of the marine algae to man has a considerable history. Of course, no one knows for certain when man first started using these algae, but the Chinese classics of about 800 B.C. refer to their use as food, and Chinese legends carry their use as therapeutics back to 3000 B.C. Confucius thought highly of them, and the Greeks used them in various ways. However, the Romans did not seem to hold them in such high esteem, and they were referred to as the *Vilega alga* ("vile algae"). However, they were used by the Romans, with Pliny the Elder recognizing their usefulness as dye, and the ladies of Rome using an extract of *Fucus* as a rouge. Marine algae have been used as fertilizer for centuries in the coastal regions of the British Isles and Scandanavian countries.

Before discussing the various categories of uses of the marine algae, it might be well to clarify a couple of terms. One of the important economic groups is called the kelps. This term has various meanings, but in general, it refers to a group of large brown algae, including *Macrocystis, Nereocystis, Pelagophycus, Alaria, Laminaria,* and *Postelsia.* The *phycocolloids,* or hydrocolloids as they are sometimes called, are almost indispensable to a number of different industries. These include furori and alginic acid from the brown algae, carragheenin (or carrageenin), and agar from the red algae.

As with the diatoms, uses of marine algae are quite extensive, so we will attempt to discuss these uses in categories.

*Food for Humans.* *Porphyra* (*nori* or *amonori* in Japan) in the Orient is dried, baked, and eaten in soups and broths; in Ireland is fried in butter; in England is eaten in soups, fried, boiled into a viscous, brown breakfast dish, and prepared as a bread; in Wales is eaten as a salad or mixed with oatmeal and fried; in Europe is eaten as a salad or after being boiled and having lemon juice, spices, and butter added; in New Zealand is used by the Maoris in soup and as a steamed dish known as *karenga. Laminaria, Arthrothamnius,* and *Alaria* (known as *kombu* in Japan) are eaten in all possible ways by the Japanese, either black, white and pulpy, powdered, or in ribbons of various lengths and widths. Some is sugar coated and eaten as a confection, while the powdered form is generally used to thicken soups. Pieces of narrow ribbons are steeped and the liquor drunk as a tea *(Cha).* Thin shavings are used to flavor fish and meat. In a suchi bar, the customer is served small balls of rice surrounded by raw fish, clam, shrimp, or sea urchin eggs, with the whole concoction wrapped in a sheet of algae. There are few meals in Japan, or other Orient countries for that matter, where algae do not accompany the food in one form or another. *Rhodymenia palmata* (dulse) has a slightly acrid flavor and the odor of violets and is used as a flavoring (eg. for soups), salad ingredient, condiment, thickening and coloring (eg. for stews). It has been used since the eighth century in Ireland, where it is eaten raw, with potatoes and fish, or boiled in milk or oil of citron. It is also eaten in Scotland, and both the Scots and the Irish use it as a chewing "tobacco." In the Philadelphia area, it is known as sea kale and is eaten there and in the New England area as a vegetable. *Chondrus crispus* (Irish moss) and *nori* are probably the most popular of the algae eaten in the United States, especially along the East Coast. It can be purchased dried and used in making blancmange and soft jellies or it can be eaten like popcorn. As its name implies, it is also a popular eating-algae in Ireland. Many other algae are used as food in various parts of the world.

We will mention just a few of these. *Nereocystis leutkeana* stipes are processed into a confection called Seatron. The red alga *Suhria vittata* has a novel preparation in South Africa: it is first washed thoroughly, then boiled to a pulp (allowing the fumes to escape during boiling), the pulp is strained and lemon or orange juice, brandy or sherry is added to it, along with cloves, cinnamon, or lemon peel if desired. The resulting dish, *Chinchow,* sounds delightful. The *Limu* (seaweeds) of Hawaii are widely known, with the algae either eaten raw, cooked with fish, squid, or octopus, or used in soup.

*Food Processing.* The phycocolloids, agar, carrageenin, and alginic acid have essentially the same properties and can be used interchangeably in most food processing. They are used in the baking industry as stabilizers, moisteners, or smoothing agents; in confections as stabilizers, fillers, moisteners, or smoothing agents; in the dairy industry to smooth and stabilize ice creams and sherbets, to prevent the formation of ice crystals, shorten the whipping time, add body, lessen melting, emulsify cream, and stabilize the chocolate in chocolate milk; in mayonnaise and various salad dressings, they are used as emulsifiers, stabilizers, or smoothing agents; in jellies, jams, marmalades, sauces, desserts, and soups as thickeners; in soft drinks to add body and sharpen the flavor; in beers and wines for clarifying and, for many years, for fining (removing suspended proteins).

*Food for Domestic Animals.* Seaweed has been fed directly to domestic animals for centuries. Although it was and still is used directly, it is now generally ground into a meal and used as a supplement. While the protein content is not high, it does have a large number of inorganic minerals and vitamins.

*Medicine.* As mentioned earlier, recorded use in medicine goes back almost 5000 years in China. Its most widespread use in medicine is still in China. *Porphyra coccinea* was recommended in 300 B.C. for the treatment of goiter, a practice therapeutically sound as this alga contains iodine, which is used in present-day treatment of the condition. In the South Pacific, natives wrapped wounds with algae to prevent infection; Pliny recommended it for the treatment of gout; Dioscorides recommended it as helpful in treating burns, scurvy, rashes, and intestinal disorders; the ancient Greeks prepared a vermifuge from *Alsidium,* and other genera have been used in like form by other peoples; extracts from *Iridea flaccida, Chondrus cripus, Delesseria sanguinea, Gelidium,* and *Iridea* seem to have anticoagulant properties; *Rhodymenia palmata* is reported to be a laxative; intestinal and bladder disorders were treated with *Gracilaria lichenoides;* in the Himalayas, *Laminaria saccharina* was thought to be a cure for syphilis, while in China, species of *Laminaria* were used to relieve menstrual cramps; more recently, the Russians have used a *Laminaria* derivative in treating trichomonad vaginitis; stipes of *Laminaria cloustoni* were used to open wounds, facilitate healing, and dilate the cervix of the uterus during labor. A most unusual use involves the nests which passerine birds build of mucus and the alga *Gracilaria*. These nests are collected and used as the ingredient for bird's-nest soup, reputed to be a strong aphrodisiac. Although many of the historic direct uses of algae in medicine were based upon superstition and folk customs, some have been shown by modern medicine to be basically sound.

*Fertilizer.* Used widely throughout the world as green manure; some coralline red algae used as substitute for lime; dried kelp found to contain 15 percent potash.

*Chemicals.* Kelp ash used for recovery of iodine, potassium salts, crude washing soda (alkali); distillation of kelp produces calcium acetate, acetone, and mixed ketones, although this has not been done commercially; active decolorizing carbon (called kelpchar), kelp oils, creosote, ammonia, and bromine have also been produced along with the potassium salts and iodine.

*Cosmetics.* Phycocolloids used as stabilizers, binders, emulsifiers, and smoothing agents in face creams, shaving creams, soaps, shampoos, hair-curling fluids, toothpastes, sunburn lotions, hand lotions, hand jellies, deodorants, rouges, compacts, pigment pencils, and other products.

*Textile Production.* Phycocolloids may be used in place of starch as sizing compound or finishing substance, in manufacture of artificial silk, as a hardener or adhesive, in making flame-resistant material.

There are many other uses of the phycocolloids, as in dyes and mordants, in rope making, in paper sizing, in pharmaceuticals as bases for ointments, suppositories, and lubricating jellies, in many pharmaceutical preparations as emulsifiers, smoothing agents, and flavor maskers; in paints and varnishes as suspending, emulsifying, and bodying agents; in glass and ceramic making; in plywood bonding and in other adhesive functions; in ornament or curio making; in mine sealing; in cartridge primers and plaster as a binder; in making charcoal briquettes; and many others uses.

## Dinoflagellates

The dinoglagellates, or Pyrrhophyta, are mostly unicellular biflagellates, with both chlorophyllous and colorless species known. They are abundant in both fresh and marine waters, where they may form an important constituent of the plankton. The cell walls probably contain cellulose and are composed of regular platelike segments (Figure 7.22). Some, however, lack a wall, with the protoplast being surrounded by a fibrous pellicle. The protoplast generally contains a single nucleus, which is unusual in that the chromosomes remain condensed and clearly recognizable throughout the

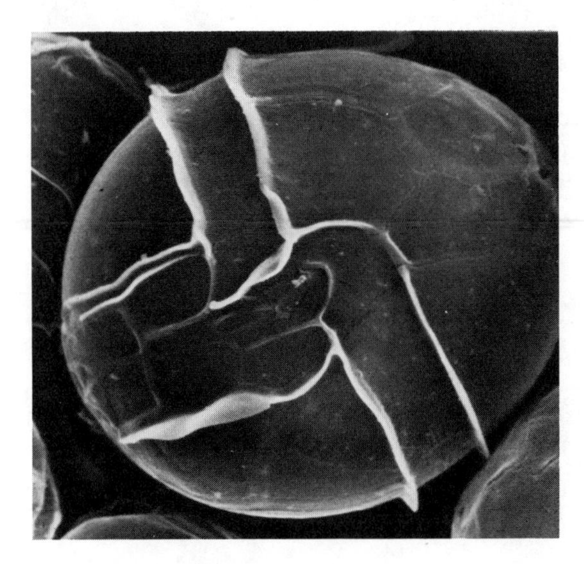

*Figure 7.22*
*Peridinium;* a dinoflagellate. Note the plates comprising the cell wall and the median groove surrounding the cell. One of the flagella lies in this median groove and completely circles the cell. [Courtesy Michael T. Postek].

cell cycle, and similar to prokaryotes, no histone, RNA, or protein component has been found. Some investigators refer to this condition as *mesokaryotic.* It has recently been reported (1971) that some dinoflagellates have two nuclei within their protoplast, one which is mesokaryotic, the other eukaryotic. The plastids are discoid, and yellow-brown to dark brown in color. Pigments include chlorophyll *a* and *c*, $\beta$-carotene, and several xanthophylls including two, *peridinin* and *dinoxanthin,* which are peculiar to the group. The storage product may either be starch or oil. The arrangement of the flagella is unusual. Although both are of the tinsel type, one is ribbonlike and lies in the groove which completely encircles the organism, while the second is elongate and attaches at a single point in the groove, extending backward with reference to direction of movement.

Reproduction is by cell division and sexual reproduction was once thought to be rare. However, extensive investigations by Von Stosch and by Pfiester seem to indicate it may not be as rare as once thought.

Some dinoflagellates have the property of being luminescent. When large blooms of dinoflagellates occur, the sea can literally be lit up at night. In times past, seafaring men described the sea as being "on fire at night" or "burning." Two dinoflagellates play a rather important role in human affairs. The so-called shellfish poisoning is caused by the dinoflagellate *Gonyaulax. Gonyaulax* is ingested by the shellfish, accumulating in their tissue. The dinoflagellate produces a potent toxin which is released when humans eat the shellfish. The toxin acts as a neurotoxin and can result in death. A second dinoflagellate is *Gymnodinium brevis.* Vast blooms of this organism can cause widespread death of fish and respiratory problems to people breathing the salt spray containing these organisms. The two organisms, and some others, are associated with what is commonly referred to as "red tide," the former generally occurring off the east coast, particularly the more northern area of the U.S., the latter off the Florida Coast, particularly in the Gulf of Mexico.

## *Euglenoids*

The euglenoids, the Euglenophyta, are primarily unicellular, except for one genus. They are all flagellated, bearing a long emergent flagellum and a short, nonemergent one. The protoplast contains a single nucleus, and as in the dinoflagellates, the chromosomes of this nucleus remain permanently condensed. The chloroplasts are variable in number and form, and with the exception of the small discoid type, all contain a pyrenoidlike body. However, this pyrenoid is not surrounded by starch. The principal storage product is a substance called *paramylon*, which is stored outside the chloroplasts. Pigments include chlorophyll *a* and *b*, β-carotene, and several xanthophylls. Many forms also have a stigma; these are phototaxic. Unlike the stigma of *Chlamydomonas*, that of *Euglena* lies outside the chloroplast.

The protoplast is bounded by a plasma membrane, inside of which is a series of ridged or grooved, helically arranged proteinaceous strips, which together with the plasma membrane form the *pellicle*. The pellicle is apparently elastic and in some forms fairly rigid. But in a number of species of *Euglena*, the pellicle is nonrigid, and the organisms undergo marked changes in shape, a phenomenon called *metaboly*. Some euglenoids live within a pectic shell or *lorica*. The lorica is often colored, due to deposition of iron salts in the pectic material. Reproduction in the euglenoids is by cell division, with sexual reproduction unknown in the group.

Many euglenoids are colorless, and even those having chlorophyll can often exist heterotrophically. It has also been possible experimentally to produce colorless euglenoids from pigmented forms. Although normally green, due to the predominance of chlorophyll, under certain conditions *Euglena* can become red in color, due to the pigment hematochrome.

The euglenoids have little or no significance to humans. Although, if you see water that is loaded with these little flagellates, you can be relatively certain that the water is rich in organic material. An interesting story regarding *Euglena* is recounted by C. E. Taft in his book *Water and Algae: World Problems:*

> On August 17, 1947, an extensive bloom caused by a species of *Euglena* . . . occurred in the Scioto River at Columbus, Ohio. The first evidence of something amiss was a reddish tint to the water as seen in the slanting rays of the afternoon sun. By the next dawn the mass had consolidated into a blood-red surface film which extended eight miles upstream . . . Its color in the early morning light recalled words that were written centuries ago:" . . . and all waters that were in the river were turned to blood. And the fish that was in the river died; and the river bank

stank, and the Egyptians could not drink of the water of the river; . . ." (Exodus 7:20-21). It does not require much imagination by the reader to visualize the reactions of persons who experienced those mystical happenings.

## Charophytes

The charophytes, sometimes called the stoneworts or brittlewort, are placed in the division Charophyta. Often included with the green algae, the morphological complexity of this group raises serious doubt if these organisms are algae at all.

Certain species have the capacity to precipitate calcium carbonate from the water, becoming covered with a calcareous surface layer, thus the common name stonewort or brittlewort (the term *wort* is an old one meaning plant or herb).

Unlike the majority of freshwater algae, the charophytes are macroscopic plants, consisting of a main axis which is differentiated into *nodes* and *internodes.* Its ontogeny involves a geometrically regular pattern from an apical cell. These features suggest those of certain vascular plants. From the nodes arise whorls of branches of determinate growth, commonly called leaves. In some species, true branches may also arise from the nodal region. From the lower portion of the axis arise rhizoids, which serve both to anchor the plant and for vegetative propagation. Protoplasmic streaming is a distinctive characteristic of internodal cells. These internodal cells generally elongate considerably, and may be covered by cells which arise from the lower node, elongating upward, and those which arise from the upper node, elongating downward, thus meeting in the central region of the internode. Such covering cells are termed *corticating cells,* and the internode thus covered is *corticated.* In the genus *Nitella* and a few species of *Chara* corticating cells are not produced.

The sex organs of charophytes are quite distinctive, being multicellular, and the sex organs are sheathed by sterile cells. Some botanists argue that the sex organs of charophytes are not truly multicellular nor truly surrounded by sterile cells. In Chapter 16, you will have the opportunity to compare these organs with those of liverworts and mosses and draw your own conclusions.

## Slime Molds

The slime molds are considered in two divisions, the Myxomycota, or plasmodial slime molds, and the Acrasiomycota, or cellular slime molds. Although we shall be considering these together, they are probably not very closely related.

The plasmodial slime molds are principally characterized by the nature of their vegetative phase: a naked, noncellular, multinucleate, amoeboid mass of protoplasm called the *plasmodium.* Most of these occur in moist, dark situations, often in woods. As they generally are within or beneath the organic substrate, they are usually overlooked by the casual observer.

The plasmodium may be colorless, or may be colored white, gray, black, violet, blue, pink, or orange. A unique characteristic of the plasmodium is the rapid cytoplasmic streaming that occurs, first in one direction, then in the opposite.

After a relatively long vegetative phase, the plasmodium, apparently induced by environmental changes, enters the reproductive phase producing *sporangia*. These sporangia have a distinctive form characteristic of the species. The *spores* formed within these sporangia have cell walls composed of cellulose and chitin. At maturity, the sporangial wall ruptures and the spores are carried away by air currents. If they land on a favorable substratum, they almost immediately germinate. In some forms, the product or products of the spore may be flagellated; in others, amoeboid, apparently depending upon whether free water was on the substratum (if so, the cells would be flagellated, if not, amoeboid). Apparently, these motile products of spore germination are gametes, for in some species it has been demonstrated that they unite in pairs to form an amoeboid zygote. Many of these zygotes aggregate together and their protoplasts fuse forming a new plasmodium.

The cellular slime molds, or Acrasiomycota, differ from the plasmodial slime molds in that a true plasmodium and flagellated cells are not produced. The product of spore germination is an amoeboid cell.

Investigations of the aggregation phenomenon of these amoebas produce a fascinating account. Apparently, it begins when individual amoebas begin to emit pulses of the chemical cyclic adenosine monophosphate (AMP) into their environment at intervals of approximately every five minutes. An amoeba perceiving such a signal pulse moves toward it, pauses, and emits a pulse of its own. Consequently, each amoeba is attracted to its nearest neighbor. The amoebas form streams, flowing toward the center of a growing aggregate in a series of waves, and then sticking together. The amoeba at the center of this mass is called the pacemaker, and is the one that emits pulses of AMP at the fastest rate.

After a time, isolated amoeba differentiate into stalk cells, and as the stalk grows, it is strengthened with cellulose fibrils, with the whole mass of amoebas abruptly rising up in a fruiting body (the *sorocarp*) with those at the pinnacle becoming transformed into elongate spores. When the latter germinate, each gives rise to a single amoeba. There is no evidence at present that this is in any way a sexual process.

## *Chytrids*

The Chytridiomycota consist of uniflagellated cells which have traditionally been placed with the fungi. (However, the true fungi do not possess motile cells.) They are aquatic organisms, many of which are parasitic on or in algae, fungi, or higher aquatic plant parts. The cell walls of these organisms are composed primarily of chitin. In some, an aggregation of hyphae (tubular filaments) occurs to form mycelia as in the fungi, other members do not form mycelia. The mycelium, when present, is coenocytic. Fruiting bodies called *sporangia* are formed and produce zoospores which upon germination produce a new thallus (a vegetative body lacking roots, stems, or leaves). Sexual reproduction also occurs, which the gametes being formed in specialized structures called *gametangia.*

### Oomycetes

The Oomycota range from unicellular forms to greatly branched coenocytic mycelia. Members of the oomycetes produce zoospores having two apically or laterally placed flagella, one of which is of the whiplash type, the other of the tinsel type. The cell wall is unusual, being constructed of D-glucose units having a different linkage than that for cellulose. In addition to asexual reproduction via zoospores, hormonally controlled sexual reproduction also occurs in this group.

Many members of this group are commonly known as the *water molds,* including *Saprolegnia* and *Achlya.* Members of the genus *Saprolegnia* parasitize fish, being a particular problem in aquaria. Many others, however, are terrestrial and may cause a variety of diseases. Members of the genus *Albugo* cause white rust on a number of different hosts, including *Amaranthus,* morning glories, sweet potatoes, mustards and others. Members of the genus *Pythium* cause a disease called *damping off* in seedlings. The great potato famine in Ireland, mentioned in Chapter 1, was due to a disease called *late blight,* which is caused by the oomycete *Phytophthora infestans.* The *downy mildew* of grapes caused by the oomycete *Plasmopora viticola* threatened to wipe out the whole wine industry in France in the late eighteen hundreds—a true calamity. The greatest scientist of the period worked frantically to find a cure, but it was just an ordinary vineyard man that provided the answer, not because he was trying to cure the mildew, but because he was trying to keep passersby from picking and eating his grapes which were growing along the road. To discourage this pastime, he applied to the vines, the most evil-looking concoction he could think of, a mixture of blue copper sulfate and lime. A professor from the University of Bordeaux happened to pass by and noted that the vines thus treated were free from the mildew. After getting the formula for the mixture from the vineyard man, he prepared and tested it himself, found it did in fact control the disease, and thus, the first fungicide for the treatment of plant diseases, Bordeaux mixture, was discovered.

## FUNGI

The true fungi are comprised of five divisions: the conjugate fungi, or Zygomycota; the sac fungi, or Ascomycota; the club fungi, or Basidiomycota; the imperfect fungi, or Deuteromycota; and the lichens, or Mycophycophyta. These organisms are characterized by their chitinous cell walls, lack of flagella, and branched mycelial body plan.

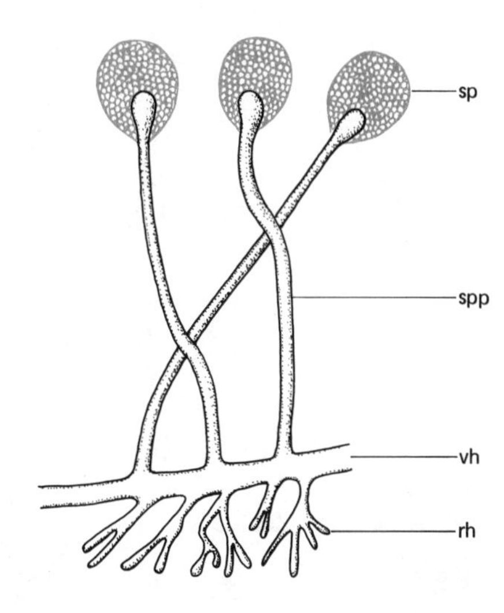


*Figure 7.23*
*Rhizopus.* A portion of the mycelium with sporangiophores; *vh.*, vegetative hyphae; *rh.*, rhizoids; *spp.*, sporangliophore; *sp.*, sporangium.

## *Conjugate Fungi*

Members of the Zygomycota are characterized by production of aplanospores which germinate directly into nonseptate (without cross walls) mycelia and by sexual reproduction featuring *gametangial contact*, or *conjugation.*

Hyphae modification occurs by development of portions of the prostrate hyphae into rhizoidal branches which penetrate the substrate and serve both for anchorage and absorption and by development of elongate upright hyphae, the *sporangiophores*, which bear enlarged sporangia at their tips (Figure 7.23). If two sexually compatible strains develop in close proximity to each other, sexuality is soon manifest by the series of events illustrated in Figure 7.24. The resulting zygote secretes a thick wall and is called a zygospore. Upon germination, the zygospore produces an upright sporangiophore which produces spores, the latter germinate forming vegetative hyphae.

The genus *Pilobolus* has a very unusual method of spore dissemination. Due to the build-up of excess turgor pressure in an enlarged portion called the subsporangial swelling (located just below the sporangium), the sporangiophore literally blows its top, shooting the sporangium and its spores as far as six feet away.

## *Sac Fungi*

The Ascomycota are characterized by having their hyphae generally septate (divided by cross walls). These cross walls are perforated however, and thus a protoplasmic connection is maintained between cells. Cytoplasmic material, including the nucleus, can pass through, and the cells thus delimited might be uninucleate or multinucleate. As the result of sexual reproduction, a saclike structure, the *ascus* (*asci*, plural), is formed.

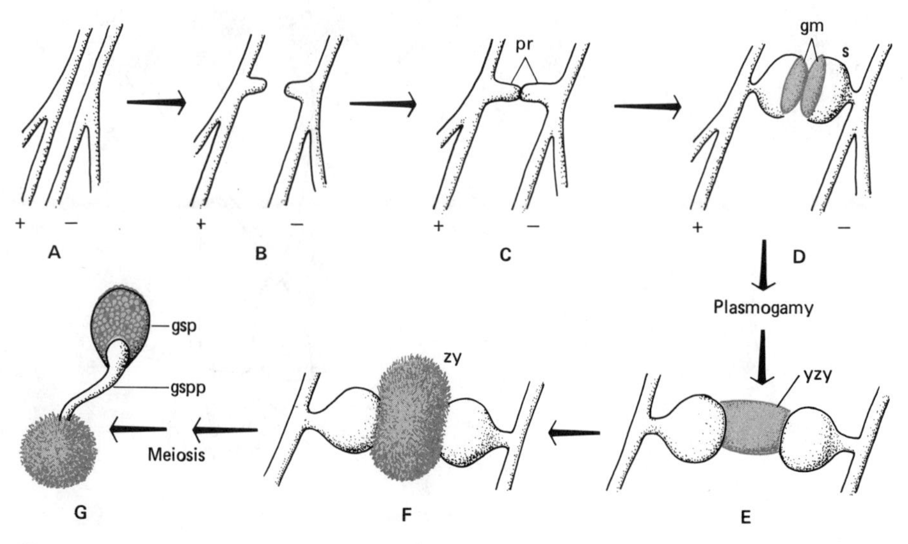


*Figure 7.24*
*Rhizopus;* sexual reproduction. *A.* Physiologically different vegetative hyphae. *B.* Vegetative hyphae began to proliferate in vicinity of each other. *C.* Progametangia *(pr)* formed. *D.* Progametangia become walled off forming gametangia *(gm)* and suspensor *(s).* The wall between the two gametangia undergoes dissolution and plasmogamy occur. *E.* The young N + N zygospore *(yzy) F.* The zygospore *(zy)* enlarges and the wall becomes greatly thickened and warty in appearance. Karyogamy occurs, followed almost immediately by meiosis. *G.* The resulting meiospore germinates to form a germ sporangiophore *(gspp)* with a germ sporangium *(gsp).* The germ sporangium produces spores which germinate forming new vegetative hyphae.

Typically eight *ascospores* are produced within each ascus, sometimes arranged in a linear order (Fig. 6.3). The asci may be single and scattered or they may be aggregated in a specialized structure called an *ascocarp.* The ascocarp may be open and more or less cup-shaped, closed and spherical in shape, or flask-shaped with a pore. Some representative members are shown in Figure 7.25. In the majority of forms, asexual reproduction is by specialized spores called *conidia,* which are cut off from the tip of modified hyphae called *conidiophores.*

The sac fungi are quite important to human endeavors. Beadle and Tatum won the Nobel prize in 1959 for their work on biochemical mutants of *Neurospora,* the pink bread mold. The key to the usefulness lies in the fact that the young asci of *Neurospora* are binucleate, with the two nuclei representing descendants of two different strains originally brought together. These eventually fuse in the ascus, followed by meiosis, which in turn is followed by a single mitotic division. Because of their linear arrangement in the ascus, the products of each divisional step is known, and the ascospores can be individually picked out of the ascus, placed in individual culture dishes and the cultures analyzed. The significance of this to Beadle and Tatum's work was discussed in Chapter 6.

A B C

*Figure 7.25*
Some representative ascomycetes. *A.* Sclerotium of *claviceps purpurea* (Ergot). *B.* Hyphae of *Ceratocystis ulmi* (Dutch Elm disease). *C.* Effect of Dutch Elm disease. *D. Morchella* sp., a morel. [Photographs courtesy of U.S. Department of Agriculture.]

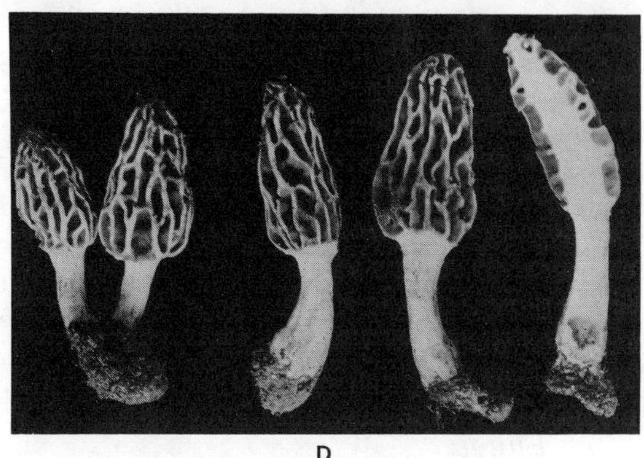

D

The activities of another sac fungus, *Saccharomyces cerevisiae* or *S. ellipsoides* commonly called yeasts, are well known in products enjoyed by many of you. I am, of course, referring to the fact that these are utilized in producing leavened dough from which bread and various pastries are made. They are also used in the brewing industry, with beer, wine, and saki being the result of direct fermentation by yeast. Whiskies, vodka, and gin are produced by distillation of the fermented substance.

A number of ascomycetes can be eaten including the common morel, or sponge mushroom, of the genus *Morchella* (Figure 7.25D). The little cup fungus, *Sarcosypha,* is also edible, but the greatest delicacy of all is the genus *Tuber*, which produces an underground ascocarp called a truffle. *Tuber* seems to develop a symbiotic relationship with the roots of certain trees, especially the oaks, and it is under these trees that truffle hunters, often with the aid of trained dogs or pigs, who locate the truffles by their smell, will search.

Another group of economically important sac fungi are the powdery mildews. They infect plants such as lilacs, roses, cereal grains, and many others. The black smut of rye is caused by the organism *Claviceps purpurea* (Figure 7.25A), and eating of such diseased grain produces a disease called ergotism. In the mild form, it can cause hallucinations; in severe cases, insanity and even the literal rotting off of limbs by gangrene. It was recently postulated that ergotism may have been the principal factor involved in the famous Witch Hunt in early Salem, Massachusetts.

The molds, particularly *Aspergillus* and *Pencillium*, seem clearly to be ascomycetes, but in most, sexual reproduction is not known; thus, they are included in the imperfect fungi. But we will discuss some of the more important ones here. *Aspergillus niger* causes the black mold on bread, *A. tamarii* is used in the orient for producing soy sauce, *A. oryzae* is used in the initial stages of sake production, and *A. fumigatus* produces the respiratory disease aspergillosis in humans. *Penicillium notatum* and *P. chrysogenum* are used in the production of the antibiotic penicillin, while *P. roquefortii* and *P. camenbertii* are used respectably in the production of roquefort and camembert cheese. The yeastlike organism, *Candida albicans*, causes several diseases in humans, including thrush and fungal vaginitis.

## Imperfect Fungi

There is little that can be said about the Deuteromycota that has not been already mentioned. They are a large group, comprising some 25,000 species in which sexual reproduction is unknown. Most have the morphological characteristics of ascomycetes while a few are morphologically similar to the basidiomycetes. An imperfect fungus, *Trichophyton*, produces various skin infections including athlete's foot and ringworm.

## Club Fungi

The Basidiomycota contain a divergence of types, including rusts, smuts, jelly fungi, pore fungi, mushrooms, and puffballs. Estimates of the number of species range from 12,000 to 25,000. The basidiomycetes are characterized by septate mycelia and lack of motile cells (as in the ascomycetes), by the production of a club-shaped reproductive body called a *basidium* in which *basidiospores* are produced, and by division of secondary myceliand by formation of clamp connections (Figure 7.26). These secondary mycelia are formed by fusion of primary hyphae from different mating types, with the primary hyphae having been produced by a germinating basiodiospore. The secondary mycelium gives rise directly to fruiting bodies *(basidiocarps)* which constitute the so-called tertiary mycelium.

The morphology of the club fungi differ considerably with the different groups. The rusts and smuts do not form basidiocarps, but some of them do have complex life cycles, utilizing several different hosts and producing as many as five different kinds of spores (Figure 7.27). The jelly fungi, pore fungi, coral fungi, mushrooms, and puffballs do form basidiocarps. Figure 7.28 illustrates some representatives of the basidiomycetes.

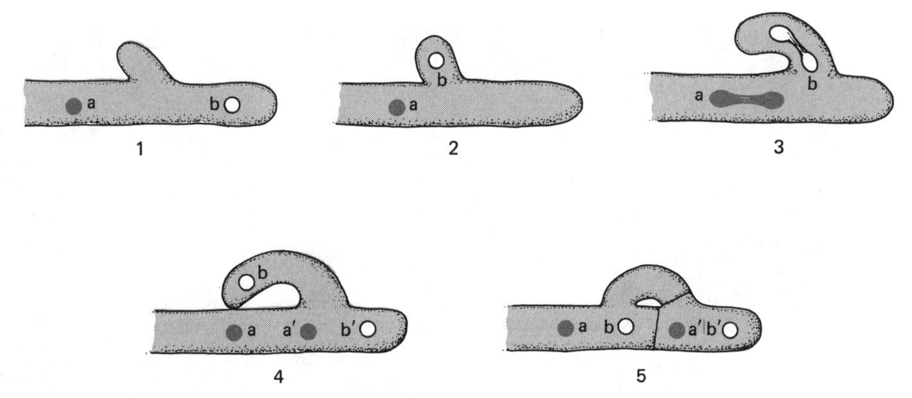


*Figure 7.26*
Sequential stages 1–5 in formation of a clamp connection; *a*, nucleus a of dilcaryon; *b*, nucleus b of dikaryon.

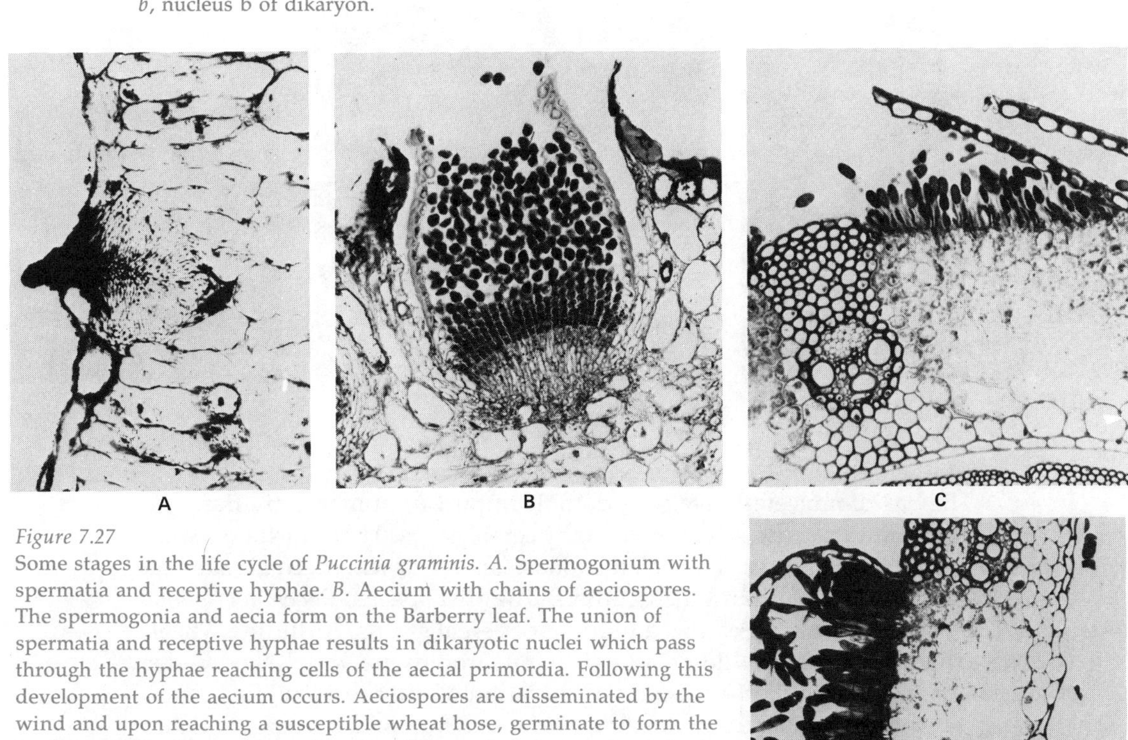


*Figure 7.27*
Some stages in the life cycle of *Puccinia graminis*. *A*. Spermogonium with spermatia and receptive hyphae. *B*. Aecium with chains of aeciospores. The spermogonia and aecia form on the Barberry leaf. The union of spermatia and receptive hyphae results in dikaryotic nuclei which pass through the hyphae reaching cells of the aecial primordia. Following this development of the aecium occurs. Aeciospores are disseminated by the wind and upon reaching a susceptible wheat hose, germinate to form the uredium. *C*. Uredium with uredospores. The uredospores may germinate forming other uredia, or, as the season progresses, telia may be formed. *D*. Telia with teliaspores (teleutospores). The teliospores overwinter and in the following spring germinate to form the basidia. Following meiosis, basidiospores are formed which are carried by the wind to a susceptible Barberry plant.

The basidiomycetes have considerable import in human activities. The rusts and smuts cause infection of many cereal crops. These include *Ustilago zeae*, the corn smut; *U. tritici*, the "loose smut" of wheat; *U. avenae*, the "loose smut" of oats; and many others. The rust species *Puccinia graminis* and its many races parasitize a number of cereal grains, including wheat, rye, barley, oats, and wild grasses. The mushrooms are widely known for eating qualities and other aspects. Cultivation of the common mushroom, *Agaricus bisporus*, has become an important commercial enterprise. Other fungi, such as the jelly fungus and puffball, are also edible, but many are not particularly palatable. The fact that some mushrooms are poisonous is also well known. Most of these are members of the genus *Amanita*, including the notorious *A. muscaria*, the fly agaric, and *A. phalloides* (or *A. verna* to taxonomists), the destroying angel. One tiny bite of the latter is said to be fatal in at least 50 percent of the cases. There is no antidote for mushroom poisoning, although the compound thioctic acid has shown some promise, and contrary to popular opinion, there is no simple way to tell

G

H

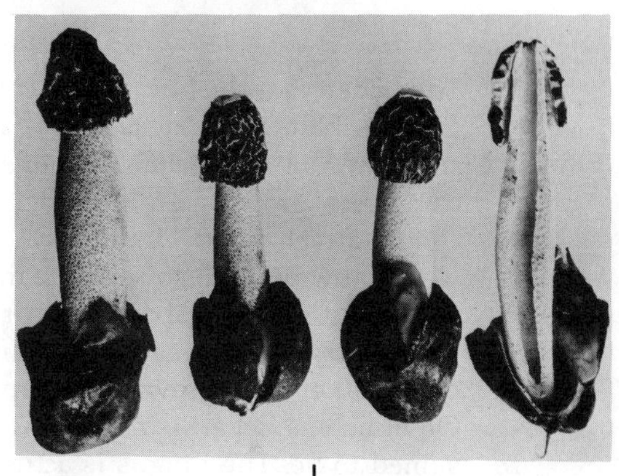

I

*Figure 7.28*
Some representative basidiomycetes. *A. Gymnosporangium* sp., cedar apple rust with telia. *B.* Gelatinous structures bearing teleutospores of *Gymnosporangium. C.* A tooth fungus; a member of the Hydnaceae family. *D.* Mushroom "fairy ring." *E.* The fly agaric, *Amanita muscaria. F.* Bird's nest fungus. *G. Calvatia gigantea*, the giant puffball. *H.* An earthstar. *I.* Ithyphallus impudicus, stinkhorn fungus. [*A., B., C., E. G.,* and *I.* Courtesy of U.S. Department of Agriculture; *D.* and *F.* Courtesy of Carolina Biological Supply Co.; *H.* From a color photograph by Dee Ann Houston.]

the poisonous mushroom from the edible one except to eat a little (not a particularly good way, in my opinion, unless someone else does the eating).

The relationship of mushrooms to hallucinogens is also well known and is tied to the religion, culture, and history of many peoples. I do not intend to get into a discourse here concerning the social or medical implications of hallucinogenic plants; this has been discussed (and cussed by some) in many books, journals, public meetings, pulpits, and other forums. But we do want to mention that certain mushrooms, particularly in the genus *Amanita* and *Psilocybe* are hallucinogens. Several species of *Psilocybe*, mentioned in the introduction, were the "sacred mushrooms" of the Aztecs, and *P. mexicana* is used today in religious ceremonies among the Indians in Mexico. The effects of *Amanita muscaria* is quite different from those of *Psilocybe*. The user initially has a trembling or twitching of the limbs, followed by a general euphoria, delusions of grandeur, and visions of the mystical or supernatural. There is also a distortion of perspective so that big things seem larger, and small things seem even smaller. It is said that Lewis Carroll had this mushroom in mind when he wrote Alice in Wonderland.

As we complete this discussion of the major saprophytic groups of fungi, we should mention that, similar to the saprobic bacteria, the fungi are extremely important in the decay of organic materials. Unfortunately, they often decay things that we do not want them to, such as living trees, books, clothing, leather, etc. We should also mention the interesting and rather remarkable predacious fungi, especially the nematode-trapping ones such as *Arthrobotrys*. Despite their frail-looking nature, these fungi are able to trap organisms that are many times larger, are powerful for their size, and are enormously active.

## Lichens

The lichen is a dual organism composed of an ascomycete fungus and a green or blue-green alga. Some basidiomycete lichens also exist, as does a single zygomycete lichen. Lichens are widespread and often are important colonizers, growing on bare rock or bare soil. Some are very tiny, others comparatively large (Figure 7.29).

It is now possible to separate the components of a lichen and grow each separately. The fungal component has a much different form then it had in the lichen and requires a large number of complex carbohydrates. The alga on the other hand grows more rapidly then it did in the lichen association. Obviously, a lichen is not the symbiotic association that it was once assumed to be. The fungus is actually exercising a controlled parasitism of the alga.

Resynthesis of the fungal-alga components proved to be far more difficult than separation had been. It was finally discovered that the lichen association will only be formed under conditions that were unfavorable to growth of both the fungus and the alga. It was also shown that during "lichenization," the fungus damaged the algal cells, sometimes killing them. The lichenologist, Ahmadjian, proposes that when spores from the fruiting body of a lichen germinate in the presence of free-living algal cells under environmental conditions that are unfavorable for the growth of either, they parasitize the alga. Algal cells that can withstand this association live, and a new lichen is formed. If not, then both the alga and the fungus die.

Lichens may be important in soil building as they produce various kinds of acids that will dissolve certain types of rock. They also have the ability to concentrate certain heavy metals, and analysis of the heavy metal content of lichens has been used as an aid to finding commercially valuable mining areas. This capacity to concentrate heavy metals also has its drawbacks. After the heavy atmospheric testing of atomic weapons following World War II, investigators were testing levels of radioactivity worldwide to determine the pattern of fallout. They found, much to their surprise,

*Figure 7.29*
Early stage of primary succession on a gabbra boulder. Lichens, mosses, and small ferns may be noted. Crustuse liverwort *(a)*, Foliose liverwort *(b)*, and Fructicose liverwort *(c)*. [Photograph courtesy of U.S. Forest Service.]

that the Laplanders of Scandanavia and Eskimos of Alaska had much higher levels of radioactivity in their bodies than had been predicted. The amount of fallout reaching the poles should have been very low. Continuing investigation indicated that the lichen *Cladonia subtennuis* (reindeer "moss") had absorbed isotopes from the atmosphere and had concentrated them many times higher than their atmospheric levels. As the common name suggests, this lichen is a primary food of reindeer, which in turn were the chief source of food for the Laplanders and Eskimos. Thus, the radioactive substance was simply being passed upward through the food chain. Lichens are also extremely sensitive to air pollution, especially sulfur dioxide, thus have been used to monitor air pollution.

## PLANTAE

The so-called true plants are placed into nine divisions (phyla): the Bryophyta (liverworts, hornworts, and mosses), Lycopodophyta (club mosses, spike mosses, and quillworts), Sphenophyta (horsetails), Filicinophyta (whisk fern and true ferns), Cycadophyta (cycads), Ginkgophyta (ginkgos), Coniferophyta (conifers), Gnetophyta *(Gnetum, Ephedra, Welwitschia),* and Angiospermophyta (flowering plants). The major characteristics of the plant kingdom are that most are autotrophic green plants, exhibit advanced tissue differentiation, and develop the diploid phase from an embryo.

## *Liverworts, Hornworts, and Mosses*

We will be considering these three admittedly distinct groups as a single division, the Bryophyta. All exhibit a similar life cycle, and some botanists believe there is a line of evolution between the groups (hence the single division). However, the great German morphologist Goebel wrote in 1905: "Between Hepaticae (liverworts) and Musci (mosses) there are no transition-forms; as there are none between Bryophyta and Pteridophyta (vascular plants), and as there never were such transitions their absence is not caused by their having died out." This viewpoint is supported by many botanists, and Bold, in his popular *Morphology of Plants,* has summarized comparatively a list of 14 diversities between the groups and on the basis of this and the lack of fossil evidence, questions the placing of these groups in the same division. Steere, a supporter of the common phylogeny between the groups, qualifies this somewhat by stating: "Even the earliest fossils yet known show no approach of mosses to hepatics; their separation was very ancient indeed."

Many botanists believe the origin of the bryophytes was from the green algae and cite common pigmentation, storage of starch, existence of cellulose cell wall, and formation of cell plate during cytokinesis (otherwise found only in a few green algae and the vascular plants, with the exception of a single brown alga). Despite the above, on the basis of evidence we shall be discussing in a later chapter, Steere states: "In spite of all that has been written to the contrary in texts and the botanical literature, we have no evidence that bryophytes have evolved directly from algae, or that they present an evolutionary waystation between algae and higher plants."

Let us not digress further at this point on the evolutionary aspects of the bryophytes, but rather let us consider their characteristics. They are small plants, some less than a centimeter long, with most less than 20 cm. Although most common in warm, moist areas, they are not confined to such areas, with some even being found on dry exposed rocks. Many are aquatic, although none are truly marine.

The bryophytes are distinguished from other groups in the plant kingdom by their lack of vascular tissue, *xylem* (the water-conducting tissue) and *phloem* (the food-conducting tissue). However, in certain mosses there is a central strand of water-conducting tissue clearly delimited from a food-conducting tissue. It has been stated by some botanists that little water is actually conducted in this central strand, but recent evidence indicates that in some mosses the water-conducting tissue is quite functional. In addition, phloem cells have been found in the food-conducting tissue of certain mosses. Thus, the concept of lack of vascularlization in mosses may change. Without such vascularlization, the organs of bryophytes cannot be referred to as roots, stems, or leaves, as such a distinction requires that they contain vascular tissue. However, it is a common custom to refer to the leaflike and stemlike structures of bryophytes by the terms leaf and stem, and we shall do so here. It is up to you to keep the conceptual distinction of true stems and leaves in mind. Rootlike structures are lacking in bryophytes, rather there is an evagination of epidermal cells in elongated structures called *rhizoids*.

A second distinctive characteristic of bryophytes is their life cycle. The gamete-producing plant, the *gametophyte*, is larger and always nutritionally independent, whereas the spore-producing plant, the *sporophyte*, is permanently attached to the gametophyte and may be dependent upon it to a greater or lesser degree. It has been commonly stated that the sporophyte is parasitic upon the gametophyte. This concept is based upon the assumption that 1) the sporophyte lacks chlorophyll and 2) it can absorb water and minerals only from cells of the gametophyte. In fact, the sporophyte of most bryophytes contains chlorophyll, and it has been demonstrated that water and minerals can be absorbed directly through the walls of the sporophyte. Thus, the sporophyte is not parasitic but rather epiphytic upon the gametophyte.

The group of liverworts called thallose are characterized by the lack of differentiation of stem and leaves in the gametophyte (Figure 7.30). The

*Figure 7.30*
A thallose liverwort, *Marchantia* sp. [Photograph by David R. Williams].

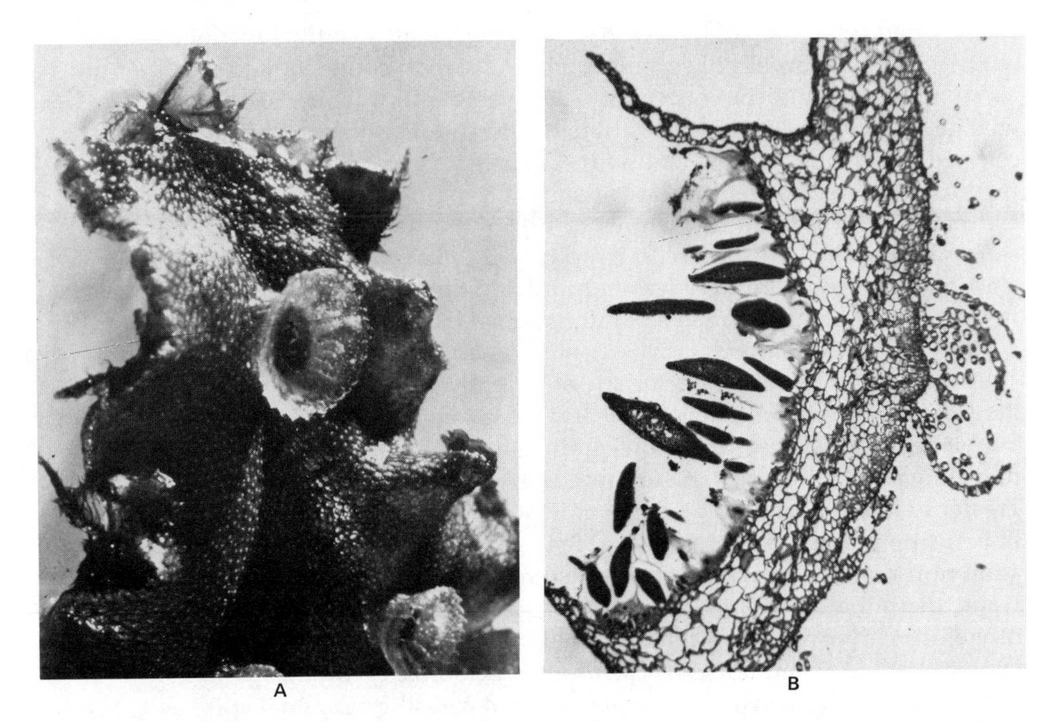


*Figure 7.31*
*A.* Dorsal surface of *Marchantia polymorpha.* Note gemma cups (almost all gemmae shed).
*B.* Transection in region of gemma cup; note gemmae. [*A.* Courtesy Carolina Biological Supply Company.]

gametophyte is dorsiventrally flattened, with the upper surface differentiated into an outer epidermal layer with pores. Each pore leads into an air chamber contained within the underlying photosynthetic tissue. Beneath the layer of photosynthetic tissue is a rather compact layer of parenchyma cells which function largely in storage. From the lower, or *ventral*, epidermis emerge multicellular *scales*, one cell layer thick and often purple in color as the result of pigmentation of the cell walls, and *rhizoids*. The rhizoids may be all of one type, or in some, two types of rhizoids are formed, smooth walled and tuberculate (pegged).

The thallose liverworts can reproduce vegetatively by fragmentation of the thallus or by specialized structures, called *gemmae*, produced in *gemmae cups* (Figure 7.31). Sexual reproduction involves the differentiation of specialized sex organs within the gametophytic thallus. The male sex organs are called *antheridia*, the female, *archegonia* (Figure 7.32). The male and female sex organs may arise on the same plant, if so the plant is termed *bisexual*, or the male sex organs may arise on one plant, the female sex organs on a different plant *(unisexual)*. The sperm cells are flagellated; thus, water is required to effect fertilization. The sperm is probably attracted to the egg by a chemical stimulus. The zygote undergoes development

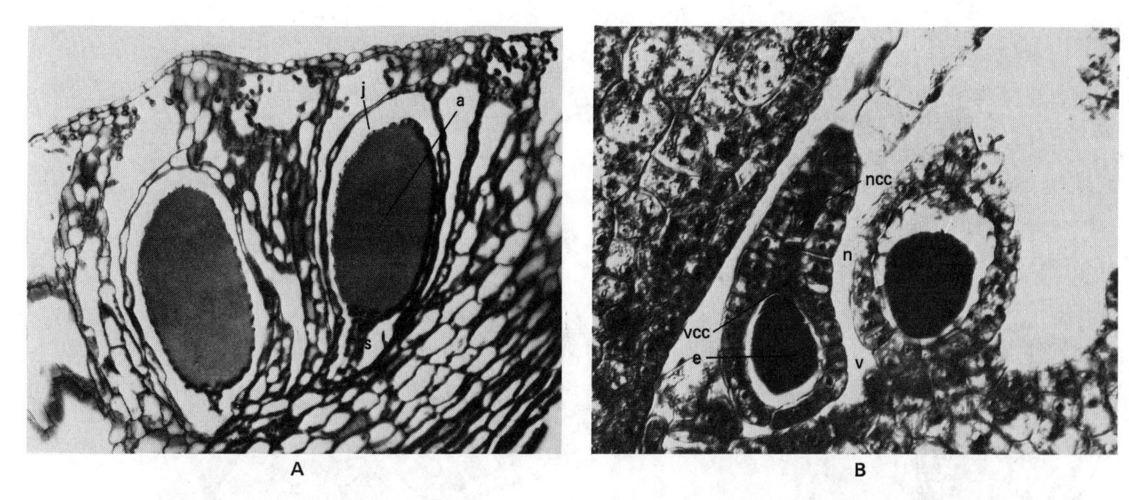


*Figure 7.32*
*A.* Longisection of antheridial disk of *Marchantia*. Note antheridia borne in antheridial chambers. The antheridia are borne above the thallus on a stalked structure termed the antheridiophore. The antheridium is a globular body borne on a short stalk; *s*, stalk; *j*, jacket; *ac*, androgonial cells. *B.* Longisection of archegonial disk of *Marchantia*, like the antheridia, borne above the thallus on an archegoniophore. The archegonium illustrated here is in median longitudinal section; *e*, egg; *v*, venter; *vcc*, ventral canal cell; *ncc*, neck canal cells; *n*, neck.

within the archegonium, with the enlarging archegonium being called the *calyptra*. The more or less spherical mass of diploid tissue within the enlarged archegonium is the sporophyte. Cells of the sporophyte undergo meiosis forming *spores*. The sporophyte of some thallose liverworts is little differentiated, while that of others is differentiated into three regions: the *foot*, which retains contact with the gametophyte; the *seta*, or stalk; and the spore-bearing region, or *capsule* (Figure 7.33). Certain cells of the capsule may not undergo meiosis but become much elongated, with spirally arranged thickenings on the inner surface of their walls (Figure 7.34). These are called *elaters* and at maturity are sensitive to slight changes in atmospheric pressure. They aid in dissemination of the spores following dehiscence (opening) of the capsule.

The leafy liverworts are a diverse group, but are characterized, in general, by being well branched and bearing two rows of oppositely arranged, lobed leaves, with a third row of reduced leaves on the ventral surface.

The hornworts are generally found in moist, shaded habits. The gametophyte superficially resembles that of a thallose liverwort but has less internal differentiation. The chlorenchyma cells contain a single massive chloroplast like those of green algae, rather than the numerous, discoid chloroplasts present in other bryophytes. The chloroplast contains a proteinaceous body which apparently is involved with starch synthesis, thus

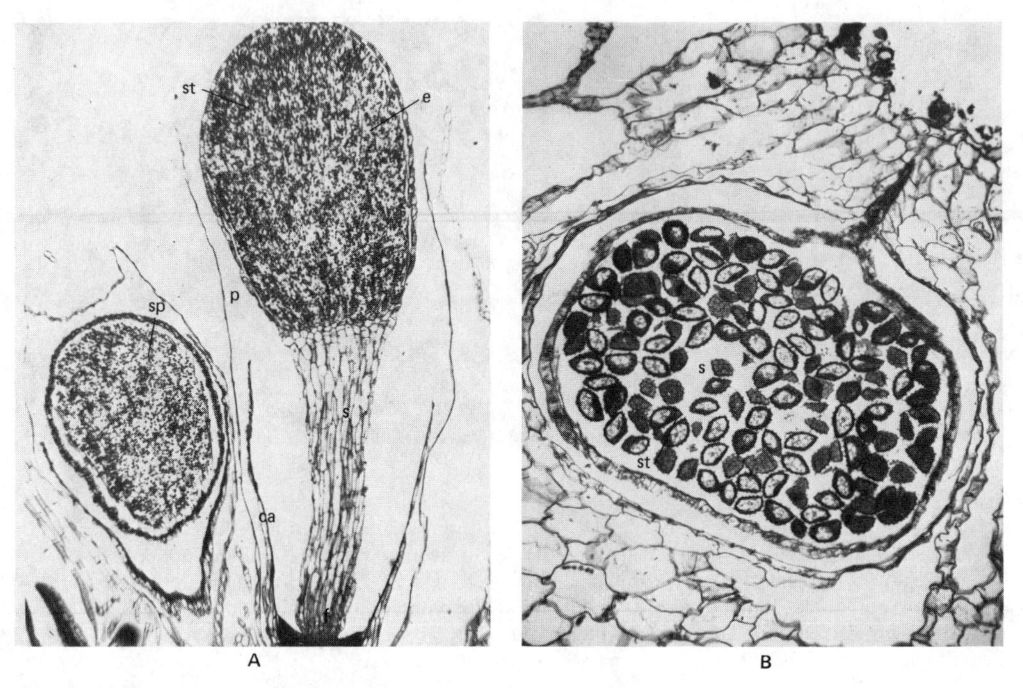


*Figure 7.33*
Stages in the development of the sporophyte. *A. Marchantia; f*, foot; *s*, stalk or seta; *ca*, calyptra; *p*, perianth; *st*, spore tetrad; *sp*, spores; *e*, elaters. *B. Riccia.* Note absence of foot, seta, and elaters in the *Riccia* sporophyte.

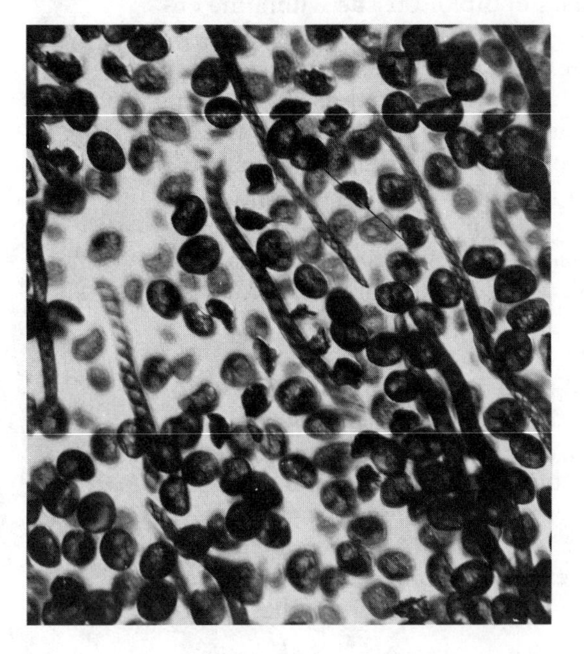

*Figure 7.34*
Enlarged view of interior of *Marchantia* capsule. Note the pointed, elongated elaters with their spirally arranged thickenings on the inner surfaces of their walls.

being analogous to the pyrenoids of green algae. Whether or not this is a truely homologous structure is debatable, but it has generally been designated a *pyrenoid body*.

Antheridia and archegonia are sunken into the dorsal surface of the thallus, and members may either be unisexual or bisexual. Mature antheridia are recognizable as small, orange-yellow dots, with the color resulting from transformation of chloroplasts in the outer layer of the antheridium into chromoplasts. The zygote develops into a much-elongated sporophyte. This elongation is accomplished by formation of a meristematic region just above the foot. This region functions as an intercalary meristem and as the result of its activities, the cylindrical sporophyte pushes through the gametophytic tissue. The sporophyte is perhaps the most complex in the bryophytes. The central region remains sterile and is called the *columella*. The epidermis is cutinized and develops stomata similar to those of higher plants. Surrounding the columella is a tract of cells which develop into sporogenous tissue, with the differential development of these into *sporocytes* (spore-producing cells) beginning at the tip of the sporangium. Each sporocyte produces a tetrad (group of four) of spores as the result of meiosis. Some of the potentially sporogenous cells remain sterile but may divide and elongate. These are called *pseudoelaters*. Dehiscence of the capsule proceeds from the top toward the apex. The sporophyte is photosynthetic, and in one species, rhizoidlike outgrowths develop from the foot penetrating through the gametophyte. The sporophyte is relatively long lived, with the gametophyte becoming progressively more disorganized. Thus, we very nearly have a completely free-living sporophyte as is characteristic of the vascular plants.

The gametophytes of mosses are characterized by being upright, leafy structures. The leafy gametophyte arises from a branched, filamentous structure termed the *protonema*. Thus, in the moss gametophyte, there are two distinct phases of development. The protonema arises directly from the germinating spore, with some of its branches becoming colorless rhizoids, others developing budlike structures which develop into the leafy gametophyte. The protonema looks superficially like a branched filamentous green alga, and this feature had been used as an argument in support of an algal origin for the bryophytes. But the cell walls of the protonema are distinctly different from those of any known algae.

Sex organs are borne either at the apex of the main axis or at the terminus of short, lateral branches. In *Sphagnum* (peat moss), the antheridium is somewhat similar to those of liverworts (although its development is different), but in the so-called true mosses, it is an elongated, somewhat cigar-shaped structure with a number of antheridia being grouped together, interspersed with sterile structures called *paraphyses* (Figure 7.35). The archegonia are frequently long stalked with an only slightly enlarged *venter* (egg chamber) and an elongated neck (Figure 7.36). The sporophyte produced from the zygote is generally elevated above the gametophytic

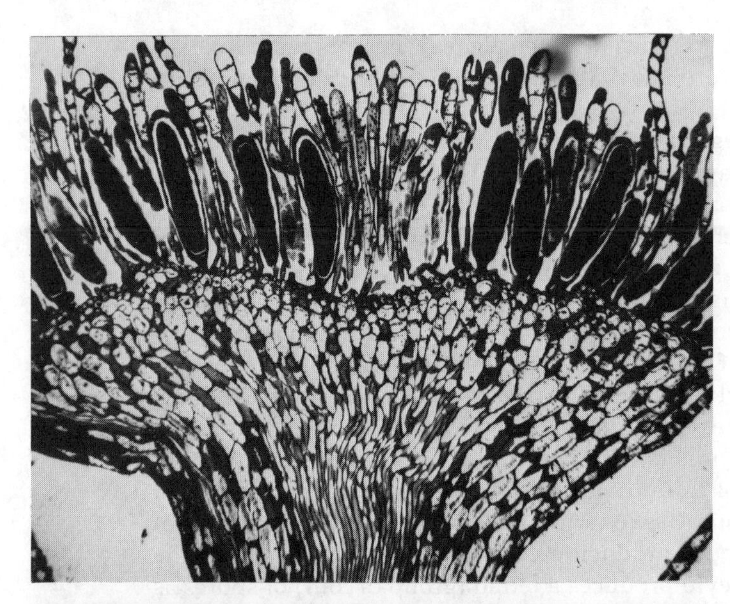

*Figure 7.35*
*Mnium.* Longisection through apex of male plant; note antheridia and paraphyses.

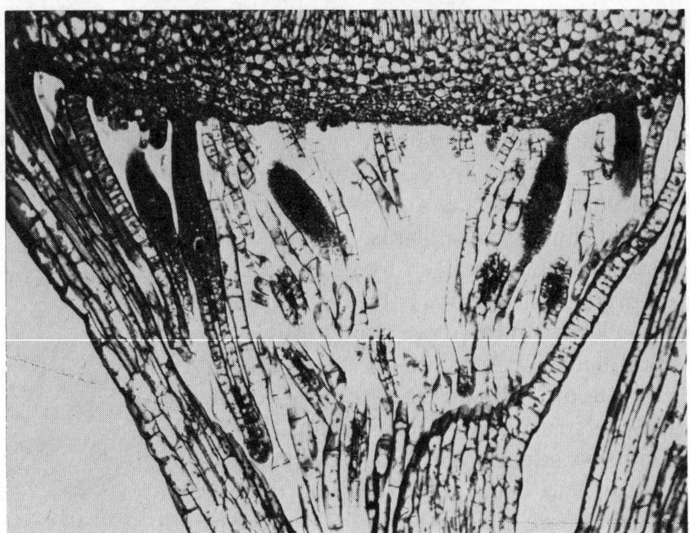

*Figure 7.36*
*Mnium.* Longisection through apex of female plant. Note mature archegonium just prior to fertilization and paraphyses.

thallus. In *Spagnum*, this is accomplished through elongation of the stem bearing the sporophyte, but in most mosses, it occurs by elongating of the seta of the sporophyte itself (Figure 7.37). The capsule is surrounded by the enlarged and modified archegonium, the *calyptra*, and it has been shown that this calyptra is absolutely necessary for normal development of the capsule. In mosses other than the peat mosses and granite mosses, the capsule is topped by a lid, called the *operculum*. Beneath this lid in many mosses is a row of *hygroscopic*, teethlike structures, called the *peristome teeth*. These aid in dissemination of the spores (Figure 7.38). Mosses may reproduce vegetatively by fragmentation or by production of gemmae.

*Figure 7.37*
Attached moss sporophytes *(arrow)* on moss plants growing among tree roots. [Photograph by David R. Williams.]

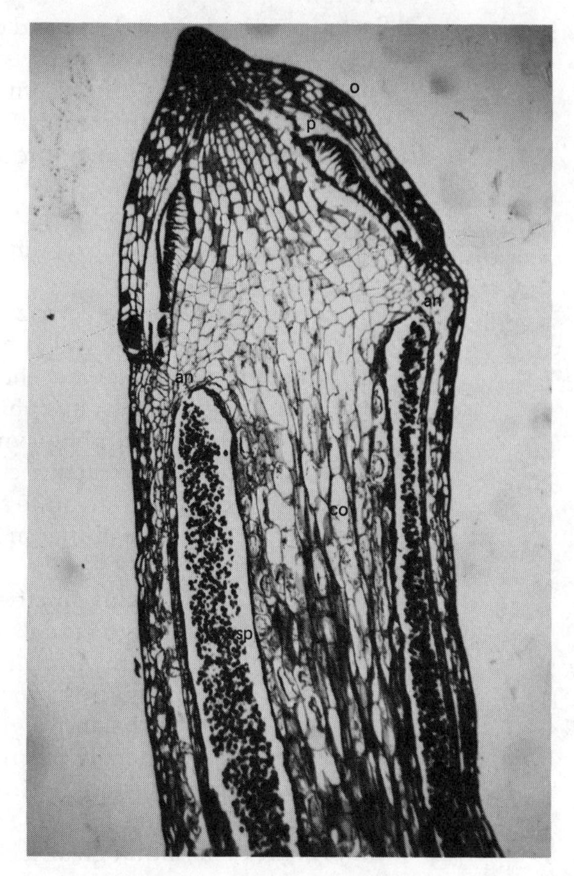


*Figure 7.38*
Longisection through tip of moss capsule; *an*, annulus; *o*, operculum; *p*, peristome; *co*, columella; *sp*, spores.

The bryophytes obviously have considerable ecological significance but, other than the peat mosses, have little significance for humans. Peat is used for packing, for mulch, for ion exchange capacity, and in some areas for building material and fuel.

## SUMMARY

1. Virus particles, by themselves nonliving, can infect living cells causing various metabolic changes in the cell and the production of new viral particles.
2. The kingdom Monera is composed solely of prokaryotic organisms and includes the bacteria and blue-green algae.
3. The bacteria are primarily heterotrophic; those which are photosynthetic do not produce oxygen.
4. The bacteria have three basic forms: cocci, bacilli, or spirilli.
5. Many bacteria are motile by means of flagella.
6. Most bacteria are asexual, although genetic recombination can occur in a few.
7. Bacteria are important decay and disease-causing organisms.
8. The blue-green algae bear many similarities with the bacteria.
9. The blue-green algae differ from bacteria in that they evolve oxygen, have greater cellular differentiation, but lack flagella.
10. Some blue-green algae may produce toxins, and many are problem-causing organisms in water.
11. The Protista are eukaryotic organisms which are characterized by their lack of true tissue and embryonic stages in their development but have sexual reproduction.
12. The green algae are characterized by their pigment which includes chlorophyll *b*, a cellulose wall, and starch as stored photosynthate.
13. The golden-brown algae are characterized by the pigments chlorophyll *c* and xanthophylls, morphologically dissimilar flagella, chrysolaminarin as the stored photosynthate, and asexual reproduction by means of a specialized structure called a statospore.
14. The yellow-green algae are quite similar to the golden-brown but have chlorophyll *e* and flagella that are always of unequal length and different construction.
15. The diatoms have a highly silicified cell wall composed of two overlapping halves and undergo sexual reproduction with the zygote becoming an auxospore. Motility is by means of a raphe.
16. The brown algae are macroscopic, almost entirely marine, store laminarin, and have the substance alginic acid in their walls.
17. The red algae are generally macroscopic, also primarily marine, have phycobilin pigments, store floridean starch, and have the substances agar-agar and carrangeenin in their cell walls. Many also exhibit a complex life cycle, although all lack flagellated cells.

18. Both the red and brown algae are important in food processing, cosmetics, and many other industries.
19. The dinoflagellates have a unique type of flagellar insertion, have several unique xanthophylls, and may have their walls composed of plates of cellulose. Some are luminescent, while others are toxin producers known as "red tide."
20. The euglenoids have a single emergent flagella, no cell walls, store paramylon, have pigments similar to the green algae, and reproduce by cell division only.
21. The charophytes are macroscopic plants, with the plant body divided into nodes and internodes and bearing "leaves," branches, and multicellular sex organs.
22. The slime molds produce a naked, multinucleate plasmodium, and a sporangium and spores having cellulose walls.
23. The cellular slime molds are aggregates of amoeboid cells, which can be transformed into spores.
24. The chytrids are colorless, uniflagellated cells which may form coenocytic mycelia, and form sporangia in which zoospores are produced.
25. The oomycetes produce biflagellate zoospores with one flagellum placed apically, one laterally. The cell wall is constructed of D-glucan units.
26. The Fungi are characterized by chitinous cell walls, flagellated stages lacking, and a branched mycelium. The various divisions are characterized by the production of zygospores, asci, basidia, or a lack of sexual reproduction.
27. The lichen is a dual organism, consisting of a fungus and an alga, with the fungus apparently parasitizing the alga.
28. The fungi are important decay organisms. Some are used for food, others for their hallucinogenic properties.
29. The bryophytes show tissue differentiation and have embryonic development. They are divided into three groups; the liverworts, hornworts, and mosses, with the distinctions between the groups based primarily on sporophyte characteristics.

# Chapter 8

## THE VASCULAR PLANTS—I. ROOTS

Roots are thought of normally as the underground structures of vascular plants. Such a concept may lead to several misconceptions: 1) all roots are found underground, 2) all underground structures of vascular plants are roots, and 3) all vascular plants have roots. The majority of vascular plants do have an underground root system and, although roots and root systems may vary considerably in form, they have in common their association with water and mineral uptake from the soil. They also typically serve to hold the plant in place.

## ROOT MORPHOLOGY

Roots differ externally from stems by an absence of nodes (leaf-bearing regions) and internodes (the portions between nodes). Thus, roots do not normally produce leaves, nor do they possess lateral buds. Shoots may form from roots in certain plants. The growing point of a root, the *apical meristem*, is not truely apical as in the stem tip. Rather, it is *subapical* in position, being covered by a mantle of parenchymatous cells in various stages of differentiation called the *root cap.* A little distance behind the apical meristem (from 700 μm to 15,000 μm, more or less, from the *root tip*), *root hairs* develop. Root hairs are simply outgrowths of epidermal cells (Figure 8.1). Older hairs are normally sloughed off. Occasionally however, they may be persistent. Not all vascular plants possess root hairs. For example, aquatic plants typically lack them as well as some terrestrial plants, such as certain gymnosperms. Anywhere from several millimeters to centimeters behind the root apex, *lateral* or *secondary roots* emerge. These have their origin from an internal tissue of the main root and in turn may bear other lateral roots.

*Figure 8.1*
Root hairs. *A*. Pea seedling. Very fine fibers about 200 nm in diameter, called *rhizoplane fibers*, extend into the soil from the surface of the root hair cells. These root hairs are normally short-lived. *B*. Cross-section of radish root through root hair zone.

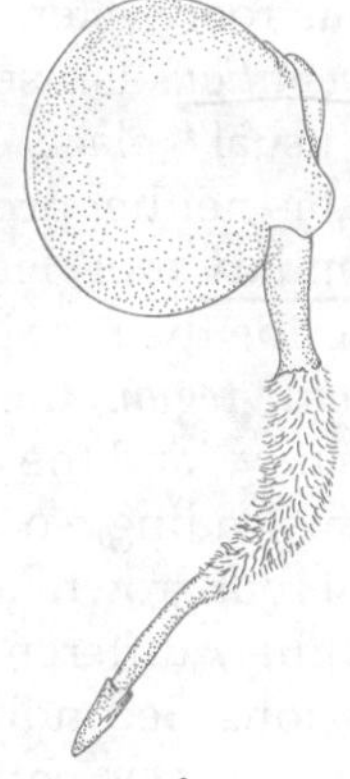


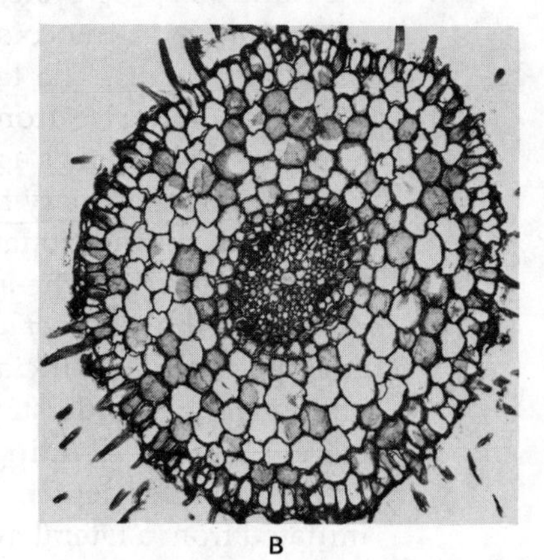

*Figure 8.2*
Types of root systems. *A.* Taproot system. *B.* Fibrous root system. *C.* Adventitious root; prop roots of Pandanus. [*A.* and C. by David R. Williams; *B.* Courtesy of U.S. Department of Agriculture.]

## ROOT SYSTEMS

All underground roots of a plant constitute the *root system.* Some plants, at maturity, bear large main, or principal, roots which produce somewhat smaller lateral roots which in turn produce still smaller laterals, and so on. This pattern is characteristic of a *tap root* system, found in most, but not all dicots. If there is no single main root, but rather the roots are all essentially the same size, the system is said to be *fibrous.* The fibrous root system is found in most monocots. Fibrous root systems are normally composed of *adventitious roots.* The term *adventitious* is used to describe any structure that does not arise from the "usual" place. Thus, an adventitious root might be described as a root that is neither primary nor secondary, arising from neither a primary nor secondary root (Figure 8.2; see also Figure 4.7).

It has been traditional to describe the root in terms of zones or regions: the *root cap region*, the *meristematic region*, the *region of cell enlargement or elongation*, the *region of differentiation*, and the *region of maturation.* Such an attempt is quite artificial and misleading, so serves little purpose in providing a better understanding of root growth and development. In the first place, cells are dividing, enlarging, differentiating, and maturing to a greater or lesser degree in all regions. Secondly, once secondary growth is initiated from a lateral meristem, the same pattern of zones, except for that of the root cap, might be postulated in a lateral direction. Thus, in that

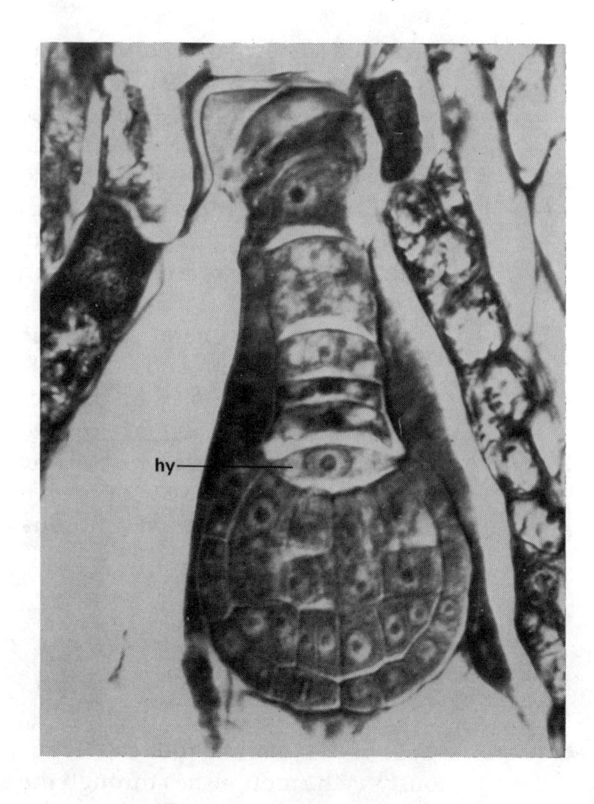


*Figure 8.3*
*Capsella* embryo. Note the hypohysis *(hy)* cell. This cell plus the cells immediately above it contribute to the development of the embryonic root.

zone termed the region of maturation, one would find cell division, enlargement, differentiation, and maturation occurring in derivatives of the lateral meristem.

## ROOT DEVELOPMENT

The initiation of an embryonic root, like that of the hypocotyl, cotyledons, and embryonic shoot, arises through divisions of the zygote and its derivatives. It differs from the other structures, however, in that a portion arises from divisions of the hypophysis, the apical cell of the suspensor, while the rest arises from cells of the embryo proper (Figure 8.3). In general, the apex of a root is less complicated than that of the stem. This is because in most plants at least, lateral organs do not arise in the apical meristem. Growth is thus considered to be relatively uniform. Even at that it is virtually impossible to construct a single classification of roots based on meristem organization. Perhaps the simplest organization to understand, and the one that will serve our purposes in attempting to get some concept of apical meristem organization and tissue differentiation, is a modification of the histogen theory first put forward by Hanstein in 1868. It was later modified somewhat by Janczewski in 1874 and Treub in 1875. This theory attributes a specific destiny to derivatives of regions in the apical meristem, termed *histogens.* I must emphasize at this point that plant cells *are not predistinated* to be any particular cell type. Every cell has the potential to develop or differentiate into any cell type, a feature termed *totipotency.* A cell's potential is determined strictly by its genetics, and neither origin nor environment (location) can influence the potentialities of a cell. But both origin and environment can and do determine the specific destiny of the cell.

Although the general pattern of root tip organization is the same in primary, secondary and adventitious roots, the specific type of histogen organization may be different in all three, even in the same plant. In most dicots, the primary root, that root which develops directly from the embryonic root, is present throughout the life of the plant. In many monocots, the primary root is *ephemeral,* that is, short lived, and is followed by development of a fibrous adventitious root system.

## DIFFERENTIATION OF PRIMARY TISSUES

### *Root Cap*

Cells of the terminal root cap (Figure 8.4) are being sloughed off continuously as the root pushes through the soil. The life of an individual root cap cell is relatively short. The root cap has always been thought of as a protective structure for the apical meristem, but it may have an equally important physiological function. If the root cap is experimentally removed, that root will no longer react to the effect of gravity. For example, if placed in a horizontal position, the root tip will not bend downward. But following removal of the root cap, cells of the quiescent zone (a zone of relatively inactive cells in the apical region) divide rapidly and a new root cap is regenerated. The root will then begin to grow downward again. These observations may suggest answers to two questions: 1) why is the quiescent center quiescent; 2) why does a root grow downward and not upward like the stem?

Acknowledgement that removal of the root cap inhibits the response to gravity and stimulates cells of the quiescent center to again divide does not *prove* its effect on either phenomenon. But it might suggest some hypotheses. In regard to *geotropism* (response to gravity), another piece of information is required. Growth-regulating substances called *auxins,* when applied to only one side of a root or shoot, have been shown to cause bending. Thus, we might postulate that the root cap somehow influences auxins in the root tip.

We might suggest two ways that this could come about: 1) the root cap may be the site of auxin production, or 2) the root cap may somehow control the lateral distribution of auxin. As auxins are growth-regulating

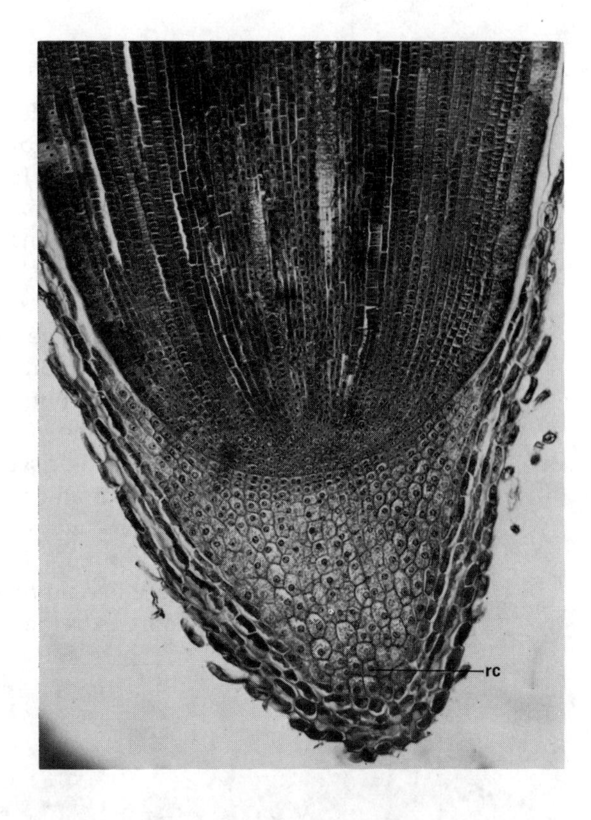


*Figure 8.4*
Longisection of onion root tip. Note cells of the root cap *(rc)*, some of which are sloughing off. This is what occurs normally as the root grows through the soil.

hormones, if they were being produced by the root cap, removal of it would result in complete loss of auxin to the root. If this occurred, root elongation would most certainly slow or cease entirely. But this is not the case. Root elongation continues in the absence of the root cap at essentially the same rate or faster than when it was present. This leaves us with the second alternative. The root cap somehow controls the movement or lateral distribution of auxin. In regard to the effect of the root cap on the quiescent center, we might suggest that the root cap is the source of a mitotic inhibitor. This hypothesis was also suggested by experiments in which only half the root cap was removed. Cells in that portion of the quiescent center just above the removed portion of the root cap began to undergo division.

Root cells subjected to geotropic stimulation show a displacement of amyloplasts to the lower side of the cells. Experimentation with a mutant plant having very small amyloplasts indicated that it showed 30 to 40 percent less downward distribution of these plastids, a 40 to 80 percent decrease in lateral transport of auxin, and considerably diminished geotropic curvature. These observations support the hypothesis that amyloplasts serve as gravity sensors. They do not offer proof of this relationship and thus not all botanists will accept this hypothesis as an adequate explanation for geotropism in roots.

## Epidermis

I have mentioned previously the development of root hairs from epidermal cells. In some plants, every epidermal cell produces a root hair. In other plants root hairs are formed from special epidermal cells called *trichoblasts*. These cells can be recognized easily even before root hairs form, by their distinctive size, shape, and densely staining cytoplasm. Other trichoblasts are distinctive, not only in appearance, but also in genetic makeup. While regular epidermal cells have the usual diploid amount of DNA, the trichoblast cell may have multiples of this normal amount. The trichoblast cells thus have polyploid nuclei.

The epidermal layer of most roots is composed of a single layer of cells. In certain epiphytic orchids, however, a multiple-layered epidermis is formed in the aerial roots. This tissue is called the *velamen* (Figure 8.5). Cells of the velamen are dead and thus devoid of protoplasm, and some may have lignified secondary walls. For a long time, it was thought that the velamen functioned to protect the underlying tissues, a function thought necessary as these aerial roots would be subjected to much more rapidly changing and more extreme environmental conditions. But, close

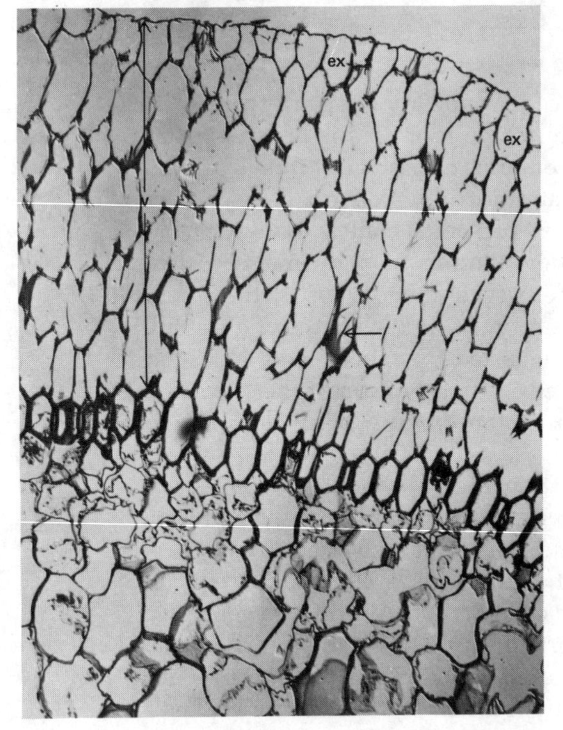


*Figure 8.5*
Cross-section of the outer portion of the aerial root of an epiphytic orchid. Note the velamen layer *(v)* and the exodermis *(ex)*. Also note remnants of the striplike wall thickenings *(arrow)*.

examination of these roots show that the velamen does not even cover the apical region of the root, and it is here that the most delicate tissues are found and thus the most protection would be needed.

This phenomenon, along with many others found in plants, illustrates the danger of attributing plant structure to need. If a particular plant has the genetic potential to develop a particular structure, then under appropriate environmental conditions the structure will develop. Under different environmental conditions the structure may not develop. So again I want to emphasize, and I can not overemphasize this point, all plant structure and function is the result of the interaction between a plant's heredity and its environment. Thus, a plant cannot adapt to its environment, in the sense that it undergoes change to meet some need. Modifications occur only in so far as it is within the plant's genetic potential for them to occur. The environment cannot cause mutation, except through certain phenomena such as X-radiation, and then these mutations are usually always harmful rather than beneficial to the plant, or they have no readily apparent effect at all.

Why then, did a tissue such as the velamen arise in certain plants? Who knows? It is not even a particularly appropriate question. What we should be interested in is the relationship of this tissue to others and to physiological processes of the plant. We may even find it has no particular function at all. Just because a tissue is specialized structurally does not indicate automatically that it must possess some specialized function. This is not to say that the velamen, for example, has no function.

## *Cortex*

This region, which is rather extensive in most roots, can be considered a single tissue on the basis of its origin from the meatic meristem. Although the cells of this region are mostly parenchymatous, a variety of cell types can be found in certain roots, and thus cortex can be considered a complex tissue. Normally, cells of the cortex are loosely packed with many intercellular spaces, a condition particularly prevalent in aquatic plants.

On the basis of differentiation, the cortex consists actually of several different tissue regions. The outermost layer, or layers occasionally, lying just beneath the epidermis may be differentiated as a *hypodermis*, or if the cells are quite thick walled, this layer is often termed *exodermis*. The innermost layer of the cortex is differentiated as an *endodermis*. Between these two layers is the rather extensive *cortical parenchyma* (Figure 8.6).

## *Endodermis*

The endodermis is a particularly interesting layer. Van Fleet indicates that as many as fourteen different types of endodermis exist. The layer is characterized by the development of a band of suberin in young endodermal

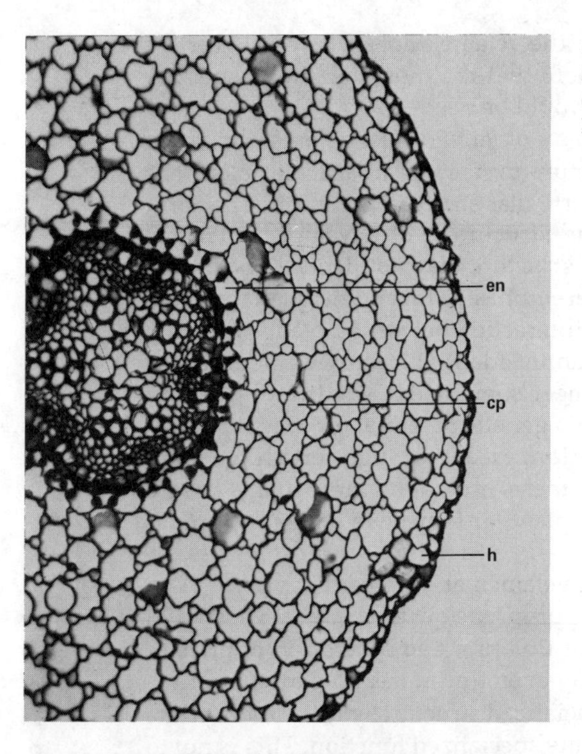


*Figure 8.6*
*Pyrus* root cross-section. Note endodermis *(en)*, cortical parenchyma *(cp)*, and hypodermis *(h)*.

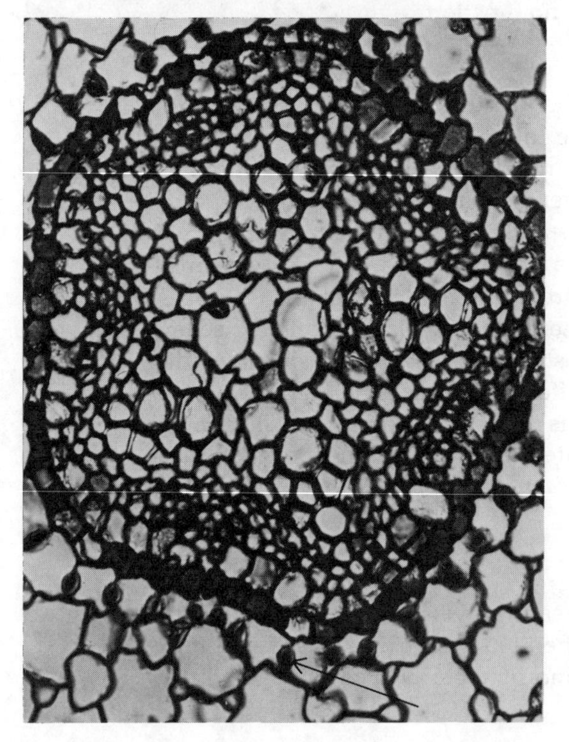

*Figure 8.7*
*Pyrus* root cross-section. Endodermis with evident Casparian dot *(arrow)*.

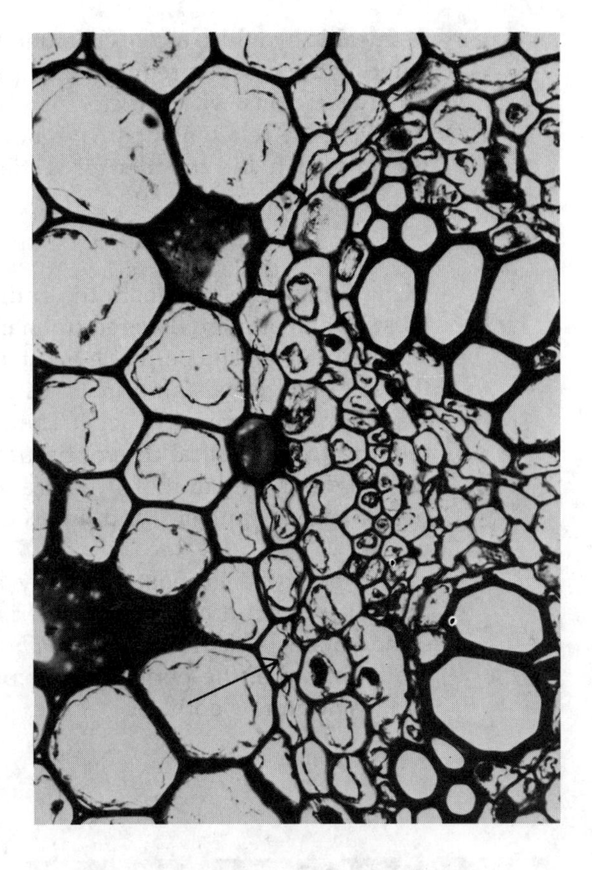

*Figure 8.8*
*Podophyllum* root, showing a portion in cross section. Note the endodermis with the protoplasmic strip extending across the cell *(arrow).*

cells, the *Casparian strip*, which runs radially around the cell and thus can be found only in the radial and transverse walls (Figure 8.7). The protoplast of the cell is attached to the strip. When the cell is plasmolyzed (the protoplast contracted away from the wall due to water loss), the protoplast moves away from all areas of the cell except in those regions of the Casparian strip. Thus, in transverse section, the protoplast can be seen as a strip extending across the cell (Figure 8.8).

The walls of endodermal cells may become thickened at maturity, and in some plants, extensive thickenings develop on the inner tangential and radial walls. Often the endodermal cells directly opposite the xylem arms do not become thick walled; these have been called *passage cells.* This term reveals what was thought to be the major function of the endodermis, that is to control movement of materials into the central cylinder, or *stele*, of the root. It was thought that the Casparian strips would prevent water movement through the middle lamellae and cell walls, areas in which water moves readily in other regions of the root. Thus, the water would have to move through the plasmalemma of the endodermal cells, and therefore, the endodermal cells, by the fact that the plasmalemmas are differentially permeable, would exercise considerable control over substances leaving or entering the vascular region of the root. The passage cells were thought to provide regions where materials could move readily through. Unfortunately, these relationships have not been demonstrated conclusively, so the function of the endodermis is still relatively obscure.

## *Pericycle*

Lying immediately beneath the endodermis is either a single-layered (in most dicots) or many-layered (in many monocots) tissue called the *pericycle* (Figure 8.9). The pericycle is often described as the outermost layer of the stele and consists normally of parenchyma type cells, although they may become sclerenchymized. The term *stele*, whose root means column, refering to the central core of tissues, was conceived by Van Tieghem in 1886 to describe a unit of structure consisting of the pericycle, stelar cambium, xylem, phloem and interfascicular regions, gaps, and pith (if present).

The pericycle is most generally associated with the origin of lateral roots. It retains a capacity for meristematic activity, and for this reason some botanists call it the *pericambium.* In addition to giving rise to lateral root primordia, the pericycle may produce, in part, the stelar cambium, cork cambium or both.

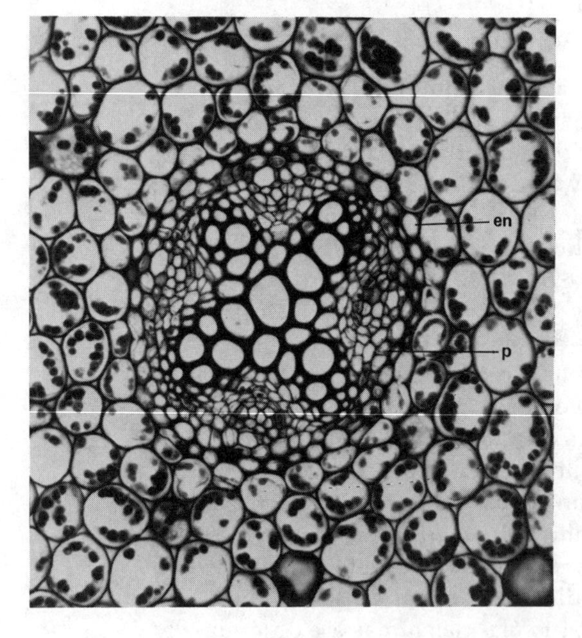


*Figure 8.9*
*Ranunculus* root cross-section. Note single-layered *(uniseriate)* pericycle; *en,* endodermis; *p,* pericycle.

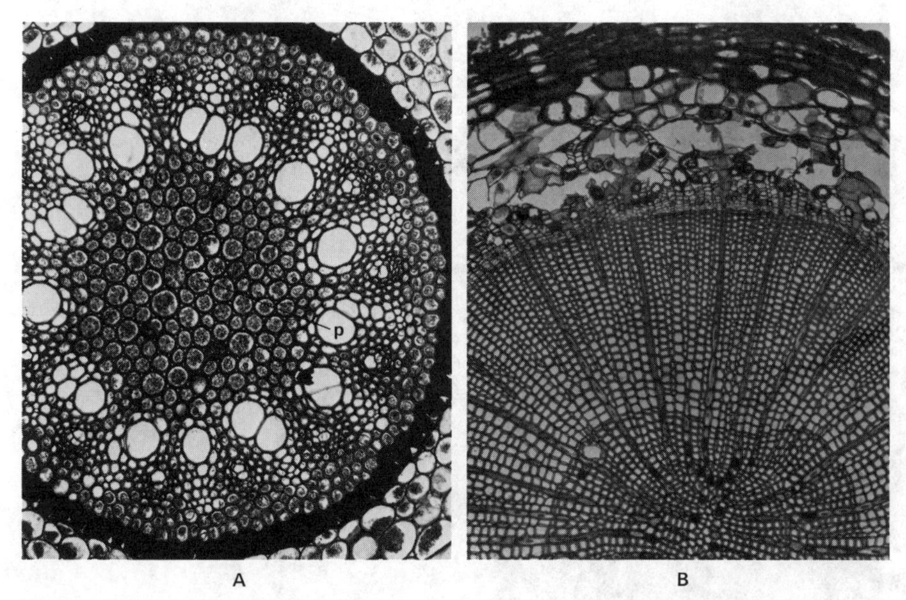


*Figure 8.10*
*A. Smilax* root cross-section. Note central core of pith (*p*) with xylem radiating outward in many arms. *B. Picea* root cross-section.

## Primary Xylem

The *primary xylem* of the root consists of one to many arcs of thick-walled, lignified cells. In dicots and gymnosperms there are usually one to five such arcs, while monocots usually have seven to many arcs of xylem. The xylem may form a solid central core (most dicots and gymnosperms) or may surround a core of pith (most monocots (Figure 8.9 and 8.10).

The xylem is a complex tissue and may be composed of as many as five different cell types. These include vessel members, tracheids, fibers, parenchyma, and ray cells. The first two cell types are associated with the conduction of water and minerals. Vessel members differentiate normally end to end, with the end wall being digested away forming continuous tubes, the *vessels*. Tracheids, vessel members, and fibers are all characterized by thin areas in the wall called *pits* (see page 53). The pits in fibers are usually either small or lacking, while those of vessel members and tracheids may be variable in form. The walls of the vessel may also contain characteristic thickening of the secondary wall. Not all plants contain all the different types of xylem cells. Vessels are found in nearly all angiosperms but are lacking in most other vascular plants. The types of cells found in different organs of the same plant may also differ.

Primary xylem elements may belong either to the *protoxylem* or to the *metaxylem*. The prefix *proto* means "first," *meta* "After" or "behind;" thus,

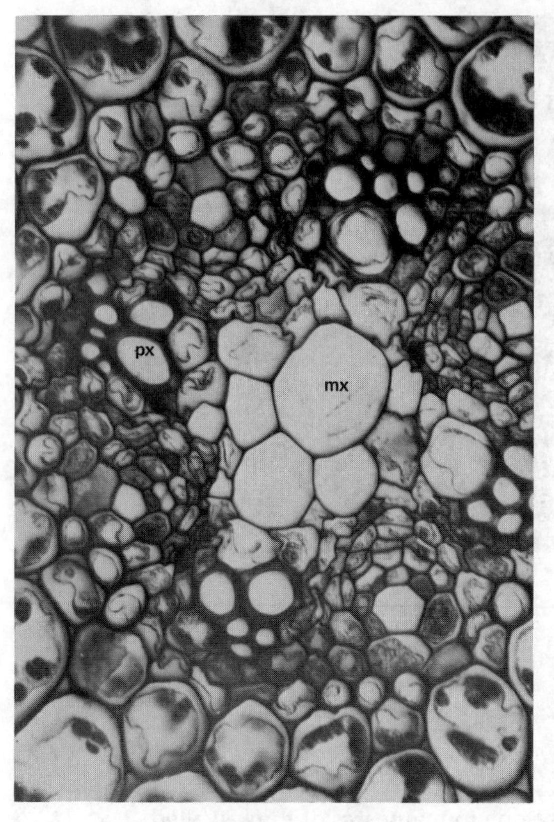


*Figure 8.11*
*Ranunculus* root cross-section. Note thick-walled protoxylem *(px)* and thin-walled metaxylem *(mx)*. This is a young root and given time the metaxylem cells would also have become thick-walled, as in Fig. 8.9.

one might logically expect that the protoxylem is the first-formed xylem with metaxylem being formed later. This is not so. In many plants, the cells are laid down at essentially the same time, with the first vascular elements to become distinguishable being the metaxylem vessel elements. The protoxylem cells, however, are the first elements to become lignified (Figure 8.11).

## *Primary Phloem*

Located between the arms of the xylem are clusters of *primary phloem* cells (Figure 8.12). The phloem may consist of several cell types and therefore is also a complex tissue. These types of cells are sieve elements or sieve cells, companion cells, parenchyma, fibers, and ray cells. The sieve elements are of two kinds, *sieve tube cells,* common to angiosperms, and *sieve cells* which occur in other vascular plants. These two constitute the cells through which organic substances are conducted. They are quite remarkable in that at maturity many have no visible nucleus. This aspect has been discussed earlier, and recent evidence indicates the possibility of

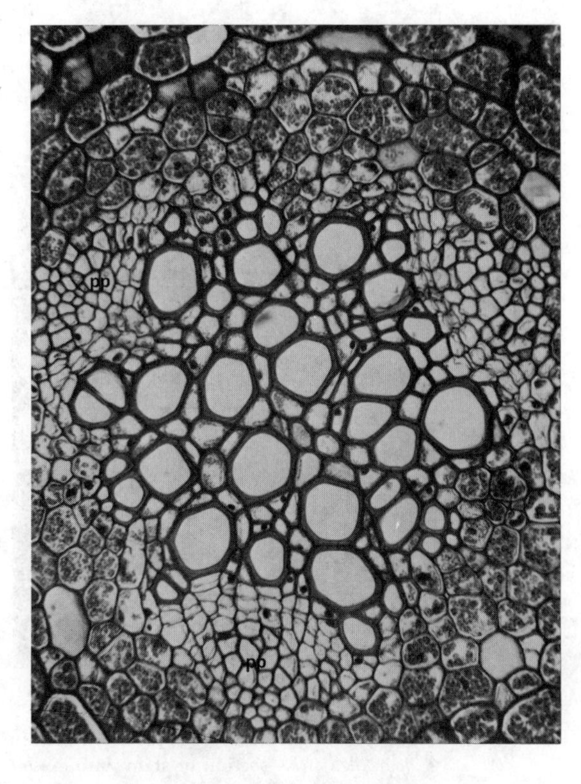


*Figure 8.12*
*Actaea alba* root cross-section. Note clusters of primary phloem *(pp)* between the primary xylem arms.

very diffuse nuclei in some plants. Sieve tube cells differ from sieve cells in that they are arranged end to end, usually are not tapered, and have a region of pores called a *sieve plate* between one cell and the next. Sieve plates may be simple or compound and connecting strands may pass through the sieve plate pores from one cell into the next (Figure 8.13). In most angiosperms, companion cells are found adjacent to the sieve tube cell and, in fact, form from the same mother cell (Figure 8.14). Companion cells are nucleated and it was thought that they somehow controlled the function of the sieve tube cells. Although companion cells do probably form some sort of a complex function system with the sieve tube cells for transport of solutes, it certainly has not been proven they are necessary to such transport. Certainly sieve cells are quite capable of transport without having associated companion cells, as companion cells are typically lacking in vascular plants other than angiosperms.

## Stelar Cambium

The *stelar*, or *vascular*, *cambium* forms between the xylem and phloem and is thus lateral in position (Figure 8.15). At first, the cambium is simply an arc of cells but may subsequently join up with cells of the pericycle adjacent to the protoxylem to form a continuous tissue. The stelar cambium produces secondary xylem toward the inside, secondary phloem toward the outside. Monocots do not normally develop a stelar cambium.

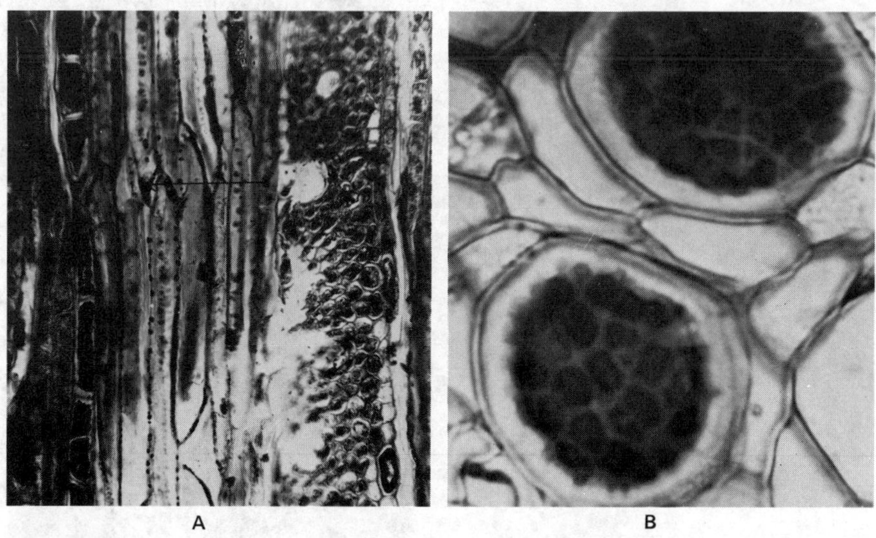


*Figure 8.13*
Sieve elements. *A. Parthenocisscus* longisection of stem showing sieve plate *(arrow)*. *B. Ibervillea* cross-section of stem with evident sieve plate.

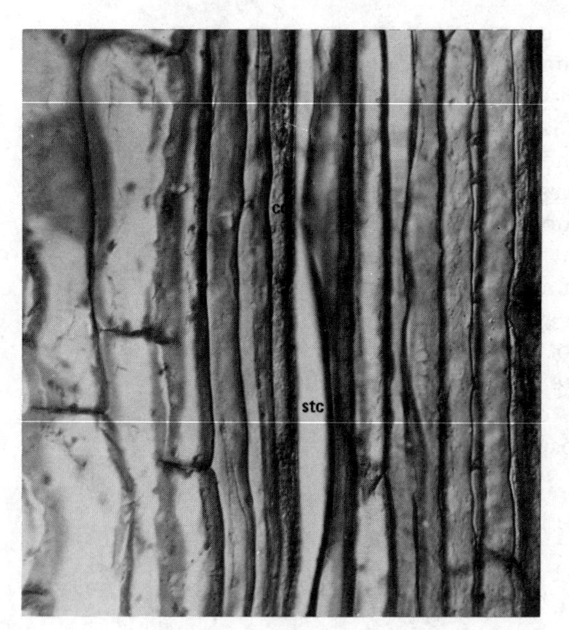


*Figure 8.14*
Note sieve tube cell *(stc)* with slime plug and smaller nucleated companion cell *(cc)* adjacent to it.

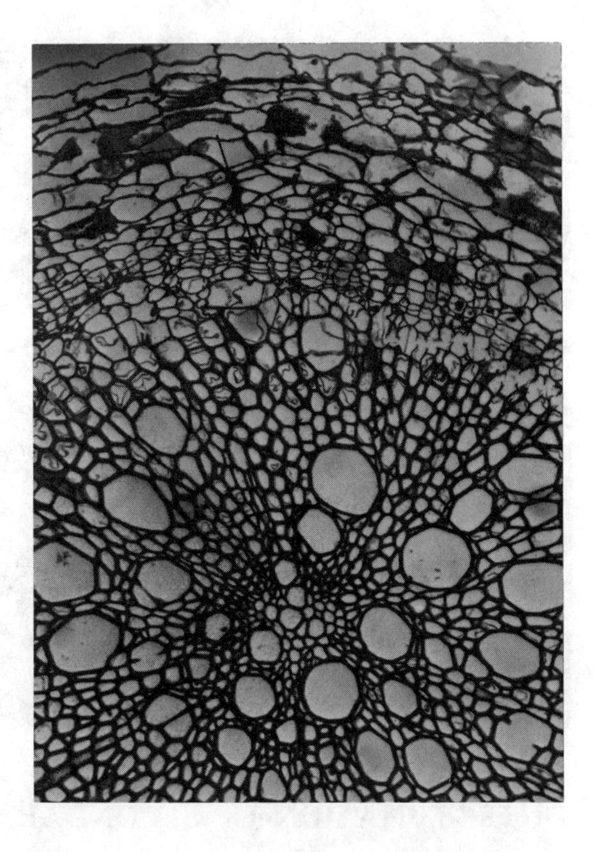

*Figure 8.15*
*Nicotiana* root cross-section. Note vascular or stelar cambium *(arrow)*. Secondary xylem is produced to the inside, secondary phloem to the outside.

## Cork Cambium

The *cork cambium*, or *phellogen*, of roots is produced frequently from cells of the pericycle, although it may differentiate in outer tissue layers, principally the hypodermis and subcortical layers. The cork cambium normally produces the bulk of its derivatives on the outside; these form the *cork*. Later, derivatives may be produced on the inside forming *phelloderm* (Figure 8.16).

## Pith

*Pith* is common to most monocot roots as well as to a few dicot roots. Unlike pith of the stem, which arises from the same meristematic region as the cortex, pith of the root arises from the procambium. Although parenchyma typically, it may also be sclerenchymized in some plants (Figure 8.10A).

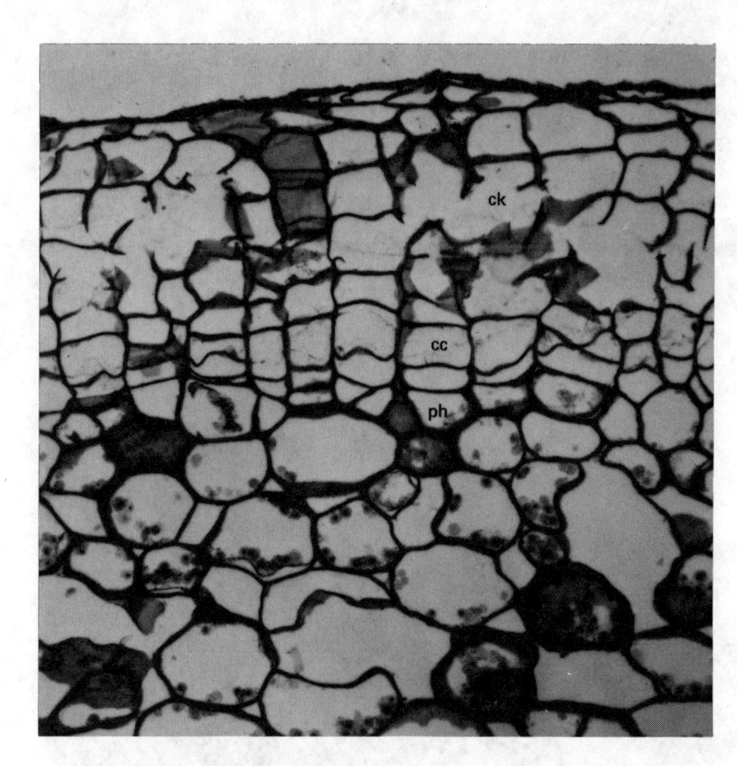


*Figure 8.16*
*Liriodendrom* root cross-section. Note cork cambium *(cc)*, cork *(ck)* and single layer of phelloderm *(ph)*.

## DIFFERENTIATION OF SECONDARY TISSUES

Mature woody plants, both angiosperm and gymnosperm, may show considerable secondary growth in their roots. In fact, the annual rings of growth associated normally with stems can also be found in roots. Roots several inches in diameter are not uncommon. Most herbaceous dicots undergo a more limited secondary growth, while some may undergo little or none. In most monocots, secondary growth does not occur, but some, even though not forming a stelar cambium, may produce a cork cambium, although cork production is normally quite limited.

As both the stelar and cork cambia are lateral in position, they contribute to growth in diameter of the root. We can recognize, generally, secondary tissue by its alignment in radial tiers from a typically narrow, rectangular, flat-sided cell having a relatively large nucleus: the cambial cell. Normally, much more secondary xylem is produced than secondary phloem. If considerable secondary growth occurs, all the outermost tissues of the root may be sloughed off.

Cork cells formed towards the outside by the cork cambium are usually short lived and soon become thick walled and suberized. The phelloderm, when produced, is less extensive than the cork and composed of parenchymatous cells. The phelloderm is sometimes referred to as the *secondary cortex*.

*Figure 8.17*
Anomalous secondary growth in the Baobab tree. Baobab trees are found in Australia only in the far North, such as the Kimberley district at the top of Western Australia. Another type of Baobab occurs widely in Africa, and is cultivated ornamentally in India and Ceylon. Madagascar has nine separate types. Theories on how the species came to Australia include continental drift, but the most popular is that the trees may have traveled as seeds on the ocean currents. [Australian Information Service. Photograph by Don Edwards. Courtesy of the Australian News and Information Bureau].

### *Anomalous Secondary Growth*

In some roots secondary growth is not of the usual type. In other words it is *anomalous* (Figure 8.17). In the carrot, the pattern of secondary growth is normal, but an unusually large amount of xylem and phloem parenchyma is formed. In the beet, the stelar cambium differentiates as usual. But then, additional cambia are formed from the pericycle and phloem outside the normal vascular cylinder. These cambia are arranged concentrically and are called *primary thickening meristem* (ptm). The ptm produces considerable parenchyma tissue, called *conjunctive tissue*, in which groups or strands of xylem and phloem differentiate. These strands are termed *desmogen strands.* Still another type of anomalous growth is found in roots of the sweet potato *(Ipomea batatas).* The xylem, which arises in the usual manner, contains a large amount of parenchyma. Accessory cambia, called *interxylery cambia*, form in the parenchyma around tracheary elements. These cambia produce some xylem toward the vessels, some phloem toward the outside and a considerable amount of parenchyma in both directions. It is this anomalous growth that accounts for the increased diameter of carrots, beets, radishes, sweet potatoes, turnips, and other roots.

## ION ABSORPTION BY ROOTS

The absorption of water and minerals from the soil occurs basically through cells of the root. The mechanism of water absorption, however, is quite different from that of solute (dissolved substance) absorption, and different solute particles are absorbed independently of water and of each other.

Minerals may be absorbed either as molecules or as ions. It has been rather clearly shown that molecules can be absorbed much more readily than ions, and the greater the degree of ionization, the less readily the substance is absorbed. But both ions and molecules are absorbed into root cells. This will occur regardless of the concentration of a substance inside the cell as opposed to its concentration outside the cell. How do solutes enter root cells?

It has been suggested that, in general, two systems must be considered: one associated with the plasmalemma and active at low concentration of a particular solute particle in the soil; and a second associated with the tonoplast and active at high concentration of a particular solute particle in the soil. The latter mechanism would entail the absorption of ions into the vacuole through the tonoplast, thus lowering the concentration of that substance in the cytoplasm. If this substance is at a high concentration in the soil, it could then move through the plasmalemma and into the cytoplasm by simple *diffusion.* We will be discussing the phenomena of diffusion in more detail later on, but very simply, it can be described as the *net* movement of a substance *from a region of its higher concentration to a region of its lower concentration.* The process is dependent upon the kinetic energy of molecules. Thus, no cellular energy would be expended in the process. This mechanism would function best for molecules possessing little or no charge, as the plasmalemma, which bears charges, is not readily permeated by highly charged ions or molecules, regardless of the concentration gradient.

How can ions move through the plasmalemma? One possible mechanism has been termed *exchange diffusion.* This hypothesis suggests that ions bound to the plasmalemma might be exchanged with other, very similar ions in the soil. Once bound to the membrane, the ions move through it simply by their own kinetic energy. A second hypothesis, called *facilitated diffusion*, postulates that certain molecules in the plasmalemma may have an affinity for certain solute particles and thus form a complex with them. Such membrane substances are termed *carriers.* The carrier-ion complex then moves through the membrane by its own kinetic energy with the dissociation of the complex on the internal side of the membrane. As you can see, neither of these mechanisms would involve the expenditure of energy by the cell. A third hypothesis concerning the absorption of ions into the root against a concentration or diffusion gradient is based upon the following observations: 1) the absorption depends upon a carbohydrate supply to the root from the shoot, 2) it is sensitive to inhibition of metabolism, 3) it will not occur under anaerobic conditions, and 4) it is almost uninfluenced by variation in transpiration rate from the shoot. (You will better understand the significance of the latter observation when we discuss transpiration.) These observations support two assumptions: 1) this absorption process involves the expenditure of metabolic energy, and 2) the rate determining step is located, not in the epidermis, but in tissues between the epidermis and the xylem. This mechanism of transport has been called *metabolically linked transport.* Metabolically linked, or active transport, has also been postulated to occur through the mediation of a carrier but at the expenditure of ATP.

All these mechanisms postulate the movement of ions from the soil to the xylem. As cells abut each other in the formation of tissues, two pathways to the stele could be postulated: one occurring through the so-called free space (the middle lamellae, cell walls and intercellular spaces), the second from cell to cell through the plasmalemma (Figure 8.18). According to the concept of free space, water and ions would move through the free space to the endodermis, at which point, due to the Casparian thickenings, the water and ions would be forced to move through the plasmalemma of the endodermal cells and thence from cell to cell until reaching the xylem tracheary elements. Many botanists, however, do not

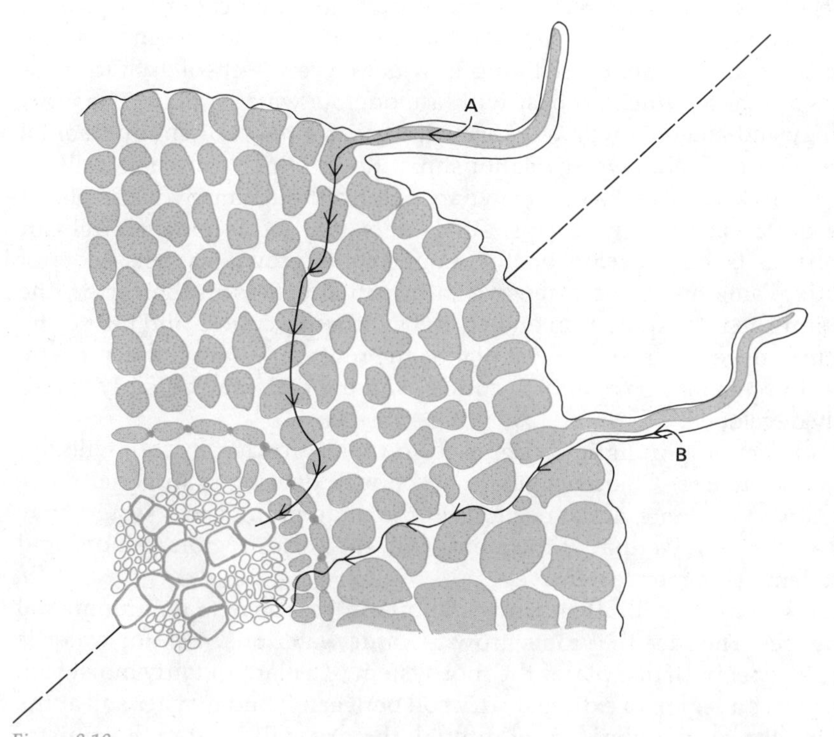


*Figure 8.18*
Diagrammatic cross-section of a root illustrating two possible mechanisms for lateral movement of ions and water. Pathway A involves osmotic activity of diffusion of water and/or ions from cell to cell across the plasmalemmas. Pathway B involves the diffusion of both water and minerals through the free space. In both instances it is postulated that when the endodermis is reached, the substances must pass through the plasmalemma and cytoplasm of these cells before reaching the stele.

agree with the concept of free space, while others believe that water and minerals need not reach the xylem but could move upward through the intercellular spaces of the cortex.

How important are root hairs to this process of absorption? There have been many conflicting views. Ion absorption has been clearly shown to occur below the root hair zone, while water absorption occurs primarily in the root hair zone. Of course, some plants do not form root hairs, yet water and mineral absorption does occur. I suppose all the conflicting data could be summarized by stating that when soil water is not limited, root hairs are relatively unimportant in absorption, but when soil water is limited and thus bound more tightly to soil particles, the root hairs do appear to play a significant role in uptake.

## ENVIRONMENTAL EFFECT ON ROOT DEVELOPMENT

Plants are classified in a variety of ways, but one classification is based on water relationships. *Xerophytes* are those plants able to live in very dry places, *hydrophytes* are those living in water or very wet soil, while *mesophytes* are those which do best with a moderate water supply. The morphology and anatomy will, as expected, vary as a result of environmental influences, including water relationships. Roots which grow normally in soil may lack a root cap when grown in water, although many water plants have quite extensive root caps. Root hairs may be lacking entirely in aquatic roots but develop on the same plants grown in soil. Roots of aquatic plants may have extensive intracellular spaces in the cortex, the epidermis may be quite thin and entirely without a cuticle, and little or no cork may develop. In xerophytic plants, the roots may have few air spaces, the epidermis may have a thick cuticle, and the cork may be fairly extensively developed.

Roots growing in light, as opposed to those growing in soil, will also show differences. Chlorenchyma may be well developed in aerial roots, and in woody plants, aerial portions of the underground root system may look very much like that of the stem, with the cortex region reduced and the xylem being more extensive.

Roots indicate still other growth habits in response to environmental influences. The idea that roots grow towards water or search for water is quite erroneous. If one places the root system of a plant in fairly moist soil, then places a region of extremely dry soil between it and moister soil at the bottom, the roots will not grow through the dry soil to get to the moister soil (Figure 8.19). But when a moisture gradient exists, a portion of the root system will extend from drier regions into more moist regions simply because roots grow better in moist soil. The growth of roots down, out, and into the soil, is of course due to apical or subapical growth associated with the meristem. The root cap cells will be sloughed off by this apical movement in the soil and replaced continuously under appropriate growing

*Figure 8.19*
Effects of dry soil (lighter color) on the growth of roots.

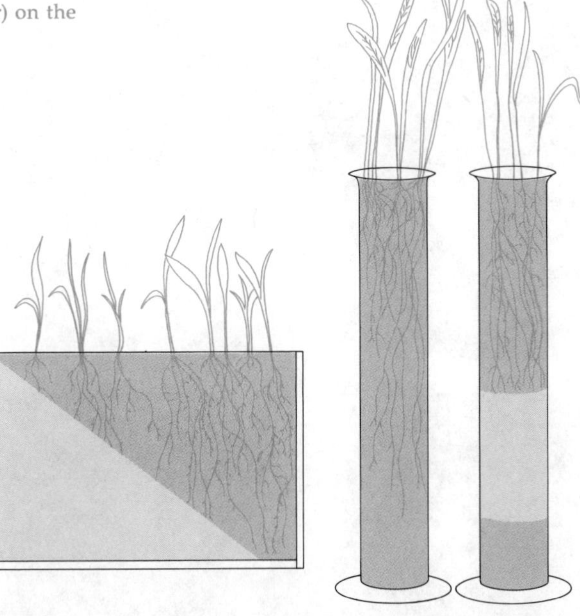

conditions by cells produced by the calyptrogen. There is absolutely no evidence that the root cap cells die when they reach a source of "nourishment" as stated in *The Secret Life of Plants.* Plant nourishment is not obtained from the soil, but is produced in the green portions of the shoot and translocated downward into the root. As mentioned earlier, minerals may be absorbed from the soil but these are not to be confused with nourishment, that is, food. Nor do root cap cells die simply because of the presence of minerals, unless they are in excessive amounts. Then, the whole root may die.

## ADDITIONAL CONSIDERATIONS OF ROOTS

For those of you that would like to consider roots in somewhat more detail, there are two general areas that bear more discussion. The first involves the ontogeny of tissues in roots, and the second, consideration of various types of steles.

### Ontogeny of Tissues

According to the theory of Hanstein as modified by Janczewski and Treub, four specific histogens have been determined. Each may be one to several cell layers thick. These have been termed the *calyptrogen*, *dermatogen*, *periblem* and *plerome* (Figure 8.20). These histogens comprise a zone of the api-

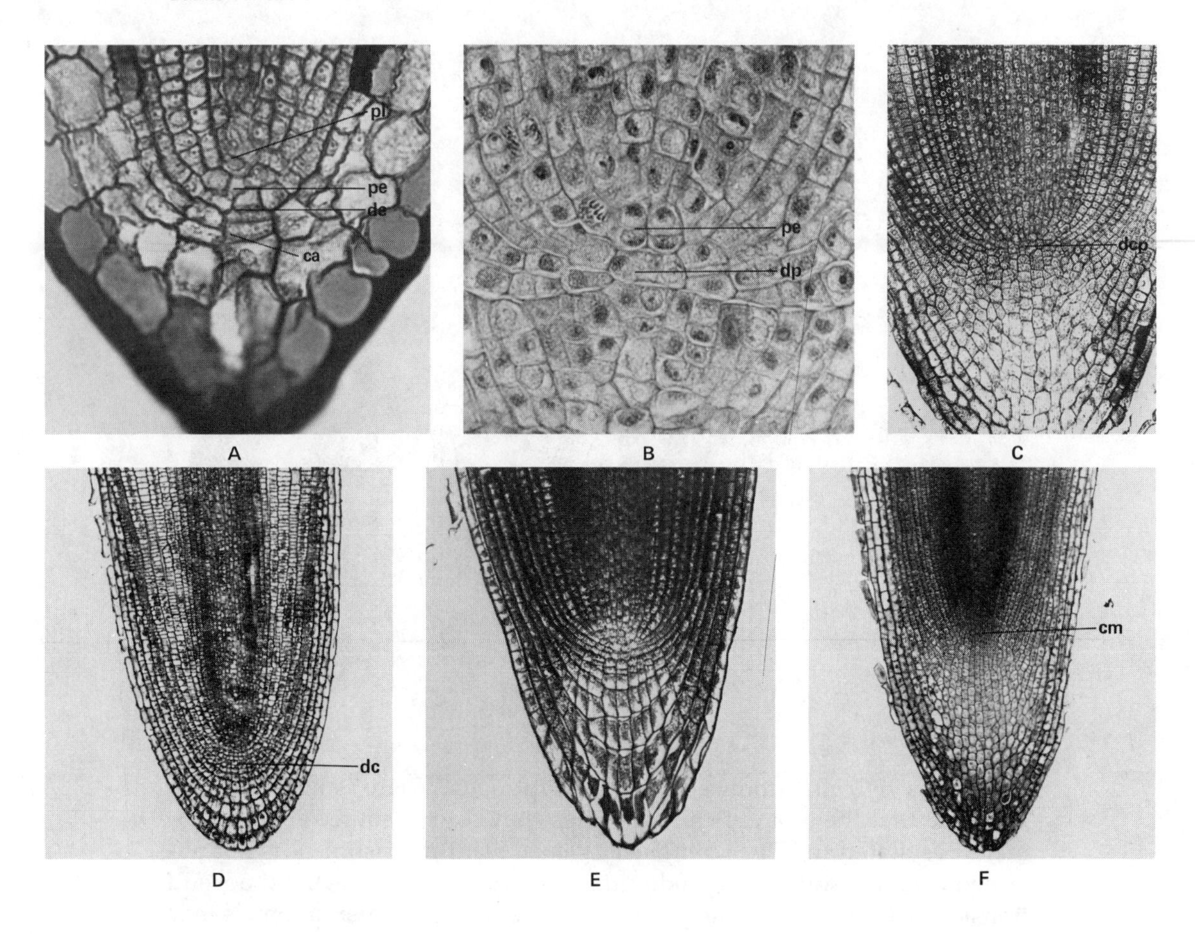


cal meristem called the *promeristem*. The histogens produce, respectively, the root cap, epidermis, cortex, and stele (central vascular cylinder). They may also combine in various fashions. For example, a single cell layer termed the dermatocalyptrogen would produce both the root cap and the epidermis. In fact, a root apex consisting of four individual histogens is rare, being found primarily in just a few lateral and adventitious roots. Engard (1944) formulated a developmental concept of zonal organization which consists of three zones: 1) the *promeristem* which, as mentioned above, is the zone composed of histogens giving rise to 2) the *fundamental tissues*, a zone of meristems with dividing, enlarging and elongating cells, which in turn produce 3) the *primary permanent tissues*, which are the physiologically mature tissues. Within zone 1 are a. the *plerome*, which produces within zone 2 the *procambium*, which gives rise in zone 3 to the *primary xylem*, *primary phloem*, *pericycle*, *stelar cambium* (in some plants), and *pith* (in some plants); b. the *periblem* which produces the *meatic meristem* within zone 2, which in turn produces the *cortex* (including *endodermis*,

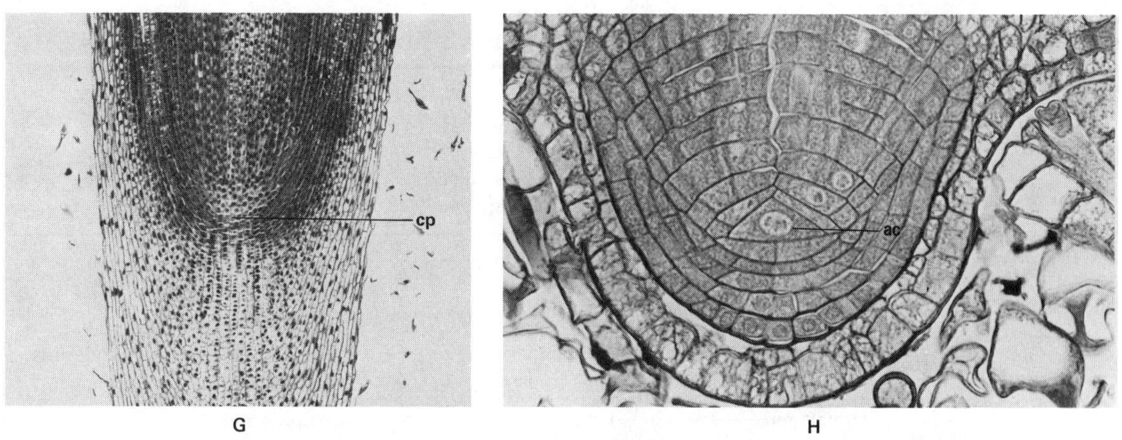


*Figure 8.20*
Longisections of root tips to illustrate root promeristem organization according to Janczewski (1874) and others. *(Opposite page.) A. Pistia*—Type I; *pl*, plerome; *pe*, periblem; *de*, dermatogen; *ca*, calyptrogen. *B. Hordeum*—Type II; *dp*, dermato-periblem. *C. Pyrus*—Type IIA; *dcp*, dermato-calyptro-periblem. *D. Raphanus*—Type III; *dc*, dermato-calyptrogen. *E. Linum*—Type IIIA; note that periblem consists of two layers, one which gives rise to hypodermis only, the second which gives rise to the rest of the cortex. *F. Curcubita*—Type IV; *cm*, central meristem. *(Above.)* *G. Pinus*—Type V; *cp*, calyptra-periblem. Note that the dermatogen later is absent, thus no epidermis is formed on pine roots and some other gymnosperms. *H. Marsilea*—Type VI; *ac*, apical cells. No histogens are present in the roots of many pteridiophytes.

*cortical parenchyma*, and *hypodermis*) within zone 3; c. the *dermatogen* which produces the *protoderm* in zone 2 which in turn produces the *epidermis* in zone 3; and d. the *calypotrogen* which produces in zone 2 the *calyptra* which in turn gives rise in zone 3 to the *root cap* (Table 8.1). The above tissues, derived from the apical meristem, are called *primary tissues*.

Although the region of the promeristem and fundamental tissues is one of the active cell division, Jensen, Clowes, Jensen and Karaljian, Pillai and Byrne, and Heimsch have shown that the cells of the central region of the promeristem have a very low mitotic activity. This region is called the *quiescent center*. The name is somewhat a misnomer as it suggests total lack of activity. This is not so. The initials situated on the periphery of the quiescent center become further distant from the center during development, and their former position is taken by new initials which must have been produced in the quiescent center. In addition, the quiescent center may be the site of hormone synthesis and may be important in maintaining the geometry of the apical meristem.

Table 8.1
**A DEVELOPMENTAL CONCEPT OF ZONAL ORGANIZATION OF ROOT TIPS***

| | | | |
|---|---|---|---|
| ZONES | Promeristem (a zone composed of histogens) | →Fundamental tissues (a zone composed of meristems; cells enlarging, elongating, and dividing) | →Primary permanent tissues (a zone composed of physiologically mature tissues) |
| ORGANIZATION WITHIN THE ZONES | Plerome | →Procambium | →Primary xylem, stelar cambium, phloem, pericycle, and pith (if any) |
| | Periblem | →Meatic meristem | →Cortex, including endodermis and hypodermis |
| | Dermatogen | →Protoderm | →Epidermis |
| | Calyptrogen | →Calyptra | →Root cap |

*Source:* After Popham, *Laboratory Manual for Plant Anatomy*, St. Louis: Mosby Co.
*Derived from Engard (1944).
†The example shown is for root tips exhibiting Janczewski's type I promeristem organization.

## The Concept of Stele

DeBary (1884) and other early plant anatomists considered the individual vascular strand the primary unit of vascular construction. Later, the continuity of the vascular system came to be emphasized and, in 1886, Van Tieghem and Douliot recognized the vascular system and associated fundamental tissue to be a unit, regardless of the arrangement of those tissues. This unit was termed the *stele*, from the Greek word meaning column. From this interpretation, the stelar theory arose, postulating that the primary bodies of stem and root are essentially alike, that both consist of a core (the stele) surrounded by cortex. Because of structural variation within this core, various stele types came to be recognized.

Although most plant anatomists now accept the concept of stele, the phylogeny of the system and the classification and interpretation of its types are still open to question. In general, the simplest form of stele, and that recognized as being the most primitive phylogenetically, is one containing a solid column of vascular tissue with no enclosed pith. This is called a *protostele* (termed an *axile bundle* or *monostele* by some). A protostele in which the xylem is a smooth column surrounded by phloem is termed a *haplostele* (Figure 8.21A), characteristic of the fossil plant *Rhynia* and extant plants such as *Selaginella kraussiana* and certain fern rhizomes such as those of *Hymenophyllum polyanthus*, *Gleichenia* (some species), and the stems of *Lygodium palmatum*. An *actinostele* (Figure 8.21B) is a type of protostele characteristized by fluted or stellate xylem (in cross section) and is characteristic of the roots of most dictoyledonous species, certain portions of the rhizome and aerial stem of *Psilotum*, and of the fossil plant *Asteroxylon*. The third type of protostele is the *plectostele* in which xylem

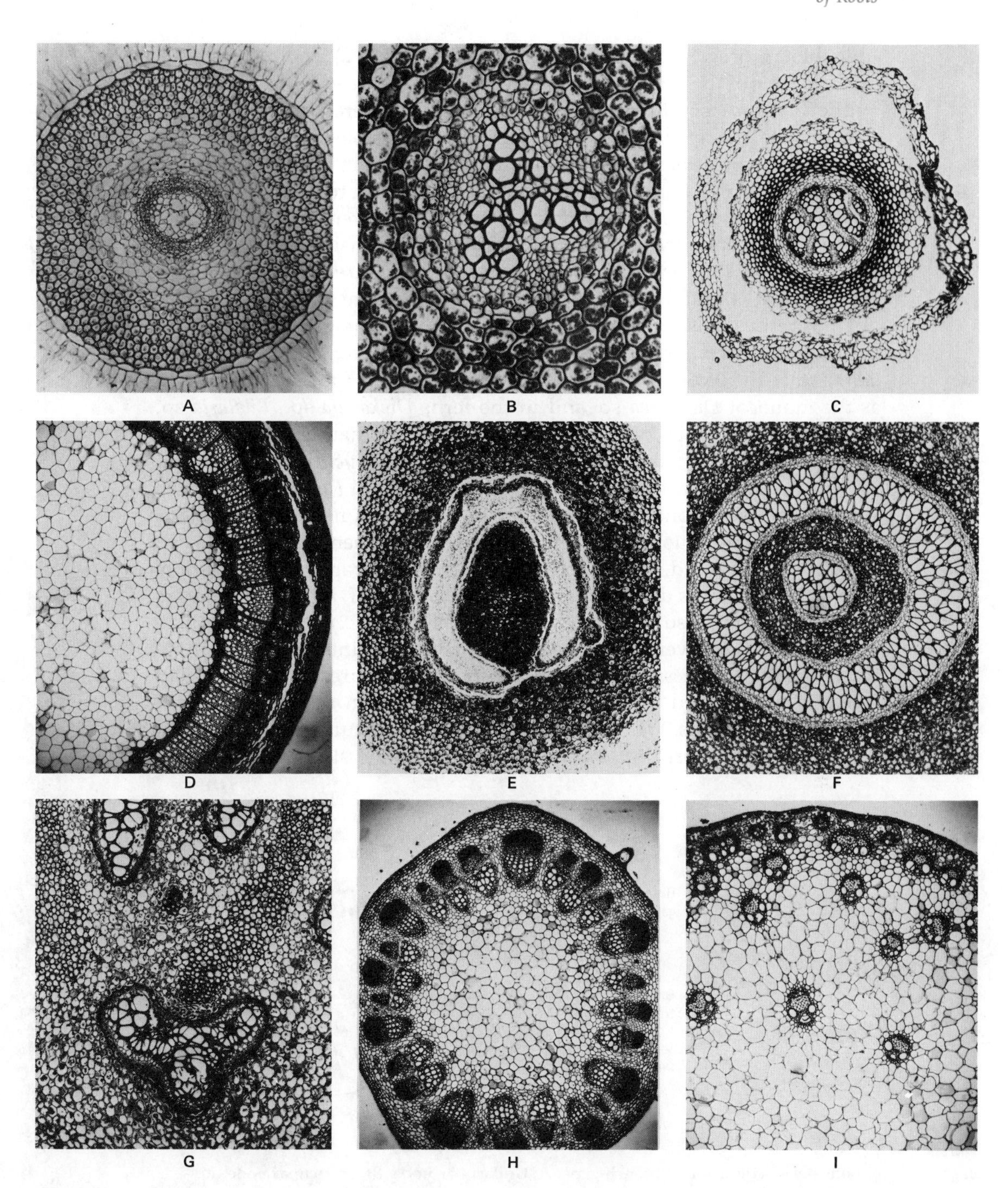


*Figure 8.21*
Stele types. *A. Selaginella kraussiana*, haplostele. *B. Actaea alba*, actinostele. *C. Lycopodium*, plectostele. *D. Liriodendron*, ectophloic siphonostele. *E. Adiantum*, amphiphloic siphonostele. *F. Matonia*, polycyclic. *G. Pteridium*, dictyostele. *H. Helianthus*, eustele. *I. Zea*, atactostele.

and phloem occur in alternating plates (Figure 8.21C) The plectostele is characteristic of the rhizomes of most *Lycopodium* species.

The presence of pith in the central column differentiates the second form of stele, the *siphonostele.* Two types of siphonosteles are distinguished based on the position of the phloem and xylem: 1) an *ectophloic siphonostele* in which phloem occurs external to xylem (Figure 8.21D), and 2) an *amphiphloic siphonostele* in which phloem is both external and internal to xylem and the endodermis occurs both outside and inside the vascular tissue, bordering the cortex and pith, respectively (Figure 8.21E). The former type of stele is found in all but the terminal portions of most dicot and gymnosperm stems, in the stems and rhizomes of some ferns such as *Botrychia* spp., *Helminthostachys* spp., *Gleichenia pectinata,* and *G. microphyllum,* and in the lycopod *Isoetes* spp. The latter type is found in plants such as the monocot *Dioscorea* sp. and in the ferns *Dicksonia* sp., *Pilalaria* sp., and *Adiantum* sp. If phloem and xylem occur in alternating concentric cylinders (Figure 8.21F), the stele type is termed a *polycyclic siphonostele* and is characteristic of such ferns as *Matonia pectinata* and *Marsilea quadrifolia.* A more complex siphonostele is one consisting of a network of bundles rather than a continuous cylinder. The regions between these bundles are parenchymatous and, when occurring in the stele above the positions where the leaf traces diverge from the stele to the leaf, are called *leaf gaps. Branch gaps* may also interrupt the stele (Figure 8.22). An amphiphloic siphonostele with overlapping gaps and the bundles interconnected in the internodes *(anastomosing),* is called a *dictyostele* (Figure 8.21G). The dictyostele is often called the fern type as it is found in the stems and rhizomes of many ferns. Where the vascular bundles occur in one circle, as in the stem tips of most dicots and gymnosperms, the stele type is termed

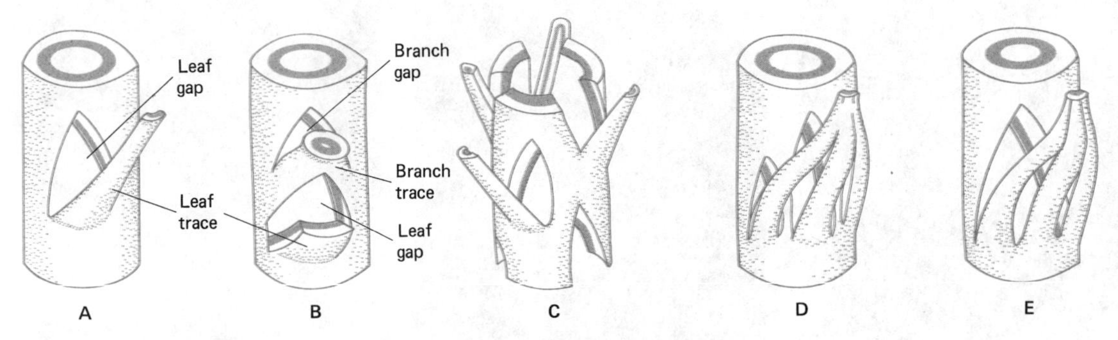


*Figure 8.22*
Diagrams of siphonosteles with leaf and branch gaps. *A.* Unilacunar node. *B.* Unilacunar node with associated branch trace and gap. *C.* Overtopping gap; stele forms a network of bundles. *D.* Trilacunar node with three leaf traces. *E.* Unilacunar node with three traces. [From A. Fahn, *Plant Anatomy,* 2d ed., Pergamon Press, New York, 1974.]

an *eustele* (Figure 8.21H). If the vascular bundles are scattered, as in most monocot stems, the stele is called an *atactostele* (Figure 8.21I). The concept of stele is one which emphasizes essentially the unity between stem, root, and leaf.

## SUMMARY

1. Roots are formed by most vascular plants and may be primary, secondary, or adventitious.
2. Root systems may be either fibrous or tap.
3. Roots differ externally from stems in lacking nodes and internodes.
4. The root apical meristem is actually subapical, being covered by a root cap.
5. Many plants have a region of epidermal outgrowths, called root hairs, which develop relatively close to the tip. These structures are not found normally in older regions of the root. They may or may not be significant in water and mineral absorption.
6. The root apical meristem may be thought of as consisting of discrete regions called histogens, each producing derivatives to specific tissue regions.
7. The endodermis is the innermost region of the cortex and is characterized by development of a suberized band in the radial and transverse walls, the Casparian strip.
8. The pericycle is the outermost cell layer in the stele and may produce lateral roots, cork cambium, and stelar cambium (in part).
9. In dicots and gymnosperms the primary xylem forms typically the central core of the stele and in cross section has anywhere from one to five radiating arms.
10. In most monocots, pith occupies the center of the stele with the primary xylem having from seven to many radiating arms.
11. Primary phloem arises generally in discrete groups between the arms of the xylem.
12. The stelar cambium, when present, may form secondary xylem and phloem, while a second lateral cambium, the cork cambium, may develop and produce cork and phelloderm. Monocots lack generally a stelar cambium and have little or no secondary growth.
13. A differential absorption of water and minerals occurs, with the substances passing through both the free space of the root and the plasmalemma of root cells. The process of absorption may occur with or without the expenditure of metabolic energy and with or without the mediation of carrier molecules.

# Chapter 9

## THE VASCULAR PLANTS— II. STEMS

We think generally of a stem as the cylindrical, upright, aboveground portion of the plant. But stems may not be cylindrical; they may be horizontal rather than upright, and they may be found underground. If we can not recognize a stem on these bases, how then can we recognize a stem? Generally, leaves and buds will be present at specific locations along the stem. The region at which these two lateral appendages arise is called the *node*, the area between two nodes is called the *internode*. Although there are many other distinctions between stem and root, or stem and leaf, it is basically this aspect of nodes and internodes which characterizes a stem.

## STEM MORPHOLOGY

Both stem morphology and stem anatomy differ among the various groups of plants. Thus, we could discuss woody versus herbaceous, monocot versus dicot, or angiosperm versus gymnosperm stems. For the purpose of establishing some general concepts of stem morphology, we shall be concerned primarily with woody dicot stems.

The first stem develops from the *embryonic bud*, or plumule, in the embryo. All new stems and leaves (collectively termed *shoots*) develop from *buds*. If the bud is at the terminus of a branch, it is called a *terminal bud*, if in the *axil* of a leaf (the upper angle formed by the attachment of a leaf to the stem), it is called an *axillary* or *lateral bud*. If we were to look at a typical woody stem in winter condition, we might note the characteristics shown in Figure 9.1. The terminal and axillary buds are prominent, and the location of the fallen leaves are noted by the presence of *leaf scars* just below the axillary buds. As certain of the vascular tissues of the stem are continuous with those of the leaf, the *vascular bundle scars* will be evident within the leaf scar. If the leaf bore stipules (leaflike structures extending on either side of the leaf base), *stipule scars* may also be present (Figure 9.2). At intervals, the stem will appear to be encircled by one or more rows of scars (Figure 9.1). These ringlike markings are the scars left by bud scales when they fell off the opening terminal bud. These markings are called *terminal bud scale scars* and may be used as a means of locating the position of the terminal bud of some previous year. Thus, by counting terminal bud scale scars, we can ascertain the age of a particular stem segment. By measuring the distance between two successive terminal bud scale scars, one can compute annual growth rate of different plants during the same year, or variations in the rate of growth of the same plant in different years. Scattered over the surface of the internode are numerous, small, pustular-looking ougrowths called *lenticels*. When young, they are composed of loosely arranged cortical cells and thus might provide for efficient gas exchange between cells of the stem and atmosphere. Later, however, cork develops beneath the lenticel and gas exhange is greatly curtailed. Lenticels may be round, elongated vertically to a greater or lesser degree, or elongated horizontally, depending on the type of plant (Figure 9.3).

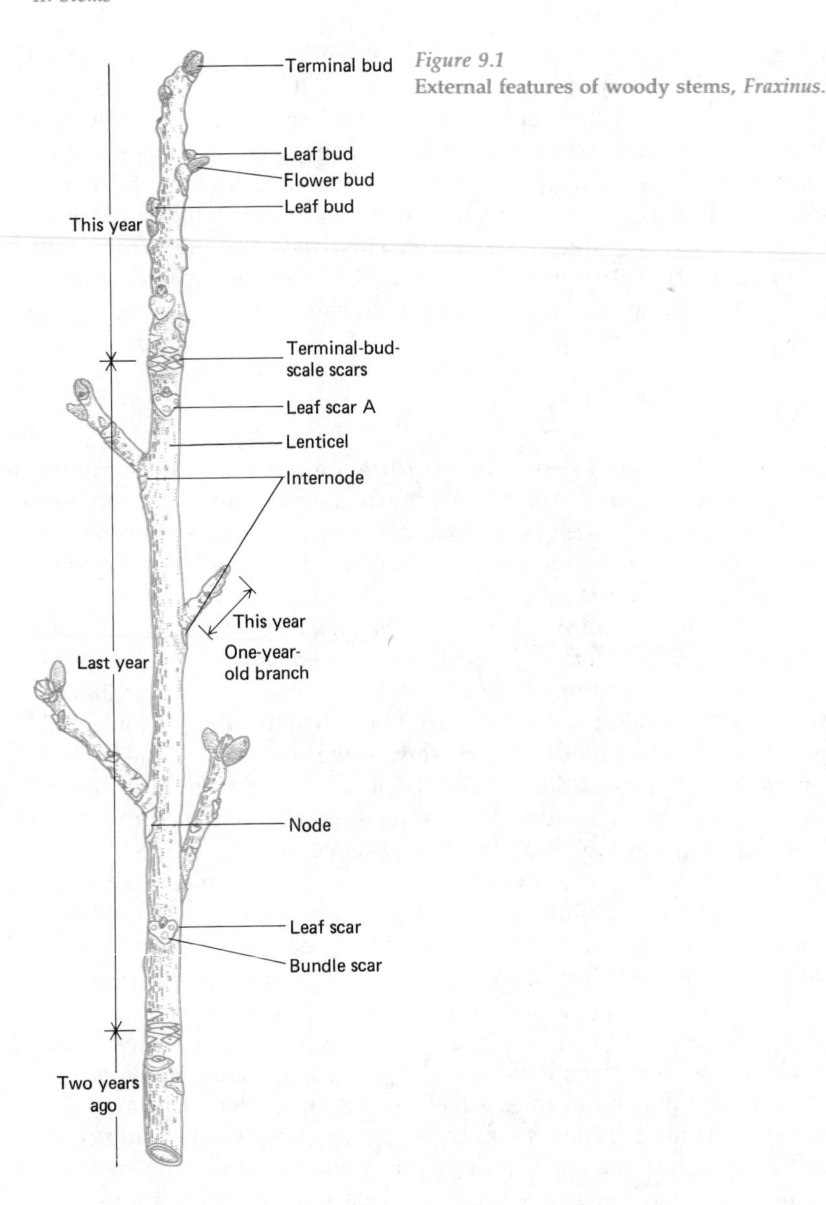


*Figure 9.1*
External features of woody stems, *Fraxinus*.

## Buds

Not all woody plants bear true terminal buds. In many, such as willow or elm, the terminal bud dies and drops off; thus, the last lateral bud formed appears to be terminal in position. Whether terminal or lateral, buds may be classified according to what they produce. Some buds produce only vegetative structures and are called *leaf buds*. Others produce only repro-

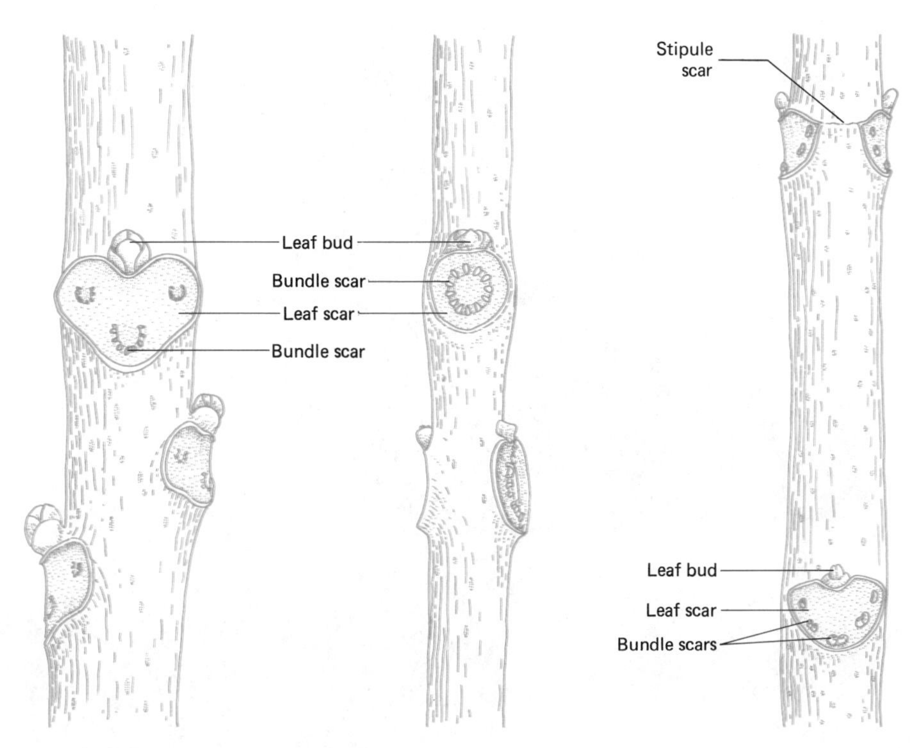


*Figure 9.2*
Leaf scars, bundle scars, and stipule scars on a woody twig.

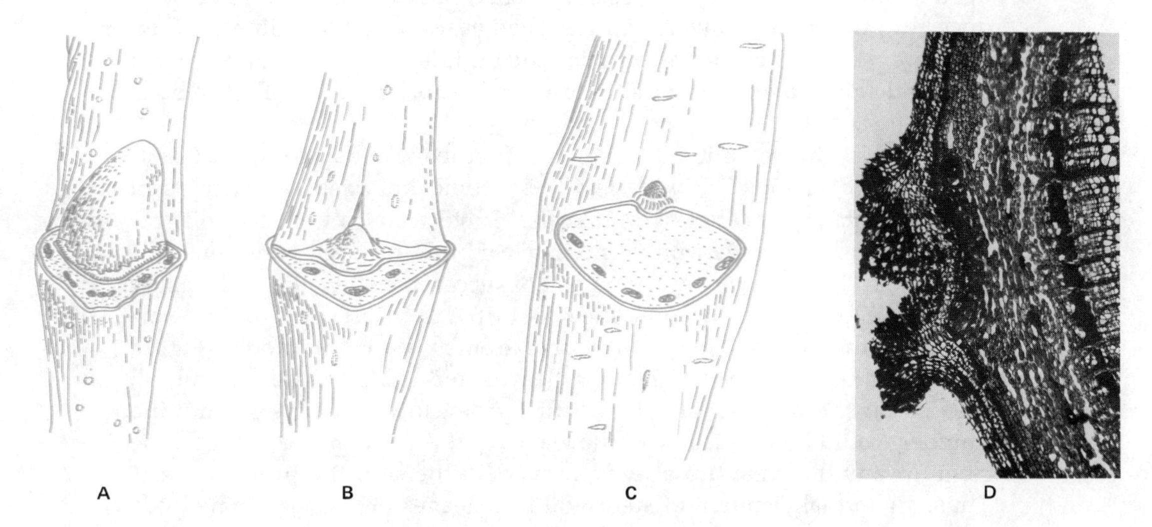


*Figure 9.3*
Lenticel types: *A.* round, *B.* vertical, *C.* horizontal. *D. Sambucus* stem cross-section with lenticel. Note corky layer.

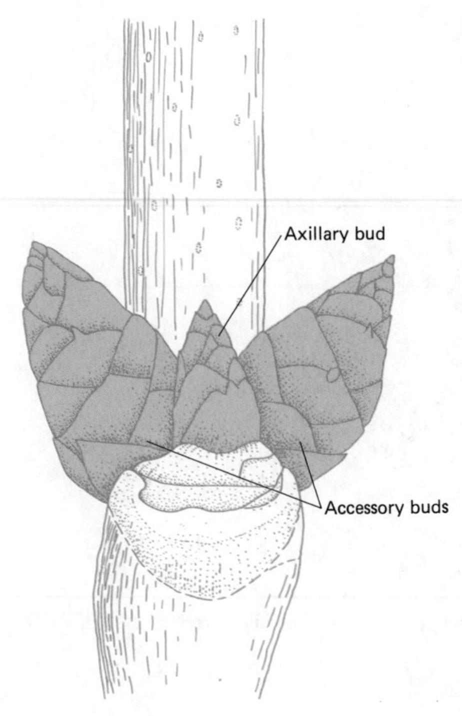


*Figure 9.4*
Axillary bud and accessory buds.

ductive structures, or flowers, and are called *flower buds.* Still others produce both shoots and flowers and are called *mixed buds.* Sometimes, several buds are produced at the same node. The middle, or axillary, bud is a leaf bud, and the buds on either side, called *accessory buds,* are flower buds (Figure 9.4).

Bud scales are considered to be modified leaves and are termed *cataphylls* (*cata,* "against," *phyll,* "leaf") by some botanists. Most buds of woody plants are covered by bud scales. If we were to very carefully remove the bud scales, we would find in a leaf bud, the largest and thus oldest of the rudimentary leaves. Removal of successively smaller and smaller leaves will leave one with only the shoot tip and its attached leaf primordia. Thus, the bud is just a shoot with extremely short internodes (Figure 9.5). If one could count rudimentary leaves and leaf primordia in a bud, then count the number of leaves actually developing in the spring, the number would be the same. All the leaves that are going to appear on the stem for any one year are already formed in the bud the previous year. Thus, an annual segment of stem will bear leaves only once. Some buds will consist only of a stem tip and will not contain leaf primordia. Such is the case with the eyes of potato tubers.

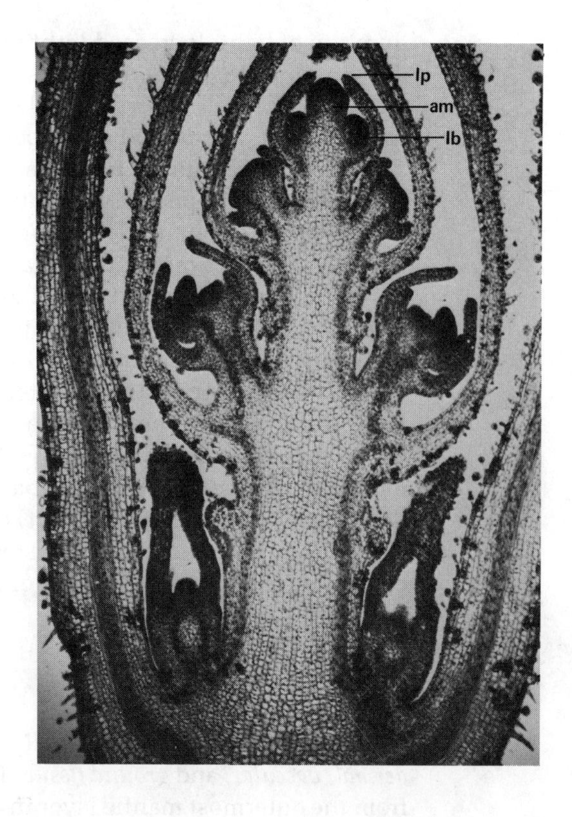


*Figure 9.5*
Longisection of shoot tip of *Salvia*. Note apical meristem region (*am*), leaf primodria (*lp*) and lateral or axillary bud primordia (*lb*).

## STEM DEVELOPMENT

The *shoot apical meristem* develops at the tip of the embryonic shoot, or plumule, in the developing embryo. Unlike the root apical meristem, the shoot apical meristem is truely apical, produces lateral appendages and undergoes considerable changes in shape, size, and rate of growth during the growth of the plant, particularly during the flowering phase. During leaf initiation, cells are displaced from one region of the apex to another. Marked seasonal changes may also occur in the apical meristem, particularly in woody plants. The apex forms a terminal bud and then ceases growth ( becomes *dormant*). In others, the whole shoot tip may abort, or *absciss* (fall off) through formation of an *abscission layer*. The apical meristem may become inactive through formation of tendrils, thorns, or flowers. Once a shoot or lateral outgrowth ceases to grow, they are said to be *determinate*, while those with an actively dividing apical meristem are said to exhibit *indeterminate* growth.

Various environmental factors can affect development of the shoot apical meristem. The most profound effects involve either the intensity or duration of light. Increased light intensity may cause an increase in apical growth in some plants, in others total darkness may enhance apical elongation, while still others will show no effect to either increased or decreased light. The period of exposure to light and darkness (the *photoperiod*) may induce a plant to flower, and for flower induction, the required length of the light period versus the dark period in any twenty-four hours will differ for different plants. We will talk more about floral induction later. Temperature, moisture, sugar concentration, hormone type and concentration may also affect apical growth.

Although we state that shoots elongate by apical growth, the apical meristem is responsible primarily for the differentiation of the primary tissues and the production of lateral organs, the leaves and axillary buds. Elongation occurs primarily in a region just behind the apical meristem, the *subapical region*. Stem elongation in some plants, including many monocots, includes not only subapical elongation, but also elongation of derivatives of *intercalary meristems*. These meristems, in monocots, are transversely oriented across the stem at the base of the internode and thus contribute to internodal elongation.

## DIFFERENTIATION OF PRIMARY TISSUES

The stem is considered generally to consist of three tissue systems, the *dermal*, *vascular*, and *ground tissue*. The dermal, or epidermis, differentiates from the outermost mantle layer in angiosperms, the vascular tissue differentiates from the *procambium*, while the ground tissue, pith, cortex, and pith rays (when present) differentiate from the peripheral and pith meristems.

The procambium may be formed in a variety of ways. Eventually, it will differentiate into primary *xylem* and *phloem* and in most dicots and gymnosperms, it will also give rise, in part, to the *vascular cambium*. In young dicot stems, near the tip of the shoot, vascular tissues occur as separate bundles, the *primary vascular bundles*. A bundle consists of primary xylem and primary phloem, while some may have the development of a meristematic region, called the *fascicular cambium*, between the xylem and phloem. The term *stele* is applicable to the central cylinder, which consists of vascular tissue, pith, and pith rays. But unlike in the root, the pericycle does not generally form the outer boundary of the stele in stems but may be present in some stems.

### Epidermis

As in the root, the epidermis is usually a single layer of cells. The epidermal layer of most plants is devoid of chloroplasts, although it may contain specialized chlorophyllous cells, called *guard cells*, which occur in pairs. Between the pair of guard cells is an opening called a *stoma* or *stomate*. The walls of epidermal cells may frequently be rather thick or lignified, or may proliferate, forming a variety of epidermal hairs, or *trichomes*. The wall may be covered with *cutin* forming a thick superficial layer called the *cuticle*. The cuticle may be particularly thick on xerophytic plant stems and thin on aquatic plants.

## Cortex

The cortex is a complex tissue and may contain parenchyma, chlorenchyma, collenchyma, sclerenchyma, or secretory cells. Usually one does not find all these different tissues or cell types in the cortex of any one plant. The cortex of the stem is generally a much smaller region than that of the root, and with secondary growth, it may be lost entirely. Hypodermis and cortical parenchyma regions of the cortex are normally present in stems, but the endodermis is less common, occurring in some underground stems and occasionally in aerial stems. Even when present, the endodermis is usually less prominent than in roots (Figure 9.6).

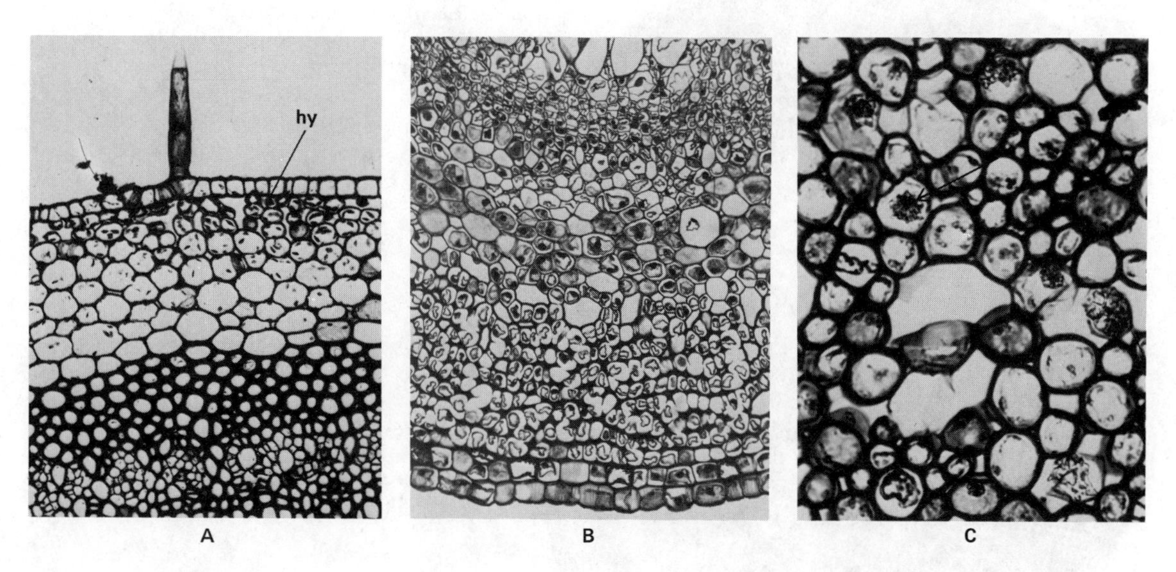


*Figure 9.6*
*A.* Young *Pelargonium* stem, cross-section. Note chloroplasts in outer cortex. Also note the trichrome on the epidermis. Both endodermis and hypodermis (*hy*) are not very prominent here. *B. Sambucus* stem, cross-section. Note the collenchyma tissue in the outer cortex, starch-containing parenchyma in the inner cortex. *C. Nerium* stem, cross-section. Note the abundance of crystal-containing cells (*arrow*).

## *Primary Vascular Tissues*

The vascular tissue of the stem is quite variable and much more complex generally than in the root. In dicots, it forms generally a discrete ring of separate bundles, with phloem and xylem lying on the radius rather than alternating with each other as in the root (Figure 9.7). The phloem may lie on one side of the xylem, both sides of the xylem, or completely encircle the xylem. In some instances, the xylem may completely encircle the phloem (Figure 9.8). A vascular cambium differentiates normally between the primary xylem and phloem. A pericycle may lie outside the phloem, but rarely is found in stems. In monocots, many vascular bundles are present generally, and these are not usually arranged in a cylinder. Thus, there is no clearcut boundary, no real distinction, between pith and cortex. The parenchyma tissue is best described as *ground tissue*. The vascular bundles are scattered throughout this ground tissue. Many monocot vascular bundles are characterized by two large metaxylem vessels which can be seen in cross section to lie opposite each other. Lying at the point of a triangle formed by it and the two metaxylem vessels is a large air space,

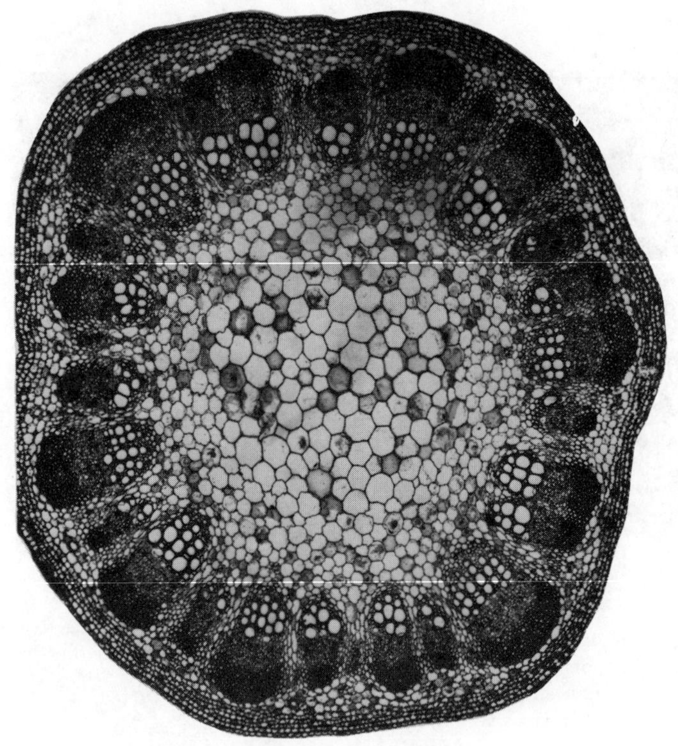

*Figure 9.7*
*Helianthus* stem, cross-section. Note ring of vascular bundles surrounding the central pith.

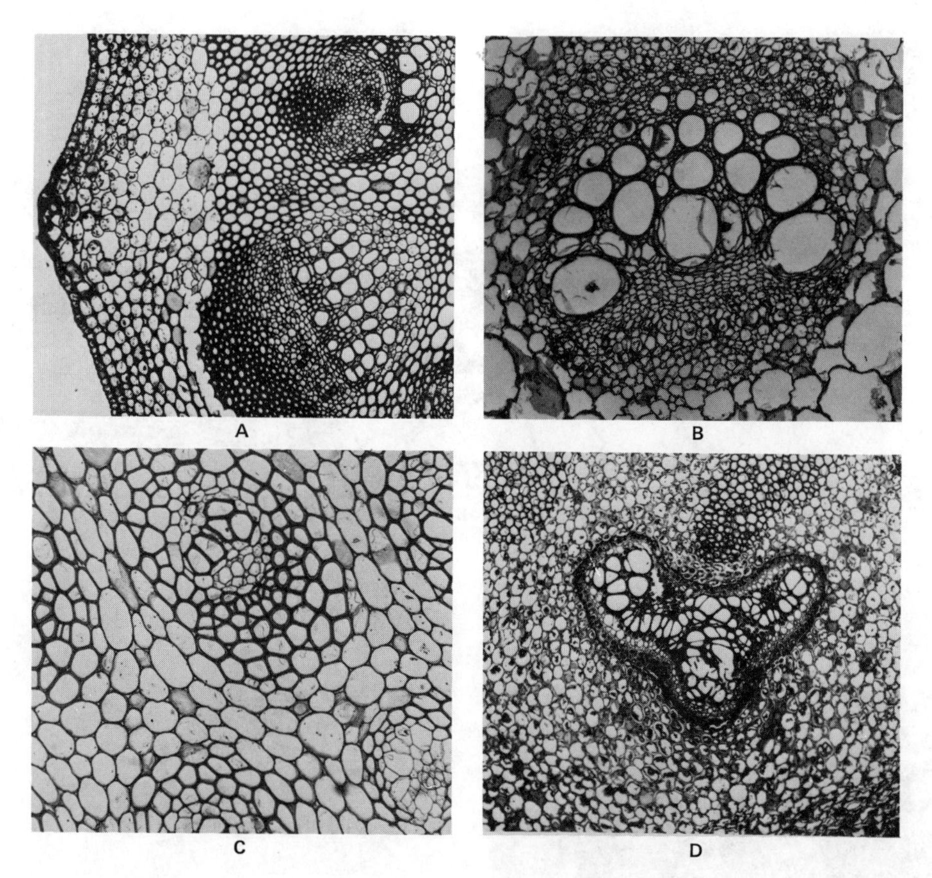


*Figure 9.8*
Vascular bundle types. *A. Helianthus* stem with collateral bundles. Note that the phloem lies on only one side of the xylem. *B. Curcubita* stem with bicollateral bundles. Note that the phloem lies on both sides of the xylem. *C. Dracaena* stem with amphivasal bundles. Note that the xylem completely surrounds the phloem. *D. Pteridium* rhizome with amphicribral bundles. Note that the phloem completely surrounds the xylem.

formed by disintegration of protoxylem elements. Some people have described these bundles as looking like a monkey face. As you would expect, these vascular bundles lack a cambium (Figure 9.9).

Vascular bundles of both monocots and dicots often have a conspicuous cap of fibers, and these sclerenchyma cells may encircle the bundle as well. Various other cell types may also be present, including xylem and phloem parenchyma, xylem tracheids and vessels, and phloem sieve tube cells and companion cells.

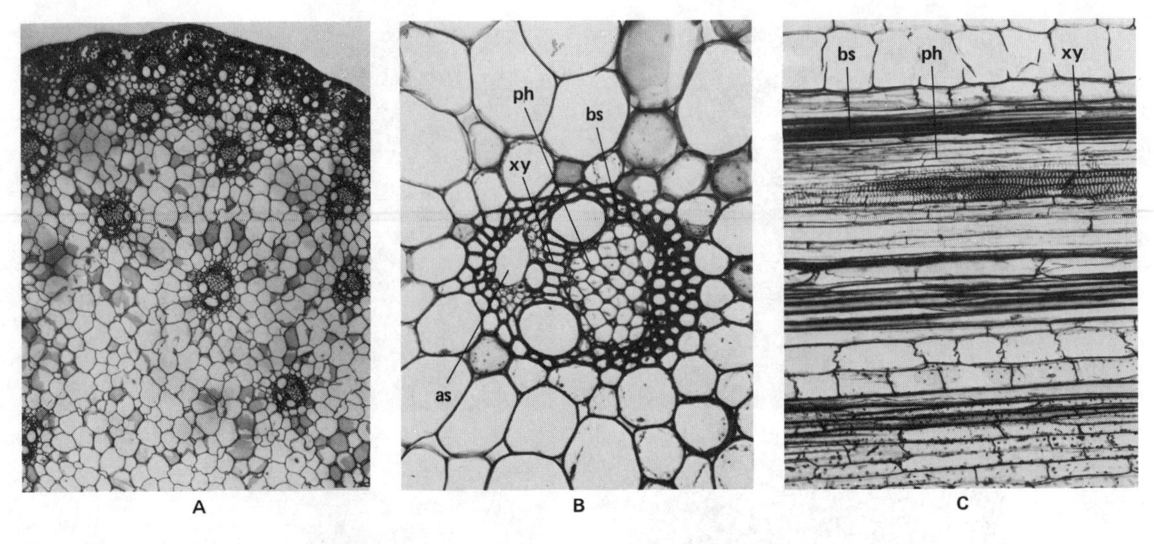


*Figure 9.9*
*A.* Corn stem cross-section. Note scattered vascular bundles. *B.* Vascular bundles of corn stem; *as*, air space; *xy*, xylem; *ph*, phloem; *bs*, bundle sheath. *C.* Longisection of stem.

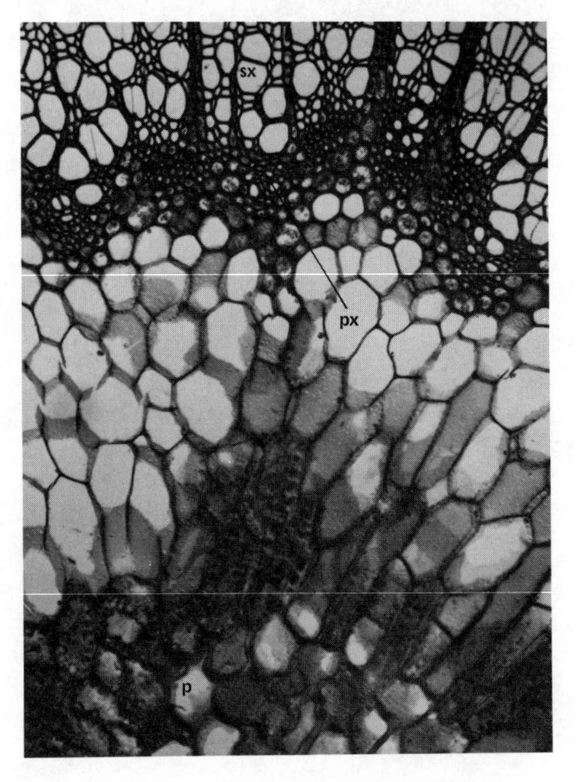


*Figure 9.10*
*Liriodendron* stem. Note variety of cells in pith; *p*, pith; *px*, primary xylem; *sx*, secondary xylem.

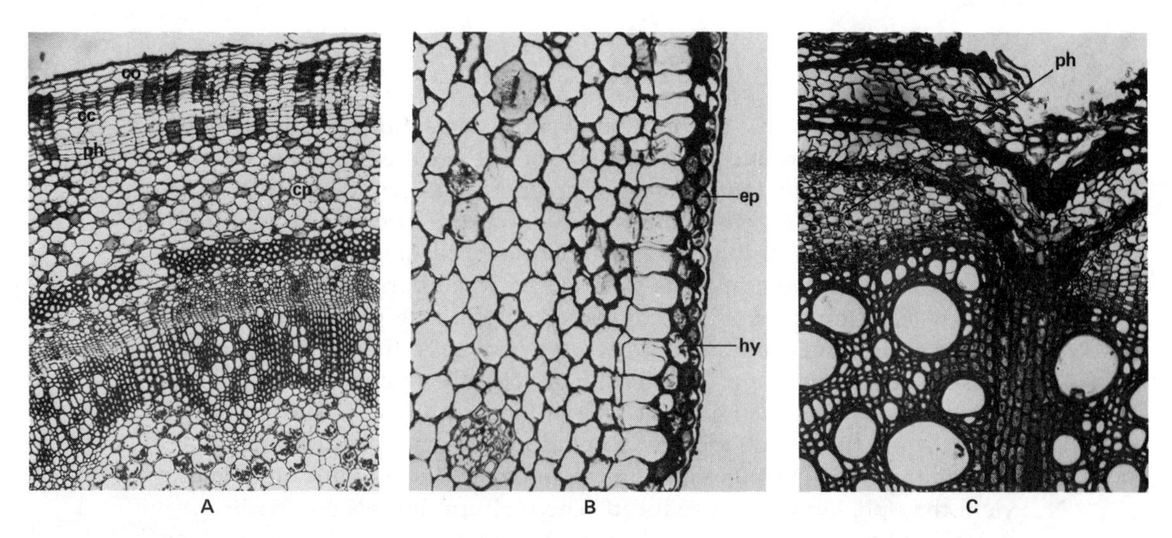


*Figure 9.11*
Origin of cork cambium. *A. Pelargonium* stem cross-section. The origin of the cork cambium is in the hypodermis; *cp*, cortical parenchyma; *ph*, phelloderm; *cc*, cork cambium; *co*, cork. *B. Anthurium* stem cross-section. The origin of the cork cambium is in an outer layer of the cortical parenchyma; *hy*, hypodermis; *ep*, epidermis. *C. Clematis* stem cross-section. Note that the cork cambium originates in successively deeper layers with the last one originating in the outer phloem (*ph*).

## Pith

Pith is common to dicot stems and may be either parenchyma or chlorenchyma, although occasionally it may be sclerenchymized. In the mature stems of some plants, the pith may be destroyed entirely due to elongation of the internodes. Often, however, the pith is retained in the nodes. Pith and cortex may be quite similar in terms of their inclusions: starch, crystals, sclereids, and others (Figure 9.7 and 9.10). Parenchyma cells may extend between the vascular bundles from the pith to the cortical parenchyma. These are called pith rays.

## Cork Cambium

The cork cambium of stems arises frequently in the hypodermis, or outermost cortical parenchyma layer, but may arise in the epidermis, endodermis, pericycle, phloem, xylem, or, very rarely, pith. Some stems have successive cork cambia, so the cambia originating in deeper tissue may not be the first cork cambium to arise in that stem (Figure 9.11).

## DIFFERENTIATION OF SECONDARY TISSUES

Secondary growth occurs primarily in woody stems, but even herbaceous annual plants may produce secondary tissues before the end of their growing season. The secondary tissues form continuous cylinders rather than being confined to discrete bundles. How does this occur? The cambium formed within vascular bundles, the *fascicular cambium*, joins with a cambium formed in the region of the pith rays between bundles, the *interfascicular cambium*, thus forming a complete cambial ring (Figure 9.12). This vascular cambium is composed of two types of cells, the *fusiform* and *ray initials.* The fusiform initials divide periclinally (parallel to the outer wall), normally, producing radially oriented files of cells. Those derivatives toward the inside differentiate as secondary xylem, while those towards the outside differentiate as secondary phloem, with much more xylem than phloem being produced. The fusiform initials do divide anticlinically (at right angles to the outer wall) occasionally which adds girth or circumference to the cambium. Derivatives of the ray initials differentiate either as xylem ray or phloem ray cells (Figure 9.13).

In perennial plants, cambial activity is usually greatest in the spring of the year. The abundant moisture, food supply, and growth hormones available at that time result in the formation of relatively larger, thinner-walled xylem vessels than those formed in mid or late summer. Thus, there is an alternation between larger, thinner-walled xylem elements and

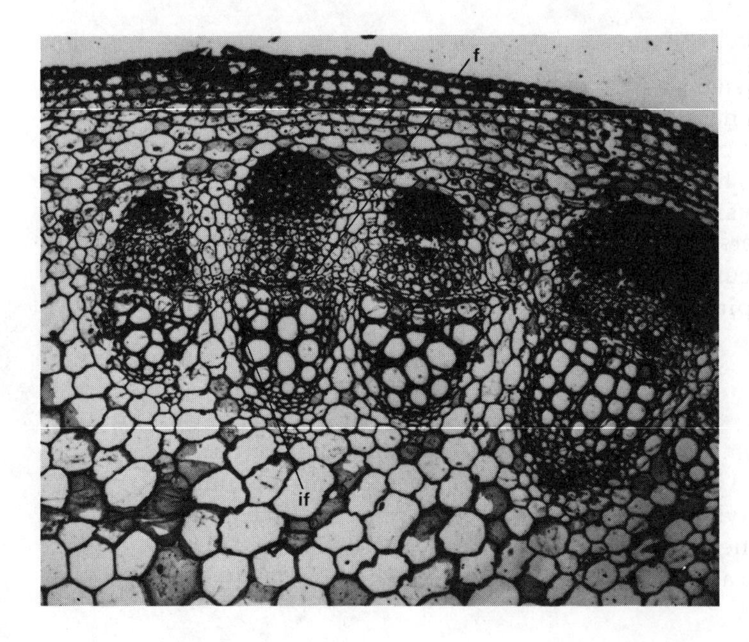


*Figure 9.12*
*Helianthus* stem cross-section. Note fascicular (*f*) and interfascicular (*if*) cambia.

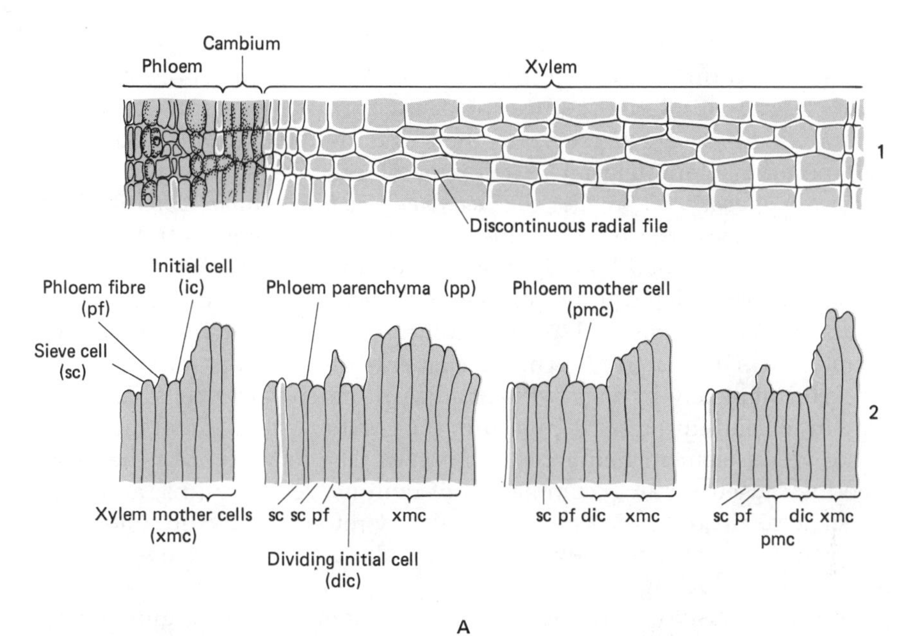


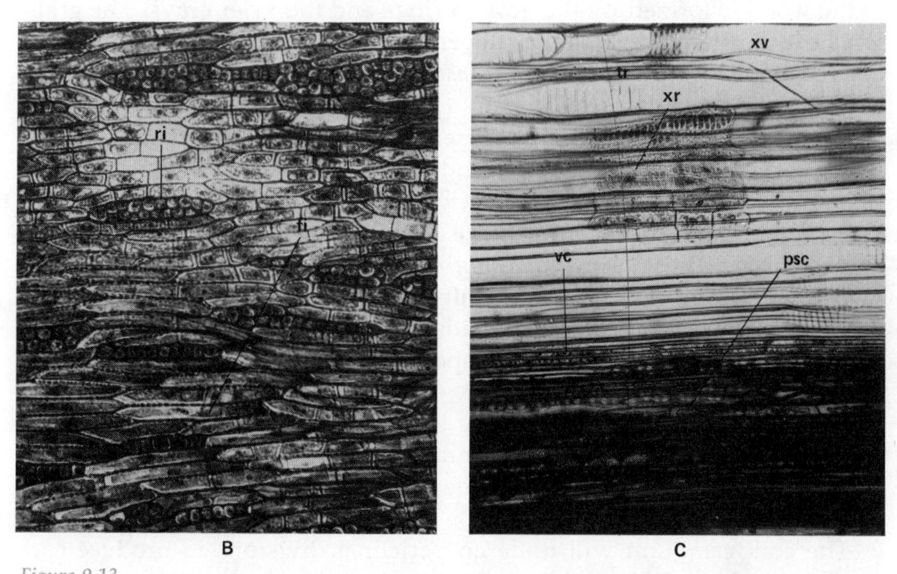


*Figure 9.13*
*A.* Cross-section through cambial region showing various stages in initiation of vascular tissue: cambial initial, phloem initial, xylem initial, phloem; xylem. *B.* Tangential section through cambial region of *Robina.* Note the narrow, tapered fusiform initials (*fi*) and isodiametric ray initials (*ri*). *C. Liriodendron* stem, longisection; *xv*, xylem vessel; *tr*, tracheid; *xr*, xylem ray, *vc*, vascular cambium; *psc*, phloem sieve tube.

smaller, thicker-walled xylem elements. This demarcation between the two regions forms the *annual rings* of the stem (Figure 9.14). As several spurts of growth might occur during a single growing season, if environmental conditions are appropriate, the rings might better be called *growth rings* rather than annual rings. The region of larger cells is usually referred to as *spring wood,* the smaller cells *summer wood* or *autumn wood.* In older stems, the xylem can be seen to be composed of a darker central area, the *heartwood,* surrounded by a light outer area, the *sapwood.* The heartwood is generally considered to be composed of dead cells, although in certain plants, xylem ray cells and xylem parenchyma may remain alive for 100 years or so. The inactive tracheary elements of the heartwood are often blocked completely by ingrowths of adjacent parenchyma cells. These ingrowths extend through the pits, with the resultant blockage of the lumen of the vessel element being called a *tyloses.* The dark color of the heartwood is caused by an accumulation of tannins, resins, gums, and other substances. The outer sapwood is frequently referred to as the conducting xylem, but studies have shown that many plants have only a narrow zone of actively conducting xylem immediately adjacent to the cambium.

The xylem tissue constitutes the *wood* of a stem, with the region outside of the xylem (from the vascular cambium to the surface) being termed *bark.* The type of cells which comprise the xylem plus their size and arrangement is extremely important to carpenters and cabinet makers. The *grain* of the wood is formed by the growth rings and the xylem rays. The grain will have a different appearance depending upon the plane of view. A cut at right angles to a long axis is a *transverse* section, a cut at right angles to the radius, a *tangential* section, and a cut which passes through both the center of the stem and the circumference, is a *radial* section. Lumbermen would term these cuts cross, slab, and quarter-sawed, respectively. Quarter-sawed lumber exhibits the most beautiful grain, but is also the most expensive. Lumbermen or furniture dealers may also talk of *hardwoods* and *softwoods.* These terms have nothing to do with the hardness or softness of the wood. In fact, some softwoods are harder than hardwoods. Softwoods consist largely of tracheids, whereas hardwoods have numerous vessels. In most cases, gymnosperm wood is softwood, angiosperm wood, hardwood.

Secondary phloem is quite similar to primary phloem, except, of course, in origin. In a woody dicot, phloem rays may be prominent, and much of the previously formed primary phloem may differentiate into phloem fibers (Figure 9.15).

The cork cambium will undergo periclinal divisions to produce cork to the outside and phelloderm occasionally to the inside. The cork cells become generally somewhat box-shaped, their walls become impregnated with suberin, and the cells die. The corky region is referred to commonly as the bark, but remember it is only the outer portion of the bark (Figure 9.16).

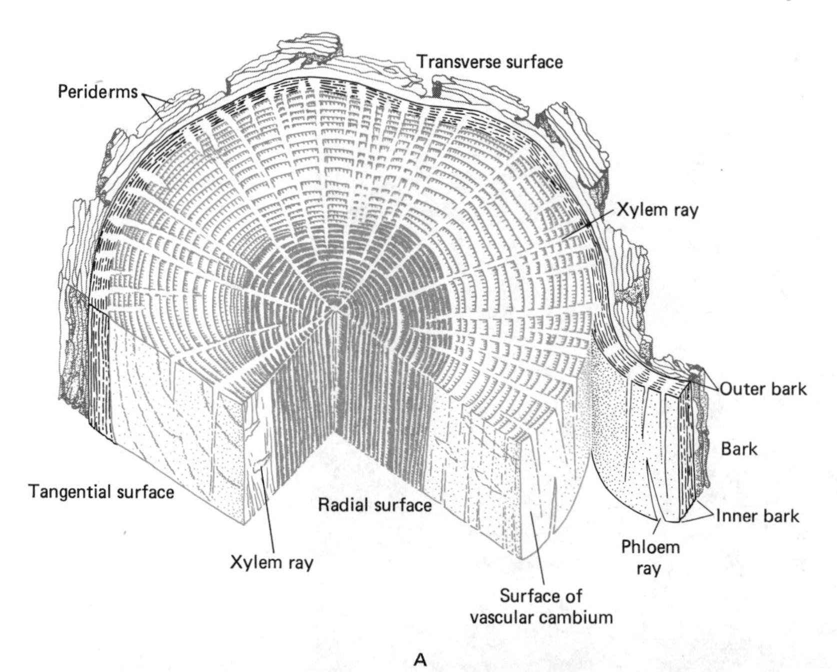

*Figure 9.14*
*A.* Portion of a cross-section of a woody dicot stem. Note wood rays, annual rings, heartwood and sapwood. *B. Liriodendron* stem cross-section. Note smaller, thicker-walled xylem cells and larger, thinner-walled xylem cells which constitute an annual ring of growth. *C.* Douglas fir stem cross-section entirely enclosing an Oregon oak tree. [*C.* courtesy of U.S. Forest Service.]

*Figure 9.15*
*Tilia* stem cross-section; *vc*, vascular cambium; *pr*, phloem ray; *pf*, phloem fiber; *psc*, phloem sieve tube cell; *pp*, phloem parenchyma; *pcc*, phloem companion cell.

### *Anomalous Secondary Growth*

Many dicots, and even some monocots, show unusual patterns of secondary structure. This anomalous growth may be the consequence of 1) the formation of accessory cambia, 2) the formation of an abnormally situated cambium, or 3) a normal cambium which produces an abnormal arrangement of the secondary xylem and phloem. This anomalous growth is an important way for certain monocots to increase in diameter, and although these plants do not have a true vascular cambium, they may produce anomalous secondary xylem and phloem.

## COMPARATIVE STRUCTURE OF STEMS

We have mentioned several features in our previous discussion by which dicot stems differ from monocot stems, woody stems from herbaceous stems, or angiosperm stems from gymnosperm stems. For comparative purposes it might be well to summarize those differences (Table 9.1).

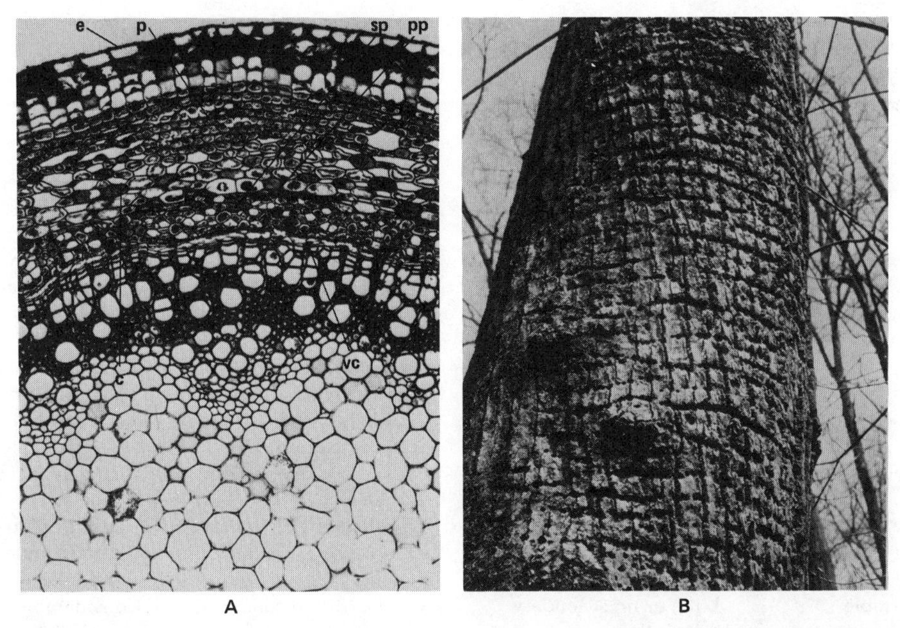


*Figure 9.16*
A. *Sambucus* stem cross-section. The bark constitutes all the tissue outside of the vascular cambium. Note that the epidermis is still intact in this stem. With additional production of cork, the epidermis will be lost as these cells are mature and will split due to expansion of the stem by secondary growth. *Vc,* vascular cambium; *pp,* primary phloem; *sp,* secondary phloem; *c,* cortex; *p,* periderm; *e,* epidermis. *B.* The tulip tree, *Liriodendron tulipifera.* Note the fissured outer bark (cork cambium outward). Note also the regular pattern of holes produced by a sapsucker. [*B.* From a Kodachrome by David R. Williams.]

## ADDITIONAL DISCUSSION OF SHOOT ORGANIZATION

For those of you that would like to consider in more detail how the tissues of the shoot may arise, let us consider two differing viewpoints.

Although it is difficult to study shoot tip organization, considering the variability of the apex during development, various theories of apical organization have been proposed. Most have since been discarded because they simply did not apply to a great many plants and were, in fact, unrealistic or invalid. In 1951, vegetative shoot organization was grouped by Popham into seven classes.

Type I (Figure 9.17A) consists of a group of plants having a *single pyramidal apical cell.* Derivatives of this cell produce all the tissue of the shoot.

Type II includes a group of plants with *two to five prismatic apical cells.*

Type III (Figure 9.17B) shoot apex exhibits three zones. The *surface meristem* consists of cells dividing in various planes producing the epidermal initials of zone 2 and zone 3 and in some the leaf primordia. The second

*Table 9.1*
**SOME COMPARISONS BETWEEN CONIFERS, HERBACEOUS MONOCOTS, HERBACEOUS DICOTS, AND WOODY DICOTS**

| CONIFER | HERBACEOUS MONOCOT | HERBACEOUS DICOT | WOODY DICOT |
|---|---|---|---|
| 1. Form typically little or no cortical fibers | Cortical fibers not unusual | Cortical fibers not unusual | Cortical fibers not unusual |
| 2. Few or no fibers in primary xylem or phloem; may be fibers in secondary phloem | Primary phloem fibers | Primary xylem and phloem fibers | Primary xylem and phloem fibers |
| 3. Little xylem parenchyma | May be considerable xylem parenchyma | Has some xylem parenchyma | Considerable xylem parenchyma |
| 4. No vessels in xylem | Vessels in xylem | Vessels in xylem | Vessels in xylem |
| 5. No companion cells or sieve tube cells in phloem | Both companion cells and sieve tube cells in phloem | Both companion cells and sieve tube cells in phloem | Both companion cells and sieve tube cells in phloem |
| 6. Xylem and phloem rays usually one cell wide | No xylem or phloem rays | No xylem or phloem rays | Xylem or phloem rays usually more than one cell wide |
| 7. Considerable secondary growth | Little or no secondary growth | May be little secondary growth | Considerable secondary growth |
| 8. Vascular tissue in concentric rings in mature stem | Vascular tissue in discrete bundles in mature stem, scattered | Vascular tissue in discrete bundles in mature stem, in single ring | Vascular tissue in concentric rings in mature stem |
| 9. Pith and cortex differentiated | Pith and cortex generally not differentiated | Pith and cortex differentiated | Pith and cortex differentiated |
| 10. Intercalary meristem absent | Intercalary meristem may be present | Intercalary meristem absent | Intercalary meristem absent |
| 11. Vascular cambium present | Vascular cambium absent | Vascular cambium generally present | Vascular cambium present |
| 12. Cork generally develops | Cork rarely develops | Cork development not extensive | Cork generally develops |

zone, the *central meristem*, gives rise to either the pith or the stele and may give rise, in part, to zone 3. The third zone, the *peripheral meristem*, lies along the flanks of the central meristem. Derivatives of this zone produce the cortex and, in some shoots, the procambium and leaf primordia.

Type IV (Figure 9.17C) is characterized by five zones. The first, the *surface meristem*, consists of a surface layer of cells in which divisions are mostly *anticlinal* although periclinal divisions do occur in this layer. It contributes initials to the adjacent zone. The second zone, the *central mother cells*, lies directly below zone one and accounts primarily for an increase in volume of cells. The third zone, a *cambiumlike zone*, contributes cells to zones beneath it. Zone four, the *central meristem*, lies directly below the central part of zone three and contributes to formation of the pith meristem. The fifth zone, the *peripheral meristem*, lies subjacent to zone three.

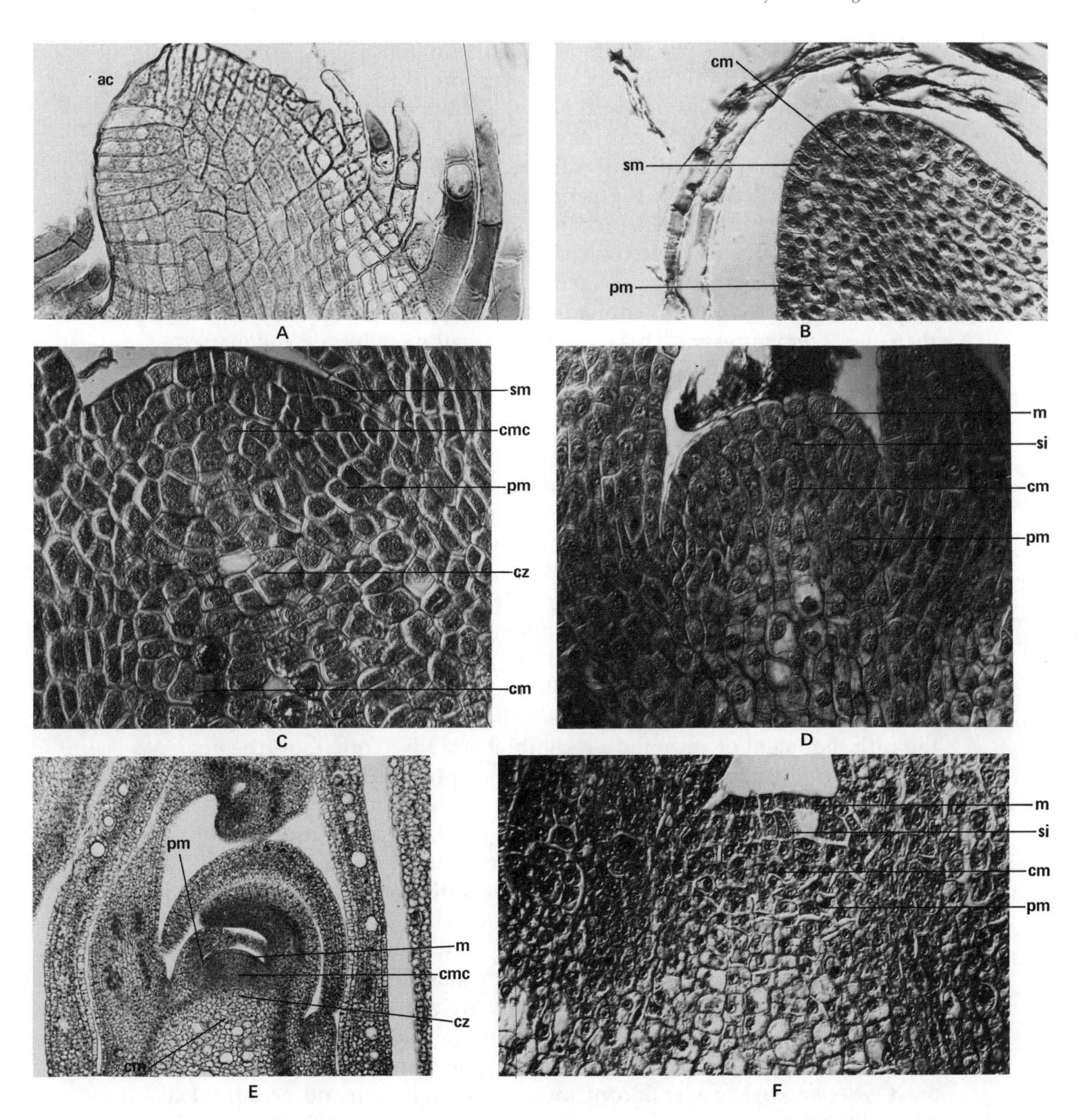


*Figure 9.17*
Vegetative shoot organization according to Popham. *A. Marsilea*—Type I. *B. Lycopodium*—Type III. *C. Ginkgo*—Type IV. *D. Sequoia*—Type V. *E. Liriodendron*—Type VI. *F. Syringa*—Type VII. *Ac,* apical cell; *sm,* surface meristem; *cm,* central meristem; *pm,* peripheral meristem; *cmc,* central mother cells; *cz,* cambium-like zone; *si,* subapical initials; *m,* mantle.

The epidermis, cortex, procambial ring, leaf primordia, and in some cases, the outer pith cells differentiate from cells of this zone.

Type V (Figure 9.17D) exhibits four growth zones. The *surface meristem* is characterized by the predominance of anticlinal divisions, but periclinal divisions do occur. Derivatives of this layer contribute to epidermis and to zone two. Zone two, the *subapical initials,* produces cells which contribute to zones three and four. The third zone, the *central meristem,* lies directly below the central portion of the subapical initials and contributes to the pith meristem. Zone four, the *peripheral meristem,* produces cells which later differentiate into cortex, procambium, and leaf primordia.

Type VI (Figure 9.17E) occurs only in angiosperms and consists of five zones. The *mantle* forms a cap over the summit of the apex and has cell divisions in one plane only, anticlinal, at least in the surface layer if more than one layer is present. This layer produces the epidermis and may contribute to zones two and three. Zone two, the *central mother cells,* contributes to all other zones except zone one. The third zone, the *cambiumlike zone,* contributes to cells in zones four and five. Zone four, the *central meristem,* may produce cells that differentiate into the pith meristem only or into pith and certain procambial strands. The fifth zone, the *peripheral meristem,* surrounds zone four and contributes cells which differentiate into the cortex, procambial tissue (all or part) and the leaf primordia.

Type VII (Figure 9.17F), which is known as the usual angiosperm type, is characterized by four zones. The first, the *mantle,* consists of one or more layers in which only anticlinal divisions occur. This zone contributes to the epidermis and may contribute some cells to zone two. Zone two, the *subapical initials,* contributes cells to zones three and four. The third zone, the *central meristem,* produces cells which differentiate either into the pith meristem or pith and procambial strands. Zone four, the *peripheral meristem,* contributes cells to the cortex, procambial strands, and leaf primorida.

This classification is somewhat complex and, although it does contribute somewhat to an understanding of the ontogeny, or development, of tissues, its usefulness is limited. This classification, as with most others, is affected by the considerable variability of the stem apex during ontogeny. A simpler system has been proposed by Newman (1965). Newman believes that no cells serve as permanent initials, but a sequence of cells function as initials over a period of time. He called these the *continuing meristematic residue* (cmr) and classified apices into three types based on the form of the cmr. The *monoplex* apex has the cmr in the superficial layer only, where a *single* division contributes to both length and breadth *(bulk growth).* The continuing meristematic residue in the *simplex* apex is also restricted to a superficial layer, but *two* divisions are required for bulk growth. In the *duplex* apex, the common angiosperm type, the continuing meristematic residue occurs in the superficial layer and at least one other layer. Only anticlinal divisions occur in the superficial layer, but divisions in at least two planes occur in cells of the inner layer or layers. Two modes of growth are provided, thus the term duplex. This classification differs from previous ones in that it does not suggest that particular derivatives will have a particular destiny.

## SUMMARY

1. Stems are characterized by the presence of nodes and internodes.
2. Buds form characteristically on woody plants.
3. Buds contain the apical meristem and may be vegetative, reproductive, or mixed.
4. The apical meristem can produce lateral appendages, leaves, axillary buds, and flower parts.
5. The production of flowers may terminate further elongation of the shoot.
6. All primary tissue is formed by derivatives of the apical meristem; thus, the apical meristem contributes to stem elongation.
7. Intercalary meristems may also contribute to stem elongation, primarily in certain monocots.
8. The primary vascular tissue differentiates from the procambium and is contained in discrete vascular bundles.
9. In dicots, a fascicular cambium differentiates generally between the primary xylem and phloem. An interfascicular cambium may differentiate between the vascular bundles.
10. In woody plants, the fascicular and interfascicular cambia join producing a concentric vascular cambium.
11. The vascular cambium produces secondary xylem and phloem and is the primary contributor to the increase in diameter of a woody stem.
12. Monocot stems lack generally a vascular cambium, have little or no secondary growth, scattered vascular bundles, and no clear delimitation of pith and cortex.
13. Herbaceous dicot stems generally have their vascular bundles in a ring. Thus, pith and cortex is delimited. They may have both fascicular and interfascicular cambia with or without the production of secondary xylem and phloem.
14. Woody stems, at maturity generally have their vascular tissue in concentric rings with considerable secondary growth.
15. Cork cambium is formed generally in dicots, rarely in monocots. It most frequently differentiates in superficial layers of the cortex and may produce extensive cork in woody plants.
16. A woody stem can be separated into three general regions: pith, wood (xylem), and bark (everything outside the vascular cambium).
17. Wood may be characterized as spring or summer wood, heartwood or sapwood, hardwood or softwood.
18. The differences in growth of cells in spring and summer wood produces growth rings which provide an estimate of the age of the stem.
19. Some stems undergo anomalous growth, and various stem modifications, both aboveground and underground are known.

# Chapter 10

## THE VASCULAR PLANTS— III. LEAVES

Leaves, as already pointed out, are lateral appendages of the stem. They typically have determinate growth and dorsiventral symmetry (broad and flat). The thinness of the leaf, plus the fact that it is composed primarily of chlorenchyma, account for the food-manufacturing process, *photosynthesis,* occurring mainly within it. The broad surface area of the leaf and the large number of stomata occurring in the epidermis account for the importance of the leaf in the evaporation of water from cell surfaces, a process called *transpiration.* Although both photosynthesis and transpiration can and do occur in other organs, the peculiar morphology and anatomy of the leaf make it a principal site for both processes.

## LEAF MORPHOLOGY

Leaves vary considerably, both in size and form. Angiosperm leaves may be over six feet long as in the banana or as much as six feet in diameter as in the floating leaf of the giant water lily, *Victoria regia,* while the entire *Wolffia* plant is only 1 to 2 mm long (Figure 10.1).

*Figure 10.1*
The banana tree, *Musa paradisiaca* var. *sapientum.* [From Kodachrome by David R. Williams.]

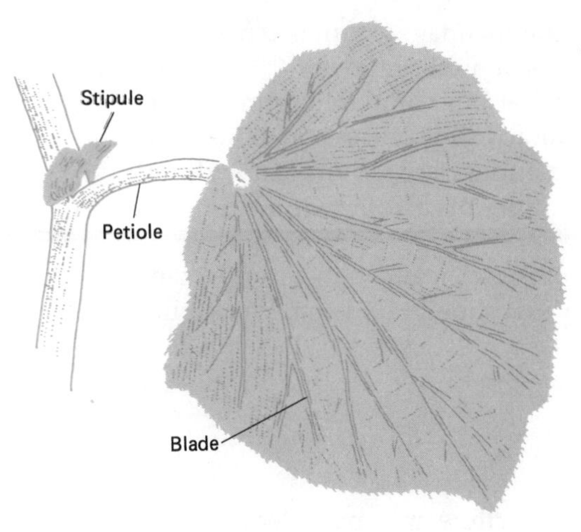


*Figure 10.2*
The dicot leaf; stipule, petiole, and blade.

As leaf morphology is so variable, it is best to consider it by plant group. Most dicot leaves are composed of two parts, the expanded *blade* and the slender *petiole.* Some leaves have a third part, the stipule, attached on either side of the stem at the base of the petiole (Figure 10.2). Some leaves consist only of a blade attached directly to the stem. They are called *sessile.*

## The Blade

If the leaf blade is composed of a single part, the leaf is called a *simple* leaf. If the blade is completely dissected so that it is composed of several separate parts *(leaflets),* the leaf is termed *compound.* If the leaflets arise in a single plane from opposite sides of a *rachis* (the continuation of the petiole), the leaf is *pinnately compound.* If the leaflets are borne on petiolelike structures *(petiolets* or *pedicels)* and arise from a common point at the tip of the petiole they are *palmately compound.* Some compound leaves may be several times compound (Figure 10.3). Leaflets may be distinguished from leaves by the facts that 1) buds will be found only in the axils of leaves, not in the axils of leaflets, and 2) stipules will be found only at the base of leaves.

Dicot leaves have a variety of shapes ranging from long slender forms to round forms. Usually the shape is described only for the blade and ignores the various lobes and indentations, determined by drawing an imaginary line around all projecting parts.

Leaf margins may be entire, or they may be wavy, toothed, or lobed in various ways. The tip of a leaf may be rounded, flattened, or pointed in various degrees. Leaf bases may also be variously formed.

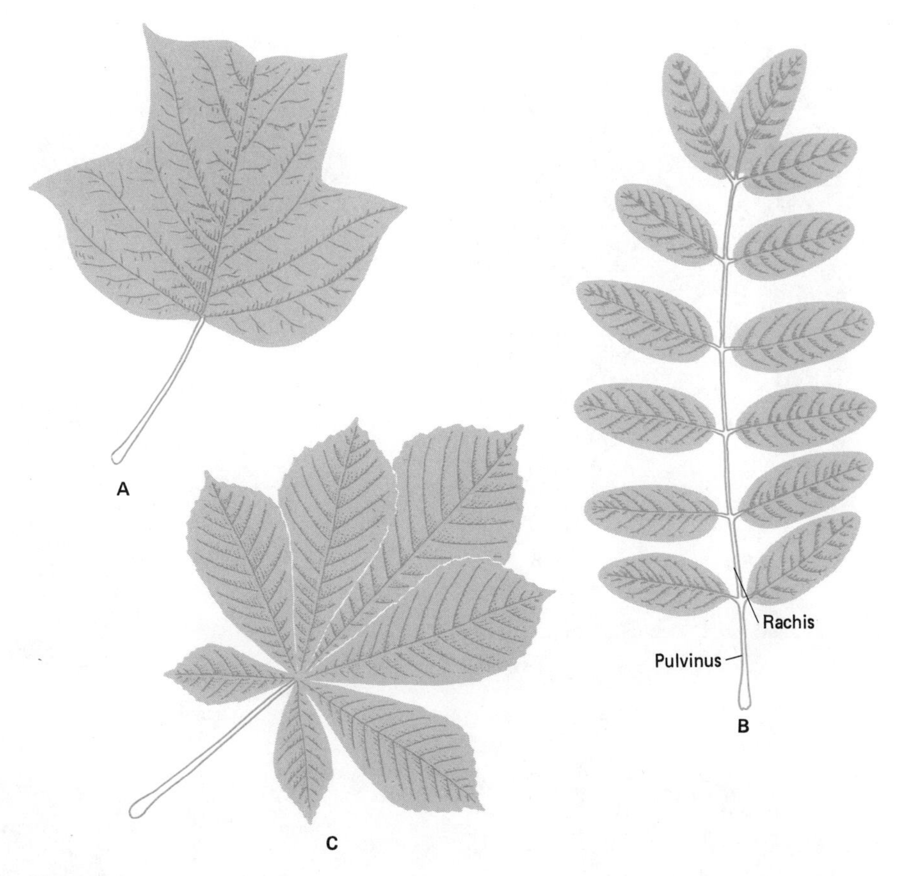


*Figure 10.3*
Simple and compound leaves: *A*. simple leaf, *B*. pinnately compound, *C*. palmately compound.

Most dicot leaves are *net veined*, consisting of one or more major veins with smaller veins branching off forming a network. A few dicots may be *parallel veined*, having numerous veins of essentially the same size running parallel to each other. This type of venation, however, is most characteristic of monocots. If the blade has a single main vein from which other smaller veins branch off, it is called *pinnately net veined*. If it has several prominent veins arising from a single point at the base of the blade, it is termed *palmately net veined* (Figure 10.4).

## Monocot Leaves

Although monocot leaves may have the same parts as found in dicot leaves, others, such as members of the grass family, have a different construction. The grass leaf is divided into two parts, the *blade* and *sheath*. The

*Figure 10.4*
Types of venation: *A*. parallel, *B*. netted, *C*. dichotomous.

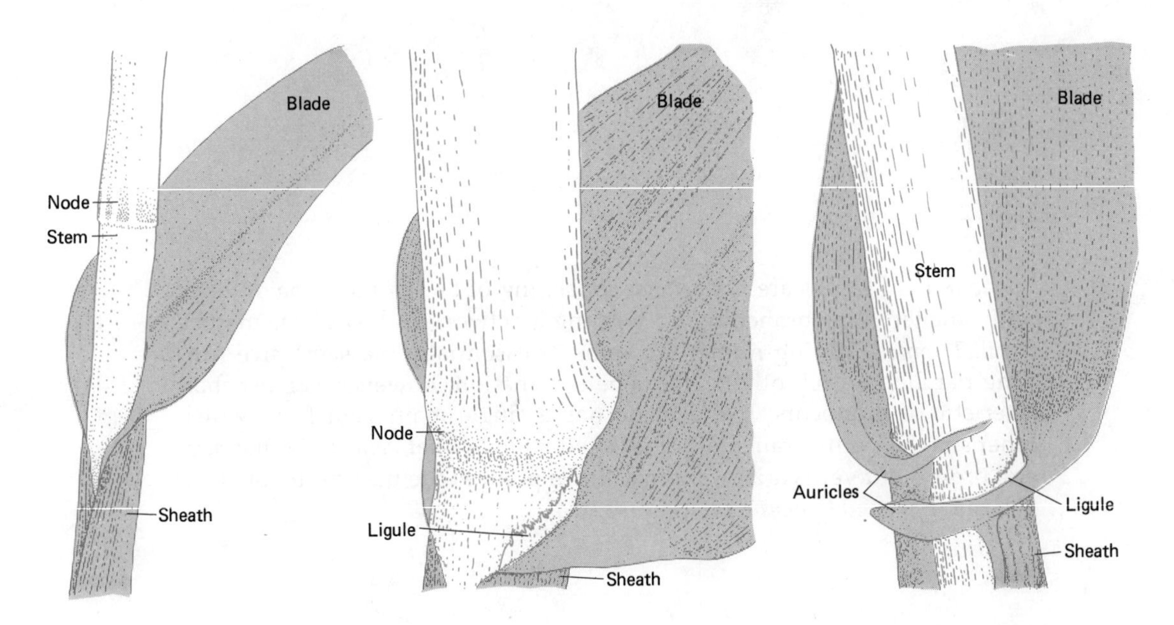


*Figure 10.5*
Leaf of a monocot; barley. Note the ligule, sheath, blade, and auricles.

sheath surrounds completely the stem and may extend over a complete internode. Extending upward from the sheath, in the region of its union with the blade, may be a small flap of tissue called the *ligule*. In some members of the grass family, the blade may wrap around the stem in two earlike projections termed *auricles* (Figure 10.5).

Most monocots have simple leaves, with the exception of the palm family. Most exhibit parallel venation, although net-veined monocot leaves are not unusual.

### Gymnosperm Leaves

The leaves of gymnosperms native to the United States are generally characterized as having needlelike, scalelike, or awn-shaped leaves. The *Ginkgo*, a native of China, has broad, flat leaves with a fan-shaped, dichotomously veined blade. A very unusual gymnosperm called *Welwitschia* is characterized by a single pair of enormous, strap-shaped leaves which may become split into several segments. A group of gymnosperms called the cycads are distinguished by having pinnately compound leaves, with the leaflets being narrow, linear, and sharply pointed (Figure 10.6).

## LEAF DEVELOPMENT

The development of leaves begins actually in many plants, in the spring, with the entire developmental process proceeding slowly through the summer and autumn months. By the end of winter, some leaves will already have their characteristic form, although in miniature. When the buds open the following spring, the continued enlargement of the leaf is due simply to the enlargement of existing cells, not by the continued production of new cells. The entire growth process entails cell division, cell enlargement, and cell differentiation. But these processes are not synchronized in each tissue of the leaf. In dicot leaves, cell division stops first in the outermost tissue, the *epidermis*, next in the *spongy mesophyll* and last in the *palisade mesophyll.* But cell enlargement continues in the epidermis long after it has ceased in palisade and spongy cells. Consequently, cells of the spongy mesophyll are pulled in a lose network with many air spaces, while cells of the palisade become separated laterally. After this period of enlargement ceases, the leaf will get no larger, and thus its growth is determinate. There are variations in this pattern, and if you have the job of mowing your lawn, you are well aware of one such exception. Unfortunately, grass leaves have an intercalary meristem in the base of the blade. Even though you cut off the top of the blade, it will continue to elongate from the base.

*Figure 10.6*
*A.* Pine. *B. Ginkgo. C. Welwitschia.*

## TISSUES OF A MATURE LEAF

At maturity, a representative dicot leaf is composed of an upper and lower *epidermis, mesophyll,* and *vascular tissue* forming the *veins.* In cross section, these tissues are seen, from upper to lower surface, as the upper or *adaxial epidermis* (the surface originally closest to the stem), the *palisade* tissue with numerous chloroplasts, the *spongy* tissue with irregularly shaped cells having fewer chloroplasts and numerous intercellular spaces, and finally the lower or *abaxial* epidermis (the surface originally away from the stem) (Figure 10.7). Vascular strands or veins are present in the central region of the leaf with the phloem located in an abaxial position, the xylem adaxial. The larger veins, particularly the midvein, may have both primary

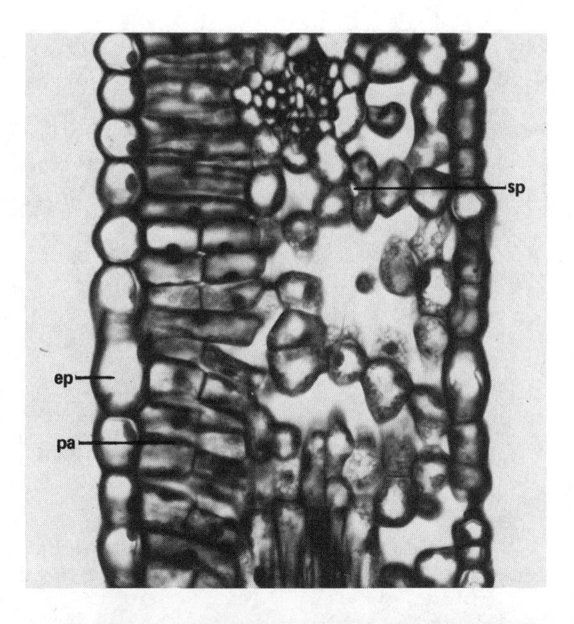


*Figure 10.7*
Tissues of a mature dicot leaf; *ep*, epidermis; *pa*, palisade; *sp*, spongy.

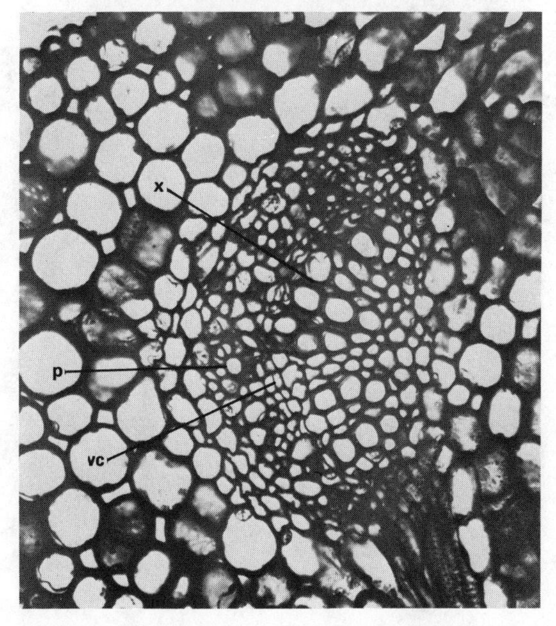


*Figure 10.8*
Mid-vein of *Ligustrum* leaf, cross-section. Note presence of cambium and secondary growth; *vc*, vascular cambium; *x*, xylem; *p*, phloem.

and secondary vascular tissue, with a variety of cell types as found in both stem and root (Figure 10.8). The smaller veins contain only primary tissue. Within them, the only xylem cells are tracheids, the only phloem, sieve elements. Near the ends of the ultimate branches, the veins may contain only tracheids and phloem parenchyma, or no phloem at all (Figure 10.9).

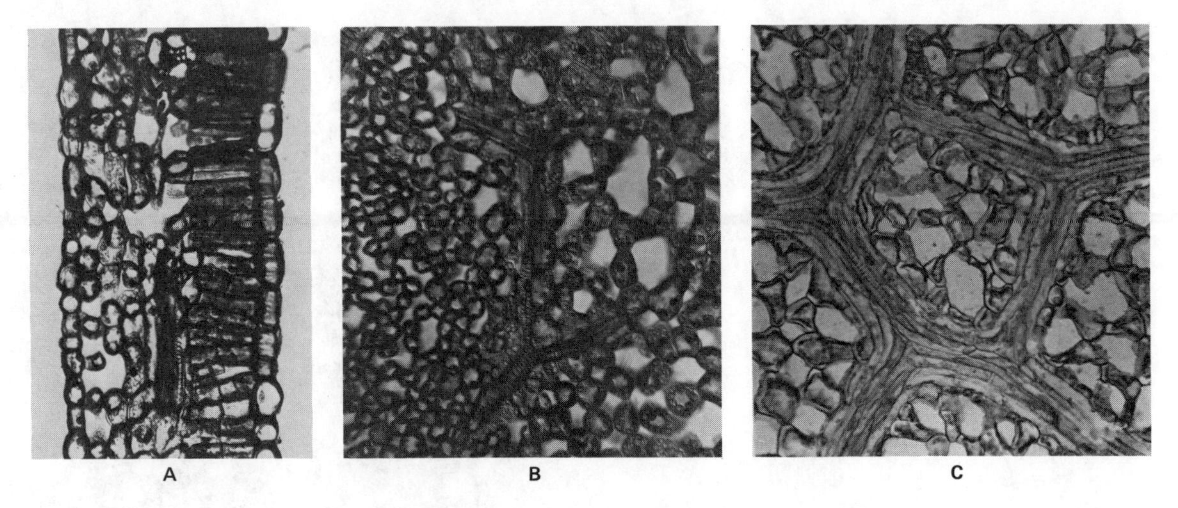


*Figure 10.9*
*A. Ligustrum* leaf cross-section, small veins. *B. Ligustrum* leaf, paradermal section to show vein endings. *C. Ilex* leaf, paradermal section showing islet net of veins.

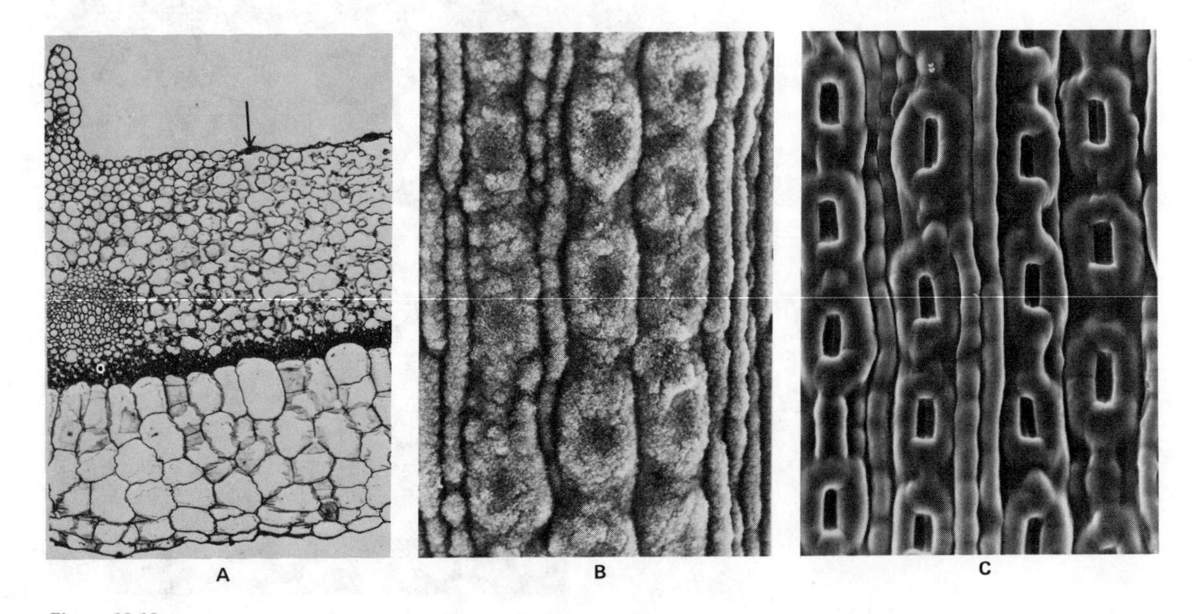


*Figure 10.10*
*A.* Stoma and guard cells (*arrow*) in *Peperomea* leaf cross-section. Note multi-seriate upper epidermis. *B.* Scanning electron micrograph of abaxial leaf surface of *Pseudotsuga* showing cuticular wax distribution in stomatal region. *C.* Scanning electron micrograph of abaxial leaf surface of *Pseudotsuga* after removal of cuticular wax. Note guard cells and stoma. [*B.* and *C.* From B. W. Thair and G. R. Lister, "The Distribution and Fine Structure of the Epicuticular Leaf Wax of *Pseudotsuga menziezii.*" Reproduced by permission of the National Research Council of Canada from the *Canadian Journal of Botany*, 53:1063–1071, 1975.]

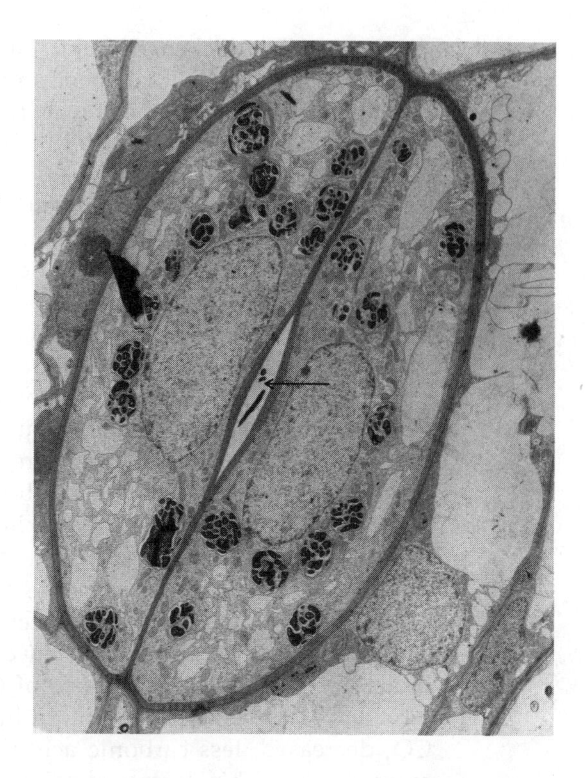

*Figure 10.11*
Guard cells. TEM of guard cells of *Opuntia*. Walls of guard cells adjacent to stoma (*arrow*) were thought to be thicker than those of opposite wall. [Courtesy W. W. Thomson, from "Studies on the Ultrastructure of the Guard Cells of Opuntia," *Amer. J. Bot.* 57(3): 309–316, 1970.]

## *Epidermis*

Usually the epidermis consists of a single layer of cells, although the leaves of some plants may contain several layers of epidermis, varying from two to sixteen. Where more than one layer is present, the upper epidermis may consist of more layers than the lower epidermis.

Pores or *stomata* (*stoma*, sing.) may occur in the lower epidermis, the upper epidermis, or both, but most commonly on the lower epidermis in terrestrial species (Figure 10.10). In the floating leaves of aquatic plants, stomata occur only in the upper epidermis. In certain submerged aquatics, they may be absent from both surfaces.

The stomata are bounded by a pair of specialized epidermal cells called *guard cells*. These have a morphology distinct from that of the usual epidermal cell. The stoma opens internally into an intercellular space, the *substomatal cavity*. When the stoma is open, it provides a pathway for gaseous exchange between the intercellular space and the atmosphere (Figure 10.10). It was once the accepted view that opening of the stoma was due to increased turgor in the guard cells and an asymmetrically thickened wall. The thinner wall was thought to push into the adjacent epidermal cell, pulling the thicker wall adjacent to the stoma with it (Figure 10.11). But Aylor, Parlange, and Krikorian (1973) postulated that a guard cell having only a ventrally thickened wall would be actually hindered from opening, and suggested instead that special micellation (the orientation of cellulosic microfibrils) in the guard cell wall, first suggested by Ziegenspeck in 1938, is the crucial feature (Figure 10.12). But why does the turgor increase in the guard cell? It was noted long ago that for many leaves the stomata were open in the light, closed in the dark. As the guard cells were observed to contain chloroplasts, the connection between light and chloroplasts was an obvious one: the opening of the stomata must somehow be related to the process of photosynthesis. It was suggested that light caused photosynthesis in the guard cells, which resulted in the formation of sugar. The sugar diluted the water in the guard cell; thus, additional water moved into the cell by osmosis from surrounding epidermal cells. This increased the turgor in the guard cells and caused stomatal opening. It was soon discovered that the effects of light were far too rapid than could be explained by the increase of sugar due to photosynthesis, and in fact, the stomata would open at a light intensity too low for any significant amount of photosynthesis to occur.

The photosynthesis theory was thus modified, based upon observations that the sap of the guard cell became less acid in light, and the amount of starch in the cell decreased. It was suggested that the decrease in acidity resulted from the use of $CO_2$ during photosynthesis. Carbon dioxide combines with water forming carbonic acid. So if the amount of $CO_2$ decreased, less carbonic acid is formed, and the solution becomes more alkaline. This decreased acidity promotes an enzymatic reaction resulting in the breakdown of starch to sugar plus phosphates. This could cause a decreased osmotic potential and water would diffuse into the guard cell. During the night, $CO_2$ is produced by the process of respiration, results in an increase in acidity which would promote the synthesis of starch from glucose. With the removal of glucose, the water potential in the guard cells would increase and water would osmose out of the cells with the resulting decrease in turgor causing stomatal closure. This so-called classical theory of stomatal function was formulated by Sayre (1923) and further developed by Scarth in subsequent years. But many observations were made that could not be explained by the classical theory. A major one is that some guard cells contain no visible starch yet still respond to light. In addition, the conversion of sugar to starch may be too slow to account for observed rates of closing, and normal $CO_2$ partial pressure in the cell cannot account for much of a pH change.

Levitt (1967) modified the classical theory by suggesting that not $CO_2$, but organic acids synthesized from $CO_2$, affect the acidity of the guard cell. This mechanism could account for the necessary changes in pH. Although Levitt remains convinced that his modified classical theory plus available factual data eliminates all arguments against the theory, other researchers remain unconvinced. In 1969, Zelitch proposed an alternate theory called

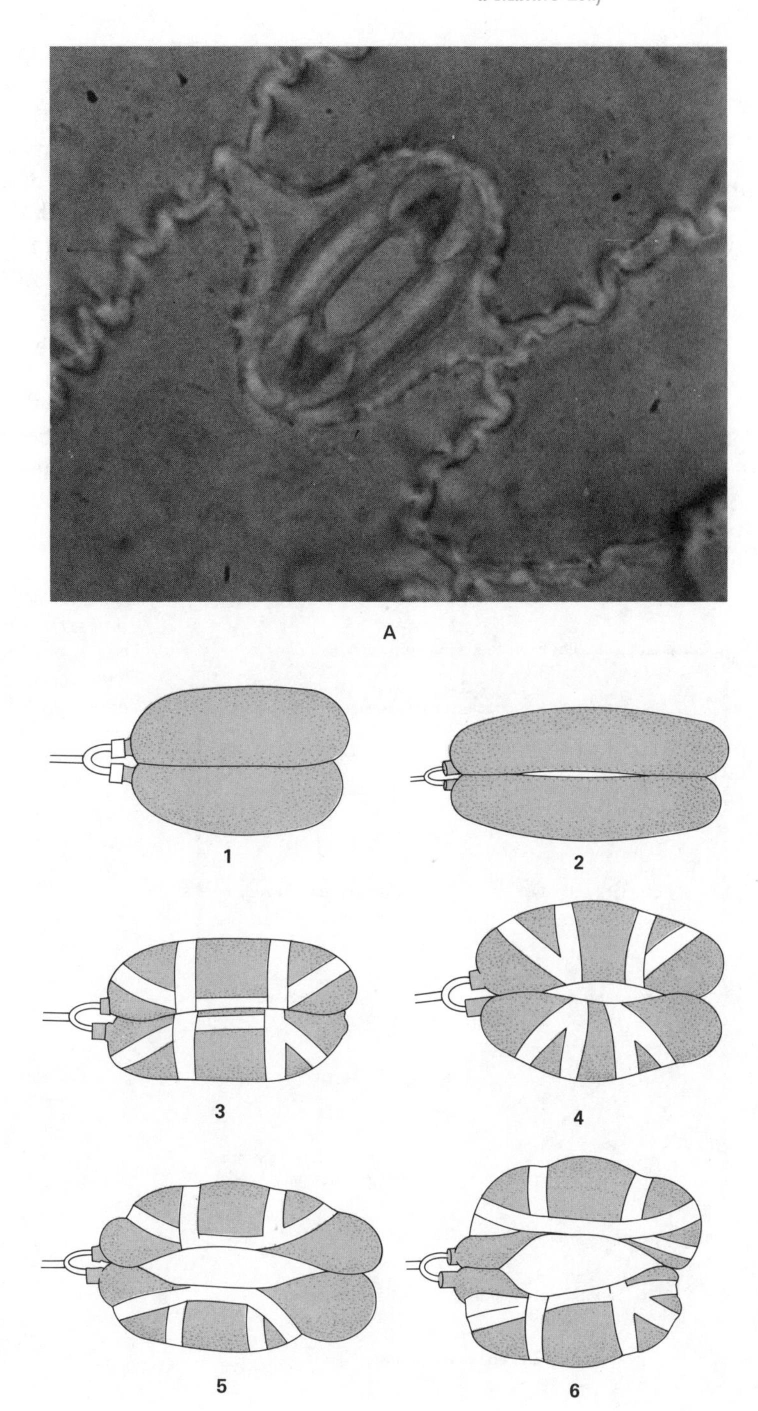


*Figure 10.12*
*A.* Open stoma from lower epidermis of corn leaf. Note thickened ends of guard cells. *B.* Schematic representation of guard cell with "radial and longitudinal micellations." [After D. E. Aylor, et al., "Stomatal Mechanisms," *Amer. J. Bot.* 60(2): 167–168, Figs. 12 and 14.]

the ion pump theory. He suggested that osmotic potential is regulated by ion pumps activated by ATP. A diagram of these two theories is shown in Figure 10.13.

You can note that Zelitch suggested the involvement of potassium ions in stomatal opening. The involvement of potassium was suggested much earlier; in fact, Macallum had observed the accumulation of $K^+$ in guard cells as early as 1905, and in 1943, Imamura suggested that $K^+$ regulated the movement of solutes in and out of the guard cells. Fujino (1959) proposed that, "Stomatal opening and closing are the result of active transport of potassium ions." More recently, Raschke (1971), using electron probe microanalysis, concluded, "$K^+$, if associated with dibasic anions, is sufficient to produce the changes in guard cell volume and osmotic pressure associated with stomatal opening." The existence of potassium fluxes in plants are now known for some 50 species.

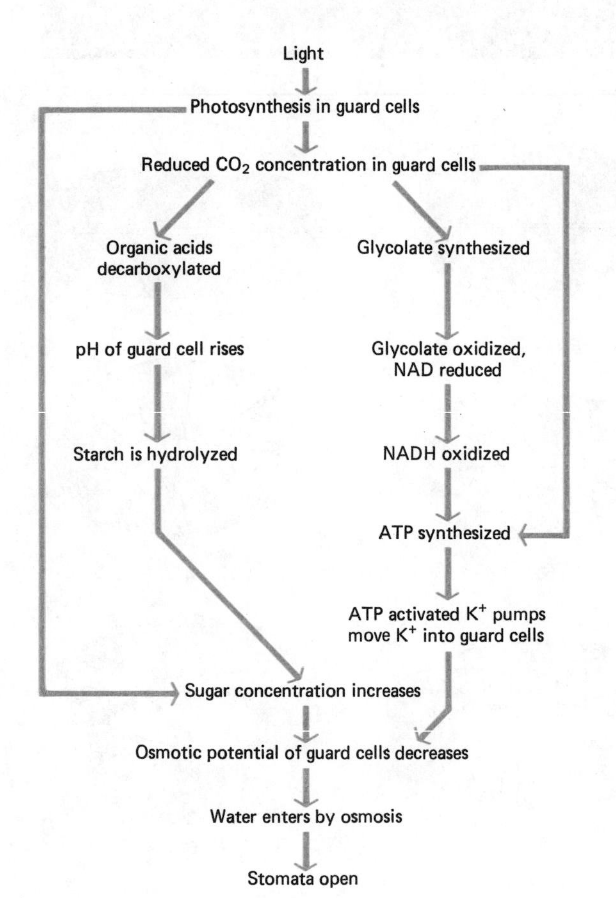


*Figure 10.13*
Comparison of the Modified Classical Theory and Ion Pump Theory of Stomatal Opening. [From R. G. S. Bidwell, "Plant Physiology," fig. 14-6, p. 303. Copyright ©1974, R. G. S. Bidwell. Reprinted by permission of Macmillan Publishing Co., Inc., N.Y.]

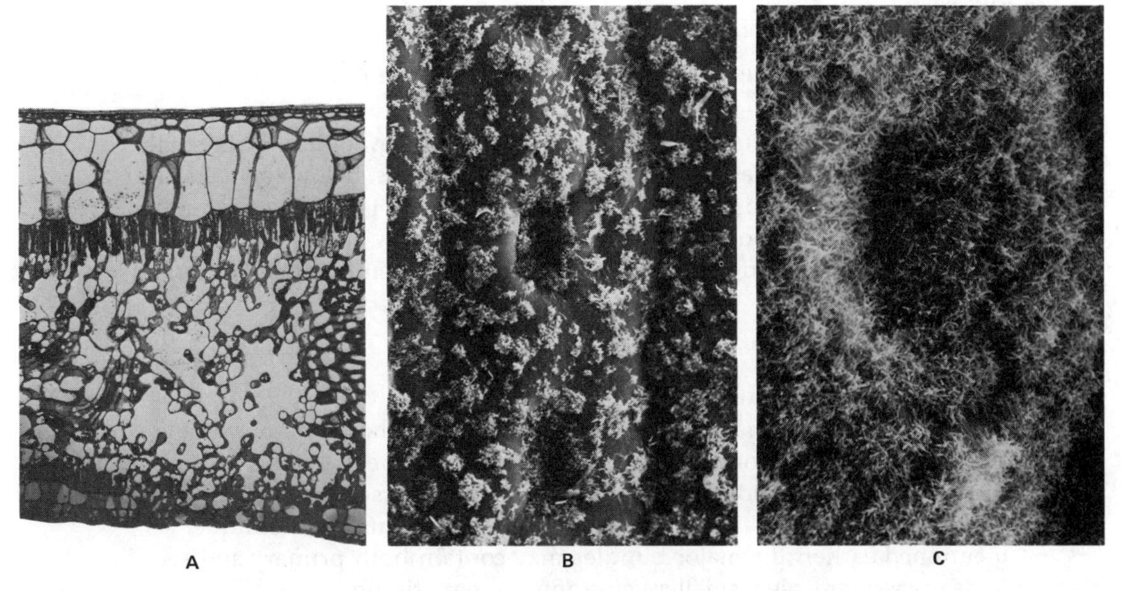


*Figure 10.14*
*A.* Note thick cuticle on leaf of *Ficus*. *B.* Scanning electron micrograph of wax distribution on adaxial surface of *Pseudotsuga* needle. *C.* Scanning electron micrograph of wax distribution of abaxial surface of *Pseudotsuga* needle. [*B.* and *C.* from B. W. Thair and G. R. Lister, "The Distribution and Fine Structure of the Epicuticular Leaf Wax of *Pseudotsuga menziezii*." Reproduced by permission of the National Research Council of Canada from the *Canadian Journal of Botany*, 53:1063–1071, 1975.]

Although both the modified classical theory and the ion pump theory are based on a number of valid observations, neither theory explains adequately all instances of stomatal activity. Many botanists now believe that both ATP and potassium are needed for stomatal opening, yet $K^+$ fluxes have not been noted for a great many plants. It is quite likely that both theories may be valid and that the phenomenon of stomatal opening and closing may differ for different plants. It would probably not be too farfetched to suppose that yet undiscovered mechanisms may exist for certain plants. But regardless of the mechanism, stomatal opening is extremely important for the influx and efflux of gases, primarily $CO_2$, $O_2$, and water vapor.

The walls of epidermal cells often contain lignin or cutin, with superficial layers of cutin, the *cuticle*, often present. The cuticle may be very thin or quite thick (Figure 10.14). Some epidermal cell walls may be projected into unicellular or multicellular hairs; some simple, others branched, some slender, others bulbous or glandular.

## *Mesophyll*

The mesophyll(*meso*, "middle"; *phyll*, "leaf") is that tissue between the upper and lower epidermis. The palisade layer, from the French *palissade* (a fence of stakes), occurs generally towards the upper epidermis. In some leaves it will occur adjacent to both epidermal surfaces. Below the palisade layer, between the palisade and the lower epidermis, is the spongy mesophyll. Cells of the spongy mesophyll have generally fewer chloroplasts than the palisade, and cells loosely packed with prominent intercellular spaces (Figure 10.7).

## *Veins*

As dicot leaves are characterized by net venation, the veins or vascular bundles form a network extending throughout the leaf. A cross section through a leaf will indicate the midvein in transverse section, with the other veins in various sections: transverse, longitudinal, and oblique. As mentioned earlier, the major bundles may contain both primary and secondary vascular tissue, smaller veins only primary tissue.

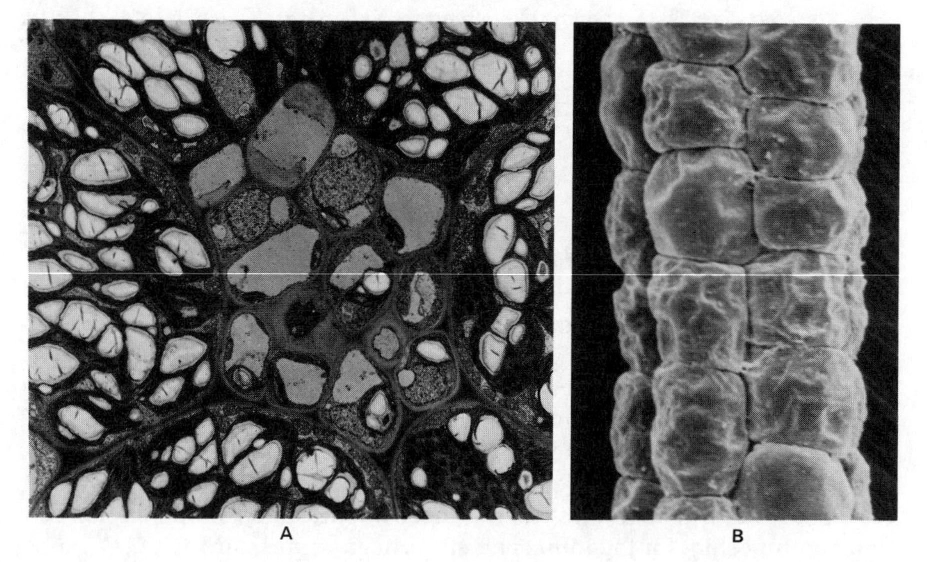


*Figure 10.15*
*A.* Transmission electron micrograph of starch sheath of *Amaranthus edulis. B.* Scanning electron micrograph of starch sheath [*A.* Courtesy C. C. Black, Jr., *B.* courtesy C. C. Black, W. H. Campbell, T. M. Chen, and P. Dittrich, "The Monocotyledons: Their Evolution and Comparative Biology. III. Pathways of Carbon Metabolism Related to Net Carbon Dioxide Assimilation by Monocotyledons," *Quart. Rev. Biol.* 48(2): 299–313. ©1973.]

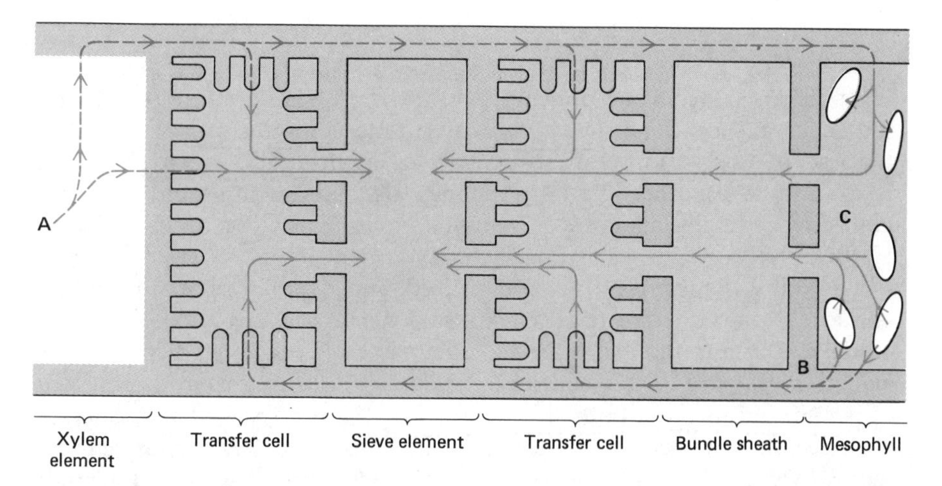


*Figure 10.16*
Transfer cells in leaf minor veins—possible pathways for solutes. Cell walls grey, cytoplasm stippled, →→ symplastic routes, -->-->--> apoplastic routes. A: solutes arriving in transpiration stream; B: solutes leaking from mesophyll to free space of leaf; C: symplastic route from mesophyll to sieve element. [From J. S. Pate and B. E. S. Gunning, "Transfer Cells," *Ann. Rev. Plant Physiol.* 23:180, fig. 2. Reproduced by permission from *Annual Review of Plant Physiology*, volume 23. Copyright © 1972 by Annual Review Inc. All rights reserved.]

Cells surrounding the vascular tissue are often distinct morphologically from other mesophyll cells. These cells constitute the *bundle sheath* and may extend from the bundles to the leaf surfaces as the *bundle sheath extensions.* Observations indicate that the bundle sheath can be considered an *endodermis,* as Casparian strips are observed occasionally. Some bundle sheath cells contain considerable starch, in which case the bundle sheath could be termed a *starch sheath* (Figure 10.15). Recent work indicates considerable structural and functional specialization of the bundle sheath cells in certain species, particularly members of the grass family, a group of monocots. We shall have occasion to discuss this aspect in much more detail later.

In minor veins, specialized cells called *transfer cells* may develop. Such cells are almost restricted entirely to dicots, mostly herbaceous species. Transfer cells are specialized cells in which ingrowths of cell wall material project into the lumen of the cell thus increasing the surface area of the plasmalemma. One of the functions of such cells is believed to be the export of photosynthates from the mesophyll. They could also allow for movements of certain solutes from the xylem tracheids to the mesophyll. Pate and Gunning (1922) suggested an exchange system as visualized in Figure 10.16.

## Monocot

The epidermal layers of monocot leaves are quite similar to those of dicots with several exceptions. The guard cells of many monocots may be below the level of other epidermal cells and are thus referred to as *sunken* (Figure 10.17). However, other plant groups may also exhibit this condition. Stomata are generally arranged in parallel rows (as in *Pseudotsuga*, Figure 10.10C).

The cell walls may be heavily lignified, and deposition of silica in the epidermal walls is characteristic of grasses. Cork cells may also occur in the epidermis of grasses. Various monocots may have specialized, large, thin-walled, highly vacuolated cells called *bulliform cells* occurring in either the upper epidermis or on both sides of the leaf (Figure 10.18). These cells are not necessarily restricted to the epidermis. Occasionally, similar cells are found in superficial regions of the mesophyll. These cells contain much water and can rapidly expand or contract. Thus, they have been associated by some investigators with the rolling or unrolling of the leaf blade. This function has not been demonstrated conclusively, however. In many monocots, the mesophyll is not differentiated into palisade and spongy tissue. The parenchyma cells are rather of a fairly uniform size and structure (Figure 10.17).

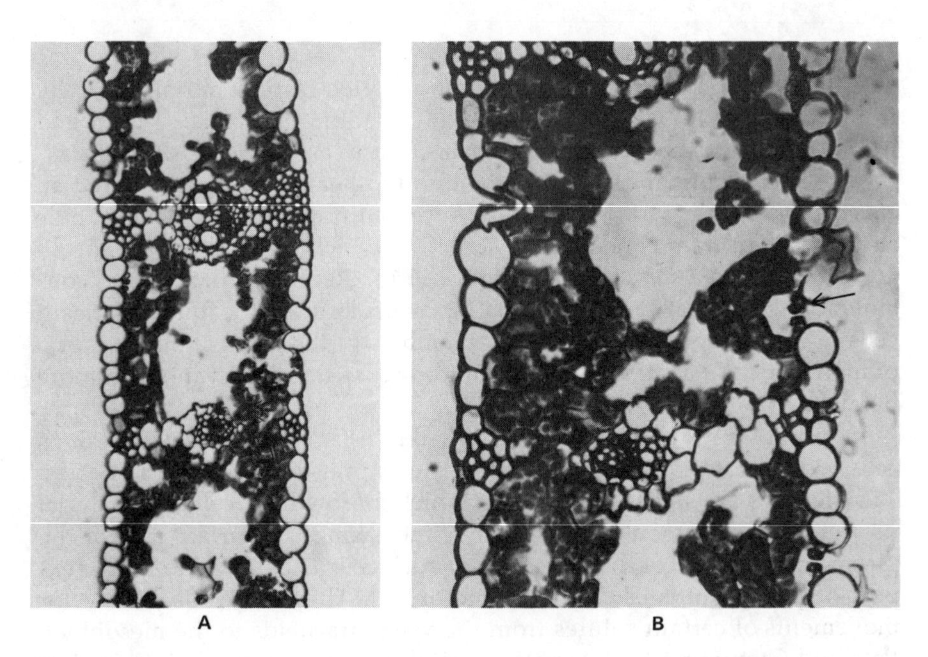


*Figure 10.17*
*Poa* leaf cross-section. Note sunken guard cells (*arrow*). *A.* Low magnification. *B.* High magnification.

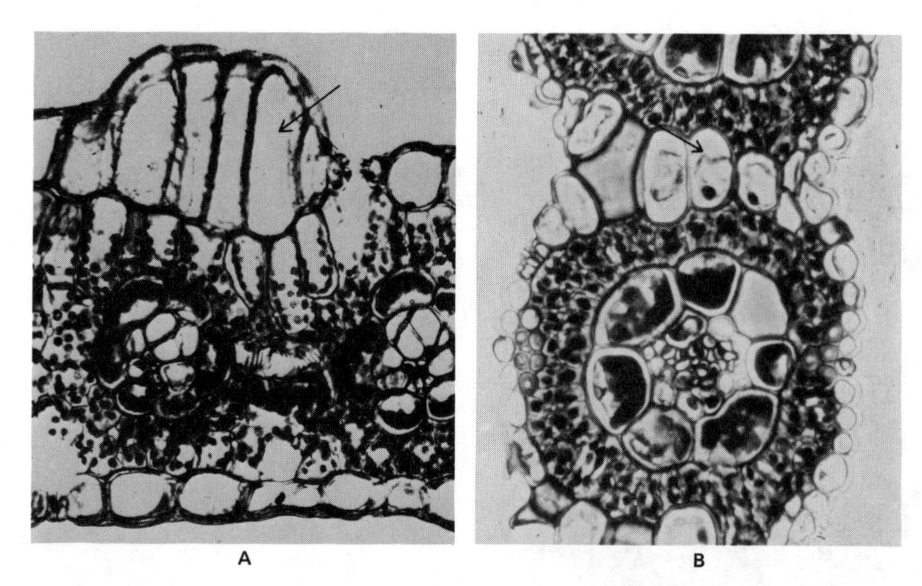


*Figure 10.18*
A. *Zea mays* leaf cross-section. Note enlarged bulliform cells (*arrow*) in upper epidermis. B. *Buchloe dactyloides* leaf cross-section. Note bulliform cells (*arrow*) in mesophyll region. [*B.* From W. M. Laetsch, "The $C_4$ Syndrome: A Structural Analysis, *Ann. Rev. Plant Physiol.* 25:28, fig. 1. Reproduced by permission from *Annual Review of Plant Physiology*, volume 25. Copyright © 1974 by Annual Review Inc. All rights reserved.]

The vascular tissue is all primary and arranged in vascular bundles that all appear in cross-sectional view in a leaf transection. This is, of course, accounted for by the fact that most monocot leaves have parallel venation (Figure 10.17). Bundle sheaths occur in monocot leaves just as in dicots. In grass leaves, these may be of two kind: 1) thin walled with chloroplasts, and 2) relatively thick walled without chloroplasts (Figure 10.19). The thick-walled sheath is termed a *mestome sheath* and, when present, occurs next to the vascular tissue with the thin-walled sheath outside of it. The mestome sheath may contain Casparian strips and thus has features of an endodermis. Monocot leaves may have a great number of fibers associated with the bundles.

## Gymnosperm (Conifer)

The needlelike leaves of pine exhibit many of the characteristics of xerophytic leaves. The leaf is semicircular or triangular in shape with the center being traversed by one or two vascular bundles. The epidermis is heavily cutinized and has thick, lignified cell walls. It bears numerous stomates in longitudinal rows, with sunken guard cells. Beneath the epi-

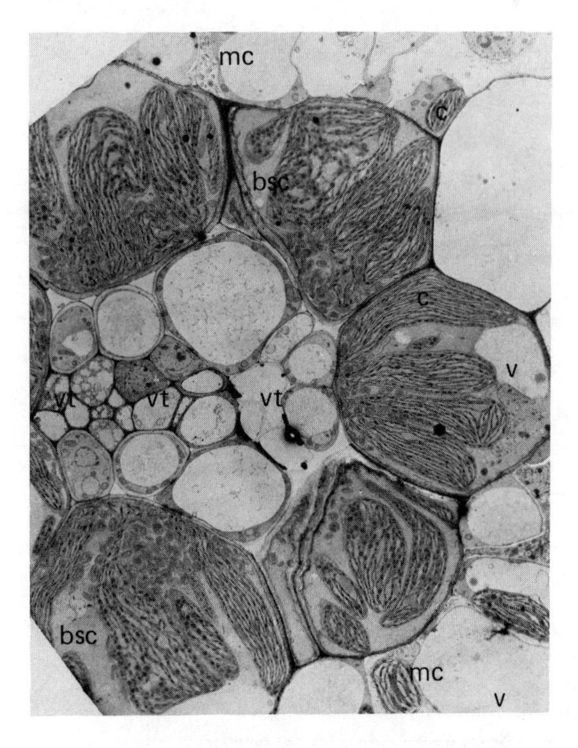


*Figure 10.19*
Transmission electron micrograph of Bermudagrass leaf. Note lack of chloroplasts in mestome sheath and presence of chloroplast in bundle sheath; *bsc,* bundle sheath cell; *mc,* mesophyll cell; *v,* vacuole; *c,* chloroplast; *vt* vascular tissue. [Courtesy Dr. C. C. Black, Jr.]

dermis is a compact hypodermis composed of fiberlike cells. The mesophyll is composed of much infolded cells having internal ridges projecting into the cell lumen. The mesophyll is not differentiated into the palisade and spongy tissue, although certain conifers other than pine show such differentiation. *Resin ducts* occur in the mesophyll and are lined with thin-walled secretory cells, outside of which is a sheath of fibers (Figure 10.20).

The vascular bundles contain primary xylem and phloem oriented in tiers like secondary tissue. After the first year, some secondary tissue is produced by the activity of a vascular cambium. (Pine needles are normally shed in three to four years.) Surrounding the vascular bundles is a tissue called the *transfusion tissue*. It consists of thin-walled parenchyma cells and thin-walled but lignified *tracheids* with bordered pits. The parenchyma cells may contain starch during their first year but later contain tannins and resins. Surrounding the transfusion tissue is an *endodermis* consisting of rather thick-walled cells. (Figure 10.21).

## ENVIRONMENTAL CONTROL OF LEAF STRUCTURE

Although leaf shape and form is under direct genetic control, with both chromosomal and nonchromosomal DNA controlling leaf development, environmental factors also play a role in leaf differentiation. Leaves are

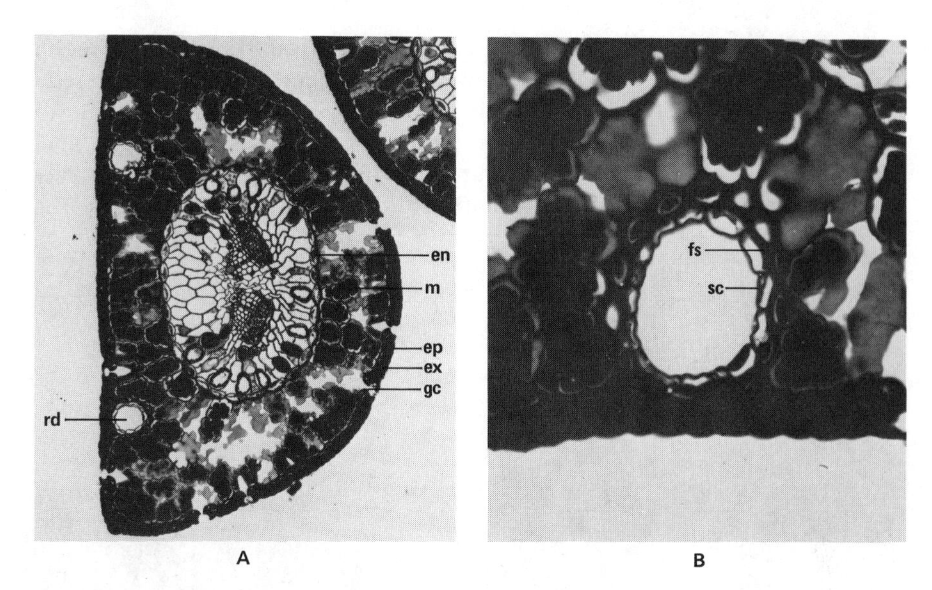


*Figure 10.20*
A. *Pinus* leaf cross-section; *ep*, epidermis; *ex*, exodermis; *m*, mesophyll; *rd*, resin duct; *gc*, guard cells; *en*, endodermis. *B*. Cross-sectional view of resin duct; *sc*, secretory parenchyma; *fs*, fiber sheath.

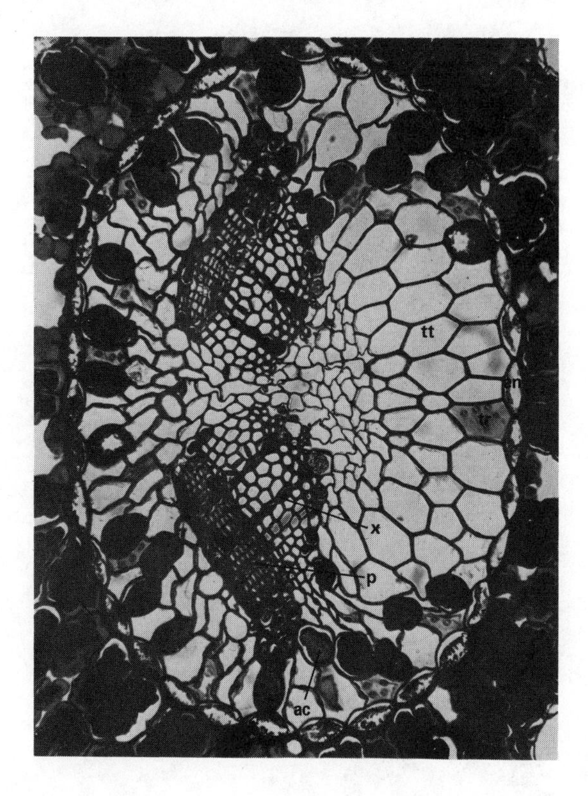


*Figure 10.21*
Central region of *Pinus* leaf, cross-section; *en*, endodermis; *tt*, transfusion tissue; *tr*, tracheid; *ac*, albuminoid cell; *x*, xylem; *p*, phloem.

subjected to various environmental conditions during their development. The most important of these are sunlight, temperature, and water relations. Internal environmental aspects include hormonal factors.

Sunlight affects size, structure, and position of leaves. In a single species, the leaves subjected to full sunlight (the sun leaves) are generally thicker, smaller, and more differentiated than those developing in the

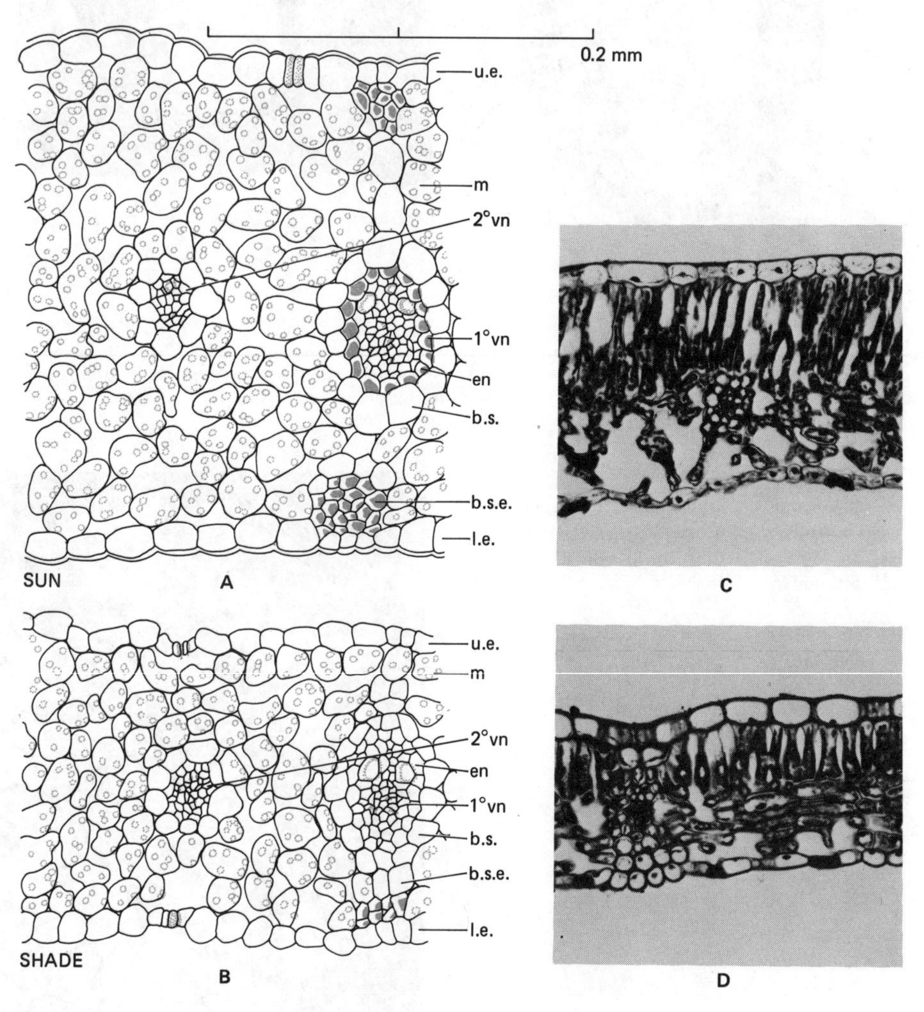


*Figure 10.22*
Sun leaf (A) versus shade leaf (B) in *Poa*, leaf cross-section; *ue*, upper epidermis; *m*, mesophyll; 2°*vn*, secondary vein; 1°*vn*, primary vein; *en*, endodermis; *bs*, bundle sheath; *bse*, bundle sheath extension; *le*, lower epidermis. Sun leaf (C) and shade leaf (D) in *Sambucus*. [*A.* and *B.* from B. F. Slade, "The Effect of Cell Elongation on Leaf Anatomy of *Poa alpina*," *Bot. Gaz.* 131(1):92, figs. 15 and 16. ©1970 by The University of Chicago. All rights reserved.]

shade (the shade leaves). In sun leaves, more mesophyll cells differentiate as palisade, and these cells are much more elongated than are the palisade of shade leaves. In shade leaves, the palisade cells are much shorter and wider. The number of palisade layers is reduced, and the spongy layer is much diffused with numerous air space (Figure 10.22). Epidermal cells of sun leaves are often smaller, more numerous per unit area, have a thicker cuticle, greater development of epidermal hairs, and more stomata per unit area than those of the same plant grown in the shade (Figure 10.23). The stomata are generally on both leaf surfaces in the shade leaf but restricted to the lower surface, or at least are greater in number there, in the sun leaves. The number of epidermal cells and stomata may not be related directly to light intensity. As cells in direct sunlight are heated more than those in the shade, greater evaporation of water would reduce the water content of these cells, thus reducing cell pressure and expansion. Therefore, temperature is also a controlling factor. Recent work by Kimball and Salisbury has indicated changes in protoplasmic components in response to low temperatures.

It has been shown that light duration can also be a factor in leaf development. A *Kalanchoë*, grown in a light regimen of 8 hours of light, 16 hours of darkness, developed small, thick, sessile leaves having an entire margin. In 16 hours of light, 8 hours of darkness, the *Kalanchoë* developed larger, thinner, petioled leaves with notched margins.

Light quality may also be a factor. Plants grown in the dark will have leaves which lack chlorophyll and are small and unexpanded. When certain plants are exposed to only two minutes of red light and then placed in darkness, the leaves will become larger and expanded. If the dark-

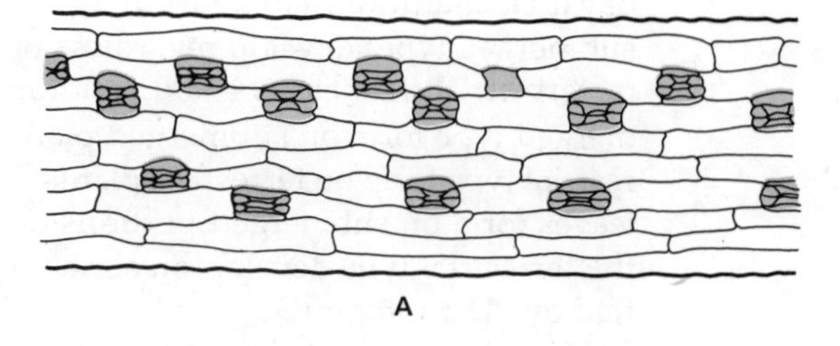


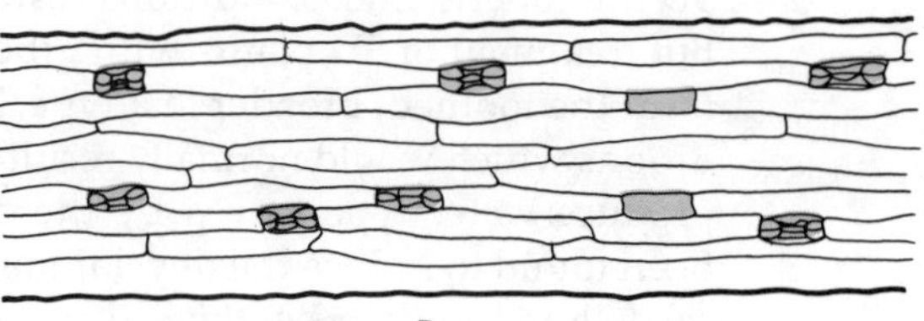


*Figure 10.23*
*Epidermal layer in Poa. A.* Sun leaf. *B.* Shade leaf. [From B. F. Slade, "The Effect of Cell Elongation on Leaf Anatomy of *Poa alpina*," *Bot. Gaz.* 131(1):93, fig. 17 and 19. ©1970 by The University of Chicago. All rights reserved.]

grown plant is exposed to two minutes of red light followed by two minutes of far red light, and then placed back in darkness, the leaves will remain small and unexpanded. Thus, it appears that far red light reverses the effect of red light on leaf expansion.

Water relations also affect leaf form. Aquatic plants, called *hydrophytes*, have certain characteristics in common. Some aquatic plants have leaves both above and below water. The submersed leaves will typically be finely divided, or dissected, thin, and without stomata, while the aerial leaves will be entire, thicker, and with stomata (Figure 10.24). If a shoot apex in *Proserpinaca* (mermaid weed) is held just below water during leaf initiation, it will form highly dissected leaves. If it is held just above water, broader leaves with a toothed margin will form. By alternately holding the stem tip first above, then below, the water level, and continuing this alternation for a period of time, one can produce a plant having successive tiers of divided and undivided leaves on the same stem. The same effects can be produced by altering the light periods.

Not only is the form of the leaf of *Proserpinaca* altered by submersion, but its internal structure also differs. The aerial leaves have a single palisade and three to four layers of spongy mesophyll, with stomata on both leaf surfaces. The highly dissected submerged leaves have generally three layers of mesophyll, not differentiated into palisade and spongy, and few or no stomata.

The effect of the period of light (photoperiod) in producing effects similar to that of submergence is an interesting one. In the aquatic plant *Ranunculus aquatilis* (common white water crowfoot), three leaf forms occur: submerged, extremely dissected leaves; lobed, more or less entire, floating leaves; and short segments of aerial, terrestrial, dissected leaves. These types normally develop under long day (LD) conditions. But under short day (SD) conditions (10 hours of light or less), highly dissected leaves (the submerged type) develop regardless of whether the plant was submerged or terrestrial. Under a 14-hour photoperiod, dissected leaves of the submerged type form on submerged plants, while dissected leaves of the terrestrial type form on terrestrial plants. Under a 16-hour light regime, entire leaves form on submerged portions of the plant while dissected leaves of the terrestrial type develop on aerial portions. Thus, water effect is modified by LD photoperiod.

Hormones have also been found to influence leaf form. *Proserpinaca* can be induced to form dissected leaves on aerial plants by subjecting them to SD photoperiod; LD conditions result in the typical toothed, broader leaf. But treatment of SD plants with gibberellic acid (GA) will induce them to form the toothed, broader leaf type. The aquatic fern *Marsilea*, under conditions which would normally result in land type leaves, can be induced to produce water type leaves by the addition of GA. Gibberellic acid has been found to induce the development of lanceolate leaves instead of typical deltoid leaves in *Xanthium* (cocklebur).

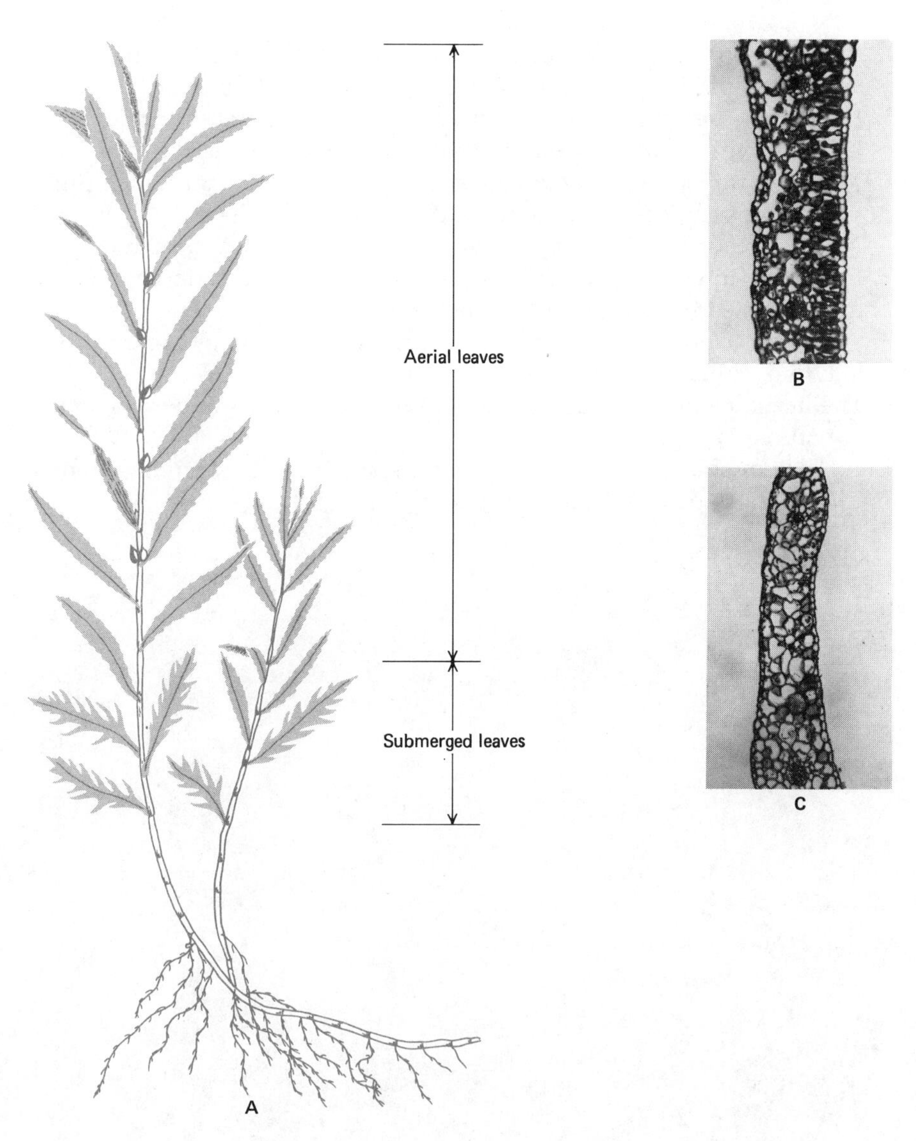


*Figure 10.24*
Air versus water leaves in *Proserpinaca*. *A*. Plant with emergent and submerged leaves. *B*. Cross-section of air leaf. *C*. Submerged leaf.

# SUMMARY

1. Leaves develop from the apical meristem of stems and are thus lateral appendages of the stem.
2. The leaf consists of a more or less flattened organ called the blade, al-

though gymnosperm leaves may exhibit a different morphology. In addition, a petiole, stipule, sheath, or ligule may be present.

3. Leaves may be simple or compound, net veined or parallel veined.
4. Leaves have variable characteristics of shape, margin, apex, and base.
5. The epidermis is perforated characteristically by stomata which function in gas exchange between the leaf and the atmosphere.
6. A unique tissue, found only in leaves, is the mesophyll, which lies between the upper and lower epidermis. It may be undifferentiated or differentiated in palisade and spongy layers.
7. Some characteristics of leaf form, structure, or physiology are transmitted from parent to offspring and are thus genetic in nature.
8. The heritable factors may be either chromosomal or nonchromosomal in nature.
9. Light, temperature, and water relations are important environmental influences on plant form and structure.

# Chapter 11

## THE VASCULAR PLANTS— A SURVEY

The vascular plants exhibit two major aspects in common; 1) they possess vascular tissue (as their name implies), and 2) the sporophyte is dominant and nutritionally independent. We think commonly of vascular plants as possessing the vegetative organs: stems, roots, and leaves. But as mentioned earlier, one or two of these organs may be lacking. In the case of *Wolffia*, the plant body appears to be completely undifferentiated. The stem appears to be the basic organ of structure, bearing lateral appendages (leaves) and roots from its base (although in higher plants a hypocotyl may be intercalated between stem and roots). Roots are thought to have evolved from an underground, rhizomelike stem, while leaves are postulated to have arisen in several ways. Two distinct morphological types of leaves exist; *microphylls* and *megaphylls*. Microphylls are relatively small leaves containing a single strand of vascular tissue and leaving no leaf gap (a region of parenchyma cells forming a gap at and above the point of departure of the leaf trace from the stele, Figure 11.1A). Megaphylls are typically larger, have branched veins, and their veins form a leaf gap, except in protostelic stems (Figure 11.1B). According to the *enation theory*, microphyllous leaves were formed simply by superficial lateral outgrowths, or enations, of the stem, with the vascular tissue eventually extending out into them (Figure 11.2). According to the *reduction theory*, microphylls were formed simply by the reduction of a megaphyllous leaf (Figure 11.3). Megaphylls were formed supposedly from branches of stem which underwent a series of steps illustrated in Figure 11.4. First the normally dichotomously branching stem developed unequal branching *(overtopping)*, fol-

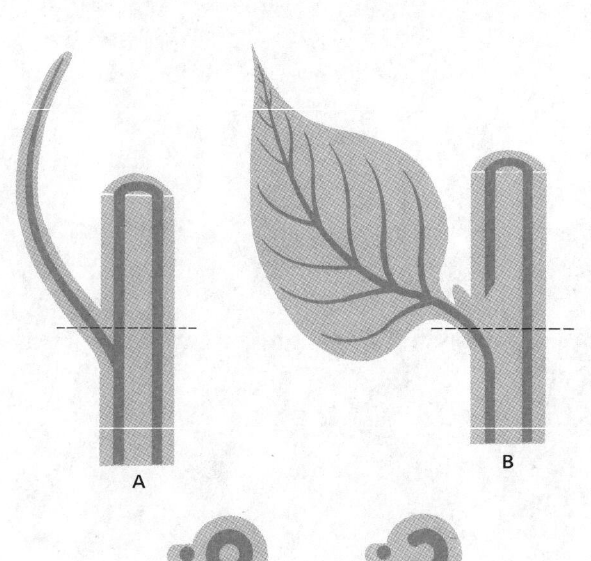


*Figure 11.1*
Segments of longisection on stems. *A*. Stem with microphyllous leaf. *B*. Stem with megaphyllous leaf. *C*. Cross-section through node of *A*. *D*. Cross-section through node of *B*. Note absence of leaf gaps in microphyll.

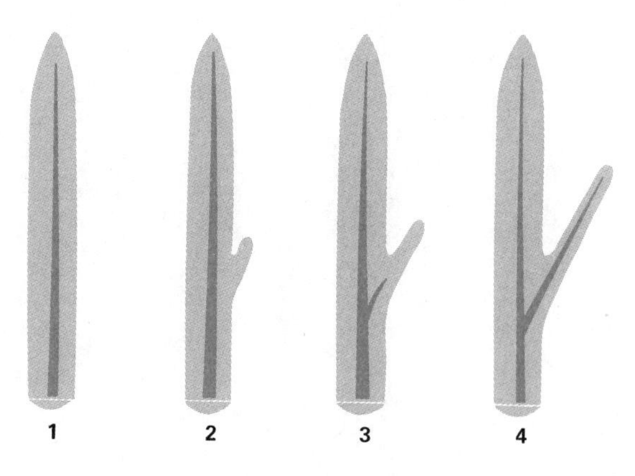


*Figure 11.2*
Evolutionary origin of a microphyllous leaf according to the enation theory.

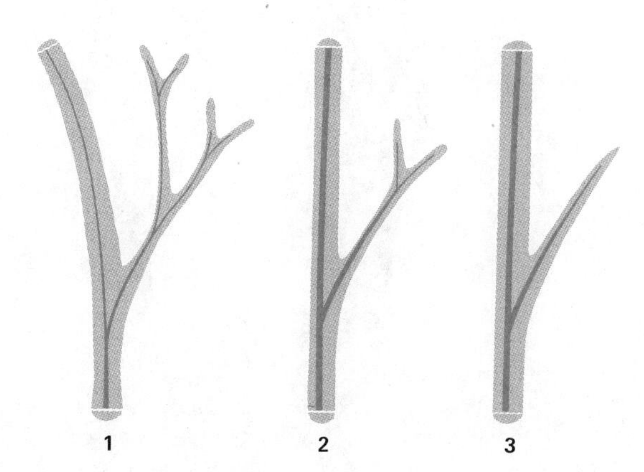


*Figure 11.3*
Evolutionary origin of a microphyllous leaf according to the reduction theory.

lowed by a flattening *(planation)* which limited the branching to one plane, and finally a fusion *(webbing)* of these separate lateral branches forming a single blade. This theory is referred to as the *telome theory.*

Another important aspect to understand is the concept of stele, discussed in Chapter 8. You recall, the two basic types of steles were the *protostele*, which lacks pith, and the *siphonostele*, which has pith. The protostele is obviously the simplest of the two and for that reason thought to be the most primitive, although the fossil record has not yet conclusively supported that assumption.

A third aspect involves the reproductive system. Up to now, all spores which were produced as the result of meiosis *(meiospores)* were of one morphological type. This is thought to be the more primitive, and such plants are said to be *homosporous* (*homo*, "one"; *spora*, "spore"). However, in our discussion of the various groups of vascular plants, we shall find that two types of spores may be produced. Such plants are said to be *heterosporous* (*hetero*, "different"; *spora*, "spores"). Of these two types, the

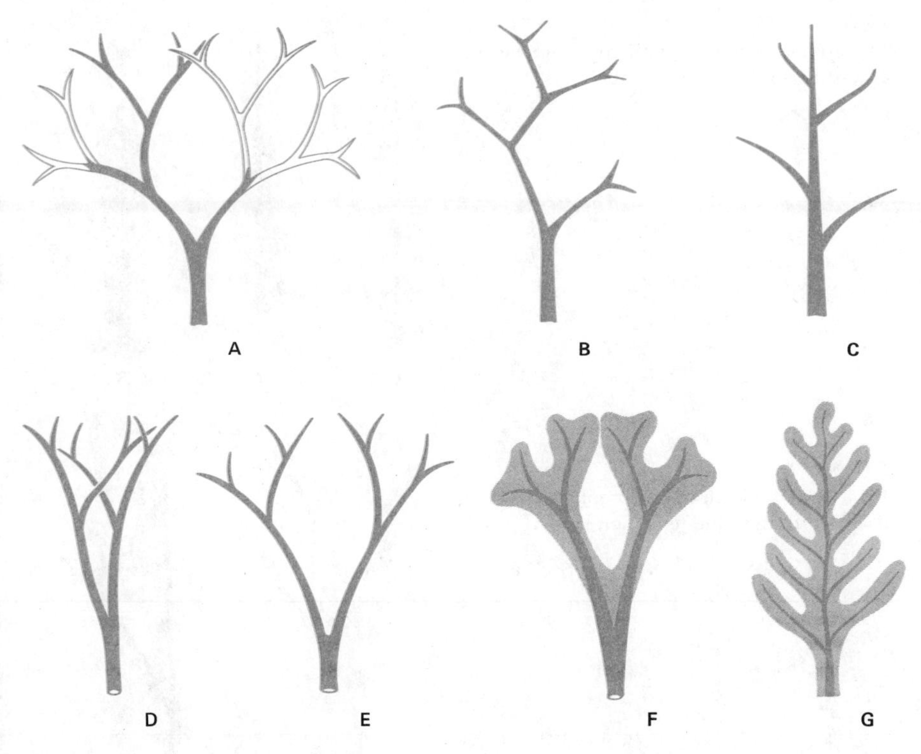


*Figure 11.4*
Evolutionary origin of megaphyllous leaves according to the telome theory. *A.* to *C.* Overtopping. Development of monopodial branch from dichotomous branches. *D.* and *E.* Flattening (planation) of the branches. *F.* and *G.* Webbing.

one producing the male gametophyte is called the *microspore*, the one producing the female gametophyte, the *megaspore*.

With the increase in complexity of the sporophyte, it might be a good idea at this point to explore some terminology so you will have a concept of the term when we use it later.

*Sporophyte* ("spore plant")—a plant that produces meiospores
*Sporophyll* ("spore leaf")—borne on the sporophyte: it is a leaf or leaflike structure upon or within which meiospores are produced
*Sporangium* ("spore case")—may or may not be borne on or in the sporophyll; it is the immediate structure within which meiospores are produced
*Sporocyte* ("spore cell")—produced within the sporangium; the typically diploid cells which undergo meiosis producing meiospores
*Gametophyte* ("Gamete plant")—normally the product of a germinating meiospore; it produces the gametes

With the aforementioned material as background, let us look at some of the diversity encountered in the vascular plants.

## CLUB MOSSES, SPIKE MOSSES, AND QUILLWORTS

Many plants have the distinction of being called mosses when in fact they are not. We encountered one such example, the reindeer moss, earlier in our discussion of lichens. Now we find two genera being called mosses, and in fact they were classified with the mosses by Linnaeus. There are five living genera in this group: *Lycopodium*, commonly called the club mosses, ground pine, or trailing evergreens; *Selaginella*, sometimes called the spike mosses; *Isoetes*, the quillwort; *Phylloglossum*, and *Stylites*. The first three are found in the United States.

This division, the Lycopodophyta, separates easily into two subgroups; those which are homosporous and eligulate (lack a ligule) and those which are heterosporous and ligulate. The term *ligule* refers to a small protuberance borne at the base of the leaf. The plants possess roots, microphyllous leaves, and a protostele.

### *Lycopodium*

The sporophyte of this genus is evergreen and consists of a branching rhizome from which aerial branches and adventitious roots arise. The roots are delicate and dichotomously branched (divided into two equal parts). The arrangement of xylem and phloem in stem and root depend upon the species and also upon the size of the organ (Figure 11.5). Stomata are

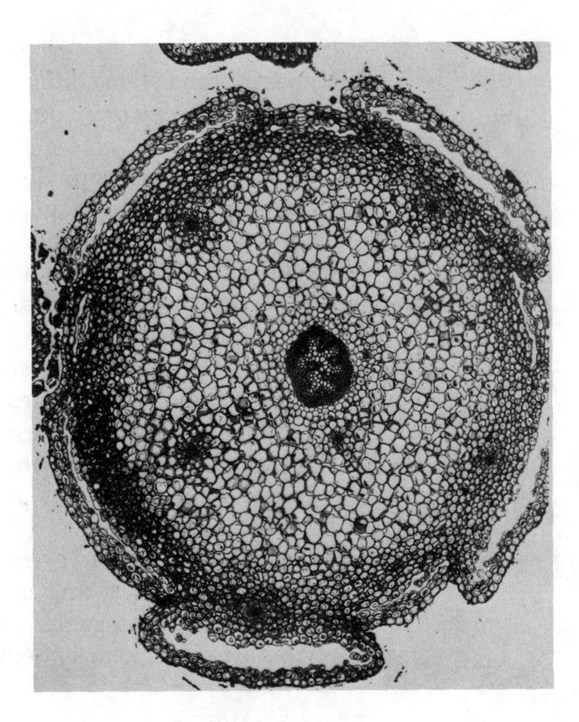

*Figure 11.5*
*Lycopodium* stem cross-section. A cresentric protostele (plectostele). The roots of older plants are adventitious and no endogenous lateral roots are formed. Root hairs are abundant and persistent (long-lived).

*Figure 11.6*
*Lycopodium lucidulum; b,* propagative gemma or bulb *s,* axillary sporangium. [From a color photograph by Dee Ann Houston].

found in both leaf and stem epidermis, with the leaves bearing stomata on only the lower epidermis or on both upper and lower, depending upon the species.

In some species, the sporophylls are similar to the vegetative leaves and occur in zones among them (Figure 11.6), while in other species they are aggregated into a terminal conelike structure called a *strobilus* (Figure 11.7). The sporangium develops into an inner mass of sporogenous tissue and an outer, three-layered wall. The innermost layer of the wall functions as a nutritive layer and is called the *tapetum.* The cells of the tapetum are digested during maturation of the sporocytes. All the cells of the sporogenous tissue develop into sporocytes, with each sporocyte dividing meiotically to produce a tetrad of spores. The spores of some species are very thick walled and may not germinate for five to seven years.

Upon germination, the spores produce a bisexual gametophyte which may either be green, somewhat heart shaped or variously lobed in form, and epiterranean (*epi,* "upon"; *terra,* "earth"); or it may be colorless, tuberous or cylindrical in form, and subterranean (*sub,* "below"; *terra,* "earth"). The subterranean gametophytes contain a symbiotic fungus. The gametophytes are small, usually being only a few centimeters in length.

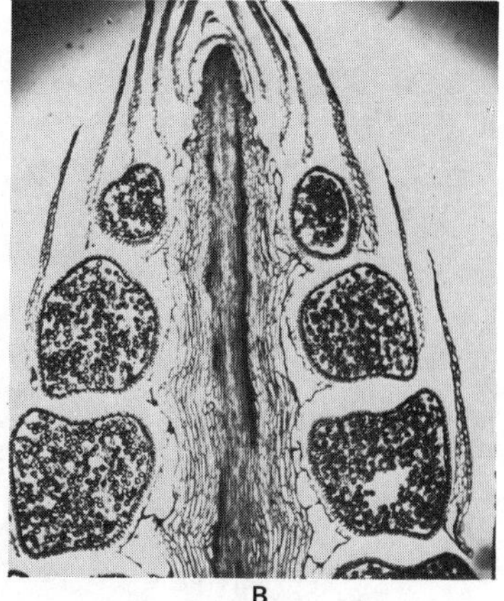


*Figure 11.7*
Strobili of *Lycopodium*. *A*. *L. complanatum*. *B*. *Lycopodium* sp., longisection of strobilus. [*A*. by David R. Williams].

Antheridia are borne on the upper surface of the gametophyte and water is required for fertilization. The biflagellated sperm swim to the archegonium and down its neck. The resulting zygote develops within the archegonium, eventually producing the free-living sporophyte. Under laboratory conditions, a gametophyte has been shown to give rise to a sporophyte directly, that is, without the fusion of gametes taking place. Such an occurrence is called *apogamy* (*apo*, "without"; *gamy*, "fertilization"), and the resulting sporophyte, assuming the gametophyte was haploid, would be haploid.

## Selaginella

Among the many species of *Selaginella*, most of which live under moist, tropical conditions, is a desert species which one can also often find in novelty shops: *Selaginella lepidophylla*, the resurrection plant.

Stem anatomy is somewhat unusual, with the cortex and tissue of the stele connected by elongated endodermal cells called *trabeculae* (*trabecula*, "little beam," Figure 11.8). In some species, the stem may contain more than one stele, a condition termed *polystelic*. Some of the cells of the xylem differentiate into vessels rather than tracheids. The microphyllous leaves bear a small tonguelike outgrowth, called a *ligule*, at their base.

All species bear their sporangia in strobili, although the strobili may be so loosely arranged that they are hardly distinct from the vegetative shoot. The sporangium wall is two layered, inside of which is a tapetal layer and sporogenous tissue (Figure 11.9). In some sporangia, a few sporocytes degenerate, but most undergo meiosis producing tetrads of spores. In other sporangia, usually all but one of the sporocytes degenerate, with the

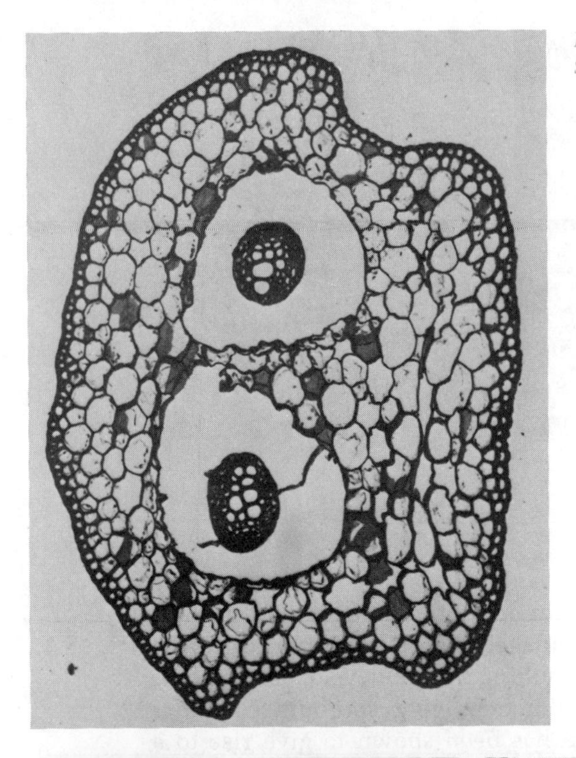

*Figure 11.8*
*Selaginella* sp., cross-section of stem.

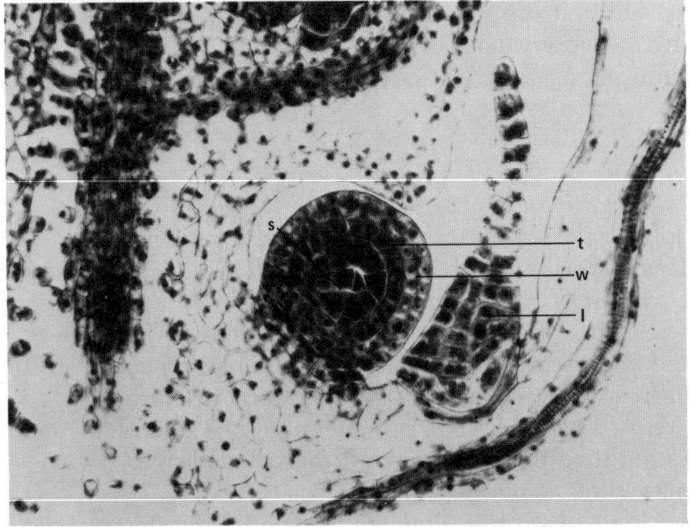


*Figure 11.9*
*Selaginella* sp., immature sporangium; *t*, tapetum; *w*, wall; *s*, sporogenous tissue; *l*, ligule.

survivor undergoing meiosis producing a spore tetrad, the members of which enlarge greatly. The enlarged spores are called *megaspores*, and the sporophylls and sporangia and sporocytes associated with them, *megasporophylls*, *megasporangia*, and *megasporocytes*, respectively. The smaller

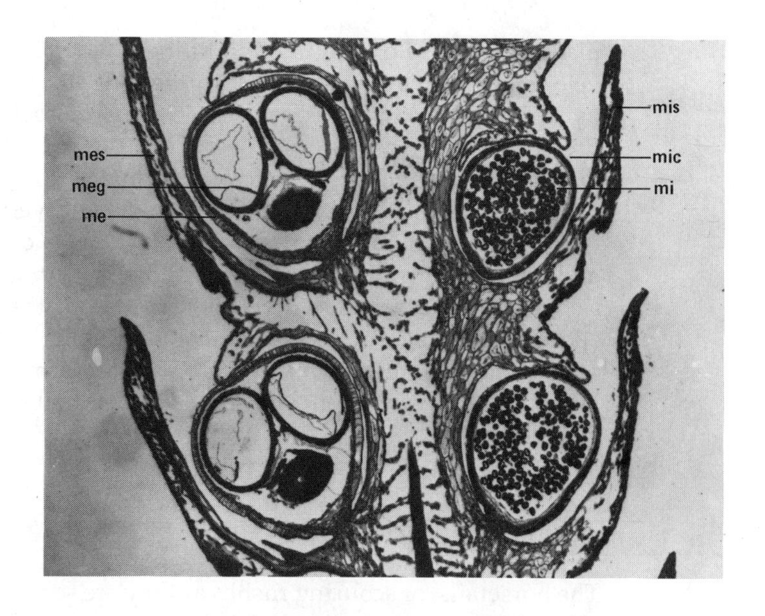


*Figure 11.10*
*Selaginella* sp., longisection of mature strobilus; *me,* megasporangium; *meg, megaspore; mes,* megasporophyll; *mic,* microsporangium; *mi,* microspore; *mis,* microsporophyll.

spores are called *microspores,* and sporophylls, sporangia, and sporocytes associated with them, *microsporophylls, microsporangia,* and *microsporocytes,* respectively. This dimorphic condition is, of course, known as *heterospory* (Figure 11.10).

Unisexual gametophytes are the result of spore germination, with the microspores producing the male, or microgametophyte, and the megaspores producing the female, or megagametophyte. Germination of the spores is termed *precocious,* in that it often occurs before the spores are shed from the sporangia. In fact, the gametophyte may become sexually mature while still in the sporangia (this rarely occurs in *Selaginella*). Union of the biflagellate sperm with the egg occurs after the spores containing the gametophytes have been shed from the sporangia. Although the megagametophyte has been reported to contain chlorophyll under certain conditions, normally both gametophytes are achlorophyllous.

## *Isoetes*

The quillworts are found generally as aquatic or amphibious plants, although terrestrial species are known. Their narrow, elongate, microphyllous leaves arise from a swollen, fleshy, basal structure that has been variously interpreted morphologically. Its upper portion has been interpreted generally as a much-shortened, vertical stem, while its lower portion bears roots. The structure has been referred to as a *corm.* It is protostelic, and outside the phloem is a meristematic region which functions as a cambium. If this is interpreted as a true cambium, then we have secondary growth occurring, an unusual situation in the more primitive vascular plants. For those of you that might like to look into the anatomy and development of this unusual plant in more depth, I can refer you to a report by D. Paollo (1963), *Developmental Anatomy of Isoetes*, University of Illinois Press.

Megasporophylls and microsporophylls contain sporangia in the spoon-shaped leaf base. Both sporophylls and vegetative leaves bear ligules. The sporangia are covered by a thin layer of tissue called the *velum* and are incompletely chambered internally by *trabeculae*. Spore germination is not precocious. After liberation from the sporangia, the microspores and megaspores germinate producing microgametophytes and megagametophytes, respectively. Gamete formation, fertilization, and sporophyte development are similar essentially to that described previously.

## HORSETAILS

The horsetails, or scouring rushes as they are sometimes called, consist of a single living genus in the division Sphenophyta (Figure 11.11). The horsetails have a number of very characteristic features. Their stems are jointed and organized into nodes and internodes, with the internodes fluted conspicuously. Minute, scalelike, microphyllous leaves are borne in whorls at each node. They are photosynthetic for only a short time, then become dry. Some species are branched conspicuously with the branches

*Figure 11.11*
Growth habit of two species of *Equisetum*. *A*. *E. telmateia*. *B*. *E. arvense*. [Courtesy of Carolina Biological Supply Company].

arising at the node and alternating with the leaves; others are unbranched. The upright, aerial stems arise from an extensive system of subterranean rhizomes. In some species, two types of aerial stems may be produced: a nonchlorophyllous fertile branch, and later, a green vegetative branch. In the development of the stem, some cells in the nodal region remain meristematic, thus constituting an intercalary meristem. These intercalary meristems are responsible for the elongation of the internode.

The central region of the stem is hollow at maturity and is surrounded by the remains of the pith. Outside of this *central canal* is a ring or circle of canals, each on a radius with a ridge of the stem. Thus, these are referred to as the *carinal* (*carina*, "keel") canals. The carinal canals are within discrete strands of xylem and phloem. In some species, an endodermis surrounds each strand. In others, there is a continuous ring of endodermal cells both to the inside and to the outside of the strands. In still others, only the outer endodermal layer is present. Just inside the endodermis is the pericycle. The cortical parenchyma, located to the outside of the endodermis, is also perforated by canals. They are opposite the grooves of the stem, thus alternate with the carinal canals. They are called the *vallecular* (*valle*, "valley") canals. The outer cortical cells are sclerenchymized, as is the epidermis layer (Figure 11.12). Silica is deposited on the inner walls

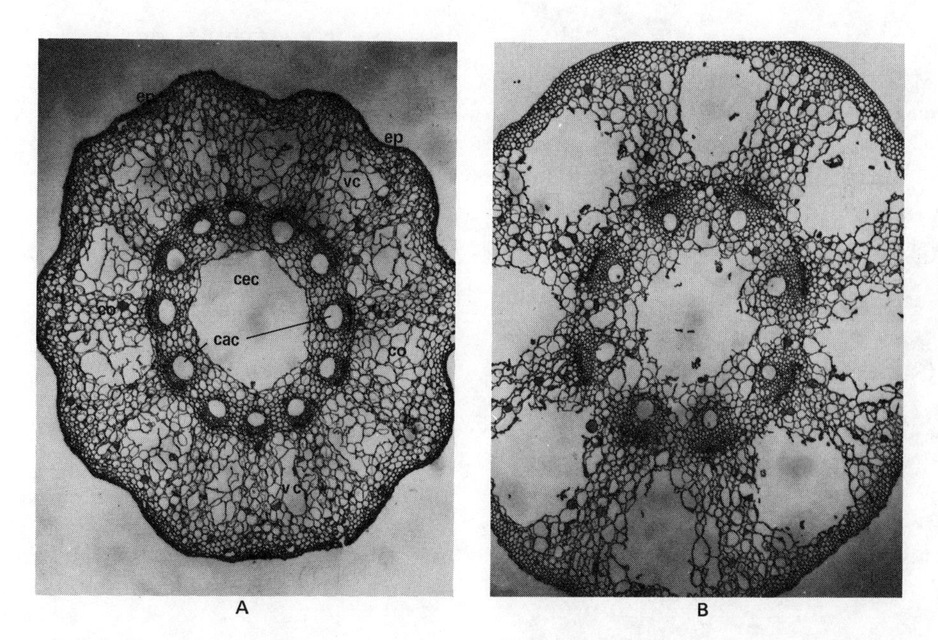


*Figure 11.12*
*A. Equisetum arvense*, stem cross-section; *ep*, epidermis; *co*, cortex; *vc*, vallecular canal; *cac*, carinal canal; *cec*, central canal. *B. Equisetum* sp., stem cross-section. Note numerous air spaces.

of the endodermal cells and in the walls of the epidermal cells, thus the usefulness of this plant to pioneer women for scrubbing pots and pans. This is why the plant is known as scouring rush.

The canals are not continuous throughout the stem, but are interrupted at each node by a thin tissue layer. Interpretation of the stele type has varied, but it is obviously of the siphonostelic type. Roots are adventitious, arising at the nodes, and have root hairs. Occasionally, branch roots develop from the pericycle.

The sporangia are borne in strobili, with the axis of each strobilus producing a series of enlargements called *sporangiophores.* The sporangiophore is generally somewhat hexagonal in surface view. Fingerlike sporangia develop from the base of each sporangiophore and are positioned inward toward the axis of the strobilus. Internal to the sporangial wall is a tapetum several layers in thickness. Digestion of these tapetal cells produces what is termed the *tapetal plasmodium* which flows around and contributes nourishment to the developing sporocytes. A number of sporocytes abort, but those that remain undergo meiosis forming a tetrad of spores. The spores are all of one type; thus, *Equisetum* is homosporous (Figure 11.13). The spore wall is complex, with the outermost layer consist-

*Figure 11.13*
*Equisetum* sp., longisection of sporangia.

ing of four spirally arranged portions which separate at dehiscence, with each bearing a somewhat spoonlike appendage at its tip. These structures are called elaters and are hygroscopic, thus aiding in spore dispersal. The spore wall contains chlorophyll. The spores germinate quickly if they encounter a suitable substrate. The gametophyte is green and free living. Some gametophytes produce antheridia only, thus are unisexual, while others are *protogynous.* The latter will produce only archegonia for a period of perhaps 35 to 80 days. When archegonia development ultimately ceases, the gametophyte produces antheridia. Sperm of *Equisetum* are multiflagellate; thus, water is required to effect fertilization.

## WHISK FERNS AND TRUE FERNS

### *Psilotum*

The whisk fern, *Psilotum*, has traditionally been placed in a separate taxonomic group. It has the least complex organization of any of the vascular plants. However, recent evidence has shown its close affinity with the ferns, thus its placement in the Filicinophyta (Figure 11.14).

*Figure 11.14*
*Psilotum nudum*, epiphytic on tree trunk. [Courtesy of Carolina Biological Supply Company.]

It has been suggested that the characteristics of *Psilotum* and its close relative *Tmesipteris* are derived rather than primitive. *Tmesipteris* is not native to the United States, so we will be discussing the characteristics of *Psilotum.*

The sporophyte consists of a dichotomously branched, upright axis produced from a horizontal underground rhizome. The rhizome bears rhizoids, and there are no roots produced. Its cells often contain endophytic fungi. The aerial stems bear small, scalelike appendages which lack vascular tissue and are called *prophylls.* The character of the stele changes at different levels of the aerial stem, but has usually been called an atactostele (Figure 11.15).

The sporangium is a globose, three-lobed structure which probably represents the fusion of three separate sporangia. Such a compound sporangium is called a synangium (Figure 11.16). The sporangial wall is several layers thick and surrounds the inner sporogenous mass. No true tapetal layer is formed; however, many of the potentially sporogenous cells are digested, forming a plasmodial mass which some botanists call the tapetal plasmodium. The remaining sporocytes undergo meiosis forming tetrads of spores (Figure 11.17).

The gametophytes, rarely larger than 2 mm in diameter, are cylindrical, lack chlorophyll, and contain an endophytic fungus. Occasionally, a small

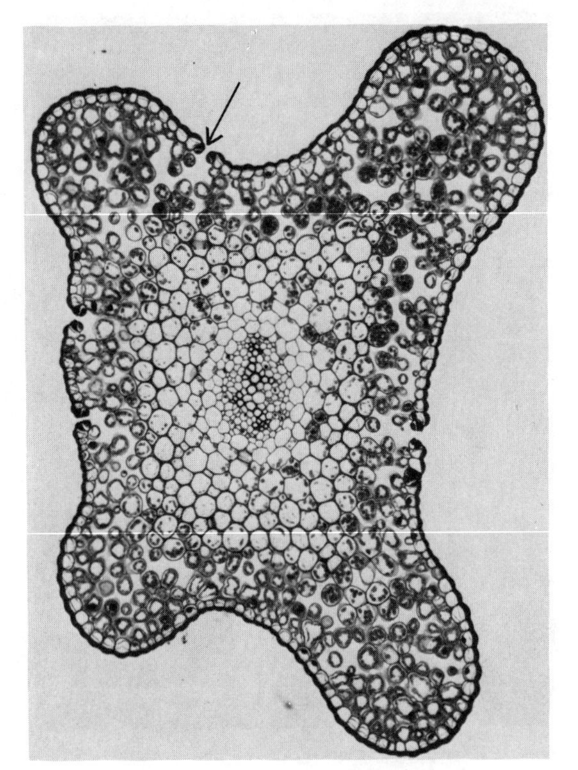

*Figure 11.15*
*Psilotum,* stem cross-section. Note the stomates (*arrow*) in the heavily cutinized epidermal layer. The stele is of the protostelic type and the cortex contains outer chlorenchyma cells and inner parenchyma cells.

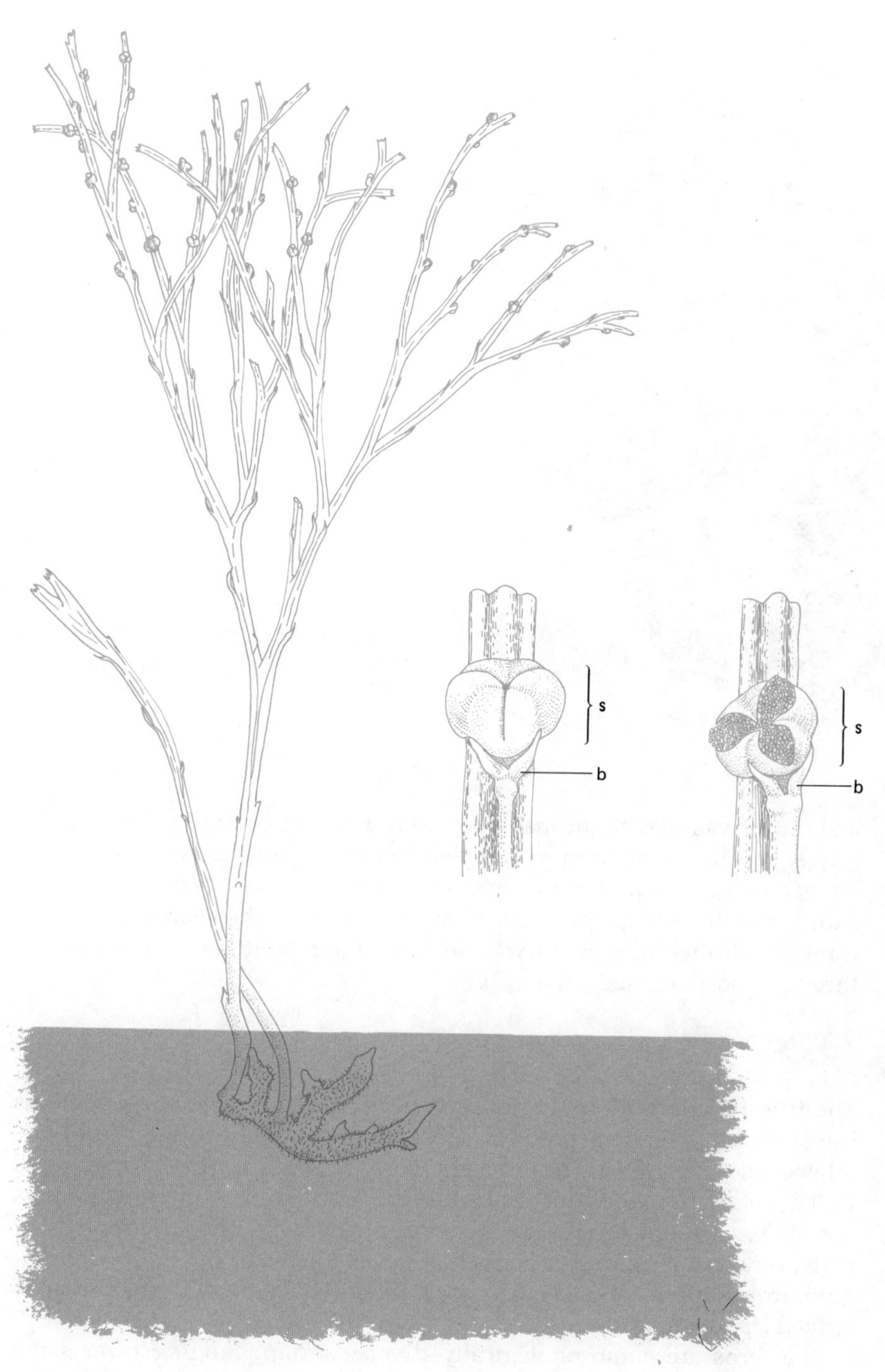


*Figure 11.16*
*Psilotum;* details of synangia (*s*) and bracts (*b*).

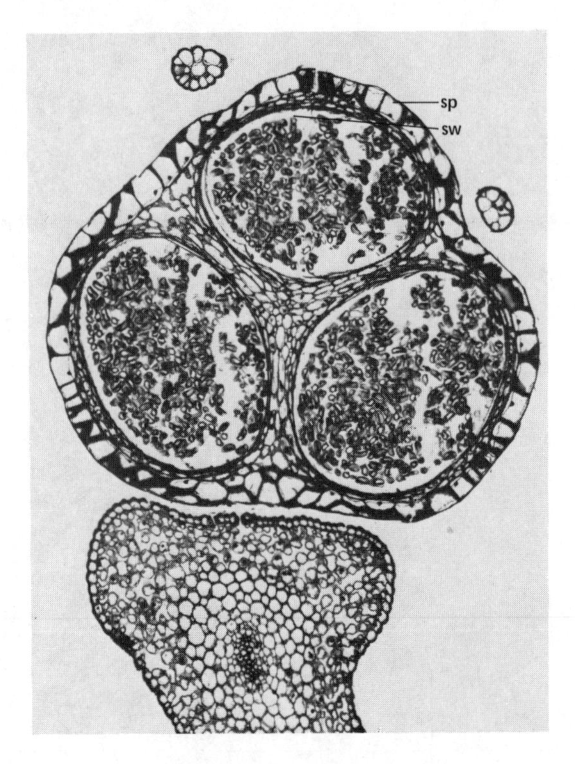


*Figure 11.17*
*Psilotum;* cross-section of sporangia; *sw,* sporangial wall; *sp,* spores.

amount of vascular tissue has been found in the center of the gametophyte. This has been interpreted by some investigators as either an incipient or a reduced stele, and such gametophytes have always proven to have a diploid rather than haploid chromosome number. The sporophyte phase from which they developed was obviously tetraploid, or they developed through apospory (see page 412).

## Ferns

The true ferns are the only nonseed-bearing plants with megaphyllous leaves. Some members produce sporangia essentially like those of the other vascular plants we have been talking about. That is, the sporangial wall develops from the superficial cells of the sporophyll, the sporogenous tissue from internal cells. Such development is termed *eusporangiate.* However, another group of ferns has development of the entire sporangium from superficial cells of the sporophyll. This type of development is termed *leptosporangiate.*

The ferns are a morphologically diverse group, ranging from small aquatic ferns about a centimeter long to large tree ferns (Figure 11.18). The most conspicuous part of a fern is its leaf or *frond.* The blade is dissected

*Figure 11.18*
Some representative ferns. *A*. The cinnamon fern. *B*. A tree fern. Other ferns can be seen in the background. [*A*. by Dee Ann Houston; *B*. Courtesy Australian News and Information Bureau.]

*Figure 11.19*
A dissected fern frond. [Photograph by David R. Williams.]

frequently into leaflets called *pinnae*, which are borne along the extension of the petiole, the *rachis*. The pinnae may be further divided into *pinnules* (little pinnae). In fact, many ferns particularly horticultural varieties, may have fronds which are many times dissected (Figure 11.19). The young

*Figure 11.20*
Fern fiddle-heads or croziers. The fern leaf exhibits circinate vernation. [From a Kodachrome photograph by David R. Williams.]

frond in most ferns is coiled up in a bud called the *fiddle head* or *crozier*, and the frond emerges through an uncoiling of the bud (called *circinate vernation*) (Figure 11.20). The leaves arise usually from a fleshy underground rhizome which, in addition, bears adventitious roots. In some ferns, these roots bear adventitious buds, thus vegetatively propagating the fern.

The vascular tissue of the rhizome may at first be protostelic, but as the stem grows older it becomes typically siphonostelic. Stele types vary considerably in the different ferns, with ectophloic siphonosteles, dictyosteles, solenosteles, and polycyclic steles being represented. The roots of ferns are protostelic. Only in the eusporangiate fern *Botrychium* is there a cambium and the formation of secondary vascular tissue. This genus also forms a periderm layer in the stem.

The sporogenous tissue may be localized in a fertile spike. More commonly the sporangia are organized in groups known as *sori*. These sori are borne on the undersurface of ordinary vegetative leaves and may be located along the veins, along the margin, or scattered (Figure 11.21A). In some ferns, the sori are covered by a flap of tissue called the *indusium* (Figure 11.21B).

Usually a vertical row of specially differentiated cells form an incomplete ring about the sporangium. The radial and tangential walls of these

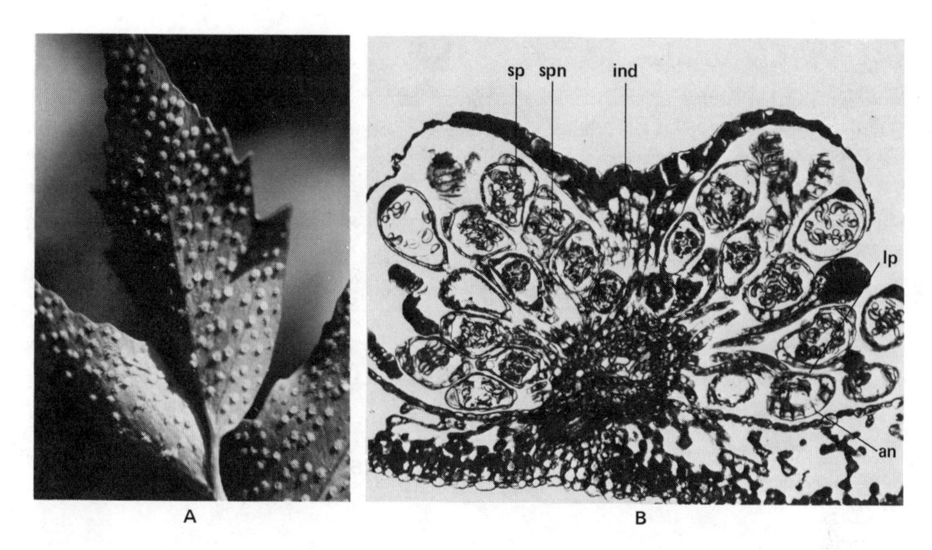


*Figure 11.21*
*A.* Sori on underside of fern leaf. *B. Dryopteris* sorus, cross-section; *ind,* indusium; *spn,* sporangium; *sp,* spores; *an,* annulus; *lp,* lip cells. [From a Kodachrome by David R. Williams.]

cells are quite thickened, and these are referred to as the *annulus.* Between the last cells of the annulus and the sporangial stalk are several thin-walled cells called the *lip cells* or *stomium.* In dehiscence of the spores, loss of water through the thin outer walls of cells forming the annulus shortens its length so the sporangial wall is ruptured in the region of the thin-walled lip cells. As the annulus continues to shorten, the outer tangential walls become increasingly concave because of adherence to water within the annulus. This water is steadily decreasing through evaporation. Ultimately a point is reached where tensile resistance of the water within the annulus cells is no longer sufficient to prevent the outer tangential walls from separating. At this point the water vaporizes, reducing the tension on the outer tangential walls, which snap back, catapulting the spores out of the sporangium.

Most ferns are homosporous with the spores developing into epiterrestrial, green, free-living, heart-shaped gametophytes, each of which is called a *prothallus.* In some, however, the gametophytes are subterranean and contain an endophytic fungus. The gametophytes are typically bisexual, although many are somewhat *protandrous* (the antheridia are produced first). Both antheridia and archegonia are borne on the ventral surface of the gametophyte, and fertilization effected by a multiflagellate sperm (Figure 11.22). Often the gametes are self-incompatable; thus, cross-fertilization is the rule. Self-fertilization can occur however in a number of species.

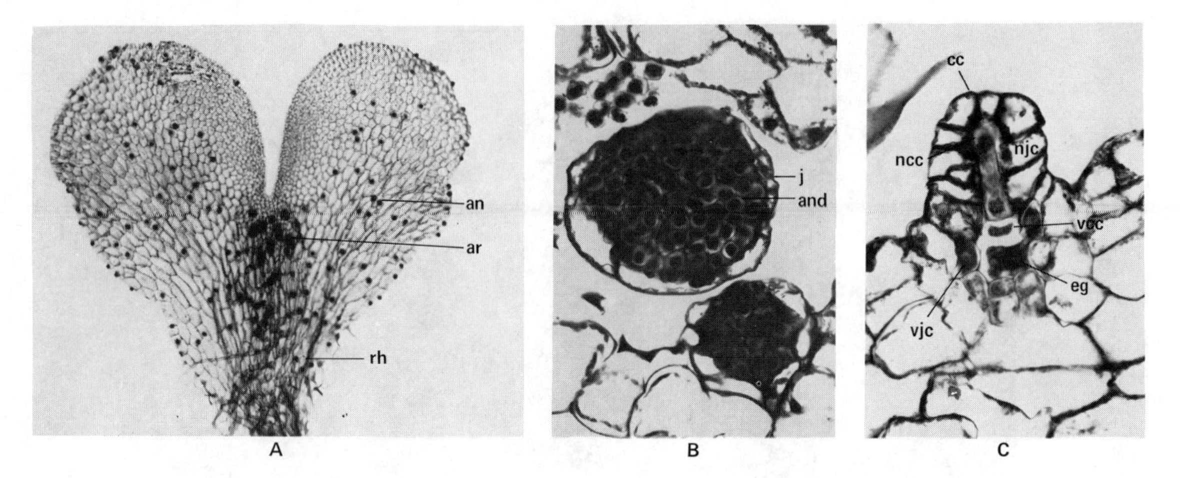


*Figure 11.22*
*A.* Fern prothallium; *rh*, rhizoids; *an*, antheridia; *ar*, archegonia. *B.* Cross-section of antheridia; *j*, jacket cells; *and*, androgonial cells. *C.* Longisection of archegonium; *njc*, neck jacket cells; *cc*, cover cells; *ncc*, neck canal cells; *vcc*, ventral canal cell; *vjc*, venter jacket cells; *eg*, egg.

Various deviations from the usual pattern of reproduction occur in the ferns, including *apogamy* (discussed earlier), *apospory* (the formation of a gametophyte directly from cells of the sporophyte without the formation of spores), and *parthenogenesis* (the development of a gamete directly into a sporophyte without fertilization having occurred). We will discuss some of these mechanisms in more detail in Chapter 16.

## CYCADS

The cycads, in the division Cycadophyta, have been called "living fossils" by some people. These plants were very numerous in Mesozoic times (see Appendix II for a geologic time table), but most are now extinct. It is predicted that all will become extinct in the future. In fact, there is good evidence that the cycad *Microcyas*, now found only in Cuba, is rapidly approaching extinction at the present time.

In their general appearance, some cycads resemble ferns; others, monocots like the palm (Figure 11.23). The stems have generally a large central pith and considerable cortical tissue. A cambium is present but relatively inactive. The roots are fleshy, although considerable secondary xylem may be produced, with certain cells of the root being quite differentiated from others and bearing endophytic blue-green algae.

*Figure 11.23*
The cycad *Zamia*. [From a Kodachrome photograph by David R. Williams.]

## GINKGO

The *Ginkgo* or maidenhair tree, is the sole living genus of a once fairly extensive group, the Ginkgophyta. It is characterized by its fan-shaped leaves with their openly branched, forking venation. Unlike most gymnosperms, the leaves are deciduous, turning a beautiful golden yellow color before they fall from the trees in the fall (Figure 11.24).

Branching in Ginkgo is dimorphic. The elongate main branches are called *long shoots*, with older portions producing short lateral branches called *spur shoots*. Each spur shoot produces a cluster of leaves (Figure 10.6B).

Like the cycads, *Ginkgo* is heterosporous. The microsporangia develop in strobili, the ovules do not, but rather occur in pairs at the tips of the spur shoots (Figure 10.6B).

The mature seeds have a rather thick outer fleshy layer and have the general appearance of a plum. This fleshy layer has a foul odor due to the production of butyric acid. For this reason, only the microsporangiate trees are generally planted.

*Figure 11.24*
*Ginkgo* tree. [Courtesy U. S. Forest Service].

## CONIFERS

The Coniferophyta are by far the largest and most significant group of gymnosperms. Most are evergreen, although there are decidious conifers as well. The leaves are highly variable in form, as mentioned in Chapter 10. This group includes the pines, spruces, hemlock, fir, juniper, larchs, cypress, redwoods, yew, Douglas fir, dawn redwood, *Araucaria* or monkey puzzle tree, and others (Figure 11.25).

Variations in both morphological and anatomical characteristics exist in the conifers. However, there is sufficient similarity that we will consider the pine as representative of the group. It is a strongly *monopodial* tree. That is, it has a well-developed main axis with smaller lateral branches. This gives pine its distinctive triangular shape (Figure 11.26). The branches are of two types, long shoots and short shoots, with the short shoots bearing clusters of two to five, needle like leaves. The short shoots actually wrap the bundles of needles at their base. Thus, the bundles are morphologically a determinate branch. Such bundles are called *fascicles.* The epidermis of these leaves are covered by a thick cuticle, below which is one or more layers of thick-walled hypodermis, with the stomata sunken below the surface (Figure 11.27). We have previously discussed the internal anatomy of pine leaves, with the veins being surrounded by transfu-

Figure 11.25
Some representative conifers. *A. Cedrus libani* (cedar of Lebanon). *B. Atlas cedar. C. Picea pungens* (Colorado blue spruce). *D. Pinus aristate* (bristlecone pine). *E. Pinus ponderosa* (ponderosa pine). [Photographs courtesy U.S. Forest Service.]

sion tissue (Figure 10.21). The roots are of the tap root type, have a protostele, undergo considerable secondary growth and become quite woody. A true epidermis is lacking in all conifer roots in that there is no dermatogen region in their apical meristem. The root hairs, unlike most, may be quite persistent with the walls becoming quite lignified. These persistent root hairs are called *secondary root hairs* by some anatomists. Within the stem a vascular strand of the eustelic type differentiates, enclosing a central pith

*Figure 11.26*
Axillary branching common to seed plants. Note that a difference in branching pattern contributes to the characteristic shape of the tree. [Photograph by David R. Williams.]

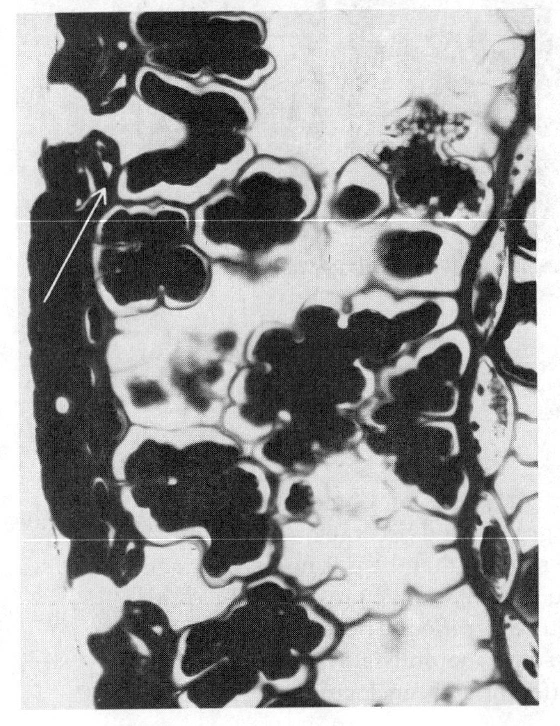

*Figure 11.27*
*Pinus*, cross-section of outer portion of leaf. Note sunken guard cells (*arrow*) and much invaginated mesophyll cells.

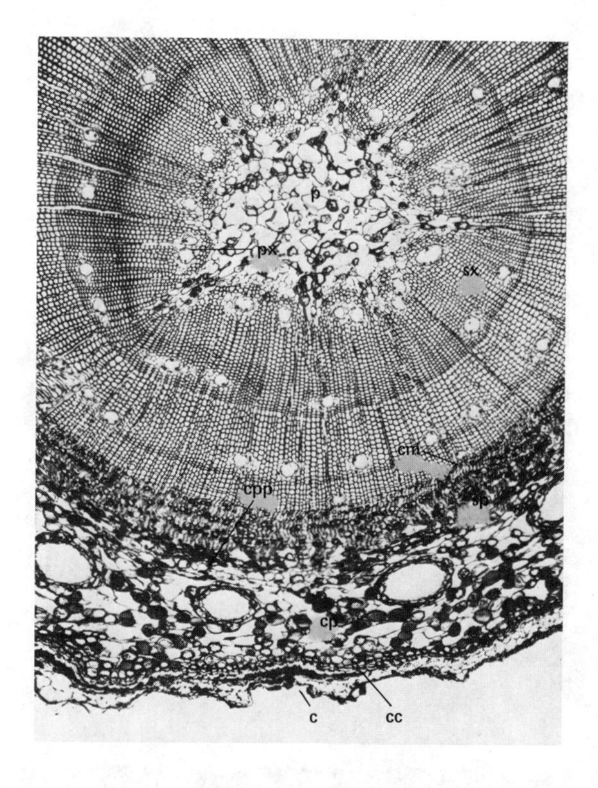


*Figure 11.28*
*Pinus;* cross-section of a young stem; *p,* pith; *px,* primary xylem; *sx,* secondary xylem; *cm,* cambium; *sp,* secondary phloem; *cpp,* crushed primary phloem; *cp,* cortical parenchyma; *cc,* cork cambium; *c,* cork.

which is surrounded by pericycle, cortex, and epidermis (Figure 11.28). An active vascular cambium is present, forming considerable secondary xylem and some secondary phloem. The xylem is characterized by being composed primarily of tracheids with prominent bordered pits. Vessels are lacking and little xylem parenchyma is formed. Narrow, generally uniseriate xylem rays extend radially through the secondary xylem, annual rings are prominent (Figure 9.14C), and *resin canals* occur in the secondary xylem and in the cortex (as well as in the root and leaf) (Figure 11.29). A cork cambium arises in the hypodermis, producing cork to the outside and phelloderm to the inside. The cork cambium may arise progressively deeper in the cortex as the stem ages, differentiating eventually in the secondary phloem. Thus, in an old stem, the bark may be composed only of periderm and secondary phloem, as all more peripheral tissues would have become sloughed off. The phloem is composed essentially of sieve cells, with sieve tube cells and companion cells lacking. Many of the primary and secondary phloem elements are crushed, except for those nearest the cambium (Figure 11.30).

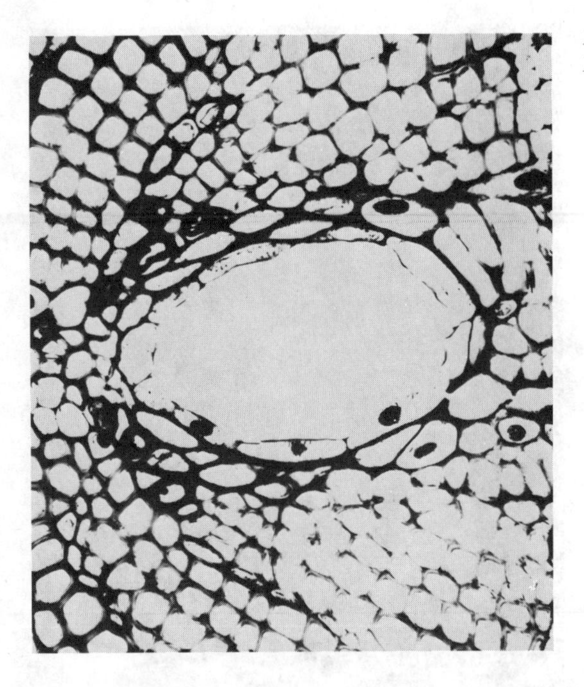

*Figure 11.29*
*Pinus,* resin canals.

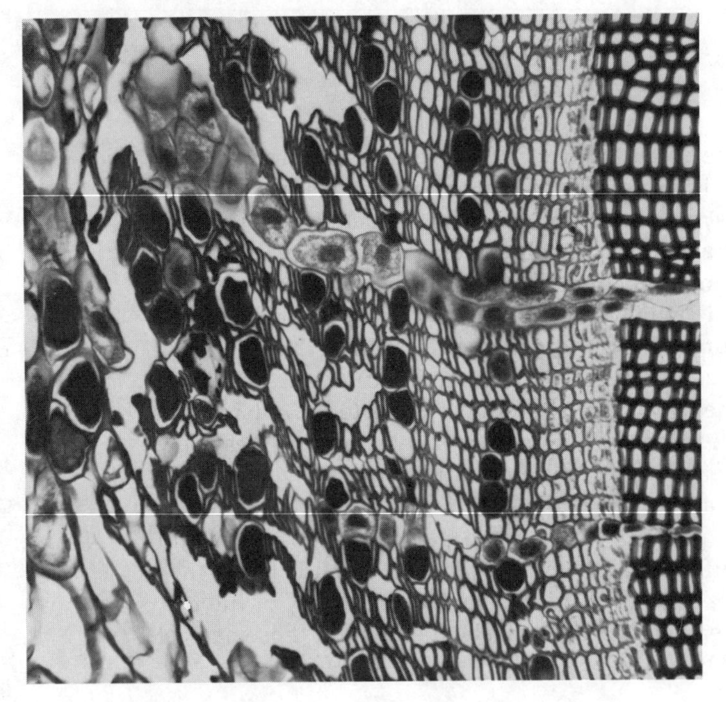

*Figure 11.30*
*Pinus;* cross-section of phloem region of stem. Note absence of companion cells.

In pine, both microstrobili and megastrobili are borne on the same sporophyte. Thus, pine is *monoecious* This is not true of all conifers, however.

The megasporangiate strobili are compound structures consisting of a central axis bearing modified scale leaves called *bracts*. In the axil of each bract is an *ovuliferous scale* bearing on its adaxial (upper) surface two *ovules*.

Each ovule consists of a central mass of tissue, the *megasporangium*, and a multicellular outer layer called the *integument*. The integument is incompletely closed, leaving an opening called the *micropyle*. A cell deep within the megasporangium will enlarge and function as a megasporocyte. This cell is present at the time of pollination, and approximately one month later undergoes meiosis producing a *linear tetrad* of megaspores. Three of these degenerate leaving a single spore as the *functional megaspore*.

The central axis of each microsporangiate strobilus bears helically arranged *microsporophylls*. Each microsporophyll bears on its abaxial (lower) surface two elongate *microsporangia*. Within the microsporangium, a number of cells function as *microsporocytes*. These undergo meiosis, forming a *microspore tetrad*. The tetrad soon separates into individual uninucleate *microspores*. The microspore nucleus divides mitotically, followed by cytokinesis, producing two cells—a larger *apical cell* and a smaller *prothallial cell*. Division of the apical cell produces a larger *antheridial initial* and a *second prothallial cell*. The two prothallial cells are termed *vestigial* structures, as they are thought to be homologous with the vegetative prothallus of lower plants. They do not divide, but soon degenerate. The antheridial initial does divide, forming a larger *tube cell* and a smaller *generative cell*. The generative cell, tube cell, and degenerating prothallial cells may be referred to as an immature *microgametophyte*, or *pollen grain*. At this stage of development the microsporophylls separate, the microsporangium ruptures, and the microspores, with their enclosed immature microgametophytes, are shed.

At about this time the internodes of the megastrobilus elongate slightly, causing the megasporophylls to separate from each other. It is through these small clefts that the windborne pollen grains sift. Cells of the megasporangial apex are digested, producing a droplet of liquid called the *pollination droplet*, and leaving a small depression, the *pollen chamber*, at the apex of the megasporangium. Pollen grains come to rest in the pollination droplet, and as it dries it contracts, carrying the pollen grains into the pollen chamber. This transfer of pollen grains from the microsporangium to the ovule is called *pollination*.

Pollination occurs in early spring, and following contact of the pollen grain with the megasporangium wall, the tube cell begins to elongate forming a *pollen tube*. The pollen tube begins a slow parasitic growth through the megasporangium. The male gametophyte overwinters in an arrested state consisting of the short and incomplete pollen tube, tube cell, and generative cell. About 12 months after pollination the generative cell divides producing a *stalk cell* and a *body cell.* About one month later the body cell divides forming two nonflagellate *sperms.*

Meanwhile, the megagametophyte has developed slowly from the surviving megaspore. The megaspore enlarges greatly, and its nucleus, and the derivatives thereof, divide many times producing a number of nuclei. Autumn cessation of growth stops the female in this *free nuclear stage,* in which it remains during winter. Shortly before sperm are produced in the pollen tube, mitosis resumes in the megagametophyte followed by cytokinesis. The mature megagametophyte consists of approximately 2000 cells. Two or three of these at the micropylar end of the gametophyte differentiate forming *archegonia.* Each archegonium consists of a very short neck and a larger venter containing a single large *egg.*

By this time, the pollen tube has completed its growth through tissues of the megasporangium and megagametophyte, penetrating the archegonium. The end of the pollen tube degenerates, depositing the tube nucleus, stalk cell, and two sperms into the egg cytoplasm. One sperm nucleus unites with the egg nucleus effecting *fertilization;* all other male cells degenerate.

The resulting zygote nucleus undergoes two successive mitotic divisions producing four free nuclei which migrate to the chalazal (opposite the micropyle) end of the egg cytoplasm. Two more mitotic divisions result in four tiers of four cells each. The lowest tier subsequently forms the *proembryo,* the next tier elongates somewhat as *prosuspensor cells.* Usually, one of the four proembryos undergoes precocious growth and only it survives. The resulting *embryo* consists of an *embryonic root, hypocotyl,* and three to twelve *cotyledons,* surrounding a small mound of meristematic tissue, the *epicotyl.* The middle layer of the integument hardens and becomes the major part of the *seed coat.* The differentiated ovule with its enclosed megagametophyte and embryo constitutes the *seed.*

Commercially, the pines are among our most important lumber and pulp wood trees and are the source of *naval stores.* Seeds of the piñon pine are collected and eaten.

## GNETUM, EPHEDRA, AND WELWITSCHIA

The Gnetophyta are an unusual group of gymnosperms. The leaves of *Ephedra* are very minute and ephemeral. Those of *Gnetum* are broad and flat with reticulate venation, much like those of flowering plants, while those of *Welwitschia* are huge strap-shaped structures which become split

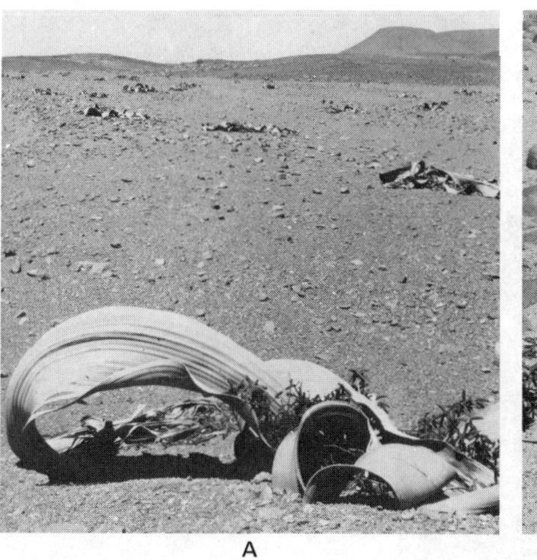

A B

*Figure 11.31*
*A. Welwitschia,* male plant. *B. Welwitschia,* female plant.

longitudinally many times as they develop. The latter have indeterminate growth from a basal meristem (Figure 11.31).

The wood is characterized by multiseriate rays and the presence of vessels; phloem contains both sieve tube cells and companion cells (Figure 11.32). These features are characteristic of flowering plants as well. The strobili are borne very much like flowers in an inflorescence, and the megagametophyte lacks archegonia in *Ephedra* and *Welwitschia.* With all these similarities to flowering plants, the literature contains numerous suggestions to the effect that the Gnetophyta are the ancestors of angiosperms. There is little good evidence for this, however, and chemical analysis has shown little resemblance between the wood of the two groups.

*Ephedra sinica* is the source of the alkaloid ephedrine. The medicinal properties of this drug were known to the Chinese as early as 2737 B.C., and they call it *Ma Huang. Gnetum gnemon* is cultivated in Java and its young leaves and sporophylls eaten. Fibers from the bark are used in making rope.

## FLOWERING PLANTS

The Angiospermophyta are separated characteristically into two groups called commonly dicots and monocots. Other than the plant kingdom survey, most of the discussion in this text has been concerned with the angiosperms or flowering plants. Thus, we can dispense with much repetitive discussion here.

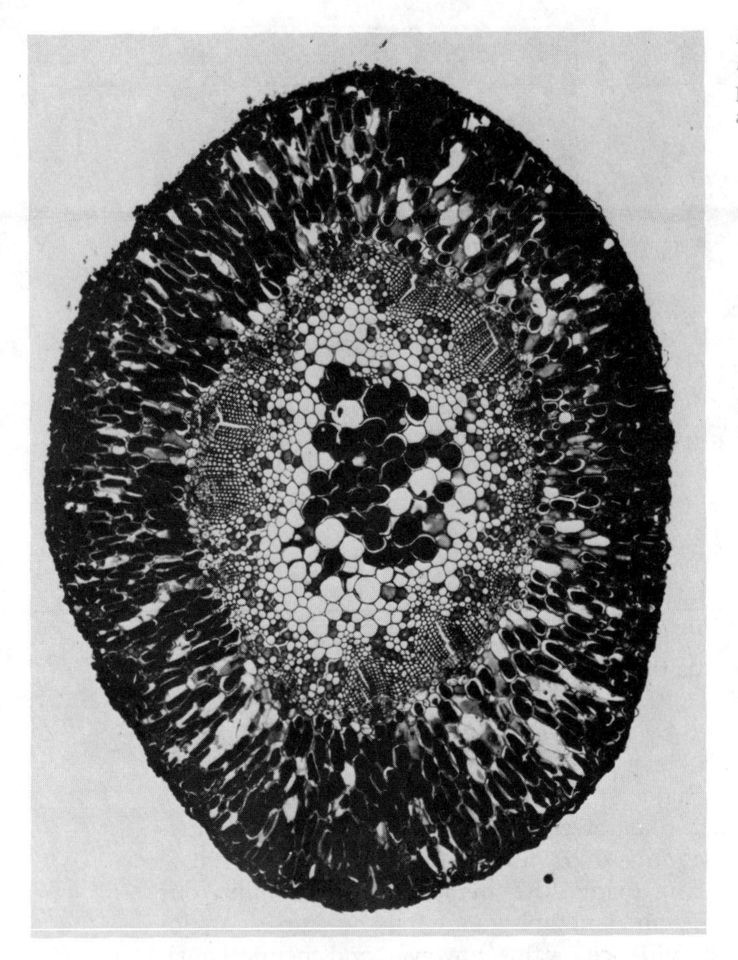

*Figure 11.32*
*Ephedra;* cross-section of stem. Unlike pine, this gymnosperm has both vessels and companion cells.

As the outstanding characteristic of flowering plants is the flower, let us consider that structure. Much has been written about the flower: its importance in the present dominance of mammals on earth; its involvement in the coevolution of flower and insect pollinator; its importance in providing food for animals, including man; its beauty; and so on. If nothing more, it has certainly proliferated a lot of terms that botany students for many years in many different classrooms have had to memorize. Flowers are obviously not all alike. But keeping this in mind, we shall consider the parts of a typical dicot flower, remembering that not all dicot flowers have all these parts.

A plant may bear only a single, solitary flower, but more often it bears a group of flowers. These aggregations of flowers are called *inflorescences* (Figure 11.33). Flowers are borne on stalks, with the stalk of a solitary flow-

*Figure 11.33*
Types of inflorescence: *A.* cyme (Sambucus). *B.* Panicle (grass). *C.* Raceme (mint). *D.* Spike (*Belaprone guttata*). *E.* Corymb (*Kalmia latifolia*). *F.* Head (Aster). *G.* Umbel (*Daucus carota*). *H.* Catkin (beech). *I.* Spadix (Jack-in-the-pulpit). [*A.*, *B.*, *C.*, *D.*, *E.*, *F.*, and *I.* from Kodachromes by David R. Williams; *G.* and *H.* from color photographs by Dee Ann Houston.]

er or an inflorescence being called a *peduncle;* that of each individual flower in the inflorescence a *pedicel.* That portion of the flower stalk to which the flower parts attach is called the *receptacle.*

Most flowers contain two sets of sterile parts: an outer whorl of *sepals,* collectively called the *calyx;* and to the inside of the sepals, a whorl of *petals,* collectively called the *corolla.* The calyx and corolla constitute the *perianth.* We often think of the sepals as being green and the petals colored, but this is not always so.

There are also two sets of reproductive parts. The *stamen* is composed of a more or less elongated, slender stalk called the *filament,* at the top of which is the *anther.* Within the anther, the microspores are produced. Thus, the anther, or perhaps more correctly, the stamen represents a microsporophyll.

The *carpel,* which is sometimes referred to as the *pistil,* is differentiated into an enlarged basal part called the *ovulary* which encloses the *ovules,* a more or less elongated middle region called the *style,* and an upper part called the *stigma.* Some ovularies are composed of several carpels which are fused together. Such *compound ovularies* commonly have a single stigma and style. But in some cases, each carpel will retain a separate stigma and style (Figure 11.34).

*Figure 11.34*
Parts of a flower; *sep,* sepal; *pet,* petal; *an,* another; *fil,* filament; *sti,* stigma; *sty,* style. The anther and filament comprise the stamen; the stigma, style and ovulary comprise the pistil. [Photograph by David R. Williams.]

The ovules may attach to the ovulary in one of three general ways: they may attach directly to the ovulary wall *(parietal);* they may attach to a central column of tissue *(free central);* or the ovulary may be partitioned, with the ovules attached near the common central point *(axile).* The cavity created by the partitions is called the *locule,* with the number of locules per ovulary representing the number of carpels which fused forming that ovulary.

The ovule consists of a central mass of tissue called the *nucellus.* The nucellus is connected to one ovulary by a stalk called the *funiculus,* and is surrounded by the *integuments* of the ovule. It is within the nucellus that the megagametophyte is formed. Thus, the ovulary is a megasporophyll, the ovule (or in the judgment of some botanists only the nucellus) the megasporangium.

If the flower contains both reproductive parts, it is said to be *perfect.* If it lacks either stamens or carpels, it is *imperfect.* An imperfect flower having stamens only is a *staminate* flower. One having carpels only is a *carpellate* flower. If both staminate and carpellate flowers are borne on the same plant, the plant is said to be *monoecious.* If staminate flowers are borne on one plant and pistillate on a different plant, the plant or, more technically, the species, is *dioecious.*

If the flower has all four parts it is a *complete* flower. If it lacks any one part it is *incomplete.* Thus, a flower can be incomplete, yet still be a perfect flower. On the other hand, all imperfect flowers are necessarily also incomplete.

If the flower parts attach to the receptacle below the ovulary, the ovulary is said to be *superior,* and the flower is termed *hypogynous.* If the flower parts seem to come from the top of the ovulary, the ovulary is *inferior,* and the flower said to be *epigynous.* In some flowers, however, the receptacle bears a cuplike extension (the *hypanthium*) around the base of the ovulary with the petals and stamens attached to the margin of it. Thus, they seem to come from around the ovulary. Such flowers are called *perigynous.* The ovulary, however, is still considered to be superior in such flowers.

In some flowers, the petals are all essentially alike and inserted equidistant from each other. In other flowers, one or more members are different. The former type of flower is said to be regular or *actinomorphic;* the latter, irregular or *zygomorphic* (Figure 11.35).

Many monocot flowers have the flower structure discussed above. However, many monocots have their floral parts in 3s or multiples of 3, whereas most dicots have their floral parts in 4s or 5s or multiples of 4 or 5 (Figure 11.36). The grass flowers are quite specialized and we will not discuss their floral morphology.

The gametophytes of flowering plants are highly reduced, the pollen grain or microgametophyte consisting of three cells at maturity, and the mature megagametophyte or embryo sac consisting in most instances of only seven cells. Neither an antheridium nor an archegonium is formed.

*Figure 11.35*
*A.* A regular or actinomorphic flower. *B.* An irregular or zygomorphic flower. (Photographs by David R. Williams.]

*Figure 11.36*
Tulip, a monocot flower. [Photograph by David R. Williams.]

Microsporogenesis and megasporogenesis is similar to that described for the gymnosperms. The microspore resulting from meiosis undergoes a single mitotic division forming a larger tube cell and a smaller generative cell. Following pollination, the generative nucleus divides forming two nonmotile sperm. In megagametogenesis, the remaining functional megaspore enlarges greatly forming an *embryo sac*. Its nucleus divides three times forming eight nuclei, four at each end of the embryo sac. One nucleus from each end migrates to the center of the embryo sac and are called the *polar nuclei*. The three remaining cells at the micropylar end of the embryo sac become walled off. One of the three enlarges forming the *egg*. The remaining two are called *synergids*. The three cells at the opposite end of the embryo sac also become walled off forming the *antipodal cells*.

The pollen grains are carried to the stigma by a variety of vectors. Wind-pollinated flowers have light, tiny, dry pollen grains, while insect pollinated flowers have larger sticky pollen grains. The deposition, by whatever means, of the pollen on the stigma is called *pollination*. If the pollen grain is compatible to the stigma, a pollen tube develops, penetrating through the tissue of the stigma, style, and ovulary. Commonly, the pollen tube enters the ovule through the micropyle, the end of the pollen tube degenerates, and the sperm are deposited into the embryo sac. One sperm fuses with the egg nucleus, the second with the two polar nuclei, resulting in a zygote in the former case, and a *triple fusion*, or *primary endosperm*, in the latter. This event is different from that of gymnosperms; as you recall, only one sperm was functional, and the other simply degenerated. The involvement of both sperm in a fusion process has been, and unfortunately still is, called double fertilization. This is an obvious misnomer. As you know, fertilization is a process involving the fusion of *gametes*. The polar nuclei are not gametes. Thus, you have, perhaps, double fusion but not double fertilization.

The process of megagametogenesis as described above occurs in perhaps two-thirds of the angiosperms that have been investigated. Slight variation in the pattern occurs in the others. In any event, the primary endosperm nucleus undergoes a number of mitotic divisions producing a free nuclear stage in the embryo sac. This is followed eventually by wall formation, forming a cellular *endosperm*. In other instances, cytokinesis follows each mitotic division. The endosperm is typically a 3N, or *triploid*, tissue and is neither sporophytic nor gametophytic in nature.

Either during or following formation of the endosperm, the zygote begins to divide forming ultimately the embryo. There is no free nuclear stage during embryogenesis, although various morphological changes occur in the surrounded ovule. Typically, the integuments harden and become the *testa*, or seed coat. Thus, as in the gymnosperms, the ovule, with its enclosed embryo and associated structures, becomes the *seed* (Figures 4.1 and 4.5). During seed development, morphological changes occur in the ovulary and, in some species, in other parts of the flower or inflorescence as well. The ovulary enlarges, frequently with the ovulary wall becoming the *pericarp* and being composed typically of three layers: an inner *endocarp*, a middle *mesocarp*, which often becomes fleshy and fibrous, and an outer, often skinlike or papery layer called the *exocarp*. This modified ovulary and, at times, associated floral parts, is called the *fruit*.

Normally, fruit formation is stimulated by the process of fertilization, occasionally by pollination alone. Sometimes fruits develop with neither fertilization nor pollination having occurred. Such fruit would obviously lack seeds and are called *parthenocarpic.* Parthenocarpy can be stimulated by hormone application, although it also occurs naturally in a number of different plants such as navel oranges, bananas, pineapples, and some varieties of watermelons and grapes.

Normally, a seed contains but a single embryo. However, in some seeds more than one embryo is produced. This may occur through fusion of sperm nuclei with both the egg cell and a synergid; or a cell of the nucellus may begin to divide, forming an embryoid (a structure morphologically identical to the embryo but not derived from a zygote). Occasionally, an embryo will develop with no fertilization having occurred; the egg cell itself developing into an embryo (or embryoid). As we pointed out earlier, this event would be called parthenogenesis. In any instance where more than one embryo is produced by a single seed, the condition is called *polyembryony.*

## SUMMARY

1. The vascular plants as a group possess two major attributes: the presence of vascular tissue and a dominant free-living sporophyte.
2. The sporophyte is differentiated frequently into the vegetative organs of stem, roots, and leaves.
3. The leaves are one of two morphological types: microphylls which have only a single strand of vascular tissue and do not create a leaf gap in the stem, and megaphylls which have typically branched vascularization and do create leaf gaps in the stem.
4. The telome theory is proposed to explain the origin of megaphylls; the enation or reduction theory, of microphylls.
5. Vascular plants may be homosporous or heterosporous, with the homosporous condition being considered primitive.
6. The Lycopodophyta are characterized by the presence of microphyllous leaves and protostelic stems and roots. Their sporangia may be borne singly or in strobili, and they may be homosporous or heterosporous. The gametophytes are typically subterranean, achlorophyllous, and contain endophytic fungi.
7. The Sphenophyta are characterized by jointed, fluted, hollow stems having silica in the epidermis and being comprised of nodes and internodes. Their leaves are minute microphylls which are achlorophyllous typically at maturity. An intercalary meristem is present forming a plate at each node and contributing to the length of the internode. The stem is perforated by canals and has a modified ectophloic siphonostele. The sporangia are borne on sporangiophores located in strobili

and are homosporous. The spores are green and possess elaters. The gametophyte is also green, free living, and either unisexual or protogynous.

8. The Filicinophyta include the primitive *Psilotum* and the true ferns. The true ferns have megaphyllous leaves called fronds. Most have siphonostelic stems. The sporangia are found frequently in clusters called sori. They are typically homosporous, and the gametophyte is epiterranean and free living.
9. The Cycadophyta have secondary growth and bear their sporangia in micro- and megastrobili. The microgametophyte is reduced to a pollen grain, and seed is produced.
10. The Ginkgophyta have large, flat leaves with openly branched venation. Their branches are of two types, long and spur, with the ovules being borne in pairs at the tips of spur shoots. The megagametophyte is chlorophyllous and the seeds are large and fleshy.
11. The Coniferophyta are characterized generally by small scalelike, awn-shaped, or needlelike leaves, have two types of branches, long and short, have considerable secondary growth, produce nonmotile sperms (unlike all the previously mentioned groups), lack vessels in the xylem, and lack sieve tube cells and companion cells in the phloem.
12. The Gnetophyta have many characters that are angiospermlike: vessels in the xylem, sieve tube cells and companion cells in the phloem, strobili borne in a flowerlike inflorescence, and lack of both antheridium and archegonium.
13. The Angiospermophyta produce flowers and fruit. Both antheridia and archegonia are lacking and the gametophytes are reduced to a three celled pollen grain and seven celled embryo sac. Two fusions occur in the embryo sac producing ultimately an endosperm and a zygote. The ovule develops into a seed, the ovulary into a fruit.

# Chapter 12

## MOVEMENT OF MATERIALS THROUGH PLANT CELLS AND ORGANS

The movements of substances into, out of, or through plants has been alluded to previously. We will now look at these processes in more detail.

The movement of any substance requires energy. All molecules have a certain amount of internal energy, and these internal energies exist as several components: rotational, vibrational, or translational. Solids have only vibrational energy and this can account for the fact that solids do not disperse. Gases on the other hand have translational energy, which means their molecules are constantly moving in straight lines until they hit something, whereupon they bounce back at some angle and move off in a different direction. Such activity results in a random dispersal of the gas molecules. The internal energy of molecules is called *kinetic energy.* The amount of internal energy can be affected in several ways. An increase in temperature affects all three components. Other energy changes, the so-called nuclear energy, can be brought about by changing the status of nuclear particles, while still other energy changes can be brought about by changing the orbitals of the electrons. Changes in both vibration and rotation accompany shifts in orbitals. Temperature cannot be raised high enough to accomplish such a shift in living cells without damage to those cells, but sunlight does possess sufficient energy to produce such changes. We will be discussing this important fact when we talk about photosynthesis, phototropism, and other light-mediated plant processes.

To understand the significance of kinetic energy, we must delve into the laws of thermodynamics. It is beyond the scope of this book to discuss such laws in any detail. One thermodynamic property is that of *free energy.* For our purposes, free energy can be described as the energy available to do work. Free energy is generally discussed in terms of a difference in free energy between an initial and a final state. If the final state has a lower free energy than the initial state, the reaction is energy releasing. If the final state has a higher free energy, the reaction is energy requiring. Free energy is usually stated in energy per mole or per gram of a substance. This free energy is called the *chemical potential* for that substance. The greater the chemical potential, the greater the tendency of a substance to undergo chemical reactions or move from one place to another.

Water has a certain chemical potential *(water potential).* If this potential varies from one place to another, water will tend to move toward a point where the water potential is lowest.

## DIFFUSION

With these points in mind, we may be better able to understand the process of diffusion. *Diffusion* is generally stated as the *net movement of a given substance due to its kinetic energy.* If we think only in terms of concentration, diffusion is a rather simple process to visualize. If we place a little perfume in a shallow pan and place the pan in a room full of students, which students will be the first to notice the odor of the perfume? Which students will be the last to notice the odor? Obviously, those closest to the perfume will notice it first; those farthest away, last. This is due to the fact that diffusion is a *directional* movement, with the *net* movement of molecules being from an area of greater concentration toward an area of lesser concentration of that substance. In time, the perfume will be entirely gone from the pan, having undergone a change in state from liquid to gas, and the molecules will be randomly and essentially equally distributed throughout the room. We call this condition *equilibrium*. When equilibrium becomes established, the *net* movement of molecules ceases (Figure 12.1). In other words, there is equal probability of the molecule going in one direction as in another. Diffusion, but not the random movement of molecules, continues only so long as a *concentration gradient* exists.

If we examine this situation in terms of chemical potential, the free energy is very high in the immediate vicinity of the perfume, lower in the room. As diffusion occurs from a region of high free energy toward a region of lower free energy, a condition will eventually be reached in which the free energy throughout the room becomes equal. At this point the change in free energy is zero and diffusion ceases.

In the above example, we get the same result whether we consider concentration or free energy gradients. But conditions may occur within a plant in which differences in free energy exist without differences in concentration existing. A more exact description of diffusion might be to state that *diffusion of a substance will tend to occur along a chemical potential gradient, from a region of high chemical potential into a region of lower chemical potential.*

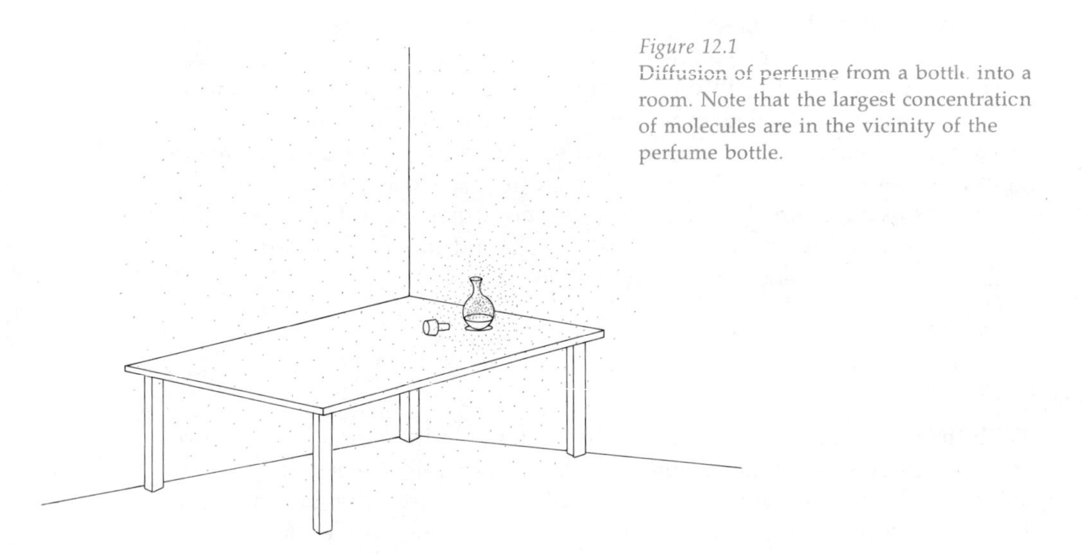

*Figure 12.1*
Diffusion of perfume from a bottle into a room. Note that the largest concentration of molecules are in the vicinity of the perfume bottle.

But even such a simple concept as diffusion can be complicated. Some botanists use the term diffusion to describe *any* movement of molecules from one place to another due to their kinetic energy and have coined the term *net diffusion* to refer to the movement along a concentration gradient. One author has stated that net diffusion is the "direction of movement of the greatest number of molecules." It is important to remember that *number* of molecules has no significance in diffusion. It is only number of molecules per unit area, in other words *concentration*, that has significance.

## Factors Influencing Diffusion

Diffusion may be influenced in several ways. In an equilibrium system, certain factors may establish a free energy gradient and thus result in diffusion. Other factors only serve to alter the rate of diffusion.

If we assume a completely enclosed system separated into two compartments by a membrane through which only water can diffuse, in which each compartment is full of water, no *net* movement of water will occur (Figure 12.2). This situation can be altered in several ways:

1. *Temperature.* If the water in one compartment is warmed, its free energy will increase and diffusion will occur toward the cooler side (Figure 12.3).
2. *Pressure.* If the water in one compartment is subjected to increased pressure, its free energy will increase and diffusion will occur toward the lower pressure (Figure 12.4).

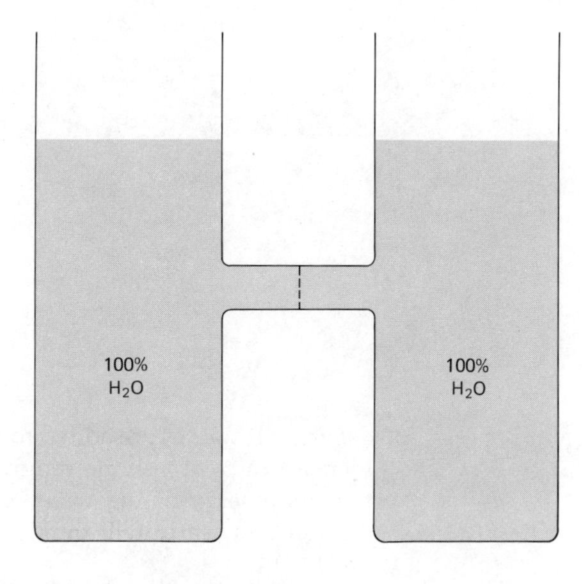


*Figure 12.2*
A system at equilibrium. No diffusion is occurring.

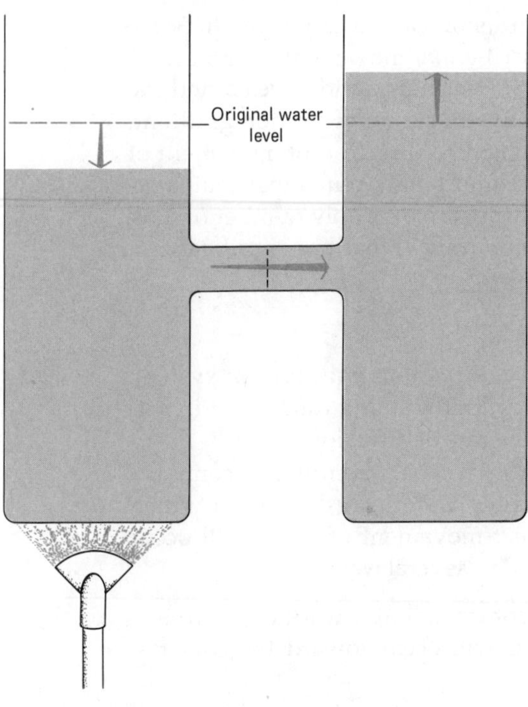


*Figure 12.3*
Increased temperature results in diffusion in the direction of the cooler area.

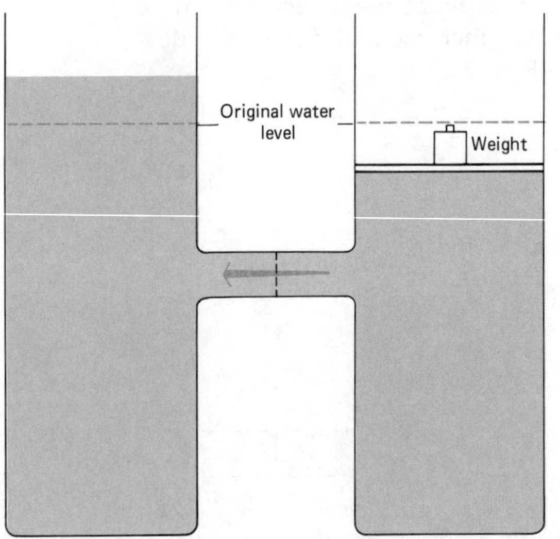


*Figure 12.4*
Increased pressure results in diffusion towards the lesser area of pressure.

3. *Solutes.* If solutes are added to the water of one container and any increase in total volume removed so pressure remains constant, the free energy of the water molecules in that container is decreased and water will thus diffuse into that compartment (Figure 12.5).

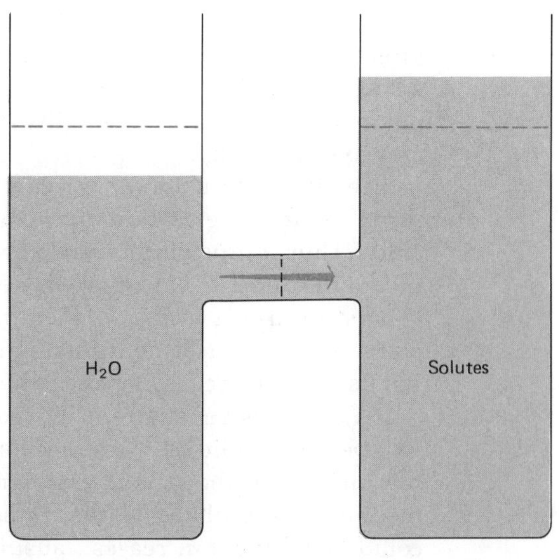


*Figure 12.5*
Diffusion occurs towards area of increased solute concentration.

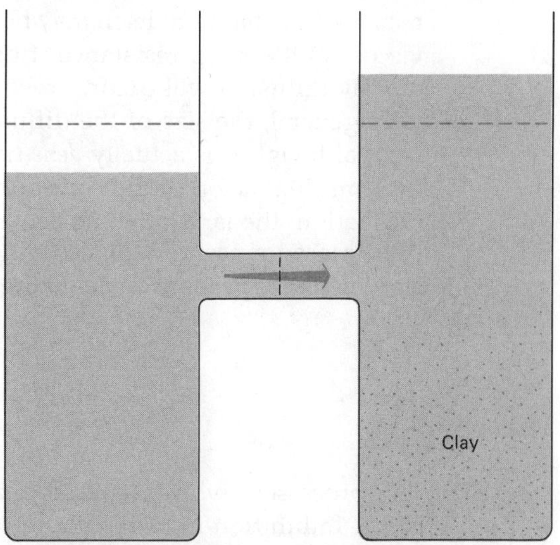


*Figure 12.6*
Adsorption of water onto clay particles in right-hand container results in diffusion of water into that side.

4. *Adsorption.* Certain molecules have water adsorbing surfaces (water molecules are attracted to and tend to collect on their surfaces). Such molecules include proteins, cellulose, starch, clay, and others. If such a substance is placed in one container, adsorption will result in lower water potential and thus, water will diffuse into that container (Figure 12.6).

The free energy aspects of a diffusion system only establishes the *tendency* for diffusion to occur. Whether or not diffusion does in fact occur, and at what rate, is influenced by other factors.

An increase in *temperature* results in an increase in the rate of diffusion. This is due to an increase in the speed, or velocity, of the diffusing particles. Occasionally, temperature may have an indirect effect on diffusion rate by affecting the permeability of the membrane the diffusing particles are passing through.

The rate of diffusion is also affected by the *concentration gradient.* This refers to the concentration of the diffusing substance over a specified distance. Thus, diffusion rate can be varied in two ways: 1) if the concentration of the diffusing substance is held constant but the *distance* over which it must diffuse is *increased,* the *rate* of diffusion will *decrease;* or 2) if the distance over which the diffusing substance must travel is held constant, but its *concentration* is *increased,* the *rate* of diffusion will *increase.*

If a diffusing substance is moving through a perfect vacuum, diffusion will be quite rapid as the molecules will be moving in a straight line. As the number or the size of "foreign" molecules increase in the diffusing medium, the probability that molecules of the diffusing substance will collide with them increases, causing the rate of diffusion to decrease. The nature of the medium itself may be a factor. The thicker the medium (more viscosity) the more resistance it offers to diffusing particles and the less rapidly diffusion will occur.

In general, the size of the diffusing particle affects the rate of its diffusion, although it is actually *density* (i.e., mass per unit volume) that is the determining factor. But for our purposes, we can make the following generalization: the larger the molecule the slower will be its rate of diffusion.

Increased pressure will also influence diffusion rate. The rate of diffusion increases proportionally to the amount of pressure increase.

## IMBIBITION

Two processes involve a special case of diffusion. The first of these is *imbibition.* Imbibition involves the uptake of water into dry solids. This process can best be demonstrated in dry wood. Water will first move into the intercellular spaces of the wood and into the xylem vessels. This movement is due to surface tension (capillarity). The water then moves into the cell walls and middle lamellae, forcing the wall particles apart and causing the wood to swell. The water molecules actually adhere to the wood particles forming a thin film around them. The force created by imbibition is considerable and before explosives, stone was quarried by drilling holes a few inches apart, pounding a piece of dry wood tightly into each hole, and then pouring water over the wood. The resulting imbibitional forces would actually split the rock. Another well-known example is the sticking of a wooden door in damp weather.

## OSMOSIS

Another special case of diffusion involves the net movement of water through a differentially permeable membrane. Water potential within the plant is influenced by the fact that various cellular membranes (plasmalemma, nuclear envelope, chloroplast envelope, mitochondrial envelope, tonoplast) inhibit the free movement of solute molecules, while water molecules pass freely through them. Various terms have been used to describe such membranes, the most common being selectively permeable, semipermeable, or differentially permeable. The term *selectively* does not quite imply the proper meaning, as it might be construed to suggest an anthropomorphism. *Semipermeable* on the other hand does not correctly describe the nature of the membrane, it is not *half* permeable. Some particles will not pass through the membrane at all, while others pass through to varying degrees. Thus, *differentially permeable* would seem to be the most appropriate term.

If the solute concentration inside the cell is greater than that outside the cell, a water-potential gradient is established. Water will thus diffuse through the plasmalemma into the cell. This process is called *osmosis*, and it is simply a special case of diffusion in which the diffusing substance is *water*, and it is diffusing through a *differentially permeable membrane*. Only water can osmose; all other substances diffuse.

The presence of *solute* particles (i.e., substances that dissolve in water, as water is the *solvent*) creates a condition called *osmotic potential* by some investigators, *osmotic pressure* by others. The two terms are equivalent, except that osmotic potential is stated as a *negative* value, osmotic pressure as a *positive* value. Completely *pure* water would lack solutes, thus would have an osmotic potential equal to zero (OP = 0). As the number of solute particles increases, osmotic potential increases and water potential decreases.

We must keep in mind that we are talking about a potential. Accordingly, the osmotic potential of a solution is that which *would* be developed by that solution at equilibrium *if* it were placed in water enclosed within a differentially permeable membrane through which only water could osmose. Alternatively, it could be looked at as the amount of pressure that must be applied to a solution to stop osmosis of water into that solution. These two situations are illustrated in Figure 12.7. In the left figure, we are starting with a beaker of a solution consisting of 25 percent sugar and 75 percent water, which is known to have an osmotic potential of −10. Obviously, the solution is not exerting that force in the open beaker. If that solution is poured into a bag having a membrane differentially permeable only to water, the bag sealed and placed into a beaker of 100 percent water, water will move into the bag via osmosis. When equilibrium is reached, assuming the membrane is rigid and will not stretch, a force of

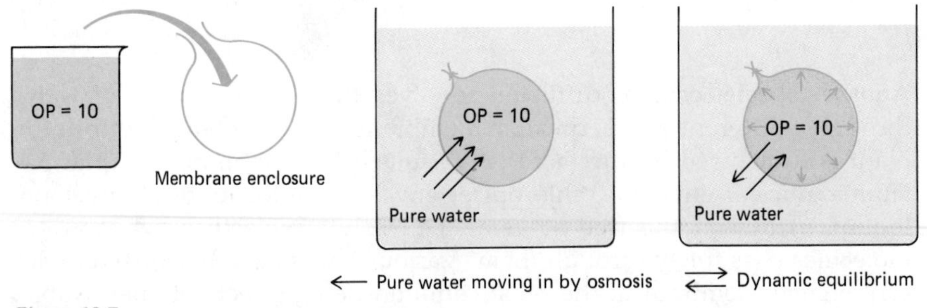


*Figure 12.7*
Dynamic equilibrium. (See text for description of figures.)

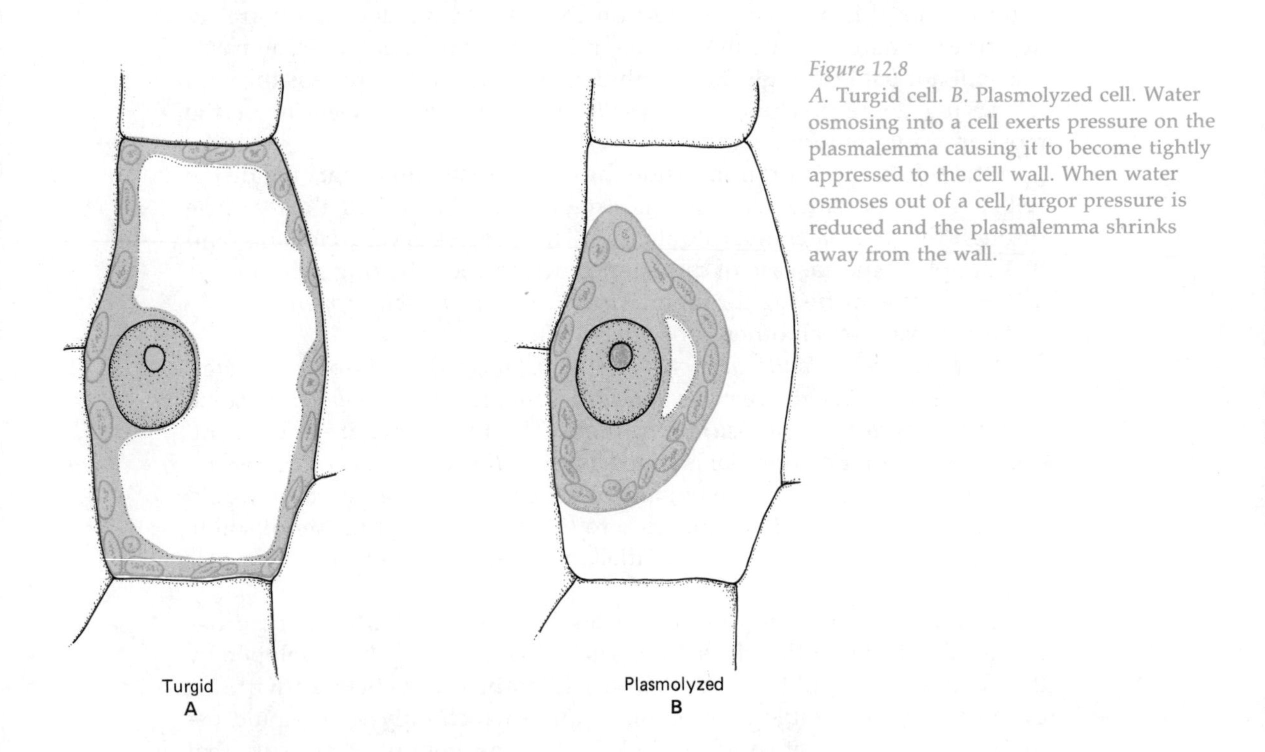


*Figure 12.8*
*A.* Turgid cell. *B.* Plasmolyzed cell. Water osmosing into a cell exerts pressure on the plasmalemma causing it to become tightly appressed to the cell wall. When water osmoses out of a cell, turgor pressure is reduced and the plasmalemma shrinks away from the wall.

−10 units (whatever units of pressure you wish to use) will be exerted on the walls of that enclosing bag.

The pressure created in a cell as the result of osmosis is called *turgor pressure*, and the cell is said to become *turgid.* According to Newton's third law of *equal* and *opposite* reactions, the pressure of the protoplast against the cell wall *(turgor pressure)* should be balanced by the pressure of the wall against the protoplast *(wall pressure).* In our diagram shown in Figure 12.7, we have equated osmotic potential with the realized turgor pressure in the cell. Actually, this is a gross oversimplification of the relationship between the two. It might be analogous to that of a racing car. The driver may tell you, "This car has the potential to run over 300 miles per hour." Rarely, if ever, would that potential be met, due to restrictions imposed by the track, by weather, by the driver's skill, etc.

If the reverse condition occurs and the solute concentration outside the cell becomes greater than that inside the cell, water will osmose out of the cell. As water leaves the protoplast, the plasmalemma will shrink away from the wall, creating a space between the wall and the protoplast (Figure 12.8). This condition is known as *plasmolysis*. If the water-potential gradient is great enough, and the cell is left in this solution long enough, the cell will die. Such a condition might occur when a considerable amount of inorganic fertilizer is placed in the vicinity of plant roots. Although the soil may contain considerable moisture, the difference in water potential will result in plasmolysis of root cells and may result in their eventual death. This condition is known as *physiological drought*.

When differences in osmotic potential exist, special terms have been coined to denote those differences. The region of higher osmotic potential is termed *hypertonic*, the region of lower osmotic potential *hypotonic*. If the osmotic potentials of the two regions are equal, they are said to be *isotonic* to each other.

## TRANSPORT

Once in the plant, substances may move from cell to cell through the free space or through specific conducting tissue. In Chapter 8, we discussed some of the mechanisms of transport. Water will diffuse independently of solutes, and each solute will diffuse independently of all other solutes where diffusion through membranes is involved.

### *Transport Across Membranes*

The transport of particles across membranes is not simply a problem of movement through the plasmalemma. Various cell membranes exist, and thus, the cell might be thought of as compartmentalized. Particle transport can then be considered in terms of movement not only into the cell, but from compartment to compartment within the cell.

Transport across cell membranes might be considered to be of two general types. If water or chemical potential is higher outside the membrane than inside, then transport inward is *passive*. But if movement is occurring *against* the water or chemical potential gradient, the transport is *active*. Active transport occurs only at the expense of metabolic energy.

Although there are reports of water being "pumped" across membranes against a water potential gradient, water osmoses normally. Solute particles however, often move against a chemical potential gradient and thus are transported actively. But establishment of a chemical gradient does not ensure that solute particles will be transported passively. As we have mentioned, cell membranes are differentially permeable, and the permeability of most healthy membranes to solutes is low. Generally, the more highly charged the particle is, the less likely its chances of diffusing across the membrane. Thus, molecules generally do somewhat better than ions. For example, ATP formed in the mitochondria is actively transported into the cytoplasm even though it is at a higher chemical potential inside than outside the mitochondria, while glyoxylate will diffuse out of chloroplasts into the cytoplasm where its chemical potential is lower. Many substances are much more concentrated in the vacuole than in the cytoplasm. Thus, movement across the tonoplast from cytoplasm to vacuole requires generally active transport. We should consider one more point concerning active versus passive transport. If a membrane is at all permeable to a substance and if a chemical potential gradient exists, that substance will diffuse through the membrane, however slowly, and equilibrium eventually will be established. But metabolic energy may be used to increase the *rate* of diffusion. Thus, metabolic energy may not only initiate transport across membranes but may also increase the rate of such transport.

## Water Transport

As indicated in the previous section, water and substances dissolved in water may move from cell to cell or between cells by a variety of active or passive mechanisms. As the bulk of these substances are absorbed by the roots yet are required by all parts of the plant, a mechanism must exist for the upward transport of these substances. Such a movement is called *translocation.* The foregoing mechanisms would be quite inadequate both in rate and volume. How then, does translocation occur?

Of paramount importance to the consideration of vertical water transport is the process of *transpiration.* Transpiration might be described as the evaporation of water from cell surfaces. Since the water evaporates, the cell surfaces involved must obviously be in contact with the atmosphere. This loss of water vapor from the plant may occur in several ways. *Cuticular transpiration* refers to loss of water through the cuticle. Even though the cuticle is composed of a waxy substance, water vapor and other gasses will move through it. About 5 to 10 percent of the water loss of a temperate plant occurs in this fashion. The figure will be higher for tropical plants, lower for desert plants. Some water may be lost directly through the corky layer, particularly through the *lenticels. Lenticular transpiration* is normally quite insignificant. Although in certain plants under conditions of low available soil moisture, enough water may be lost by this mechanism to cause dehydration and injury to the plant.

By far the greatest amount of water vapor is lost through the stomata *(stomatal transpiration)*, with the leaves being normally the principal transpiring organs. Although a certain amount of water vapor will diffuse

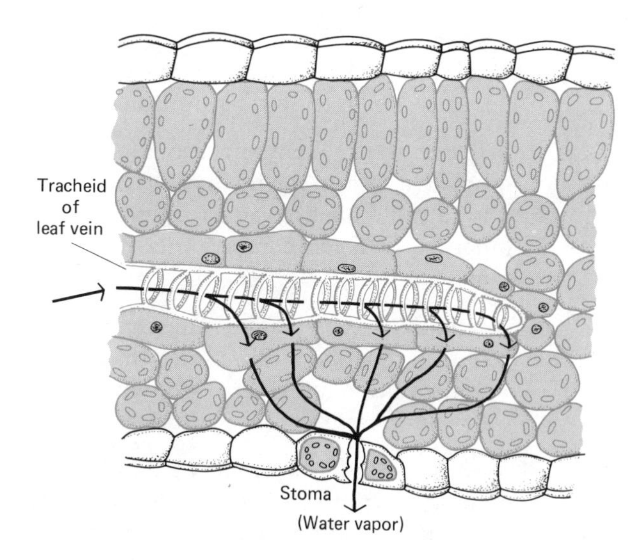


*Figure 12.9*
Water concentration pathway in a leaf. Transpiration of water creates the water deficit responsible for the directional pathway shown.

through the stomata whether they are open or closed (a stoma cannot close completely), by far the greatest amount will be lost when the stomata are open.

In living leaf cells, water fills the vacuoles, makes up a large per cent of the cytoplasm, and penetrates the cell walls and middle lamellae. In addition, it is adsorbed on the surface of cell walls. Thus, the walls of mesophyll cells bordering a substomatal air space will be moist, and the air in the substomatal chamber will be saturated. As a consequence, the water potential of the air space will be high, while that of the external atmosphere will normally be lower. Therefore, water vapor diffuses out of the substomatal cavity, which lowers the water potential of the cavity to the point that water molecules evaporate from the wet walls of the mesophyll cells and diffuse into the air space. As the walls become drier, they imbibe water from the cytoplasm and vacuole of the cell.

In this fashion, a diffusion gradient is established. The mesophyll cells surrounding the substomatal cavity, having lost water into the cavity, will have a water potential less than that of cells adjacent to them. Thus, water osmoses from these cells, and these in turn gain water from adjacent cells, eventually forming a water-potential gradient back to the water-filled tracheids in the veins (Figure 12.9). As long as transpiration occurs, water will continue to be pulled from the vein endings.

A number of factors affect the rate of transpiration, and these are elucidated below:

*Water Concentration in Plant.* The amount of water in leaf cells may effect the transpiration rate in two ways: it may act by affecting directly the water-potential gradient, or it may act by affecting indirectly the stomatal opening.

*Atmospheric humidity.* Humidity in the outside air obviously affects transpiration by its effect on the water-potential gradient. As humidity increases, the rate of transpiration decreases.

*Temperature.* Temperature affects transpiration rate in several ways. It has an affect on the relative humidity (the amount of water vapor required to saturate the air), and it increases the speed or velocity of the diffusing particles. As temperature increases, the rate of transpiration increases, up to a point.

*Light Intensity.* Light intensity also has several affects on transpiration. In some plants, it stimulates the opening of the stomata, thus increasing transpiration rate. Increased light intensity also tends to increase the internal temperature of the leaf, which results in increased transpiration.

*Air Movement.* Normally a diffusion cone exists above the surface of each stoma (Figure 12.10). If this layer is undisturbed, it will decrease the water-potential gradient between leaf and atmosphere, thus causing a decrease in the transpiration rate. Wind currènts tend to carry away this moist air layer, resulting in an increased transpiration rate. But if the wind is relatively strong, it may cause the epidermal tissues to dry out, the stomata may close, and the transpiration rate will decrease consequently.

*Soil Moisture.* Normally, the less available water in the soil, the less water will be present in the plant. Under these conditions, a high transpiration rate may result in dessication of the leaf cells, and the leaves will wilt. The water-potential gradient will be reduced, reducing consequently the rate of transpiration. If the guard cells become dessicated, the stoma will narrow or close completely, also resulting in decreased transpiration.

*Stomatal Position or Distribution.* If the stomata are borne only on the lower surface of the leaf, they will not be subjected to the direct rays of the sun. This will result in lower temperature and thus a decreased transpiration rate as opposed to leaves having stomata on the upper surface or on

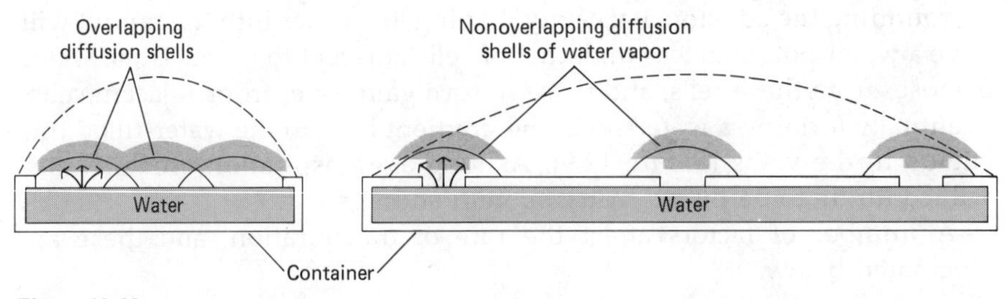


*Figure 12.10*
As transpiration occurs through stomates, a diffusion cone is established for each stoma. This effectively reduces the transpiration rate unless wind currents help to dissipate the cone.

both surfaces. In some plants, the guard cells are sunken below the surface of the epidermis. This may also lead to a decreased transpiration rate. An increase in the number of stomata per unit area of leaf surface will not necessarily increase the transpiration rate. As mentioned earlier, each stoma will have a diffusion cone associated with it. If the stoma are close enough to each other so the diffusion cones overlap, the water-potential gradient will be decreased, resulting in a decreased transpiration rate for each stoma.

Not all water lost by a plant is lost as water vapor. Certain plants, usually under conditions of high water potential in the cells and a high humidity in the atmosphere, will lose water in liquid form. This process is called *guttation.* Guttation usually occurs through specialized openings called *hydathodes*, located at the tips of the leaves (Figure 12.11). The liquid of guttation contains both organic and inorganic compounds, and although the quantity of water lost by guttation is not great, young leaves may be damaged occasionally by the drying of guttation fluid on the blade.

## Mechanisms of Water Transport

The movement of water from the roots to the top of a very tall tree creates an interesting problem for investigators. How does it occur? Where does it occur? The second question was answered with relative ease. If a ring of bark was removed from the stem (remember that "bark" refers to all tissues outside of the vascular cambium and thus would include phloem),

*Figure 12.11*
Guttation in leaves. The liquid water is exuded through special pores called hydathodes. [Photograph by F. H. Norris.]

water movement in the plant was little affected, and the top of the tree would live for some time. This suggested that the water moved through the xylem. This can be demonstrated clearly by placing the cut end of a stem into a dye solution. After a time, it can be seen that only the walls of xylem vessels and tracheids are stained by the dye.

The first question is much more difficult to answer. *Atmospheric pressure* alone could account for a rise of only approximately 33 feet. As the vessels and tracheids are relatively small in diameter, *capillarity* was suggested as a possible mechanism. But capillarity is sufficient to raise water only a short distance, certainly not to the height or in the quantities observed to occur in stems.

It has been observed that if a plant is decapitated and the soil kept well-watered, water will exude from the cut end of the stem. This phenomena is called *root pressure.* It would follow naturally that this process would be suggested as a possible mechanism for the ascent of sap. But the measurable pressure will only reach about 2 to 3 bars (a bar equals about 70.3 $g/cm^2$ pressure at sea level) by this mechanism, while it has been calculated that 10 to 20 bars of pressure are required to lift water to the top of a 300 to 350 foot tall tree. In addition, root pressure moves water rather slowly, and is minimal under conditions of low availability of soil water or high rates of transpiration. Yet water still continues to be translocated under these conditions.

If the water were in the form of *vapor* rather than liquid, then not as much pressure would be required to move it. But all evidence indicates that water is translocated in liquid form. Perhaps, some sort of *active pumping* mechanism exists in living cells of the xylem. Most evidence does not support this assumption. It is indicated clearly that water moves through nonliving rather than living cells, and the application of metabolic inhibitors seems to have little or no effect on water movement.

All the above theories rest on the basic law of physics that all forces in nature are pushing, not pulling, forces. But what if the ascent of sap does not fit this rule; what if there is a pulling force associated with it? Recall that in our discussion of transpiration, we established a water-potential gradient from the atmosphere to the xylem tracheids in leaf veins. If, as we suggested, water molecules are moving from these tracheids into surrounding cells, could they not be pulling other water molecules with them? If this mechanism is to account for movement of water from the roots to the leaves, several requirements must be met. A column of water must exist from roots to leaves, and this column must be continuous. We mentioned in an earlier chapter that some of the xylem *is continuous* from root to leaves, thus this requirement is established. But do these columns have sufficient tensile strength to prevent them from being broken? Evidence indicates that these water columns can withstand tensions greater than 300 bars without being broken, tension far greater than they would actually be subjected to in the plant. This great tensile strength of the water column results from two forces: the strong attraction of water molecules to each other *(cohesion)* and the attraction of water molecules to the walls of the xylem elements *(adhesion)*. Thus, a continuous water column is established and can be maintained. But is the force of transpiration great enough to pull these water columns all the way up a tree? It has been calculated that the water potential of air at 50 percent relative humidity is a negative 1000 bars. Thus, a considerable water-potential gradient exists between the atmosphere and the leaf cells and considerable tension is induced within these cells. The force of transpiration is therefore, much greater than necessary to account for vertical transport of water. Why do the leaf cells not collapse under pressures? Probably simply their small size prevents such a collapse.

## Phloem Transport

Girdling experiments and other evidence indicate that rapid translocation of food in the plant takes place primarily in the phloem. The cells of the phloem involved specifically in transport are the sieve elements.

Information regarding mechanisms of the movement of substances through the phloem has come from two sources: study of the exudate from aphid stylets and use of radioactive tracers. In the former, advantage is taken of the fact that most species of aphids are phloem feeders. They insert their modified mouth parts, or stylets, into a leaf or stem until they puncture a sieve tube. The turgor pressure in that tube forces sap into the hollow stylet and through the insect's digestive tract, with the unabsorbed substance emerging from the posterior as a drop of "honeydew." If the aphid is anesthetized while feeding and severed from its stylet, phloem sap will continue to emerge from the cut end of the stylet for several hours. The substance can be collected and analyzed and, in addition, can give some impression of the turgor pressure of the sieve tube. Most of the solute portion of the phloem sap consists of sucrose, although amino acids and various other nitrogenous compounds may be found, as well as certain other kinds of sugars in some plants. Glucose is not contained in phloem sap. The use of radioactive sucrose has also shown the localization of this material to be in the sieve elements.

Data obtained with the above procedures reveal that rates far in excess of those to be expected as the consequence of normal diffusion occur in the sieve elements. Rates in the neighborhood of 100 centimeters per hour have been estimated.

The sieve element was first discovered by Theodor Hurtig in 1837. He suggested that it might be the principal food-conducting element in the phloem. His suggestion was, however, ignored by his contemporaries and most of those following him. It was not until 1927 that serious thought was given to the movement of assimilates in the phloem. In that year, a *pressure-flow hypothesis* for the movement of assimilates through the sieve

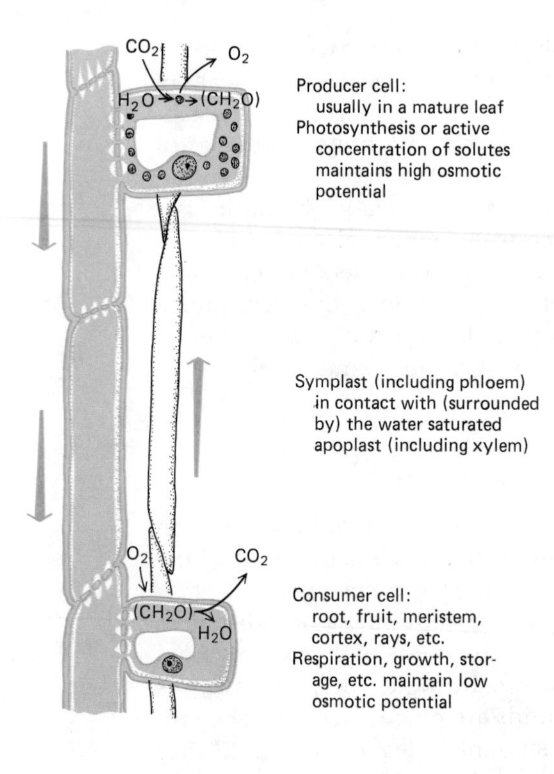


*Figure 12.12*
An illustration of phloem transport from source to sink. [From F. B. Salisbury, "Translocation: The Movement of Dissolved Substances in Plants," in W. A. Jensen and L. G. Kavaljian, eds., *Plant Biology Today: Advances and Challenges*, Wadsworth Publishing Company, Inc., Belmont, Calif., 1963, p. 82.]

elements was proposed by E. Münch. The hypothesis has been modified since its inception, but its major features are basically the same. According to this hypothesis, assimilates move through the sieve elements along a concentration gradient established between the *source* and *sink*. Sources are those areas where the food substances are being manufactured (such as photosynthesizing leaves), and sinks are areas where the food substances are being utilized, either in storage or metabolism. Examples of the latter would be the growing tips of shoot or root, or the cortical cells of the root or hypocotyl. A schematic pathway for a source to sink transport is shown in Figure 12.12.

Metabolic energy must be utilized by either companion cells or other parenchyma cells surrounding the sieve element to "pump" the assimilate into the sieve element. This decreases the water potential of the sieve element, and water moves into it from the xylem. The sugar is carried downward through the sieve element with the water to the root (or upward to the growing stem tip), where neighboring parenchyma cells "pump" the sugar out of the sieve element. This results in an increase in water potential, causing the water to move out of the cell at the sink area. As the sugar molecules are postulated as simply moving passively with the water, the pressure-flow hypothesis is an example of a *mass flow* mechanism. Several other mechanisms have been suggested to explain assimilate movement in the phloem. These are discussed in the next section for those of you interested in pursuing the subject somewhat further.

## SOME ADDITIONAL OBSERVATIONS

This section is intended for those of you that might wish to go into the aspects of translocation in somewhat more detail.

### *The Ascent of Sap*

With all the evidence mentioned earlier, it is understandable why many plant physiologists ascribe to the transpiration-pull-water-cohesion theory of water translocation. Yet many problems still exist.

In 1927, a graduate student at Ohio State University by the name of Askenasy devised an apparatus similar to that shown in Figure 12.13. With the evaporation of water from the porous clay pot at the tip, a process similar to that of transpiration, he demonstrated that mercury could be drawn

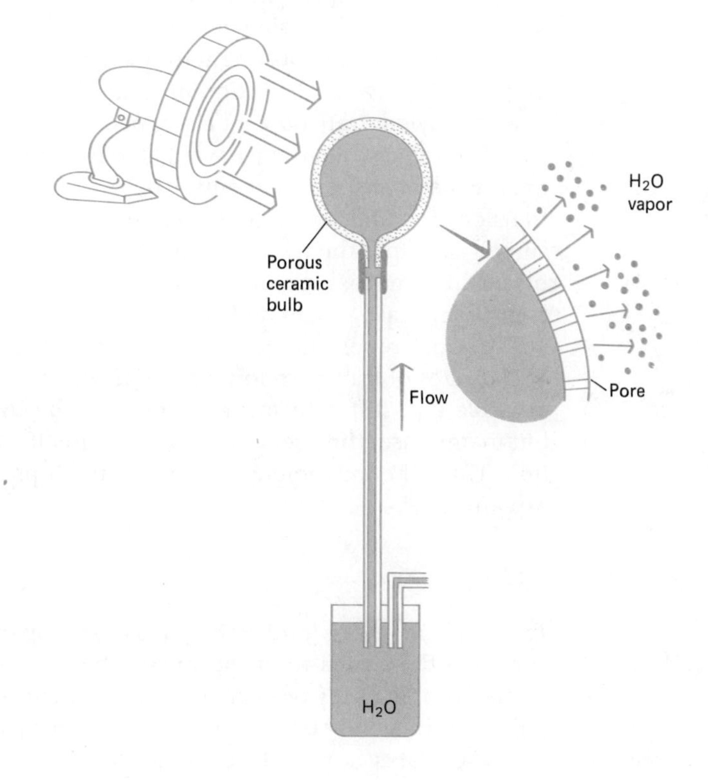


*Figure 12.13*
The Askenazy apparatus. This apparatus helped to establish the role of transpiration in movement of water and minerals up a tall stem.

up a capillary tube far beyond that ascribed normally to the effect of atmospheric pressure. It was later demonstrated that a leafy twig could be substituted for the clay pot, and the actual effect of transpiration would produce the same result. It was demonstrations of this type that led to the widespread acceptance of this theory to explain the ascent of sap to the tops of tall trees. But a single bubble of air or an improper seal would cause this system to break down. This theory makes implicit the continuity of the water column.

The question then becomes, are such water columns in the plant continuous? We have already mentioned that some xylem *is* continuous from root to stem. But what if the column is disrupted? As early as 1728, Stephan Hales, and many others since then, made transverse cuts across stems, coming in from opposite sides and overlapping beyond the midline. Such cuts would sever all vertical water columns. Yet they have little or no effect on the transpiration rate, unless the cuts are very close together (2–6 cm). Perhaps the transpiration stream is able to move laterally in the region of the cuts. Greenridge (1955), Postlethwait and Rodgers (1958), and others have shown that this does in fact occur (Figure 12.14). It has been suggested that this lateral movement is through living cells of the xylem rays. But Mackay and Weatherly (1973) have shown that lack of living tissue in the region between the cuts did not influence the transpiration rate, nor did it prevent lateral water flow.

Another type of experiment to disrupt the continuous water columns is to quick freeze a small section of twig, thus causing a block of ice to form in all the xylem transporting elements. As you know, water expands upon freezing; thus, when the ice was allowed to thaw, air bubbles were formed in all water transport elements. Theoretically, this should have prevented translocation and subsequent transpiration. It did not. Transpiration resumed as before the freezing, either by movement of water around the air blocks in some fashion or by movement of the air blocks on up the stem, being eventually dissipated at the tip.

We must report however, that some investigators have reported reduced rates of transpiration when the water columns are interrupted. It is possible that different mechanisms may be involved in different plants, but in any case, the breaks do not seem to affect seriously water translocation. Thus, the movement of water to the tops of tall trees is still not fully explained.

## Food Transport

The mechanisms of food transport are still open to considerable question, and even the structure of the sieve tube cells in relation to their function creates problems of interpretation. As mentioned earlier the sieve tube cells contain sieve plates. Electron microscopy reveals that the sieve pores in those plates appear to contain closely packed filaments which have

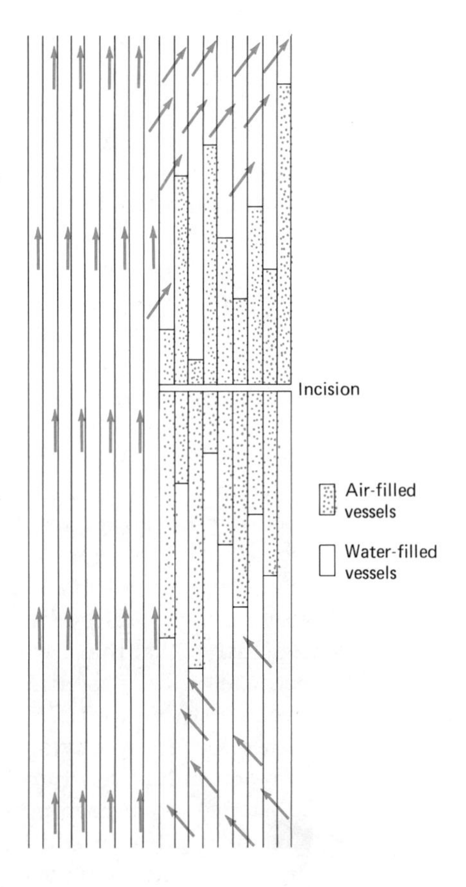


*Figure 12.14*
Hypothetical pattern of sap-and-air-filled vessels and the flow pathways after a stem incision (longitudinal section). [J. F. G. Mackay and P. E. Weatherley, "The Effects of Transverse Cuts through the Stems of Transpiring Woody Plants on Water Transport and Stress in the Leaves," *J. Exp. Bot.* 24(78):15–28. © Oxford University Press, 1973. Reprinted by permission of the Oxford University Press.]

been called *p-protein.* Other investigators, however, state that p-protein is simply an artifact of preparation. But if these filaments actually exist in the living, functioning cell, and are not simply artifacts, they would offer considerable resistance to flow. Pores also exist in the side walls of sieve tubes and plasmodesmata have been reported forming interconnections between sieve tube elements and their associated companion cells.

Among the theories proposed for phloem transport, the following probably represent the primary ones, and may be categorized as one of two types: 1) no pathway energy, the movement determined simply by concentration gradients between source and sink, although energy would be required for maintaining the system or 2) pathway energy required. The former type of system would include 1) *mass flow*, including pressure flow and 2) *interfacial flow* hypothesis of Van den Honert. The latter type would include 1) two types of mass flow: "*tubular peristaltic flow*" of Thane and Fensom and "electroosmosis" of Spanner and Fensom, 2) *activated diffusion* of Mason and Maskett, and 3) *cytoplasmic streaming*, first suggested by the Dutch botanist Hugo de Vries in 1885, developed further by Curtis, and more recently championed by Canny.

Specific metabolites can be accumulated by sieve elements against a concentration gradient, and these can be retained during translocation. In addition, the rate of flow does react to metabolic inhibitors. This information and other data seem to supply evidence for an energy requirement. The possibility of certain substances being translocated by a simple diffusion mechanism is not discounted, but the more common substances such as sucrose would not appear to be translocated by simple diffusion. Sucrose concentrations of 5 to 50 percent have been reported in sieve tubes, with concentrations of 10 to 20 percent most often mentioned. Such concentrations would result in a flow rate of some 40 cm $hr^{-1}$ for the average concentrations, with the rate reaching as high as 100 cm $hr^{-1}$ by some reports.

Not only do foods move through the phloem, but inorganic ions and even certain hormones may be translocated in the phloem sap, which would, of course, include water. There is considerable evidence that different compounds move at different speeds in the sieve tubes, with water moving most rapidly. There is also evidence, although still somewhat inconclusive, that the flow may be in both directions in the same sieve tube. Because of these aspects, cytoplasmic flow would appear to be an attractive theory (Figure 12.15). But it has been observed that as a sieve element matures, the rate of cytoplasmic streaming in that cell decreases and finally stops entirely. Even in cells where cytoplasmic streaming occurs, certain substances have been reported to be translocated quite independently of streaming rates. In addition, the rate of streaming does not begin to approach the rates of phloem transport actually observed.

There is evidence both for and against the other theories mentioned, but at present, some mechanism of bulk or mass flow (described earlier) seems to be most widely accepted. Considerable evidence for mass

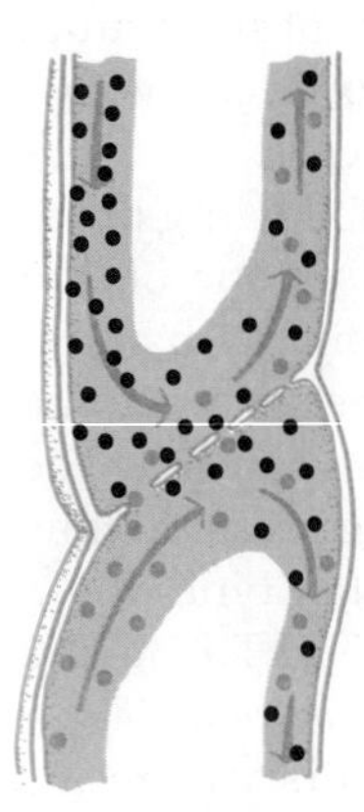

*Figure 12.15*
Schematic illustration of the cytoplasmic streaming hypothesis of solute translocation. The black and white circles represent different solute particles. Each will diffuse according to its own concentration gradients. Movement of the two materials might occur simultaneously in opposite directions. [From F. B. Salisbury, "Translocation," in W. A. Jensen and L. G. Kavaljian, eds., *Plant Biology Today: Advances and Challenges*, Wadsworth Publishing Company, Inc., Belmont, Calif., 1963, p. 81].

flow exists, including the construction of mathematical models that fit actually observed rates. But there are several problems with the mass flow theory. These include the points mentioned earlier concerning movement in opposite directions in the same cell and differential rates of movement of different substances. In fact, if these observations are true, they would be certainly incompatible with the mass flow theory. In addition, the reported presence of p-protein strands in the sieve pores, strands reported to be about 16 nm in diameter with a space between each filament of about 50 nm, would certainly cast doubt on the pressure-flow hypothesis of bulk movement. It is the reported presence of these proteinaceous strands that has revived interest in the interfacial flow hypothesis, as these would increase the area of boundary layers considerably. Some investigators (Thane, 1963) have interpreted these strands as actual tubules and have hypothesized that a bulk flow mechanism could exist through them.

As you can conclude from the above discussion, phloem transport mechanisms are still much in dispute. It may well be that different substances are translocated by different mechanisms. While we can draw few definitive conclusions regarding phloem transport, we have indicated that most investigators agree that xylem transport is associated in some way with the process of transpiration. We can always fall back on basic generalizations: water and inorganic substances are transported through the xylem; food, or organic compounds, are transported through the phloem. But phloem transports water and some inorganics, and xylem may transport some organic compounds. In fact, there appears to be a definite lateral movement of materials between xylem and phloem. If we are considering the region of *greatest* amount of transport, the above generalizations do hold.

## SUMMARY

1. All molecules have a certain amount of inherent, internal energy known as kinetic energy.
2. The amount of kinetic energy a molecule possesses determines, in part, its free energy, or energy to do work the so-called chemical potential.
3. One aspect of work is the movement of molecules. If differences in chemical potential exist between adjacent regions, molecules will tend to exhibit a net movement into the region of lower chemical potential. This movement is termed diffusion.
4. Thus, diffusion is the net movement of a substance from a region of its higher chemical potential to a region of its lower chemical potential. Stated in different terms, it is the movement of a substance from a region of its higher concentration to a region of its lower concentration.

5. When the chemical potentials (or concentrations) of a substance become equalized between the two regions, the system is at equilibrium and diffusion, but not movement, of the substance ceases.
6. The diffusion process, rate, or both are affected by temperature, pressure, solute concentration, adsorption, concentration of the diffusing substance, distance over which the substance diffuses, and the density or size of the diffusing particles.
7. A special case of diffusion is imbibition. It involves the uptake of water by a dry solid and in wood, involves the movement of water into the middle lamella and cell walls.
8. Another special case of diffusion is osmosis. It involves the diffusion of water through a differentially permeable membrane.
9. The concentration of the solute determines the osmotic potential of a solution. In the cell, osmotic potential is the potential to create turgor pressure in the cell, with an equal and opposing wall pressure.
10. When turgor pressure in a cell is fairly high, the plasma membrane is tightly appressed against the wall and the cell is said to be turgid. In the absence of turgor pressure, the plasma membrane may shrink away from the wall and the cell is said to be in a state of plasmolysis.
11. The osmotic potential of a cell may be related to that of its surroundings. The region of higher osmotic potential is termed hypertonic, the region of lower osmotic potential hypotonic. If the osmotic potentials are equal, they are isotonic to each other.
12. Transport across a membrane may be inactive or passive. That is, no cell metabolic energy is required. This would be the case in normal diffusion or osmosis.
13. Transport across a membrane may be active. That is, cell metabolic energy is required to facilitate the transport. Such instances are termed active diffusion or active transport.
14. Metabolic energy may either initiate tranport of a substance across a membrane, or it may increase the rate of such transport.
15. Transpiration is the evaporation of water from cell surfaces.
16. The greatest amount of transpiration is associated with the stomata.
17. The process of transpiration may be associated with the movement of water from the roots to the tops of plants.
18. Factors which affect transpiration include water concentration in the plant, atmospheric humidity, temperature, light intensity, air movement, soil moisture, and stomatal position or distribution.
19. Water may be lost from leaves in liquid form via a process termed guttation.
20. Vertical water movement in plants could be due to capillarity, atmospheric pressure, root pressure or transpiration pull.
21. Transpiration pull is the most widely accepted mechanism for the movement of water to the tops of tall trees and involves transpiration from the leaf and the cohesion and adhesion of water molecules.

22. The principal vertical transport of water occurs through the tracheids and vessels of the xylem.
23. Food transport occurs from source to sink and may be either upward or downward in the stem. The principal mechanism is thought to be associated with mass flow and is termed the pressure-flow hypothesis.
24. Food transport occurs primarily through sieve elements of the phloem.

# Chapter 13

## PLANT FOODS, FOOD SYNTHESIS, AND DIGESTION

# PLANT FOODS

Ask your parents or the typical person in the street what he or she considers to be the source of plant foods, and you would probably get the following list:

1. soil
2. minerals
3. water
4. air
5. carbon dioxide
6. oxygen
7. carbohydrates
8. fats
9. proteins

At one time or another, most of these were considered to be plant foods. But ask those same people to list the source of their foods, and they would say plants, animals, or plant and animal products; in other words, *organic substances.* Are plant foods different from animal foods? Are all the substances listed above really plant foods? In order to answer these questions, we obviously need to develop a concept of food. This I hope to do in the next several chapters. Keep in mind that the above *list* of plant foods does not necessarily represent my idea of foods, but rather what have been considered foods at one time or another. When you have more information and have developed a concept of food, you will probably want to come back and modify this list.

Around 350 B.C., Aristotle concluded that green plants obtained their food from the soil. This belief was accepted widely and it was not until 1648 that someone attempted to test experimentally this hypothesis. A Flemish physician and chemist, Van Helmont, weighed a bucket of soil and a small willow branch, then planted the branch in the soil. For a period of five years, he added nothing to the soil but rain water. At the end of the five year period, the willow branch had grown into a small tree, gaining some 75 kilograms in weight, while the soil had decreased by only 75 grams. Having no other evidence, Van Helmont made the rather natural assumption that the food must have come from the rain water.

In 1727, Stephan Hales, in a book hardly larger than its title, *Statical Essays, Containing Vegetable Statics, or, an Account of Some Statical Experiments on the Sap in Vegetation,* while not disagreeing with Van Helmont's conclusion, surmised that plants drew a part of their nutrition from the air. Somewhat as an aside, he also stated that he thought sunlight might also be involved in some way. No further research into these matters was reported for some fifty years.

# FOOD SYNTHESIS

## *Photosynthesis of Foods*

From the 1770s onward, work was conducted which led to increased understanding of a process called *photosynthesis.* Most introductory biology or botany students are at least somewhat familiar with the so-called general formula for photosynthesis: carbon dioxide plus water, in the presence of light and chlorophyll (in chloroplasts), produces sugar plus oxygen. This statement is represented by the equation.

$$6CO_2 + 6H_2O \xrightarrow[\text{chloroplasts}]{\text{light energy}} C_6H_{12}O_6 + 6O_2$$

Do we simply have to accept this equation on faith? Not at all. We could perform some very simple experiments that would demonstrate these relationships. As you are well aware, if we place a green plant in the dark, it will soon become very unhealthy looking and eventually will die. This does not prove, however, that light is necessary for photosynthesis. If we place two plants in bell jars, one in light and one in darkness, and then test each for oxygen evolution, we would find that oxygen is evolved only by the plant in the light. Again, this does not prove that light is necessary for photosynthesis, but it does indicate a relationship between light and oxygen evolution.

Another simple experiment utilizes the fact that an iodine–potassium-iodide solution reacts with starch forming a dark blue complex. If we place one plant in the light and a second plant in a dark room for approximately four days and then test the leaves of each plant for starch, we find starch only in the leaves which have been in the light. But how does this indicate that light is necessary for photosynthesis? It could logically be inferred from this demonstration that light is necessary for starch synthesis. We can modify the demonstration somewhat by removing leaves from a plant and placing some petioles into distilled water, others into a 4 percent glucose solution. Both groups of leaves are then placed in a dark room. After four days, each group of leaves are tested for starch. Those in the distilled water test negative as before, but those in the glucose solution are now starch positive.

This demonstration should suggest several inferences: 1) light is *not* necessary for starch synthesis, but a source of sugar *is* necessary and 2) light *is* apparently necessary for sugar formation. Thus, these dark-grown plants have demonstrated several aspects of photosynthesis: 1) light is necessary for sugar formation and 2) light is necessary for oxygen evolution. A more graphic demonstration of oxygen evolution can be shown if we place a sprig of an aquatic plant such as *Elodea* in a funnel, place a test tube filled with water upside down over the tip of the funnel, and place the whole apparatus in a beaker of water (Figure 13.1A). If we shine light

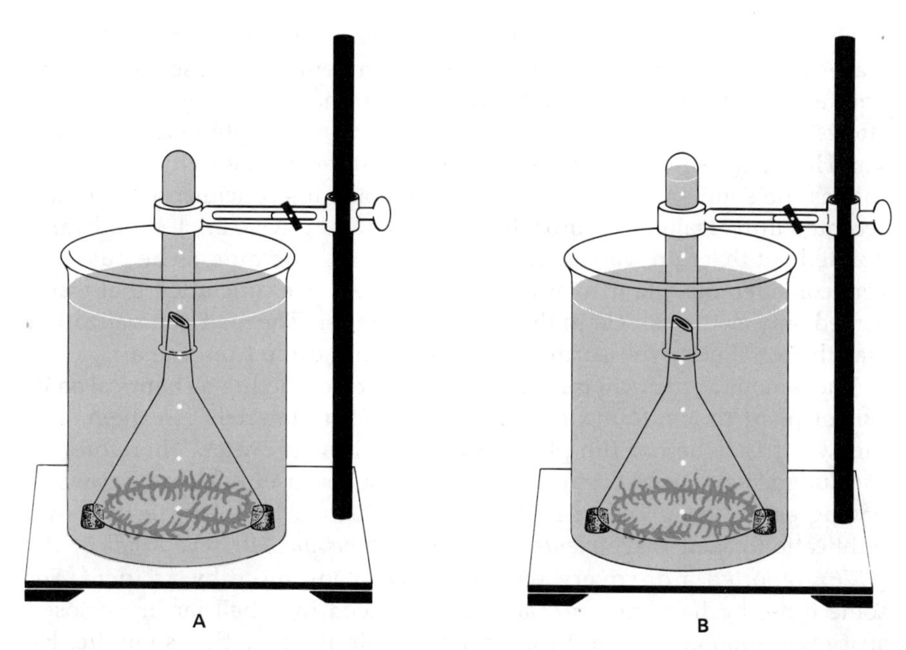


*Figure 13.1*
An apparatus to demonstrate oxygen evolution by an aquatic plant. *A.* Initial condition, with test tube completely water-filled; *B.* after some time. Note that the evolving oxygen has displaced some of the water in the end of the test tube.

on the beaker, bubbles will be seen to rise from the *Elodea* up through the funnel and into the test tube where the water at the end of the tube will become displaced by the gas (Figure 13.1B). If we test the gas in the end of the tube, we will find it to be oxygen.

We can modify the above experiment somewhat by first boiling the water in the beaker to drive off all the $CO_2$. Under these conditions, no oxygen will be evolved by the *Elodea*. If bicarbonate is then added to the water (bicarbonate is a source of $CO_2$, i.e., $HCO_3^- + H^+ \rightarrow CO_2 + H_2O$) bubbles of oxygen will begin to rise from the *Elodea* plant. This demonstration indicates the necessity for $CO_2$.

The necessity for *green* plant cells can be shown by utilizing green and albino strains of corn. The albino strains, of course, lack chloroplasts. The albino corn leaves will test starch negative, and if a green and nongreen corn plant are placed into separate bell jars, no oxygen will be evolved by the albino plant. So we can demonstrate for ourselves that the basic formula for photosynthesis is essentially correct. We cannot state whether water is actually a material used in photosynthesis as there is no simple demonstration that would indicate this relationship. It has been shown clearly to be so, however, by tracer experiments that we will mention in more detail later.

How was the general formula for photosynthesis arrived at in the first place? It evolved through a number of experiments and observations. As pointed out previously, an understanding of photosynthesis arose out of efforts to determine how plants obtained their nourishment. Although Van Helmont demonstrated that the old Aristotelian theory of humus, or soil, as the source of plant food was erroneous, his conclusions were not accepted universally. A Cambridge professor and physician, J. Woodward, pointed out that rain water is not pure but in fact contains, as he put it, "a very considerable quantity of terrestrial matter." He concluded that water served only as the vehicle to the terrestrial matter. The next generalization was that of Hales, who surmised that nutrition came from the air.

The famous French chemist Lavoisier, after conducting a chemical analysis of plant tissue, found that they had "earth materials" in them, and this was true whether the plants grew in soil or in water. Therefore, he concluded that plants "draw" their materials from either earth or water having earthy materials in it. The British chemist Joseph Priestley, in a publication called *Experiments and Observations on Different Kinds of Air* (1776), reported a discovery which he had made quite by accident. For some time, he had known that candles placed in a bell jar or enclosed space will soon go out, or mice placed in a similar situation soon die. He assumed that both the candle and the animal had somehow "damaged" the air by perhaps releasing a substance called *phlogiston.* He went on to say, "I flatter myself that I have accidentally hit upon a method of restoring air which has been injured by the burning of candles, and that I have discovered at least one of the restoratives which nature employs for this purpose. It is vegetation." Priestley had been of the opinion that *both* plants and animals would affect the air in the same way. Accordingly, he placed a sprig of mint into a bell jar inverted in a pan of water. Much to his surprise, the mint had continued to grow. He then proceeded to place a sprig

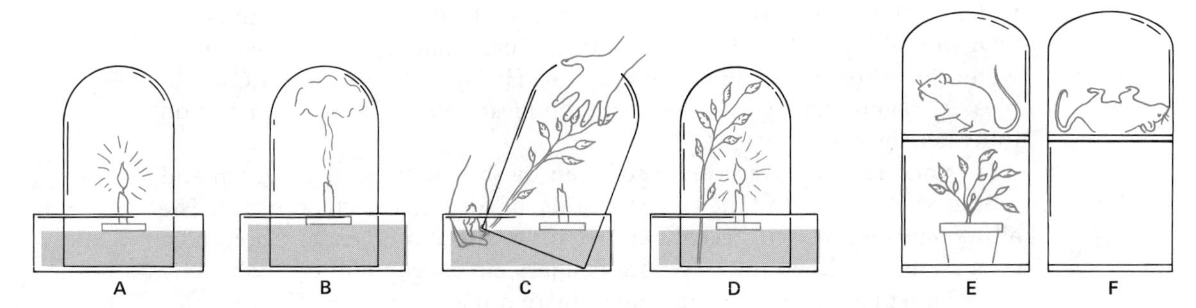


*Figure 13.2*
An interpretation of Priestley's experiments. *A.* Candle placed under an air-tight bell jar will at first burn. *B.* However, it soon goes out. *C.* A green plant placed under the bell jar. *D.* After a few days the candle will again burn. *E.* With a green plant in the bell jar, the mouse lives. *F.* Without the green plant, the mouse dies.

*Figure 13.3*
A contemporary French cartoon making fun of Priestley's ideas.

of mint into a bell jar in which a candle had burned out, and found the air would again support a candle flame (Figure 13.2). He referred to this as restored air. He found also that if a plant were placed with a mouse under a bell jar, both would continue to live. Although Priestley was quite aware of the importance of this discovery, he was quite unaware of the full implications of it. Unfortunately, Priestley had no understanding of the importance of light to the plant; thus, he was not too careful in providing light. For this reason, he often could not repeat his experiments, nor could other workers repeat them with any kind of regularity, so his findings were not accepted widely (Figure 13.3).

A Dutch physician, Jan Ingen-Housz, repeated the work of Priestley and enlarged upon it. He reported his findings in 1779 in a book titled *Experiments Upon Vegetables, Discovering Their Great Power of Purifying Common Air in Sunshine and Injuring it in the Shade and at Night.* He remarked on the rapidity with which plants "dephlogisticated" or restored air and the necessity of *light* for the process. He found that the operation varied directly with the amount of light available to the plant. He also found that only the *green* parts of plants could carry out this process, but *all* parts of the plant could "contaminate" the air, particularly at night. On the basis of his ex-

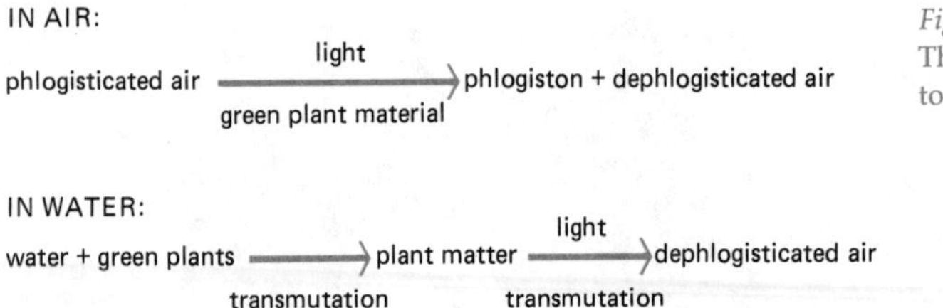


*Figure 13.4*
The two hypothesis proposed by Ingen-Housz to explain his experimental results.

periments, Ingen-Housz arrived at two conclusions, shown in Figure 13.4. Ingen-Housz really did not understand why light was important, nor did he recognize the importance of the green substance. As characteristic of his time, he believed that substances could be changed or *transmuted* into other substances. Although the observations and experimentation of Ingen-Housz were important to later investigators, his conclusions had little value.

The work of Lavoisier did much to cause the eventual overthrow of the phlogiston theory. He isolated and named *oxygen* and helped to determine some of its physical and chemical properties. Now plant scientists could talk about oxygen production by plants. The question was, where does this oxygen come from? Does it come from water, from "fixed air" ($CO_2$), or from both? The Frenchman Berthollet approached the problem by reasoning that if plants grew in a hydrogen-free atmosphere and were given only water, then any hydrogen found in their tissues had to come from the water. Berthollet conducted an experiment to test his hypothesis and, upon chemically analyzing the plants, did in fact find hydrogen in their tissues. He concluded that the plants used the hydrogen of the water, releasing the oxygen to the air.

However, another Frenchman, Jean Senebier (1782), took issue with Berthollet's conclusions. He noted that if Berthollet's conclusions were valid, then oxygen should be produced when either boiled water ($CO_2$ free) or carbonated water is used. He pointed out that his experiments did not support that assumption. Oxygen would be produced only when the plants were in water containing "fixed air." He therefore concluded that the oxygen did not come from the water, but rather from $CO_2$.

Nicolas Theodore de Saussue entered the argument in 1804 with his paper, *Recherches Chimiques Sur la Vegetation.* He made quantitative studies using varying concentrations of $CO_2$ and found that there was not a positive correlation between the amount of $CO_2$ supplied and the amount of $O_2$ produced. He found that above approximately 8½ percent, carbon dioxide actually caused a *decrease* in oxygen production by the plant and, at higher concentrations, caused the death of the plant. He made quantitative measurements concerning the *carbon* content of the plant and reasoned that the carbon in the plant must have come from the carbon dioxide. He concluded that the fixation of carbon in plant tissues resulted from the *breakdown of $CO_2$,* the *release of $O_2$,* and the *combining of carbon with $H_2O$* producing plant compounds.

Thus, by 1804, the major aspects of our basic formula were known, although the specifics were still unknown. The basic formula of that time may have looked like this:

$$CO_2 + H_2O \xrightarrow[\text{green plants}]{\text{sunlight}} \text{plant compounds} + O_2$$

In 1845, Mayer noted that this process entailed the conversion of radiant energy to the chemically bound energy of plant compounds. In addition, the carbohydrate nature of the organic matter was recognized.

With the understanding of the carbohydrate nature of the photosynthetic product, it is easy to understand why most plant physiologists of the late 1800s and early 1900s believed the oxygen to come from carbon dioxide. The equation above can now be rewritten as follows:

$$CO_2 + H_2O \xrightarrow[\text{green plants}]{\text{sunlight}} [CH_2O] + O_2$$

The formula $CH_2O$ is a general one for carbohydrates. The C obviously came from $CO_2$, the $O_2$ of $CO_2$ was set free as the liberated oxygen, and the $H_2O$ simply joined with the C forming $CH_2O$. This reasoning seems quite logical. But in 1941, Ruben and Kamen, using a heavy isotope of oxygen ($^{18}O$), were able to label water, and for the first time, the oxygen of the water could be distinguished from the oxygen of the carbon dioxide. Their results indicated that the general formula could now be written as follows, with the asterisk indicating the $^{18}O$ atom:

$$CO_2 + H_2O^* \xrightarrow[\text{green plants}]{\text{sunlight}} [CH_2O] + O_2^*$$

The oxygen from the *water* is that evolved during photosynthesis. If this is true, how can we explain the fact that carbon dioxide has *two* atoms of oxygen, yet only *one* of those end up in the carbohydrate. This dilemma can be solved easily:

$$CO_2 + 2H_2O^* \xrightarrow[\text{green plants}]{\text{sunlight}} [CH_2O] + O_2^* + H_2O$$

Thus, water is both used and produced, with the water produced being entirely new, its oxygen having come from the carbon dioxide. Our original general formula for photosynthesis can now be modified as follows:

$$6CO_2 + 12H_2O \xrightarrow[\text{chloroplasts}]{\text{light energy}} C_6H_{12}O_6 + 6O_2 + 6H_2O$$

Does our general formula describe adequately the process of photosynthesis? Some observations and experiments by the British plant physiologist F.F. Blackman offered evidence that the process was not really that simple. He noted that as he varied light intensity while measuring oxygen evolution, the oxygen evolution was proportional to light intensity only to a certain point. If he continued to increase the light intensity beyond that point, he found no further increase in oxygen evolution. He concluded that photosynthesis must consist of two phases, one which is directly dependent upon light, a second which is not. Later research would prove the correctness of his hypothesis.

Arnon and his co-workers were able in 1954 to isolate chloroplasts and show that these could perform the entire photosynthetic process. Although it had been known for some time that the chloroplast was the site of photosynthesis, as Engleman had demonstrated this in 1882 (Figure 13.5), it was now shown clearly that the process resided *entirely* within the chloroplast.

Work with isolated chloroplasts has furnished evidence that can allow for certain generalizations regarding the *light-dependent phase* of photosynthesis. When light strikes a chlorophyll molecule, the molecule becomes energized and an electron is raised to a higher energy level. Rather than this energy being dissipated, it is accepted by a compound termed an "electron acceptor." As a result, the acceptor is reduced. This compound passes the electron to an oxidized nucleotide called nicotene adenine dinucleotide phosphate, NADP for short, causing it to become reduced.

As the oxidized chlorophyll molecule is unable to lose a second electron, it must somehow become reduced if the electron transfer is to again occur.

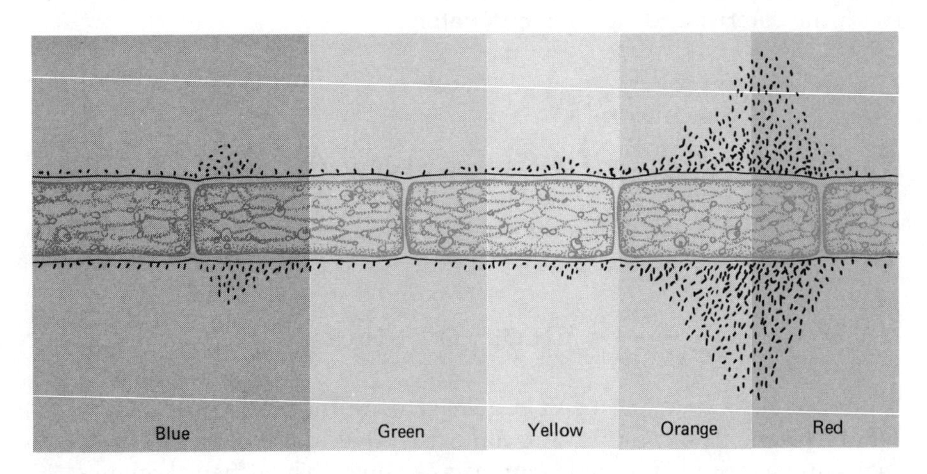


*Figure 13.5*
Engelmann's experiments illustrating the evolution of oxygen in light by algal cells. Aerobic bacteria concentrated around those areas in which oxygen was evolving.

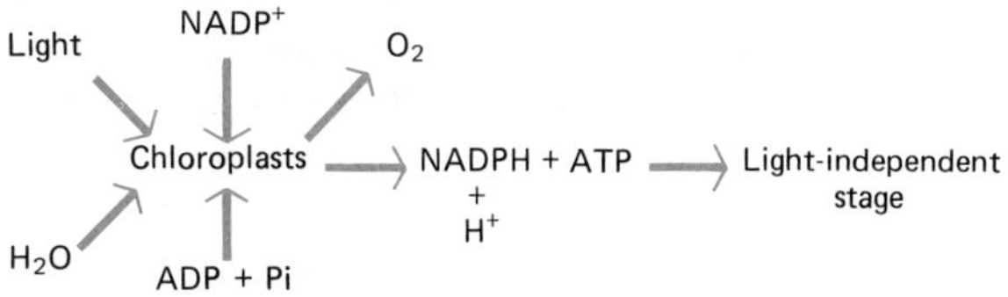

*Figure 13.6*
A simple scheme for the light-dependent stage of photosynthesis. Note that the products, NADPH + $H^+$ and ATP are reactants in the light-independent stage.

Light
$NADP^+$
$O_2$
Chloroplasts → NADPH + ATP + $H^+$ → Light-independent stage
$H_2O$
ADP + Pi

This results through the excitation of a second chlorophyll molecule by light. The released electron passes through a series of electron acceptors and ultimately to the first chlorophyll molecule, reducing it. As the electrons pass from the initial electron acceptor to more electron positive compounds, free energy is released and somehow is used, forming ATP from ADP and inorganic phosphate ($P_i$).

The second chlorophyll molecule, now oxidized, becomes reduced by somehow accepting an electron from water. The water molecule is split, utilizing light energy (photolysis), producing: an electron which goes to the chlorophyll; oxygen, which is liberated; and two hydrogen atoms which are transferred to the reduced NADP forming *NADPH* + $H^+$. Thus, three important products are formed as a result of the light-dependent phase of photosynthesis: oxygen, ATP, and NADPH + $H^+$. Thus, some of the light energy is stored as chemically bound energy in the NADPH and ATP. This reaction might be summarized as follows:

$$H_2O + ADP + P_i + NADP^+ + \text{chlorophyll (in chloroplasts)} \xrightarrow{\text{light}} \tfrac{1}{2}O_2 + ATP + NADPH + H^+$$

A simple diagram of the steps mentioned above for the light-dependent phase of photosynthesis is shown in Figure 13.6. Those who might wish to delve into this phase in more detail, can refer to the last section of this chapter.

As you note in Figure 13.6, the products of the light-dependent cycle, NADPH + $H^+$ and ATP, are utilized in a *light-independent* series of reactions. It is during this series of reactions that carbon dioxide becomes incorporated into organic compounds. The pathway by which carbon dioxide would be incorporated into organic compounds could only be guessed at until a radioactive isotope of carbon ($^{14}C$) became available following Ward War II. Calvin, Benson, Bassham, and other co-workers conducted many of the early experiments dealing with radioactive carbon tracing. They operated on the assumption that $^{14}C$ would behave little different from $^{12}C$, thus would enter all reaction normally participated in by the nonradioactive atom. They supplied $^{14}CO_2$ to algal cells in the form of $NaH^{14}CO_3$, sodium bicarbonate. The algal cells could be exposed to the radioactive carbon for varying periods of time and then killed instantly by dropping them into boiling 80 percent ethanol. The metabolites of the alga could be extracted and assayed for radioactivity. Although, the actual assay and identification procedures were quite complex and time-consuming, the basic philosophy of the procedures was quite simple. If the algae were subjected to radioactive $CO_2$ for seven seconds, only those compounds actually synthesized during that seven seconds would be radioactive. In addition, if a new carbon atom is added to an existing compound, and that atom is radioactive, then not only will the compound be radioactive, but only a single atom of that compound will be radioactive. By degrading the compound carbon by carbon, the position of the radioactive carbon can be determined.

By these procedures, $CO_2$ was found to be incorporated very rapidly. After only seven seconds, 12 radioactive compounds were formed. By reducing the time of exposure to less than five seconds, the investigators were able to identify the first product formed from $CO_2$: a three carbon acid called *3-phosphoglyceric* (PGA). Subsequently, the light-independent phase was found to be a cyclic one, fed by its own products, and apparently utilizing the products of the light-dependent stage. The reactions of the cycle can be summarized as follows:

$$6CO_2 + 18ATP + 12NADPH + 12H^+ \xrightarrow[\text{chloroplasts}]{} $$
$$\text{hexose phosphate} + 18ADP + 17P_i + 12NADP^+ + 6H_2O$$

It is now known that two other pathways of $CO_2$ fixation are found in plants. In one, the $CO_2$ combines with a three-carbon rather than a five-carbon compound; thus, a four-carbon compound is the first product formed. Because of this, the pathway is called the *$C_4$ cycle*. This pathway is found in both monocots and dicots but is particularly characteristic of certain grasses. Its major feature is that one of the enyzmes of the system has a high affinity for $CO_2$ at low concentrations; thus, the $C_4$ plant is a superior utilizer of available carbon dioxide.

In the second pathway, which occurs in many succulent plants such as cacti, four-carbon compounds accumulate in the leaves during darkness and are converted to $CO_2$ plus a three-carbon compound in the light. This cyclic series of reactions is called crassulacean acid metabolism (CAM). Plants exhibiting this cycle live typically in hot, dry climates and their stomata are closed frequently during the day, opening at night. It is thought CAM evolved because, given the pattern of stomatal opening and closing, much more $CO_2$ could be fixed during the day if $CO_2$ was available in the cell. As mentioned above, this is exactly what happens when the $C_4$ compound is converted to a $C_3$ compound plus $CO_2$ in the light. Such plants would undoubtedly have a selective advantage.

The photosynthetic rate is determined by the factors implicit in the general formula for photosynthesis, as well as by other factors not so readily apparent. Some of these factors, with a brief discussion of each, are as follows:

*Carbon Dioxide.* Only about 0.03 percent of the atmosphere is carbon dioxide, yet $CO_2$ generally diffuses rapidly into the leaf. The leaf has been estimated to be about 50 percent as efficient as inorganic $CO_2$ absorbers such as sodium or potassium hydroxide. As we normally associate the diffusion of $CO_2$ into the leaf with the daylight hours, and as the stomata of leaves are open normally during the day, it is reasonable to expect a correlation between $CO_2$ diffusion and the degree of stomatal opening. This correlation exists in many cases, but there are also some exceptions. In some plants, $CO_2$ diffuses into the leaf even when the stomata are closed. As mentioned earlier, the stoma cannot close making an airtight seal; thus, some $CO_2$ may still enter the stoma even though the guard cells are not turgid. But cuticular penetration of $CO_2$ must also be involved in this instance. In some plants, cuticular penetration may account for as much as 70 percent of the total $CO_2$ to enter a leaf even though the stomata are open.

Carbon dioxide diffuses quite rapidly through air; thus, it readily enters the leaf. As the $CO_2$ molecules strike the moist cell walls of the mesophyll or epidermal cells, the rate of diffusion is drastically decreased. Diffusion of $CO_2$ through water is some 10,000 times slower than through air. So even though the distance from the cell wall to the chloroplasts is very small, it is still sufficient enough to impose a limiting factor on $CO_2$ availability to the chloroplast.

*Water.* Water may be a limiting factor in several respects. As a raw material and supplier of electrons to the process of photosynthesis, its importance is obvious. It may also influence the rate by affecting the degree of stomatal opening. As water in the guard cells decreases, the stomata will close. (See Chapter 10 for a discussion of stomatal mechanisms.) In many plants this will reduce $CO_2$ penetration, thus limiting the rate of photosynthesis.

*Light.* Three aspects of light must be considered: 1) intensity, 2) quality, and 3) duration. Obviously, in the absence of light, there will be no diffusion of $CO_2$ into the leaf, although some $CO_2$ may actually be fixed in the dark. As light intensity increases, $CO_2$ fixation will eventually reach a level where it compensates exactly for that lost by respiration. That particular light intensity is referred to as the *compensation point* for that plant. Different plants have different compensation points; hence, this aspect would appear to be under genetic control. As the light intensity increases beyond the compensation point, there is a net increase of the ratio of $CO_2$ fixed to $CO_2$ lost by the leaves. At high light intensities, the actual degree of intensity depending upon the plant (*shade plants* being saturated at one-half or less the light required for saturation by *sun plants*), there is apparently destructive oxidation of cellular components, and then there is actually diffusion of $O_2$ into the leaf and $CO_2$ out of the leaf. The intensity of light may also affect the orientation of chloroplasts in the cells. In bright light, they are appressed generally against the radial and transverse walls, in dim light against the tangential walls (Figure 13.7).

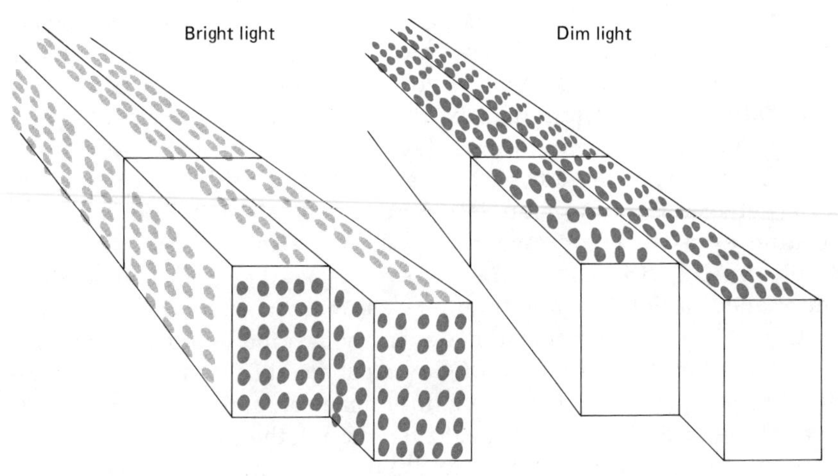


*Figure 13.7*
The influence of light intensity on chloroplast orientation.

Light quality may also affect the photosynthetic rate, and it may also affect the nature of the photosynthates. In general, blue light is absorbed most, with red light exhibiting a somewhat lower absorption peak. Greens and yellows are not generally absorbed to a very great degree. Obviously, only those wavelengths absorbed can be of any significance in photosynthesis. Although more of the blue light is absorbed, red light seems to be more effective in photosynthesis. Far red light, by itself, is quite inefficient, but if supplemented by blue light, photosynthetic rates will be increased substantially. An increase in blue light will cause an increase in protein production, while an increase in red light will cause increased production of carbohydrates.

*Chlorophyll Content.* We might suspect that the more chlorophyll a leaf has, the more rapid will be its rate of photosynthesis. But available information shows no such correlation. Partly chlorotic (nonchlorophyllous) leaves photosynthesize as rapidly as normal leaves, and sun leaves (which contain less chlorophyll than shade leaves) have been shown to synthesize more rapidly than shade leaves at light intensities above 1000 lumen per square foot.

*Oxygen.* The earth's atmosphere, having 21 percent oxygen, is highly oxidizing. When one considers that photosynthesis depends upon the production of a strong reductant, it would be quite surprising if oxygen did not affect this process. In plants that do not exhibit photorespiration, oxygen levels must reach over 60 percent before the photosynthetic rate is inhibited. In plants that do photorespire, oxygen does appear to inhibit photosynthesis, with the degree of inhibition being approximately 12 percent inhibition at 21 percent $O_2$, rising to 30 percent inhibition at 60 percent $O_2$. But a paradox seems to exist in that, even in plants that photorespire, a small amount of oxygen seems to be necessary before photosynthesis can occur.

*Photosynthetic Products.* As in any chemical reaction, the concentration of products influences the rate of the reaction. If the photosynthetic rate is not rapid, photosynthates are normally exported out of the chloroplasts, and thus, that rate will be maintained, assuming other conditions remain equal. But if photosynthesis is occurring at a rapid rate, the products can not be exported as rapidly as they are made. In such cases, they will accumulate normally in the chloroplast as either sucrose or starch. This accumulation of photosynthate in the chloroplasts will result in a decreased photosynthetic rate.

*Temperature.* The influence of temperature on the photosynthetic rate is somewhat difficult to ascertain. Its influence is quite dependent upon other factors such as light intensity and $CO_2$ availability. It is also a factor seemingly under genetic control. For most plants, under moderate light, temperature has very little influence between approximately 18 to 32°C. From O°C to 18°C an increase in temperature results in an increase in the rate of photosynthesis. Above 32°C, increased temperature results in decreased photosynthesis, with most plants completely unable to photosynthesize at temperatures above 45°C. But certain plants, termed *thermophiles,* are able to photosynthesize at temperatures of 70°C, while others, called *psychrophiles,* can photosynthesize at temperatures of 0°C or even somewhat lower. At higher light intensities or somewhat elevated $CO_2$ levels, as increase in temperature from 18°C *will* result in an increased photosynthetic rate.

Only certain of the reactions of photosynthesis are temperature sensitive, although all the reactions of the light-independent phase are temperature dependent. That temperatures above 20°C do not seem to effect total $CO_2$ assimilation is believed to be due to an increase in endogenous $CO_2$ through an increased rate of respiration.

*Minerals.* The shortage of certain minerals such as magnesium, manganese, iron, calcium, and nitrogen will cause a reduction in the rate of photosynthesis. This is due either to their incorporation into pigments or their function as cofactors in certain of the enzymatic reactions of photosynthesis.

*Leaf Age.* As the leaf enlarges, the rate of photosynthesis increases. Once the leaf has reached its maximum size, the photosynthetic rate will no longer increase, other factors being the same. In fact, the rate will begin to decrease somewhat from that point onward, and with increased senescence (old age) of the leaf, there is a more rapid decrease in the photosynthetic rate.

*Endogenous Rhythms.* Certain internal, or endogenous, factors seem to determine rhythmic increases or decreases of the photosynthetic rate in some plants. These peaks and valleys in the daily or seasonal rates of photosynthesis seem to occur in these plants irrespective of environmental influences. If the plant is brought into a greenhouse and maintained under constant environmental conditions, the rhythms will still continue as before. What controls these rhythms is not known, but they occur apparently by the synthesis or destruction of carboxylase.

*Photorespiration.* It has been known for some time that $C_3$ plants are not particularly efficient photosynthesizers. Part of this lies, as we mentioned earlier, in the fact that enzymes of the $C_3$ pathway do not have a high affinity for $CO_2$ at low concentrations. It was also found that during the light-independent phase of the $C_3$ photosynthetic pathway, a three-carbon sugar was converted to a two-carbon sugar plus $CO_2$. It has been estimated that as much as 50 percent of the fixed carbon may be reoxidized to $CO_2$ during this process called photorespiration. Thus, only one-half of the fixed carbon may be available for maintaining the pathway and forming the photosynthetic product.

The most important thing to remember about photosynthesis is what goes in and what comes out. It is perhaps not now fashionable in botany to consider things from the black box approach. However, if you remember that carbon dioxide, water, chlorophyll (i.e. green plants), and light, plus of course appropriate enzymes, are required, and sugar and oxygen are formed during the process we call photosynthesis, and if you remember some of the aspects mentioned above which influence this reaction, you do know a lot about photosynthesis. With this information, you can predict the direction of diffusion of $O_2$ and $CO_2$ and many aspects concerning the physiology of the organism.

## Nonphotosynthesis of Foods

One type of food synthesis is the light mediated process of photosynthesis discussed previously. But there are other types of syntheses apart from those of the various photosynthetic pathways. Ultimately, of course, all these synthetic pathways are dependent upon the process of photosynthesis for their raw materials, and this is true not only for plants but for all animals as well.

The carbohydrates represent an important group of compounds. They represent the most readily respirable material and thus serve as a major energy source. Certain carbohydrates represent important food reserves, while others are structural components of the cell.

One of the most basic reactions of sugars is its combination with phosphate, a step called *phosphorylation.* In almost all reactions that occur in the cell and involve sugar, the sugar is in a phosphorylated form.

You recall that one of the products of photosynthesis was a monosaccharide. Such monosaccharides may be converted to other monosaccharides; for example,

Fructose-6-phosphate⇌Glucose-6-phosphate⇌Glucose-1-phosphate

The glucose-1-phosphate could then combine with a compound similar to ATP called uridine triphosphate (UTP), forming uridine diphosphoglucose (UDPG). From this compound, the translocatable assimilate, sucrose, of which we spoke in Chapter 12, can be formed as follows:

UDPG + F-6-P→GF-6-P (sucrose phosphate) + UDP
GF-6-P + ADP→GF (sucrose) + ATP

or glucose-6-phosphate could combine directly with fructose forming sucrose and phosphate.

The important storage product, starch, is probably formed by a reaction sequence requiring a primer or starting molecule; that is,

-G-G-G-(primer) + UDPG→-G-G-G-G- + UDP

Thus, the conversion of glucose to starch is simply a chain-lengthening reaction, occurring molecule by molecule until the final starch macromolecule is formed. As starch is a branching molecule, one enzyme would be required for chain-lengthening, a different enzyme would be required for branch formation. Cellulose, the primary cell wall substance, is formed in essentially the same way as starch. The only difference is the type of linkage formed between the number one and four carbons of the glucose molecule.

Lipids are probably derived more or less directly from photosynthetic products. The usual storage lipid is one called a glyceride. Triglycerides are formed by three molecules of fatty acid combining with a single glycerol molecule; that is,

3 fatty acids + glycerol→triglyceride (lipid) + $3H_2O$

The glycerol is probably formed directly by the reduction of phosphoglyceric acid (PGA), the fatty acids by the acetyl-coenzyme A pathway. Important lipids include the waxes, which cover the outer surface of epidermal cells and are important substances of commerce; phospholipids, which are components of cell membranes; and steroids and sterols, which are the ingredients active in certain drugs and poisons, important in the synthesis of certain vitamins, and intermediate in the formation of terpenes, rubber, and carotenoid pigments.

Proteins generally do not serve as food storage products but rather are enzymatic or structural components of the cell. We discussed the mechanisms of protein synthesis in Chpater 6. The amino acids from which proteins are formed may be synthesized directly from products of photosynthesis, or they can be synthesized from a keto acid via a process called

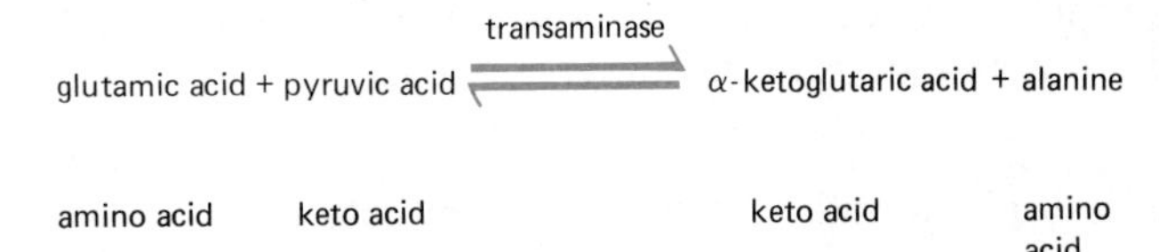


*Figure 13.8*
An over-all equation for transamination. Note that the process involves the transfer of an amino group to a keto acid forming a new amino acid.

*transamination* (the transfer of the -$NH_2$, or amino group, from one compound to a different compound) (Figure 13.8). Plants, as a group, have pathways for the synthesis of all the keto acids required to form the 20 or so common amino acids. But not every plant has each of the different keto acids; thus, some plants will be deficient in certain amino acids. Mammals, on the other hand, lack the enzymes to synthesize approximately one-half of the required keto acids. Mammals must obtain these lacking amino acids in their diet, and therefore, these amino acids are called the *essential amino acids.* Table 13.1 lists the essential and nonessential amino acids for humans. Given the amino acids, both plants and humans can synthesize these into various proteins.

## DIGESTION OF FOODS

Just as chemical sequences exist in cells for the synthesis of complex molecules from simpler ones, mechanisms also exist for breaking these complex molecules back down to simpler ones. This latter conversion process, called *digestion,* is of primary importance in that the complex molecules are *insoluble* normally and thus are incapable of being translocated out of the cell or even from one cellular organelle to another within the same cell. Digestion of such compounds results normally in the formation of *soluble* compounds.

Certain of these digestive processes use phosphorus in the process, others involve water. The former are termed *phosphorolysis,* the latter *hydrolysis.* Both types of digestive processes are reputed to occur in cells, al-

*Table 13.1*
LIST OF ESSENTIAL AND NONESSENTIAL AMINO ACIDS FOR MAN

| ESSENTIAL | | NONESSENTIAL | |
|---|---|---|---|
| Lysine | Tryptophan | Alanine | Proline |
| Isoleucine | Valine | Aspartic acid | Hydroxyproline |
| Leucine | Phenylalanine | Glutamic acid | Serine |
| Methionine | Histidine | Cysteine | Tyrosine |
| Threonine | Arginine* | Glycine | |

*Source:* After Irvin E. Liener, *Organic and Biological Chemistry*, New York: Ronald Press, 1966. Copyright © by The Ronald Press Company.

*Synthesized by man but at a rate insufficient for normal growth.

though there are some biochemists that insist that all intracellular digestion is by phosphorolysis and only extracellular digestion involves hydrolysis.

The digestion of starch demonstrates a process in which both types of digestion occur. Starch plus phosphate, in the presence of an enzyme called a phosphorylase, will result in breakage of the $\alpha$ 1, 4 linkages of the starch molecule, forming amylopectin and G-1-P. The enzyme phosphorylase is unable to attach to $\alpha$ 1, 6 linkages of amylopectin. The amylopectin portion of the starch molecule is broken down through the action of the enzyme amylo-1, 6-glucosidase yielding molecules of free glucose. An alternate system utilizes two enzymes also: $\alpha$-amylase and $\beta$-amylase. The internal $\alpha$ 1, 4 linkages are attacked by $\alpha$-amylase, while $\beta$-amylase works at the end of the chains. The activity of both enzymes results in the formation of the dissaccharide maltose plus $H_2O$. Maltose is then hydrolyzed by the enzyme maltase forming glucose plus water.

## SOME ADDITIONAL CONSIDERATIONS OF FOOD SYNTHESIS

Much conflicting data exist regarding the specific reactions of the light-dependent phase of photosynthesis, and many points of dispute have arisen. Current information would seem to suggest a model similar to that of Figure 13.9. The current view of photosynthesis regards the process as one of electron transport between two photosystems. Each photosystem appears to possess a *reaction center* surrounded by a large number of various chlorophyll molecules which constitute a so-called light trap. The light trap of photosystem I (PS I) would seem to consist primarily of chlorophyll *a* molecules which absorb light energy and pass the resultant excitation energy from molecule to molecule to the reaction center. The reaction center consists of a molecule of P 700 in association with "X" and plastocyanin. The light trap of photosystem II (PS II) is apparently somewhat similar except chlorophyll *b* molecules will be present in much higher numbers than in PS I (Table 13.2), as will certain accessory pigments. The reaction center consists of a molecule of P 682 in conjunction with "Q" and water.

*Table 13.2*
**SUSPECTED LIGHT-TRAPPING PIGMENTS OF PHOTOSYSTEMS I AND II**

| PS I CHLOROPHYLL | PS II CHLOROPHYLL |
|---|---|
| *b* 650—less than in PS II | *b* 650 |
| *a* 672—somewhat less than in PS II | *a* 672 |
| *a* 683—more than in PS II | *a* 683 |
| *a* 695—all | |
| *a* 703—all | |

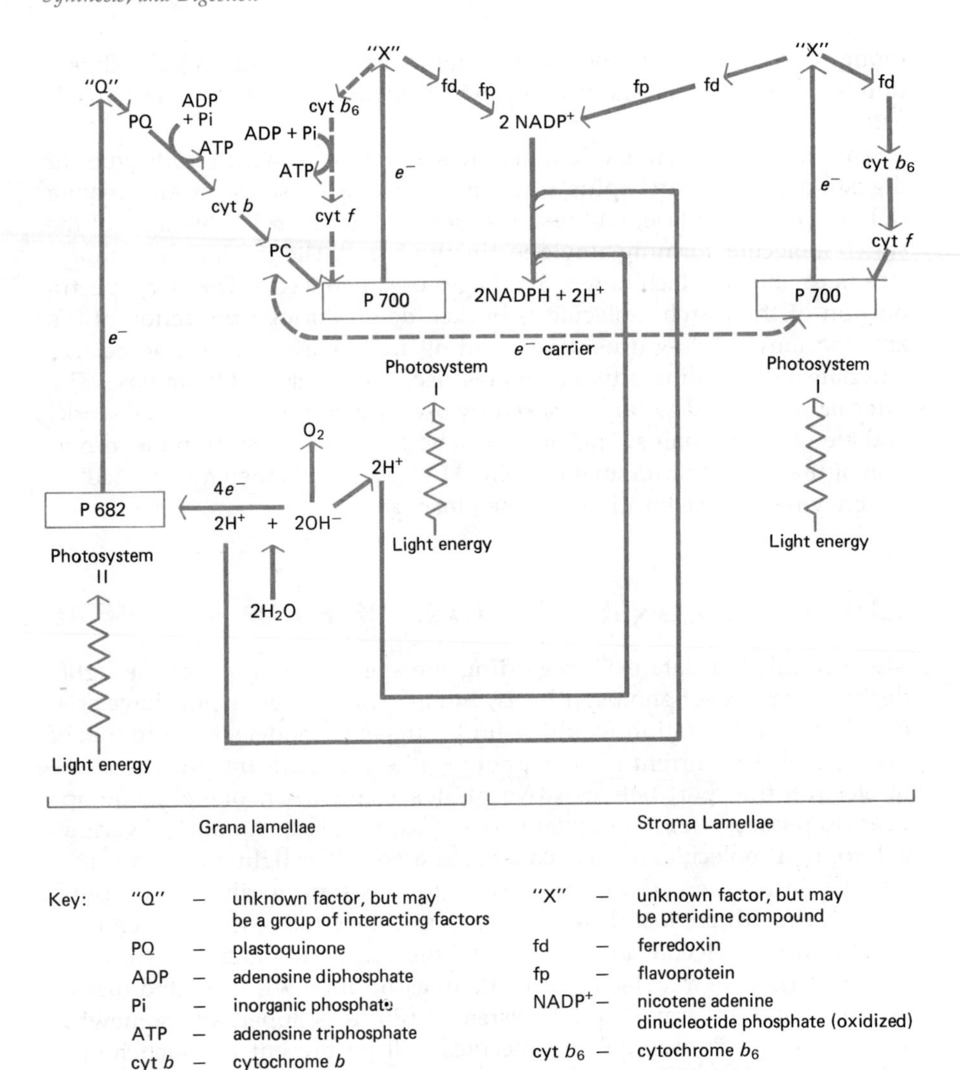


*Figure 13.9*
A model for the light-dependent phase of photosynthesis. The dotted lines indicate speculative pathways. [Modified from R. B. Park and P. V. Sane, "Distribution of Function and Structure in Chloroplast Lamellae," *Ann. Rev. Plant Physiol.* 22:395–430. © 1971 by Annual Reviews Incorporated. All rights reserved.]

The grana lamellae contain both PS I and PS II, while the stroma lamellae contain only PS I (Fig. 3.16). The two kinds of photosystems are designated PS $I_S$ and PS $I_G$. I have suggested previously that the two PS I systems may be coupled by an electron carrier, but this has not been demonstrated clearly. Electron micrographs have revealed spheres about 20 nm in diameter and 10 nm thick located only on the grana lamellae. Park (1963) has called these structures *quantasomes.* The quantasomes are an integral part of the membrane and were thought by Park to be the actual physical or morphological expression of the photosynthetic unit. But more recent studies indicate that the entire membrane must be present in order for the entire electron transport sequence to occur. Trebst (1974) further emphasized this view by presenting evidence for the sidedness of the chloroplast inner membrane. He noted that whereas most components of the electron transport system were located within the membrane, they were oriented generally either towards the inside or towards the outside (matrix side) of the membrane. Several were associated actually with the membrane surface. Thus, there is an electron flow from water to NADP across the membrane from inside to matrix side.

The pigment systems of PS I and PS II were at one time thought to consist of different pigments, and one can still find textbooks that refer to the pigment of PS I as chlorophyll *a*, that of PS II as chlorophyll *b*. More recent evidence indicates that the systems do not differ so much in kind of pigment as in the ratio of different pigments present (as mentioned earlier). Recent studies indicate that in higher plants the transformation of a protochlorophyllide fluorescing with a peak at 655 nm results in the formation of a chlorophyllide fluorescing at 675 nm, which in turn is converted to chlorophyll *a* 672. Phototransformations can occur, resulting in the formation of various chlorophyllides which in turn are converted to chlorophyll *a* molecules having different absorption maximums. Thus, the various chlorophyll *a*'s isolated from chloroplasts may be due simply to photoconversions. In addition, in some plants, the place of chlorophyll *b* is taken by chlorophyll *c*, chlorophyll *d*, fucoxanthin, or the phycobilins, phycocyanin or phycoerythrin. Although fucoxanthin is a carotenoid pigment and is about as efficient a photosynthesizer as chlorophyll *a*, the carotenoids in general are relatively poor sensitizers in photosynthesis. Many of the carotenoids, termed as a group *accessory pigements* apparently do not participate at all in the process of photosynthesis. The primary function of the carotenoids may be simply as a light screen to prevent the photooxidation of chlorophyll. The above data suggest that the primary sensitizers of each photosystem cannot be as simple as stated in earlier works, that is, PS I—chlorophyll *a*, PS II—chlorophyll *b*, but is rather a specific complex of molecules. Some investigators have hypothesized that PS II is sensitized by chlorophyll *a* molecules having a chlorophyll *b* tail.

The specific reactions of the light-independent cycle are complex, and for some time investigators searched for the two-carbon compound that could combine with $CO_2$ to form PGA. They found however, that a five-carbon compound seemed to accumulate when $CO_2$ was removed, thus suggesting that it was the precursor. Evidence indicates that $CO_2$ combines with the five-carbon sugar, ribulose-1,5-diphosphate (RuDP), forming an unstable six-carbon intermediate which is immediately broken down to two molecules of PGA. Phosphoglyceric acid is reduced to glyceraldehyde phosphate (GAP) by NADPH + $H^+$, with the phosphate coming from ATP. After isomerization of some of the GAP to dihydroxyacetone phosphate (DHAP), GAP and DHAP combine, forming a six-carbon sugar, fructose-diphosphate (FDP), which in turn forms fructose-6-phosphate (F-6-P) and inorganic phosphate ($P_i$). *One molecule of F-6-P becomes an end product of the reaction.* Other molecules of F-6-P form a two-carbon intermediate plus a four-carbon sugar, erythrose-4-phosphate (E-4-P). Erythrose-4-phosphate combines with DHAP forming a seven-carbon sugar, sedoheptulose diphosphate (SDP), which in turn forms sedoheptulose-7-phosphate (S-7-P), plus $P_i$. The S-7-P forms a five-carbon sugar ribose-5-phosphate (R-5-P) and a two-carbon intermediate. The R-5-P is then isomerized to ribulose-5-phosphate (Ru-5-P). The remaining GAP combine with the two-carbon intermediates forming xylulose-5-phosphate (Xu-5-P), which is isomerized to Ru-5-P. The Ru-5-P is then phosphorylated forming RuDP, with ATP being the donor, and the cycle is primed to continue (Figure 13.10). These light independent reactions are associated with the stroma rather than with the membranes of the chloroplasts.

The light-dependent and light-independent phases of photosynthesis are linked in that the source of energy and reducing power for the light-independent reactions are supplied via the production of ATP and NADPH during the light-dependent reactions. As much more ATP and

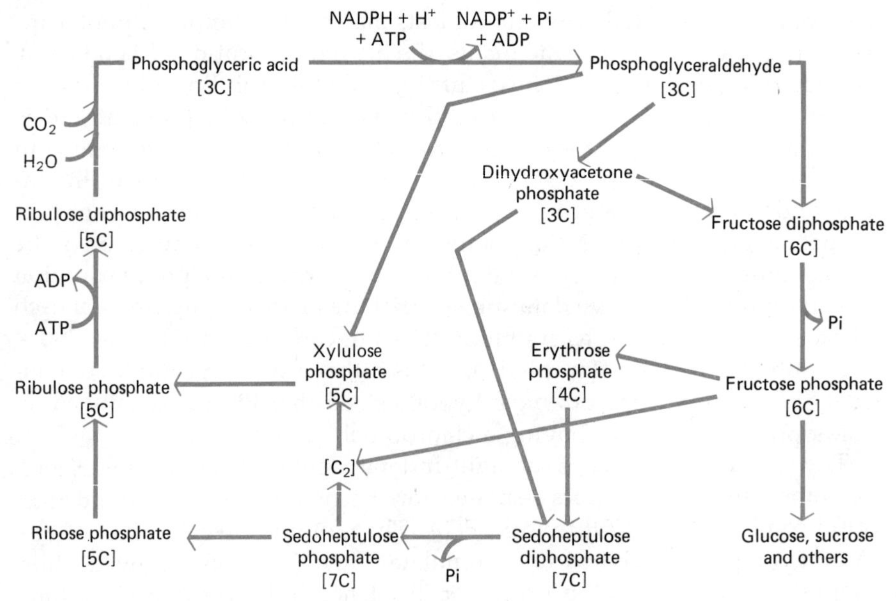


*Figure 13.10*
A model for the light-independent phase of photosynthesis.

NADPH is required for one complete cycle of light-independent reactions than is produced during a single series of electron transfers in the light-dependent phase, the two phases are obviously not synchronized. In fact, the light-dependent series of reactions occurs much more rapidly than does the light-independent series. The process of photosynthesis is somewhat analogous to a little wheel driving a big wheel; the little wheel in this instance being the light-dependent phase, the big wheel, the light-independent phase. In order for the big wheel to complete a single revolution, the little wheel must revolve many times. But even considering this difference, present information would seem to indicate that more ATPs are required than are formed in the photosynthetic process.

Although the six-carbon sugar phosphates are the principal end products of photosynthesis, there is evidence that neither G-6-P or F-6-P leaves the chloroplast. The principal compounds exported from the chloroplasts are PGA, DHAP, and glycolic acid (or glycolate). Oxaloacetate (OAA) may also be exported (Figure 13.11). There is some evidence that phosphorylated compounds do not pass readily through the chloroplast envelope; thus,

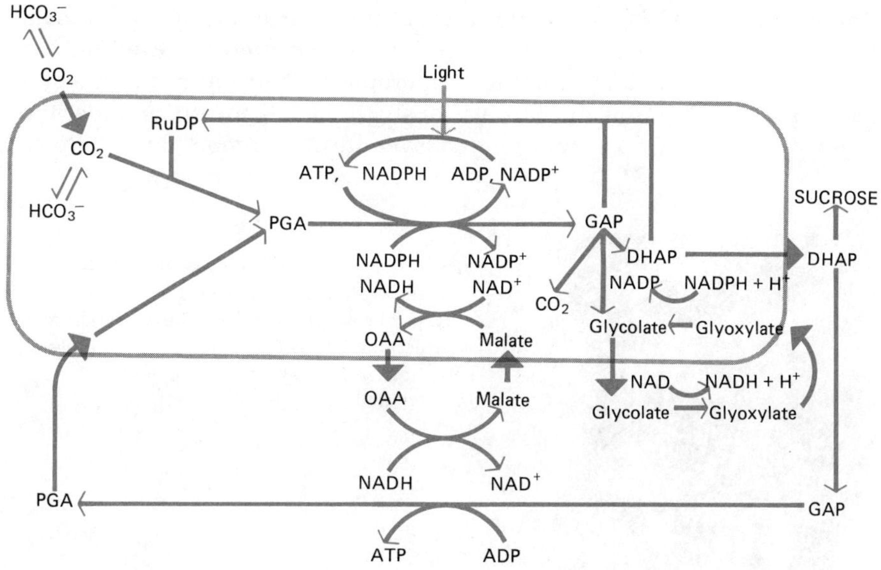


*Figure 13.11*
Export of carbon and of phosphate energy from the chloroplasts of photosynthesizing leaf cells. Sucrose is formed outside the chloroplasts from DHAP. Formation of ATP outside the chloroplasts is brought about by shuttle transfer of DHAP and PGA and of OAA and MALATE. The chloroplast envelop has a low permeability to sucrose and ATP. [After Ulrich Heber, "Metabolite Exchange Between Chloroplasts and Cytoplasm," *Ann. Rev. Plant Physiol.* 25:393–421. © 1974 by Annual Review Incorporated. All rights reserved].

PGA and DHAP may be dephosphorylated in the chloroplast and enter the cytoplasm as GA and DHA. Both sucrose and starch may be synthesized within the chloroplast. The growth of starch grains may disrupt eventually the organization of the grana and stroma lamellae, resulting in the conversion of the chloroplast to an amyloplast. Although sucrose is a major product of photosynthesis, the chloroplast envelope is essentially impermeable to it; thus, it is not exported from the chloroplast. It is possible that amino acids are also manufactured from photosynthates directly in the chloroplast and form the protein of the structure.

As mentioned earlier, the $C_4$ pathway has been found to exist in a number of plants. These plants were found to have certain aspects in common: they had distinct bundle sheaths, a low $CO_2$ compensation point (the $CO_2$ level at which photosynthesis is equal to respiration and no *net* gas exchange occurs), and high temperature optima for photosynthesis.

The characteristics of $C_4$ plants, although possessing common distinguishing features, also show differences. These differences concern the degree of grana development in the chloroplasts of bundle sheath cells, the position of the chloroplasts in these cells, and the activity of enzymes, particularly those thought to function in decarboxylation.

The leaf anatomy of $C_4$ plants is similar to that shown in Figure 13.12. The chlorenchyma is in a radial arrangement surrounding the vascular bundles. The innermost layer consists of large, relatively thick-walled, cylindrical cells containing prominent chloroplasts. This layer is termed the *bundle sheath* and has been called variously the starch sheath, endodermis, or mestome sheath. Surrounding this layer are one or more layers of pali-

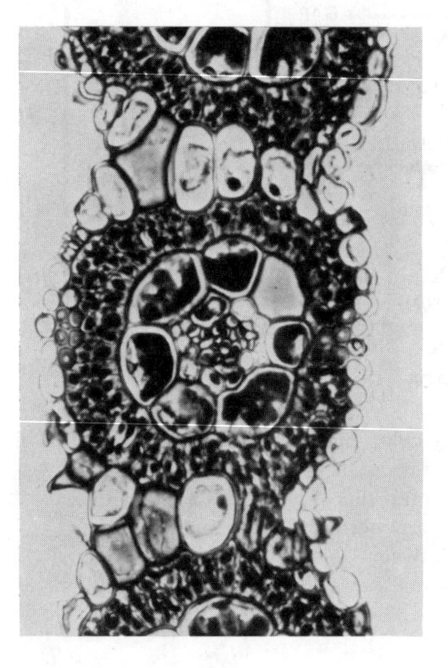

*Figure 13.12*
*Buchlor dactyloides* leaf cross-section. This is a $C_4$ plant. Note the specialized bundle sheath cells. [From W. M. Laetsch, "The $C_4$ Syndrome: A Structural Analysis," *Ann. Rev. Plant Physiol.* 25:27–52. © 1974 by Annual Reviews Incorporated. All rights reserved].

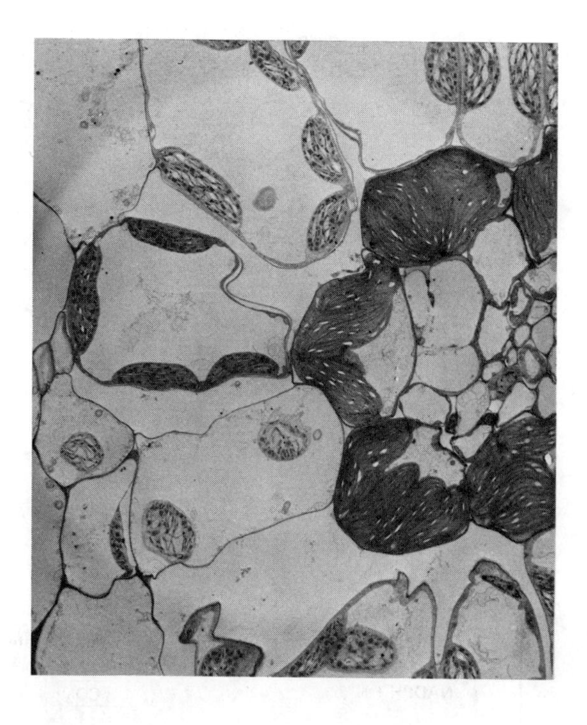

*Figure 13.13*
Transmission electron micrograph of crab grass leaf transection. Note lack of grana in chloroplast of bundle sheath cell as compared to those of mesophyll cells. [Courtesy C. C. Black, Jr.].

sadelike mesophyll cells. The German botanist Haberlandt called this anatomical pattern Kranz type. Kranz anatomy has been given as a criterion for the presence of $C_4$ photosynthesis.

Ultrastructural observations of the chloroplasts of bundle sheath cells and mesophyll cells indicate obvious differences. Early work on $C_4$ plants, particularly sugar cane, gave evidence of extreme structural dimorphism in chloroplasts from the two cell layers (Figure 13.13). But other work has indicated absolutely no dimorphism, but a gradation in between. Thus, the absence of grana in the chloroplast of bundle sheath cells is not a generalized characteristic of $C_4$ plants. In fact, the term "agranal" is used to describe the bundle sheath chloroplasts of maize and sugar cane is probably a misnomer. Thylakoid appression does occur apparently in these species, particularly in the peripheral regions of the chloroplast. Grana in these species could perhaps be termed *rudimentary*. Although structural dimorphism may not exist in $C_4$ plants, a size dimorphism does generally exist.

Current work on the $C_4$ cycle is quite controversial in its interpretation. Some investigators believe that at least three different groups of $C_4$ plants exist. Black (1973) has presented a generalized scheme, as shown in Figure 13.14, which summarizes the ideas of Andrews and others (1971) and Hatch and others (1971). It is hypothesized that compounds move between the mesophyll cells and the bundle sheath cells. Evidence would seem to indi-

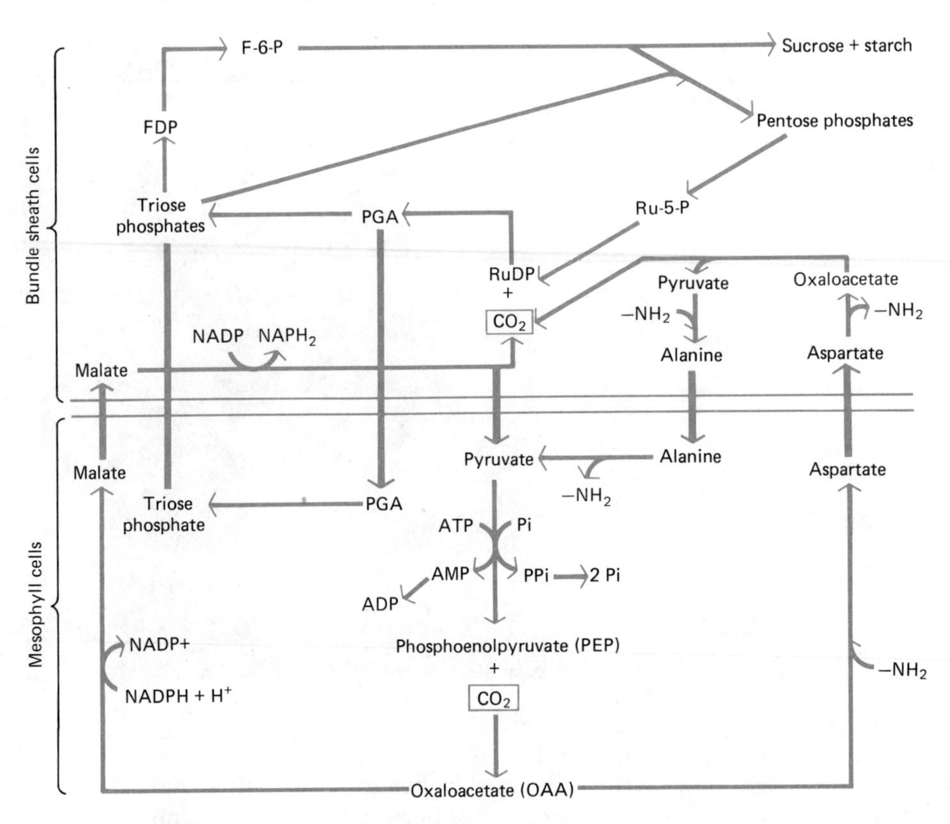


*Figure 13.14*
Probable reactions and intercellular movements of metabolites during the operation of the $C_4$ dicarboxylic acid pathway. [C. C. Black, Jr., "Photosynthetic Carbon Fixation in Relation to Net $CO_2$ Uptake." *Ann. Rev. Plant Physiol.* 24:253–286. © 1973. by *Annual Reviews* Incorporated. All rights reserved.]

cate that diffusive processes could account for the rates necessary for $C_4$ photosynthesis. In addition, the compounds postulated as moving are those which are capable of moving through the chloroplast envelope. The decarboxylation of malate in the bundle sheath cells furnishes both $CO_2$ and reduced dinucleotide. Thus, $CO_2$ is concentrated at the site of RuDP carboxylation, and the reaction proceeds quite rapidly. In the mesophyll cells, the carboxylation of PEP should be coupled to the reduction of OAA. The required dinucleotide could be produced photosynthetically as follows:

$$HCO_3^- + PEP \longrightarrow OAA + P_i$$

$$NADP^+ + H_2O \xrightarrow[\text{chloroplasts}]{\text{light}} NADPH + H^+ + \tfrac{1}{2}O_2$$

$$OAA + NADPH + H^+ \longrightarrow Malate + NADP^+$$

The CAM plants, mentioned earlier, demonstrate a night fixation involving the production of the organic acids PEP, OAA, and malate, thus indicating a correlation between this pathway and that of $C_4$ plants. The CAM plants however, do not exhibit the Kranz anatomy of $C_4$ plants, thus supporting the suggestion that $C_4$ plants are simply CAM *mit* Kranz ("CAM with Kranz"). A scheme for carbon dioxide fixation in CAM plants is given in Figure 13.15. The characteristics distinguishing the three types of $CO_2$ fixation are shown in Table 13.3.

I mentioned earlier that some plants exhibit a light-stimulated $O_2$ uptake and $CO_2$ release. This process, called photorespiration, was said to be absent in $C_4$ plants. This is probably not so. Measured in terms of $O_2$ uptake, they have considerable photorespiration. Apparently they simply do not lose $CO_2$ during the process, either because it does not escape from the bundle sheath cells, or because it is "refixed" in the $C_4$-dicarboxylic acid cycle. Some investigators believe the relative efficiency of $C_4$ plants in $CO_2$ fixation is due to this recycling of photorespiratory $CO_2$.

Most evidence seems to indicate that glycolate, mentioned earlier as one of the main products of photosynthesis to be exported from the chloroplast, is the major compound associated with photorespiration. During the oxidation of two molecules of glycolate to one molecule of serine (an amino acid), there is an uptake of one molecule of $O_2$ and the release of one

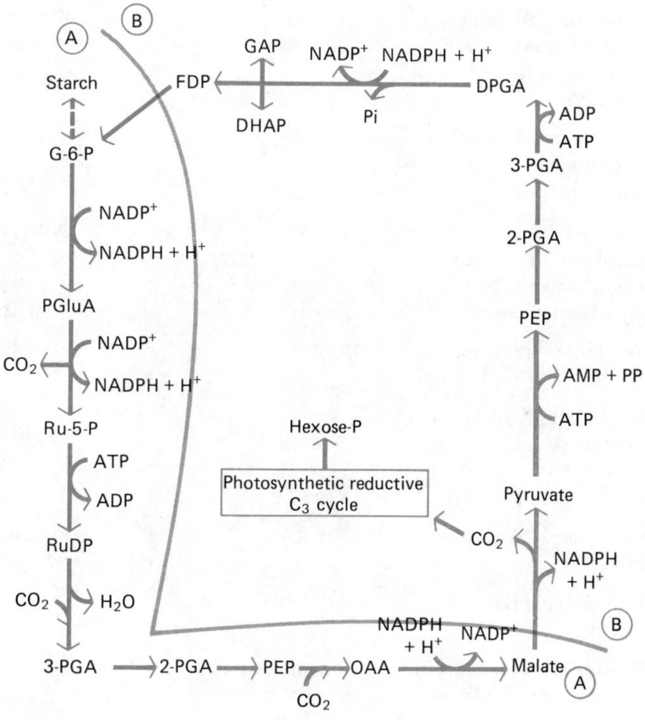

*Figure 13.15*
Scheme for carbon dioxide fixation in Crasulacean acid metabolism plants in the dark (A) and light (A plus B). [C. C. Black, Jr., "Photosynthetic Carbon Fixation in Relation to Net $CO_2$ Uptake," *Ann. Rev. Plant Physiol.* 24:253–266. © 1973 by Annual Reviews Incorporated. All rights reserved.]

*Table 13.3*
**CHARACTERISTICS DISTINGUISHING THREE GROUPS OF HIGHER PLANTS**

| CHARACTERISTIC | PRIMARY TYPE OF CARBON DIOXIDE FIXATION | | |
|---|---|---|---|
| (at 21% $O_2$, 0.03% $CO_2$) | $C_3$ | $C_4$ | CAM |
| Leaf anatomy in cross section | Diffuse distribution of organelles in mesophyll or palisade cells with similar or lower organelle concentrations in bundle sheath cells if present | A definite layer of bundle sheath cells surrounding the vascular tissue which contains a high concentration of organelles; layer(s) of mesophyll cells surrounding the bundle sheath cells | Spongy appearance, mesophyll cells have large vacuoles with the organelles evenly distributed in the thin cytoplasm. Generally lack a definite layer of palisade cells |
| Theoretical energy requirement for net $CO_2$ fixation ($CO_2$ : ATP : NADPH) | 1 : 3 : 2 | 1 : 5 : 2 | 1 : 6½ : 2 |
| Major leaf carboxylation sequence in light | RuDP carboxylase | PEP carboxylase, then RuDP carboxylase | Both PEP and RuDP carboxylase |
| $CO_2$ compensation concentration (ppm $CO_2$) | 30 to 70 | 0 to 10 | 0 to 5 in dark: 0 to 200 200 with daily rhythm |
| Transpiration ratio (g $H_2O$/g dry wt) | 450 to 950 | 250 to 350 | 50 to 55 |
| Maximum rate of net photosynthesis (mg $CO_2$/dm² of leaf surface/hr) | 15 to 40 | 40 to 80 | Usually about 1 to 4; highest reported, 11 to 13 |
| Photosynthesis sensitive to changing $O_2$ concentration from 1% to 21% | Yes | No | Yes |
| Leaf photorespiration detection: a) exchange measurements; | Present | Difficult to detect | Difficult to detect |
| b) glycolate oxidation | Present | Present | Present |
| Optimum day temperature for net $CO_2$ fixation | 15 to 25°C | 30 to 47°C | ~ 35°C |
| Maximum growth rate: (g of dry wt/dm of leaf area/day: | 0.5 to 2 | 4 to 5 | 0.015 to 0.018 |
| g/m² of land area/day | 19.5 ± 3.9 | 30.3 ± 13.8 | – |
| Leaf chlorophyll *a* to *b* ratio | 2.8 ± 0.4 | 3.9 ± 0.6 | 2.5 to 3.0 |
| Leaf isotopic ratio ${}^{13}C/{}^{12}C$ | –22 to –34 | –11 to –19 | –13 to –34 |
| Requirement for sodium as a micronutrient | None detected | Yes | – |

| CHARACTERISTIC | PRIMARY TYPE OF CARBON DIOXIDE FIXATION | | |
|---|---|---|---|
| (at 21% $O_2$, 0.03% $CO_2$) | $C_3$ | $C_4$ | CAM |
| Leaf postillumination $CO_2$ burst sensitive to oxygen levels above about 1% | Yes | Not completely inhibited | – |
| Rapid leaf postillumination release of newly fixed $^{14}CO_2$ > 5% | No | Yes | Yes |
| Translocation of freshly fixed $^{14}C$ from illuminated leaves (%/6 hr) | <50% | >50% | – |
| Optimum day temperature for growth (dry matter production) | 20 to 25°C | 30 to 35°C | ~ 35°C |
| Response of net photosynthesis to increasing light intensity at temperature optimum | Saturation reached about one-fourth to one-third full sunlight | Either proportional to or only tending to saturate at full sunlight | Uncertain but apparently saturation is well below full sunlight |
| Dry matter production | 22.0 ± 3.3 | 38.6 ± 16.9 | Extreme variability in available data |

molecule of $CO_2$. However, the term "photorespiration" associated with this particular reaction is a misnomer. Light is involved only indirectly in the formation of the photosynthate, glycolate.

There has been speculation that dark respiration (which we will be discussing in the next chapter) is reduced in the light. This has not been substantiated completely. However, glyoxylate or its condensation products have been shown to inhibit three enzymes of the aerobic pathway of dark respiration, and reducing equivalents may be drained off by pathways connected with photorespiration. Thus, it is not unreasonable to assume that dark respiration might, in fact, be inhibited somewhat in the light.

I do want to at least mention the role of the microbodies, peroxisomes and glyoxysomes, not only in the reactions we have just been discussing, but also in food synthesis, food breakdown, and respiration, which we shall be discussing next. These microbodies have some metabolic steps in common. All contain peroxide-forming enzymes and the enzyme catalase which immediately destroys the newly formed $H_2O_2$. In fact, some investigators believe there is no distinction between peroxisomes and glyoxysomes, and thus refer simply to these structures as microbodies. A whole metabolic pathway is not contained within a single microbody. Photores-

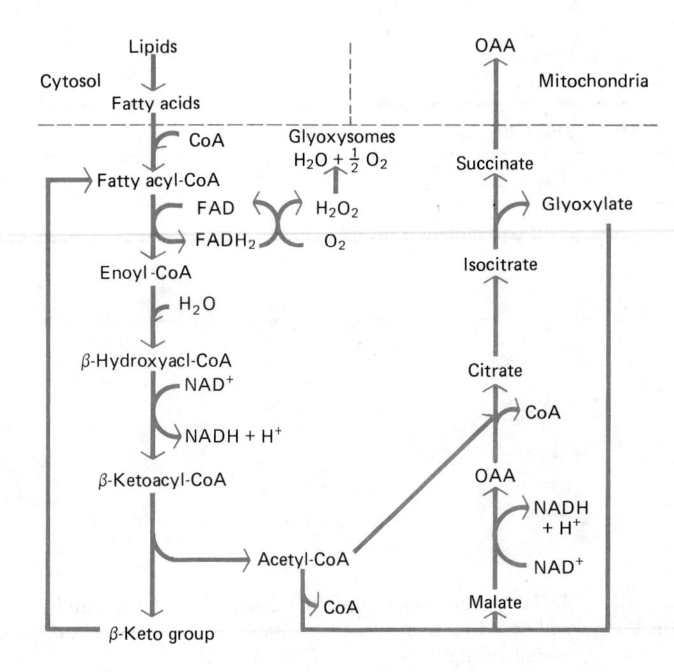


*Figure 13.16*
Postulated reactions in glyoxysomes. [N. E. Tolbert, "Microbodies—Peroxisomes and Glyoxysomes," *Ann. Rev. Plant Physiol.* 22:45–74. © 1971 by Annual Reviews Incorporated. All rights reserved.]

piration, for example, represents a metabolic series of reactions, parts of which are in the chloroplasts, peroxisomes, mitochondria, and cytosol (the cytoplasm as a membrane-bounded organelle).

Figures 13.16 and 13.17 indicate some postulated reactions in the microbodies. As you can see, the breakdown of fats to sugars, which we were discussing earlier, occurs partly in the glyoxysomes. The acetyl-CoA formed can become involved directly in the process of aerobic respiration occurring in the mitochondria.

The major aspect of the peroxisomes is their association with the process of photorespiration. So long as glycolate is supplied by the photosynthesizing chloroplasts, oxygen will be utilized in the peroxisome and carbon dioxide will be evolved in the conversion of glycine to serine in the mitochondria.

## SUMMARY

1. At one time or another, soil, minerals, water, air, carbon dioxide, oxygen, carbohydrates, fats, and proteins were all considered to be plant foods.
2. Aristotle believed plants obtained their food from the soil.
3. Van Helmont concluded that plants obtained their food from rain water.

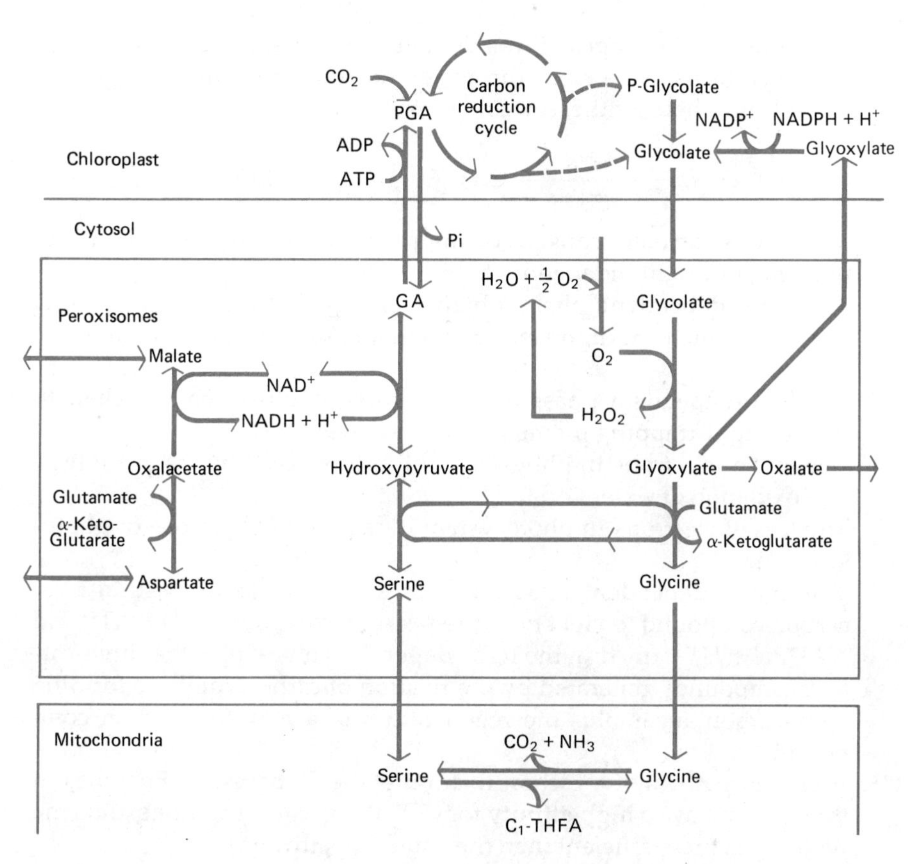


*Figure 13.17*
Postulated reactions in leaf peroxisomes, the glycolate pathway, and the glycolate-glyoxylate shuttle. [N. E. Tolbert, "Microbodies—Peroxisomes and Glyoxysomes," *Ann. Rev. Plant Physiol.* 22:45–74. © 1971 by Annual Reviews Incorporated. All rights reserved.]

4. Hales believed that plants obtained their food from the air.
5. From investigations on how plants obtained their food, an understanding of the process of photosynthesis arose.
6. Priestley discovered that plants could produce a substance that "restored injured air." He did not realize that substance was oxygen.
7. Ingen-Housz demonstrated that green plant material and light were necessary in the process observed by Priestley.
8. It later became known that the substance produced by green plants in light was oxygen and that carbon dioxide and water were utilized in the process. It was thought that the carbon of $CO_2$ combined with the water forming an organic compound, $CH_2O$, with the $O_2$ of the $CO_2$ being released.

9. With the use of isotopes, it was demonstrated that the $O_2$ produced in photosynthesis came from the water not the $CO_2$, thus the general formula for photosynthesis was

$$6CO_2 + 12H_2O \xrightarrow[\text{chloroplasts}]{\text{light}} C_6H_{12}O_6 + 6O_2 + 6H_2O$$

10. The process actually consists of the series of reactions, one light dependent, one light independent.
11. The light-dependent phase obtains energy from sunlight forming products which the light-independent phase utilizes as a source of energy.
12. The light-dependent phase consists of two photosystems, each with its own light-trapping pigments and active center.
13. The loss of electrons in photosystem II are replaced by the photolysis and oxidation of water.
14. The loss of electrons in photosystem I are replaced by those of photosystem II.
15. The light-independent phase combines carbon dioxide with a five-carbon compound to yield two three-carbon compounds. The ATP and $NADPH + H^+$ formed in the light-dependent phase plus the three-carbon compounds generated by the fixation of carbon combine forming a six-carbon sugar plus the regeneration of a new five-carbon compound.
16. A second pathway for carbon fixation is the $C_4$ pathway. Enzymes in this system have a high affinity for $CO_2$ at low concentrations, thus the pathway is more efficient than the usual $C_3$ pathway.
17. A third pathway for carbon fixation is crassulacean acid metabolism. In this pathway, four-carbon compounds accumulate during the night and furnish a source of $CO_2$ and three-carbon compounds during the day.
18. Factors which affect the rate of photosynthesis include $CO_2$ concentration, water, light (intensity, quality and duration), chlorophyll (concentration and quality), oxygen concentration, concentration of product, temperature, minerals, leaf age, endogenous rhythms, and photorespiration.
19. Photorespiration occurs in the light and results in the utilization of $O_2$ and the production of $CO_2$ through a photosynthetic intermediate compound, thus decreasing the rate of photosynthesis.
20. Nonphotosynthesis of food utilizes the products of photosynthesis and combines them forming dissaccharides, glycerol, fatty acids, amino acids, and other compounds which represent the carbohydrates, lipids,and proteins of the cell.
21. Thus, photosynthates are directly or indirectly responsible for all storage, structural, or enzymatic compounds of plants, and ultimately of animals.

22. Mammals cannot synthesize essential amino acids, thus must rely on plants as the source of these compounds.
23. The breakdown of complex molecules into simpler ones can occur by the processes of digestion, either phosphorolysis or hydrolysis.
24. Digestion is extremely important in that the end products are usually soluble.
25. The most important thing to remember about photosynthesis is the general equation.

# Chapter 14

## ROLE OF FOODS IN PLANTS— I. RESPIRATION

If we place some chips from common rocks such as limestone or marble into a flask and place a few drops of hydrochloric acid onto them they will effervese, in other words, bubble. This bubbling is due to the production of a gas. If we allow the gas to diffuse into a flask containing lime water, a white precipitate (tiny solid particles which eventually settle to the bottom of the flask) will form. We can consider what is happening from a chemical standpoint if we have several additional pieces of information: limestone and marble are substances which consist of the compound calcium carbonate, $CaCO_3$; lime water is calcium hydroxide, $Ca(OH)_2$. The reactions occurring can be shown as follows:

$$\underset{\text{calcium carbonate}}{CaCO_3} + \underset{\text{hydrochloric acid}}{2HCl} \longrightarrow \underset{\text{calcium chloride}}{CaCl_2} + \underset{\text{carbon dioxide}}{CO_2} + \underset{\text{water}}{H_2O}$$

$$\underset{\text{calcium hydroxide}}{Ca(OH)_2} + \underset{\text{carbon dioxide}}{CO_2} \longrightarrow \underset{\text{calcium carbonate}}{CaCO_3} + \underset{\text{water}}{H_2O}$$

The gas produced is carbon dioxide, which combined with lime water forms an insoluble precipitate. The whole point of this demonstration is simply to show that lime water can be used to indicate the presence of carbon dioxide.

If we exhale through a soda straw into a flask of lime water, we will note the formation of a white precipitate. Your breath, as you are well aware, contains carbon dioxide. If we flush carbon-dioxide–free air over germinating seeds and into a flask of lime water, again a white precipitate forms. The same thing will occur if we use colored flower petals or green leaves. Carbon dioxide must therefore be diffusing out of these plant parts.

Some people confuse the diffusion of $CO_2$ out of plants and the breathing of animals with respiration. This is like confusing getting fat with overeating. The fact that one is getting fat may simply be an outward manifestation of overeating. By the same token, the diffusion of $CO_2$ out of plant tissue and breathing in animals are simply outward manifestations of respiration. Respiration is an *intracellular* process.

As we have intimated above and in the preceeding chapter, respiration occurs in *all living cells*, plant and animal, pigmented and unpigmented, lighted and unlighted. If respiration ceases, the cell will die. An exception to this last statement may be found in dry seeds or spores. They may remain viable for some time without *detectable* respiration. There are some biologists which insist that even here, respiration must be occurring, even if at an infinitesimally low level.

Evidence of the occurrence of respiration can be obtained by placing a living organism or living tissue into a calorimeter. This device measures changes in temperature and gas concentrations. With such a device it will be noted that temperature and $CO_2$ concentration will increase, $O_2$ concentration will decrease. With this device, it will also be noted that the reaction will not occur unless an organic substance, *a food,* is provided. Although the calorimeter indicates two external manifestations of respiration, $CO_2$ and heat, it does not indicate the most important and most fundamental feature of respiration. That aspect is the transformation of the chemically bound energy of foods into a form of chemically bound energy readily usable in energy-requiring reactions of the cell. Thus, the major significance of respiration is that it provides the cell with energy.

Basically, the process utilizes $O_2$, water, and food and produces $CO_2$ and energy (in the form of heat, which is lost, and ATP), in addition to $H_2O$. We can summarize these reactions as follows:

$$6O_2 + 6H_2O + C_6H_{12}O_6 \rightarrow 6CO_2 + 12H_2O + \text{energy}$$

Overall, respiration is the oxidation of foods with the subsequent production of energy. Not all oxidation reactions utilize gaseous oxygen, however, nor do they result in the complete oxidation of glucose to $CO_2$ and water. Some cells, such as yeast, muscle, certain bacteria, and others oxidize glucose to some other organic compound, with or without the formation of $CO_2$. This type of oxidative process is called *fermentation.* Perhaps the best known fermentation is that of yeast. One molecule of glucose is oxidized to two molecules of ethyl alcohol (EtOH), $2CO_2$, and 2ATP. Of the 686 kcal (kilocalories) of energy bound in the glucose molecule, 630 kcal remain bound in the two molecules of alcohol, about 42 kcal are lost as heat, leaving only 14 kcal in the ATP as the energy available to the cell. This represents an energy yield of only slightly over 2 percent, a very inefficient process. In addition, the product, ethyl alcohol, is toxic to the cell. When the concentration of alcohol reaches approximately 12 percent, the yeast cells are killed. This is why naturally fermented alcoholic beverages will not have an alcoholic content above 12 percent.

When free gaseous oxygen is available, and when the cells have the appropriate enzymes to utilize it, glucose will be oxidized *aerobically.* During aerobic respiration the glucose molecule will be completely oxidized to $CO_2$ and $H_2O$, with the production of 38ATP. Thus, about 266 kcal of energy is in the ATP or approximately 40 percent of that originally bound in the glucose molecule. The remainder is lost as heat. Aerobic respiration is thus some 20 times as efficient as fermentation, and when one considers that the automobile operates at something less than 10 percent efficiency, aerobic respiration represents a quite efficient system. These energy figures are only approximate, some investigators state that glucose contains 710 kcal with each ATP producing about 10 kcal, giving an energy yield of 54 percent for aerobic respiration.

Regardless of whether the organism undergoes fermentation or aerobic respiration, the initial steps are the same. These initial series of reactions are termed the *Embden-Meyerhof-Parnas* (EMP) pathway, or glycolysis.

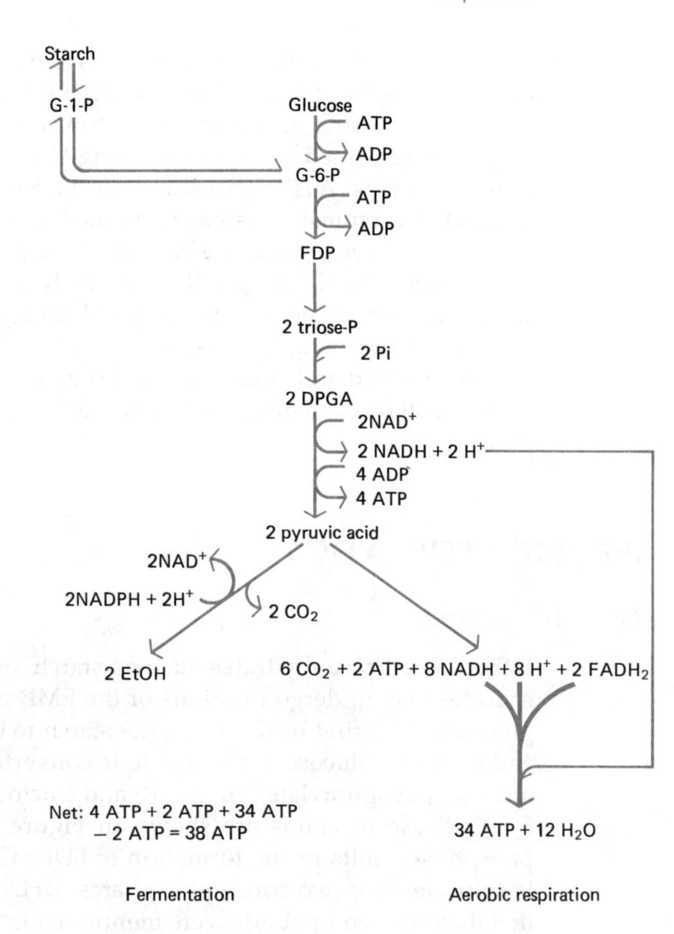


*Figure 14.1*
Simplified diagram illustrating the relationship of the EMP pathway, fermentation and aerobic respiration.

Although the term *glycolysis* is commonly used for this pathway, it is not particularly appropriate, having been originally coined for the breakdown of glycogen in animal cells.

The association of the EMP pathway with fermentation and aerobic respiration is shown in Figure 14.1. Note that molecular or gaseous oxygen is not involved in the EMP pathway. This has led some authors to describe the pathway as anaerobic. At best, however, it can only be considered *potentially* anaerobic if enzymes for fermentative oxidation are available; in most plants it is truly aerobic. The major steps in the EMP pathway include the conversion of glucose to G-6-P and finally to FDP, utilizing 2ATP in the process, the breakdown of FDP into two triose phosphates, and the conversion of each triose phosphate into pyruvic acid, with the production of $4ATP + 2NADH + 2H^+$ in the process. These reactions occur in the cytoplasm.

Under aerobic conditions, the pyruvic acid enters a series of reactions called the *Krebs cycle*. This cyclic series of reactions result in the production of $CO_2$ and $H^+$. The hydrogen ions and electrons are transferred to an electron acceptor, $NAD^+$. The $NADH + H^+$ thus formed, passes through a series of electron acceptors and ultimately both the hydrogen and the electrons are passed to oxygen forming $H_2O$. As the result of this electron transfer process, ATP is formed. Both the Krebs cycle enzymes and those of the electron transfer system are located in the mitochondria.

You now have a good working knowledge of the process of respiration, which occurs in all living cells and involves the oxidation of food, either aerobically or anaerobically, to yield energy. Aerobic respiration will result in $O_2$ diffusing into the cells, $CO_2$ diffusing out. We will now look at the process in detail, and as in the previous chapter, the information that follows will be difficult to understand unless you have had some advanced chemistry.

## AEROBIC RESPIRATION

### *The EMP Pathway*

Either stored carbohydrates such as starch and sucrose or free monosaccharides may undergo reactions of the EMP pathway. Obviously, storage products must first be digested; the starch to G-1-P, the sucrose to glucose and fructose. Glucose-1-phosphate is converted readily to G-6-P, glucose must be phosphorylated to G-6-P, and fructose can be phosphorylated to F-6-P. These reactions are shown in Figure 14.2. Addition of a second phosphate results in the formation of FDP. The fructose-1-6-diphosphate is split yielding two triose phosphates, DHAP and GAP. You may recall that these two compounds were mentioned in Chapter 13 as the photosynthates exported from the chloroplast. It is entirely possible that some of these photosynthates may enter directly the respiratory pathway at this point rather than being converted into other compounds. The DHAP is isomerized to GAP and both molecules of GAP are then oxidized by removal of two hydrogen atoms which are transferred to $NAD^+$. In the resulting compound DPGA, the phosphate of carbon-1 is transferred to ADP, forming ATP, and 3-PGA is converted to PEPA. Transfer of the phosphate group to ADP, forming ATP, and conversion of the enol structure of PEPA to a keto form results in the formation of pyruvic acid (pyruvate).

Several important side reactions involving EMP intermediates can result. Dihydroxyacetone phosphate can be reduced forming $\alpha$-glycerolphosphate, an important precursor of glycerol, necessary in fat synthesis. Some of the PEPA might be used as a precursor for the synthesis of lignin or anthocyanin pigments. Pyruvate itself can be utilized in reactions other than that of respiration. It can be converted to the amino acid alanine, for example.

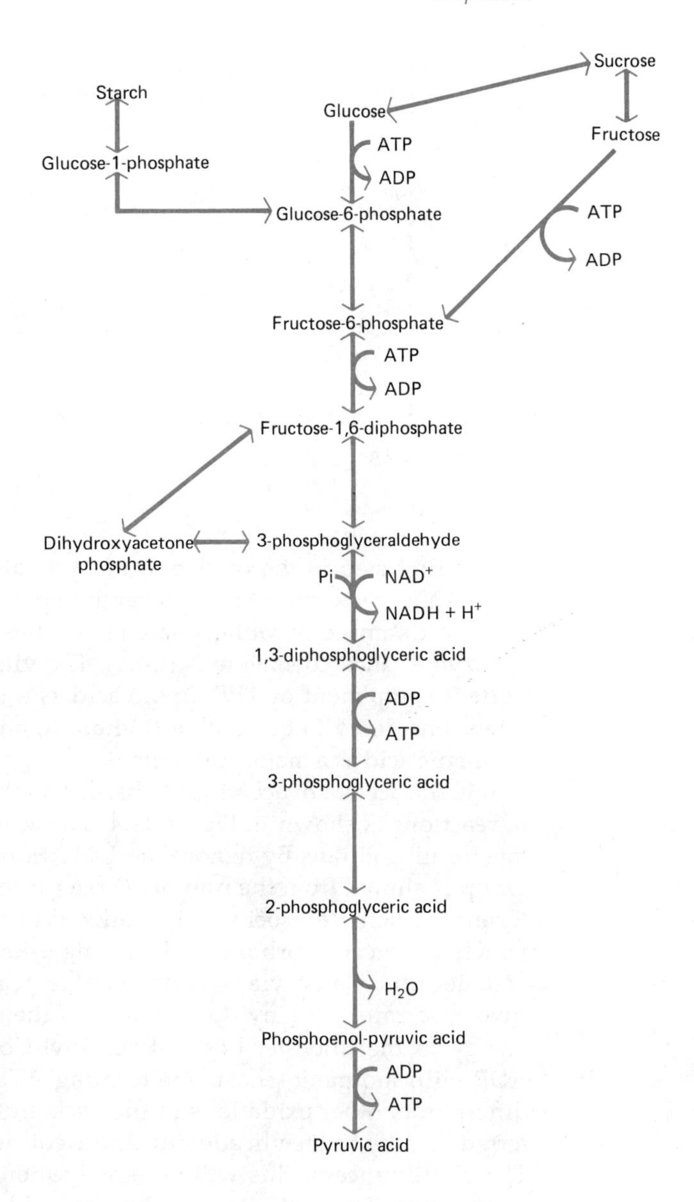


*Figure 14.2*
The Embden-Meyerhof-Parnas metabolic pathway.

## The Citric Acid Cycle

The citric acid or tricarboxylic acid cycle is known commonly as the Krebs cycle. In 1937, the English biochemist H. A. Krebs proposed a cycle of reactions to explain what happened during the oxidation of pyruvate in the breast muscles of pigeons. In the early 1950s, it was established finally that essentially the same cycle of reactions occurred in plant cells.

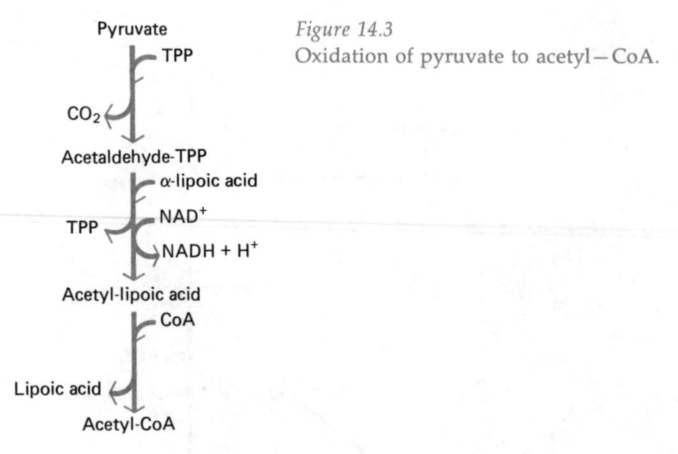


*Figure 14.3*
Oxidation of pyruvate to acetyl—CoA.

The first step in the oxidation of pyruvate is its conversion to acetyl-CoA. This reaction occurs in several steps (Figure 14.3) and utilizes the enzyme thiamine pyrophosphate (TPP), the cofactor α-lipoic acid, and a coenzyme called coenzyme A (CoA). The vitamin thiamine ($B_1$) is an important component of TPP, lipoic acid is a sulfur-containing compound often considered to be another B vitamin, and CoA has the vitamin panthothenic acid as a major component.

It is the acetate of acetyl-CoA that enters the citric acid cycle. This series of reactions is shown in Figure 14.4. The acetate combines with oxaloacetate forming citrate. By removal and addition of a water molecule an —OH group is shifted from the number three carbon to the number four carbon forming isocitrate. Isocitrate is oxidized to the keto acid oxalosuccinate, which in turn is decarboxylated forming α-ketoglutarate which is oxidized and decarboxylated via a nonreversible reaction similar to that of pyruvate, forming succinyl-CoA, which is then converted to succinate. The energy of the thioester bond of succinyl-CoA is used in the reaction of ADP with inorganic phosphate forming ATP. The oxidation of succinate differs from other oxidations in the cycle in that the hydrogens are transferred directly to a flavin adenine dinucleotide (FAD) rather than to $NAD^+$. The significance of this will be noted when we discuss oxidative phosphorylation. Fumarate, formed by the oxidation of succinate, combines with water, forming malate, which is oxidized to oxaloacetate thus completing the cycle.

## *Electron Transfer and Oxidative Phosphorylation*

As you can note from our previous discussion, only four ATP are formed directly by reactions of the EMP pathway or Krebs cycle. We term this type of ATP formation *substrate phosphorylation*. Then how are the other 34ATP

*Figure 14.4*
The citric acid, or Krebs, cycle.

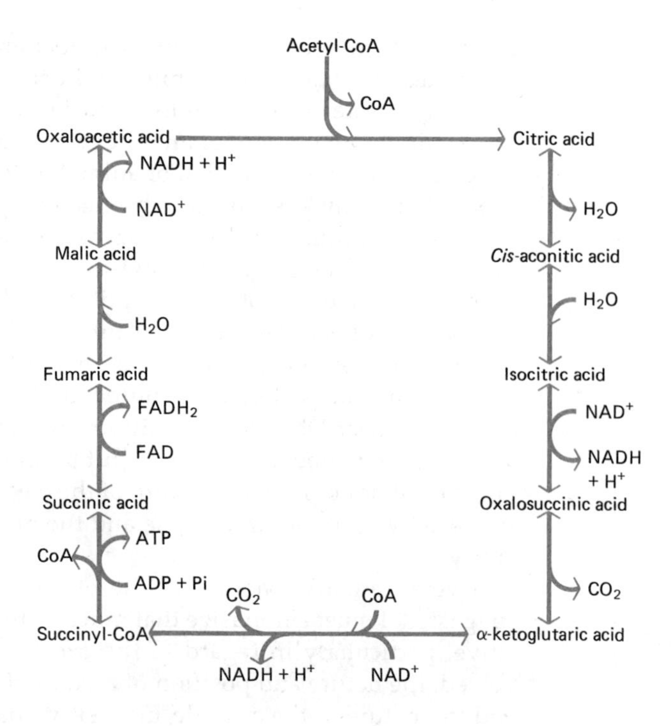


produced? In considering this question, the importance of the oxidation reactions will become evident. You recall that one oxidative step occurs during the EMP pathway, one during the oxidation of pyruvate, with the hydrogen, or electrons, being accepted by $NAD^+$ forming NADH + $H^+$. Three more molecules of NADH + $H^+$ are formed during the Krebs cycle plus one molecule of $FADH_2$. As water is formed as an end product of respiration, you might think that the electrons or hydrogen could be passed directly to oxygen forming water. But this is not biochemically possible. These electron acceptors are quite specific and can transfer their electrons only to another specific acceptor. Thus, a chain of electron acceptors is established, with the electrons being transferred from one enzyme to another in the chain. These specific enzymes constitute what is known as the *electron transport system*, with the electrons being transferred via a series of oxidation-reduction reactions.

Although electron transfer in plant mitochondria resembles that in animal mitochondria, close examination has shown that plant mitochondria exhibit many respiratory activities different from those of animal mitochondria. Ikuma (1972) cited ten differences:1) low temperature (at 77°K) spectra show three *b*-type cytochrome peaks, as compared with one cytochrome *b* in animal mitochondria, and two *c*-type cytochrome peaks different from the cytochromes *c* and $c_1$ of animal mitochondria, 2) the flavoproteins of plant mitochondria are apparently different from those of animal mitochondria, 3) the plant mitochondrial envelope appears to be more "leaky" than that of animal mitochondria, 4) respiration rates are generally greater in plant mitochondria than in animal mitochondria, 5) the respiratory pattern of plant mitochondria (ADP : O ratios) shows a less coupled appearance than that of animal mitochondria, 6) malate oxidation proceeds rapidly even in the absence of pyruvate or glutamate in plant mitochondria (this is not so in animal mitochondria), 7) plant mitochondria contain ATP-specific succinyl-CoA synthetase in contrast to the GTP-specific synthethase of animal mitochondria, 8) a "cyanide-resistant" pathway has been detected in plant mitochondria, 9) rotenone, a potent inhibitor of electron transport in animal mitochondria, only weakly inhibits electron transport of plant mitochondria, and 10) plant mitochondria contain a linear DNA of about 10 $\mu$m in contrast to animal mitochondria which carry a circular DNA of about 5 $\mu$m in circumference. Thus, differences exist not only in the nature of the electron transport system, but also in the activity of the Krebs cycle and the morphology of the mitochondria envelope.

Several tentative models of the electron transport system have been suggested. I must emphasize that much of these data are still highly speculative, particularly in regard to the exact nature of the flavoproteins involved, the nature and position of certain of the cytochromes in the chain, and the nature of the cyanide-insensitive pathway. It is also unclear how $NADH + H^+$ enters the chain in plant mitochondria. With these qualifications in mind let us look at possible pathways of electron transfer in plant mitochondria (Figure 14.5).

There is evidence that at least four flavoproteins appear in plant mitochondria that are different from those of animal mitochondria. In addition, there are three to four flavoproteins similar to those of animal mitochondria. The nature of participation of these flavoproteins in electron transport is not known with any degree of certainty. For example, in the oxidation of pyruvate, the electrons may first be transferred to a flavoprotein, then to $NAD^+$. The same might be true for the oxidation of GAP in the EMP pathway. The oxidation of fatty acyl-CoA and glycerol phosphate may also involve flavoproteins different from those of the electron transfer system shown in Figure 14.5.

The flavoproteins may pass their electrons directly to a coenzyme called ubiquinone (UQ) or, as in the case of FAD $H_2$, may transfer electrons directly to cyt $b_{553}$. Ubiquinone may act as a storage pool for reducing equivalents, thus as the concentration of cyt $b_{553}$ becomes quite high, it may push the equilibrium back towards UQ.

The most distinctive feature of the mitochondrial electron transfer system is the cytochrome components. These are iron containing enzymes. Three *b* cytochromes have been identified, $b_{562}$, $b_{557}$, and $b_{553}$. The oxidation-reduction behavior of $b_{562}$ is confusing, and the position of this cytochrome in the respiratory chain is not at all clear at this time. Cytochrome

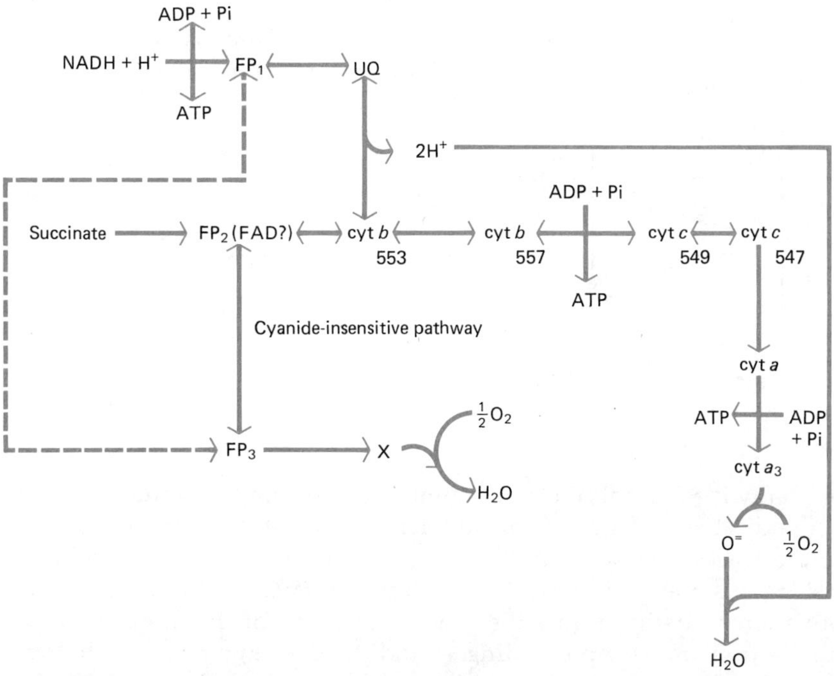

*Figure 14.5*
Possible electron transfer chain in plant mitochondria.

$c_{547}$ seems to be quite similar to the cyt *c* of animal mitochondria, while cyt $c_{549}$ resembles cyt $c_1$ of animal mitochondria, however their absorption spectra are different from those of the animal mitochondria. At this time it would appear that the *c* cytochromes of plant and animal mitochondria have some subtle differences. The *a* cytochromes (cytochrome oxidases) differ from the others in having a copper component in addition to the iron component. Again, subtle differences seem to exist between plant and animal cytochrome oxidases. It is interesting to note that both plants and animals can be killed by subjecting them to carbon monoxide or hydrogen cyanide gas. It is the latter gas that is used in gas chambers. These compounds act by combining with the iron component of cytochrome oxidase, thus rendering it unable to accept electrons. Because of this, the whole electron transfer chain ceases to function, halting respiration. The organism therefore dies of respiratory failure. It is interesting to note that some plants, such as shunk cabbage, are not killed by hydrogen cyanide. They have an alternate pathway of electron transfer that does not include cytochromes (see the cyanide-insensitive pathway of Figure 14.5).

As the result of electron transfer, water is formed. But this is not the most significant aspect of electron transfer. Electrons are transferred normally from a compound with a higher electrochemical potential to a compound with a lower electrochemical potential. As the result of this transfer,

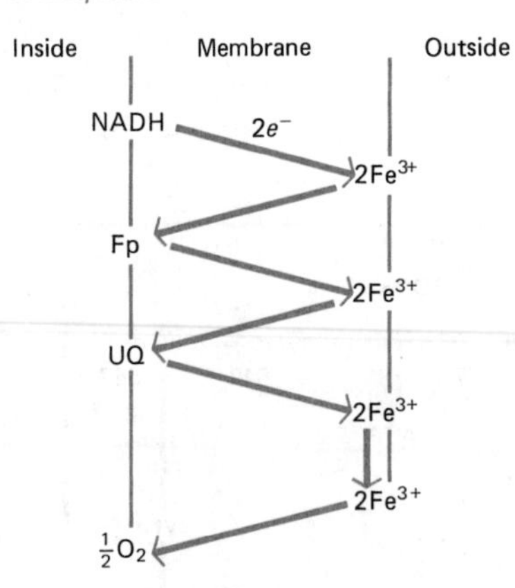


*Figure 14.6*
Location of electron carrier on inner and outer surface of inner mitochondrial membrane.

free energy is released. At certain points along the pathway (thought to be those indicated in Figure 14.5) sufficient free energy is available to catalyze the reaction between ADP and $P_i$ forming ATP. Some investigators consider this energy in terms of an electrochemical potential or proton-motive force existing across the inner membrane of the mitochondrion, with electron transfer proceeding as indicated in Figure 14.6. The total electrochemical or proton-motive force generated would be calculated as follows:

$$\Delta\mu_{H+} = \Delta H^+ + \Delta\Psi$$

| $\Delta\mu_{H+}$ | $\Delta H^+$ | $\Delta\Psi$ |
|---|---|---|
| Total electrochemical potential (proton-motive force) | Gradient of $[H^+]$ (alkaline inside) | Transmembrane electrical potential (negative inside) |

The transmembrane electric potential is greater than that for any other membrane with the exception of the inner chloroplast membrane. As the result of this gradient, energy is available to catalyze the reaction between ADP + $P_i$ forming ATP. But there are two possible ways in which this can occur. The gradient can have a direct effect on ATP formation as follows:

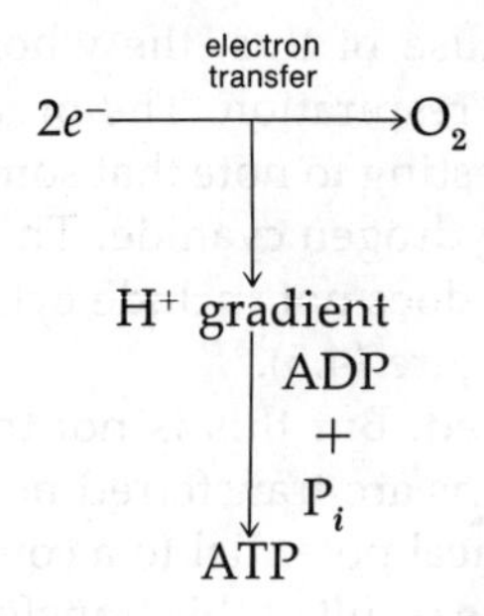

or it can have an indirect effect by affecting some unknown compound X as follows:

$$2e^- \xrightarrow{\text{electron transfer}} O_2$$

$$\downarrow$$

$$H^+ \text{ gradient} \rightleftharpoons X \sim i$$

$$\downarrow \; \begin{matrix} ADP \\ P_i \end{matrix}$$

$$ATP$$

A third theory of oxidative phosphorylation is based on the observation that hydrogen ions are secreted from mitochondria while hydroxyl ions accumulate inside the mitochondria. This would create a dehydration effect in the membrane and would drive the reaction $ADP + P_i \rightarrow ATP + H_2O$ toward the right, thus resulting in ATP formation. This hypothesis is illustrated in Figure 14.7.

Regardless of the mechanism, for every pair of electrons received from $NADH + H^+$, generally *three* ATP will be formed. This is not so for succinate oxidation, and if you look at Figure 14.5, you can see that only *two* ATP can result. Some investigators report that the oxidation of $\alpha$-ketoglutarate results in the formation of *four* ATP. If this should actually be proven to be so, we must revise our net yield of ATP during aerobic respiration upward from 38 to 40. You can determine this yield for yourself by referring back to Figures 14.2, 14.3, and 14.4 and counting the ATP formed by substrate phosphorylation plus those gained during electron transfer from $NADH + H^+$ and $FADH_2$ by oxidative phosphorylation. Remember that your total must be multiplied by two as glucose is split into two molecules of triose phosphate.

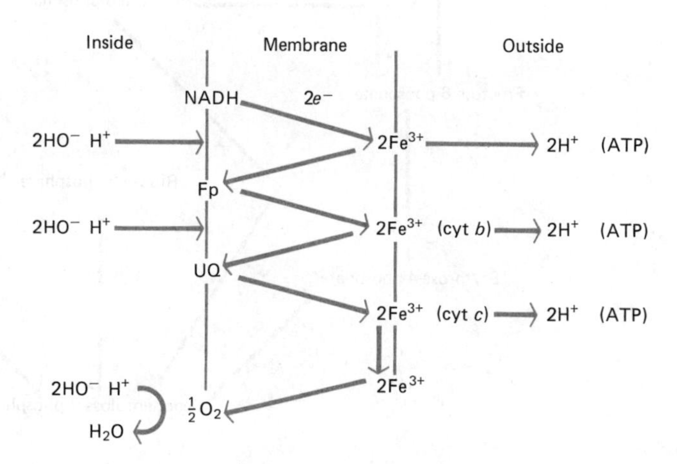


*Figure 14.7*
Possible ATP synthesis via membrane dehydration pathway.

## *An Alternate—The Pentose Phosphate Pathway*

During the 1950s, it became apparent to plant physiologists that the EMP pathway and Krebs cycle were not the only series of oxidative reactions in which $CO_2$ was formed and energy produced. Investigations indicated that five-carbon sugars were intermediaries in the pathway; thus, this series of reactions is often called the *pentose phosphate pathway* (PPP). The pathway is also referred to commonly as the hexose monophosphate shunt, a name which might give the wrong impression, as in some plants this pathway is not simply a shunt but a major oxidative pathway.

The reaction pathway is shown in Figure 14.8. The first reaction involves G-6-P, formed either by the digestion of sucrose or starch or by the phosphorylation of glucose. This compound is oxidized forming 6-phosphogluconic acid. Unlike the EMP pathway and the Krebs cycle, the electrons are accepted by $NADP^+$ rather than $NAD^+$. The 6-phosphogluconic acid is dehydrogenated and decarboxylated forming Ru-5-P. These first two steps of the reaction series are the only ones which are oxidative, and the $CO_2$ formed is all that will be formed in the cycle. The reactions that

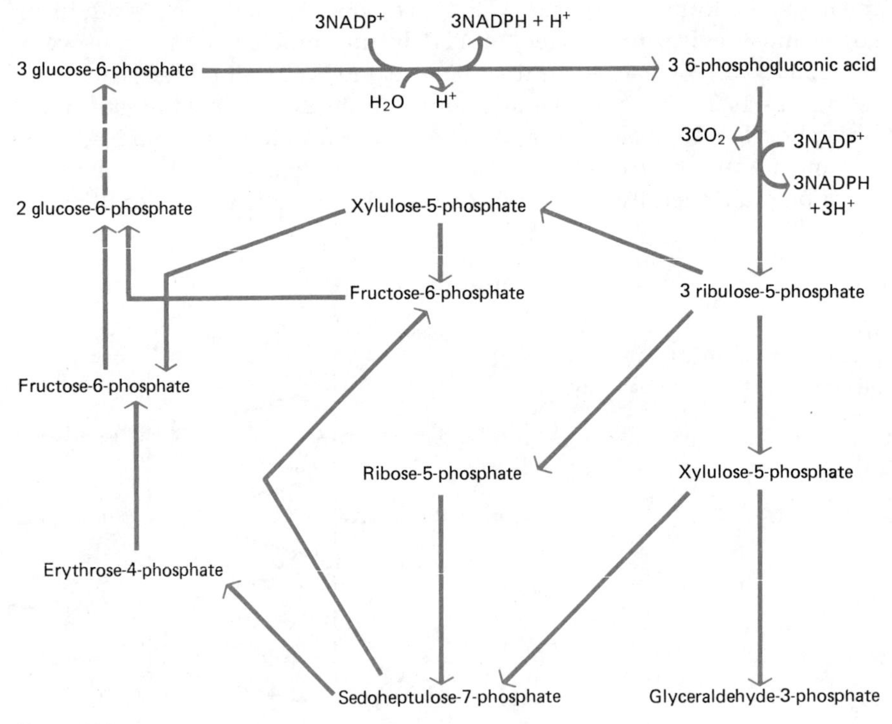


*Figure 14.8*
The pentose phosphate respiratory pathway.

follow serve to transform the Ru-5-P back to G-6-P, and in the process, several important intermediates are formed. The ribulose-5-P can be converted to either Xu-5-P or R-5-P. A thiamine-requiring enzyme facilitates the transfer of a ketol group from Xu-5-P forming sedoheptulose-7-P, with the remaining three carbons of Xu-5-P forming GAP. The upper three carbons of sedoheptulose-7-P are transferred to a three-carbon group of a second Xu-5-P forming F-6-P and E-4-P. Erythrose-4-P can accept the ketol group from Xu-5-P, yielding another F-6-P. Fructose-6-P is then isomerized to G-6-P. It would appear to require three molecules of G-6-P for one complete turn of the cycle, returning two G-6-P, six $NADPH + H^+$, and one GAP. Enzymes of the PPP cycle are located in the cytoplasm.

If one considers the possible energy yield from the PPP pathway, it can be calculated as follows. Assuming the cycle is already functioning, it requires only a single new glucose molecule for another cycle to occur, as two glucose molecules are available from a previous cycle. Thus, a single ATP is required to phosphorylate the glucose. Two molecules of $NADP^+$ are reduced for each $CO_2$ produced from glucose, resulting in a total of 12 $NADPH + H^+$ per molecule of glucose ($2 \times 6$). If we assume that each $NADPH + H^+$ goes through electron transport, three ATP will result. Three ATP times the 12 $NADPH + H^+$ available from the cycle give a total energy yield of 36 ATP minus the 1 initial ATP for a net yield of 35 ATP. Thus, this cycle would be only slightly less efficient than the EMP–Krebs cycle pathway.

From these calculations, it would seem that the PPP pathway would be a major energy-yielding pathway. But there is good evidence that the cytochromes of plant mitochondria cannot effectively use NADPH; thus, no oxidative phosphorylation can result. Of course, the PPP pathway is important as the NADPH could be used in a variety of reactions as a reducing agent. In addition, the GAP formed could enter the EMP pathway. Erythrose-4-P is an important intermediate as it is required in the synthesis of lignin, anthocyanin, and certain aromatic compounds.

## FERMENTATION

Fermentation or anaerobic respiration is common to certain bacteria or fungi such as yeast. Many green plants are also capable of fermentatively respiring glucose. In the so-called higher plants, fermentation may occur under a variety of conditions of low $O_2$ concentration. Roots of plants in waterlogged soil, initial stages of seed germination where the hard seed prevents ready diffusion of $O_2$ into the seed, and other situations where oxygen cannot readily enter the plant organ or where there is a temporary unavailability of $O_2$ are examples. Under these conditions, pyruvic acid is not oxidized, but rather is reduced, either forming alcohol or lactic acid (Figure 14.9). The products of fermentation accumulate generally in the

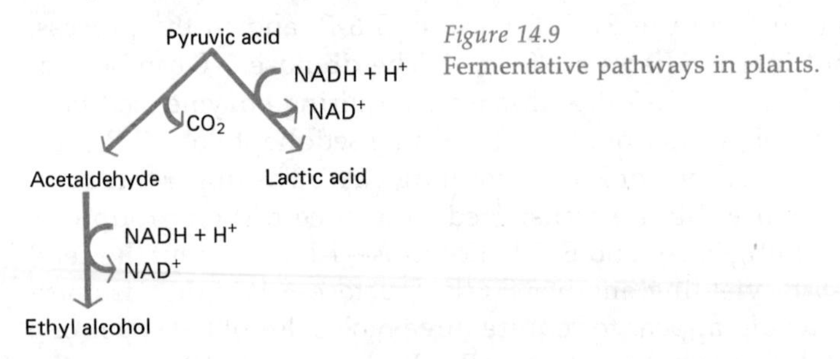


*Figure 14.9*
Fermentative pathways in plants.

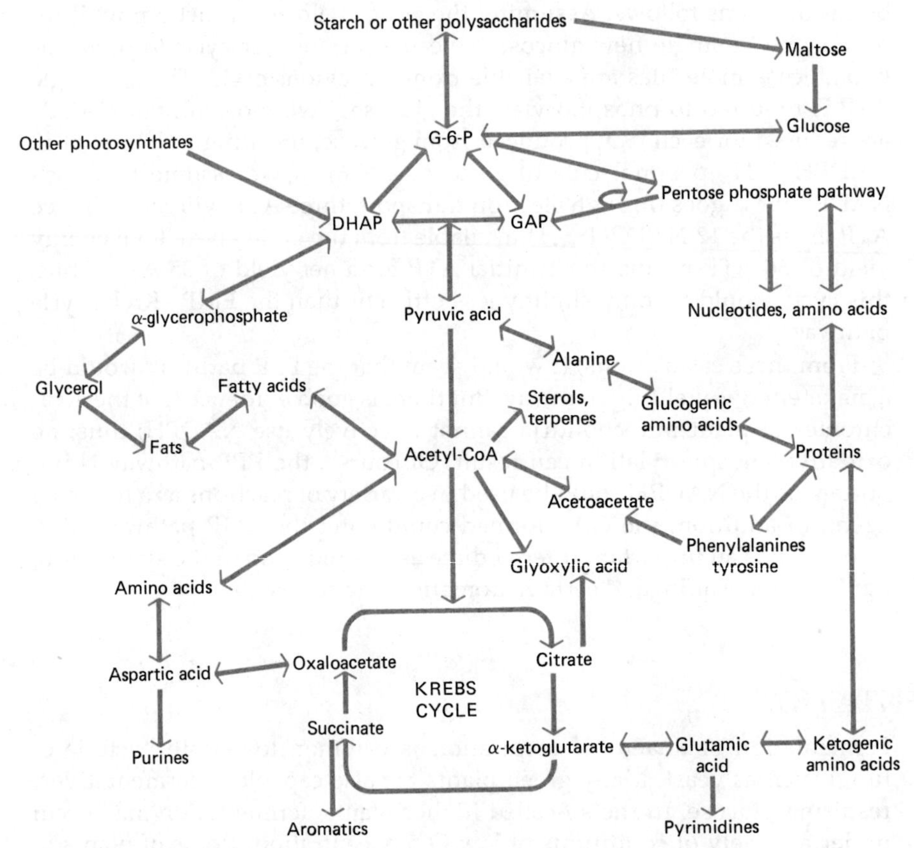


*Figure 14.10*
Relationships between aerobic respiration and other metabolic pathways.

cells as no provision for utilizing them is available to higher plant cells. This accumulation will result eventually in death of the cell if anaerobic conditions persist.

# RELATIONSHIP BETWEEN RESPIRATION AND OTHER METABOLIC PATHWAYS

The process of respiration depends upon the availability of food as a substrate. If you haven't already figured it out, this is one of the major aspects involved in the concept of food. Obviously, the source of food for plants and animals is identical. Foods are *organic compounds* which are *capable of being respired.* With this concept in mind, you may wish to go back to page 319 and revise that list of foods.

We have been discussing the respiration of carbohydrates. Does this mean that only carbohydrates are foods? Not at all. If you have been reading carefully, you will have noted that we mentioned several important side reactions involving fat or protein synthesis. Pathways also exist for the respiration of the digestive products of fats and proteins.

Normally, carbohydrates are most readily respired, but fats also serve as a major source of stored energy. We do not think normally of proteins as sources of energy, but under certain conditions, they may be digested, and the resulting amino acids respired. This will occur under conditions of acute starvation or during senescence of certain organelles or cells. The possible pathways involved are shown in Figure 14.10. We have previously discussed the importance of digestion in forming soluble, and thus respirable, compounds.

# FACTORS AFFECTING RESPIRATION

## Internal or Genetic Factors

Obviously, if carbohydrate concentration in the cell is very low, the respiration rate will be somewhat reduced. If the cell is undergoing starvation, the rate of respiration will continue to decrease until all respirable substrates have been utilized or until the rate has become so low as to no longer support life processes; then the cell will die.

Respiration is an energy-liberating reaction. If there were no controls over the process, it would run at full speed all the time until all substrates were respired. But controls are present so this does not occur. A common type of control is *feedback inhibition.* Such inhibition results normally when a product, such as ATP, NADH, or malate, is present in high concentration, thus forcing equilibrium to the left. These same compounds may exert a different effect. Rather than causing a shift in equilibrium, they may affect directly certain enzymes. This is termed an *allosteric* effect. This type of effect may also result through a process called *competitive inhibition.* This occurs when an enzyme combines with a compound resembling closely its usual substrate. As such compounds irreversibly bind the enzyme, less enzyme is available to react with the normal substrate, thus the rate of the reaction is reduced.

The availability of $NAD^+$ and ADP may also limit the reaction rate. This is termed *cofactor control.* Obviously, if $NAD^+$ and ADP are not available, NADH and ATP cannot be formed. All the oxidations of aerobic respiration are dependent upon $NAD^+$ as the oxidizing agent, while the cytochrome system seems to be coupled with ATP formation, thus requiring ADP and $P_i$ as substrates.

Reference to Figure 14.10 indicates that intermediates of the EMP pathway and of the Krebs cycle serve in the synthesis of many compounds. The draining of these intermediates in fat and protein synthesis will cause an overall increase in the respiration rate.

Age of the cell is also an important factor. Meristematic cells have normally the highest rate of respiration. As differentiation and aging progresses, the rate of respiration declines.

## External or Environmental Factors

As you might expect, an increase in temperature will result in an increased respiration rate, within genetically prescribed limits. For most plants, the respiration rate will actually decrease at temperatures above 35 to 45°C. Certain organisms can respire at temperatures above 90°C (organisms of hot springs), while others have been found growing in ice cream at temperatures below freezing.

As mentioned earlier, if oxygen is unavailable many plants are capable of anaerobic respiration, at least for a time. In fact, under such conditions the rate of respiration, as measured by $CO_2$ production, may actually increase. This is due to the inhibiting effect of $O_2$ on the EMP pathway. As oxygen concentration is increased from O to approximately 3 to 5 percent, the rate of $CO_2$ production is significantly decreased. As the oxygen concentration increases beyond that point up to that of atmospheric (approximately 21 percent), there is an increase in the rate of $CO_2$ production because of stimulation of Krebs cycle activity.

We have discussed previously the process of photorespiration. This process results in increased $CO_2$ production. There have been reports that light actually inhibits mitochondrial respiration. This is difficult to demonstrate, however, as $CO_2$ is being utilized in the light by the process of photosynthesis. As these rates are measured normally by determination of the net exchange of $O_2$ and $CO_2$, and as respiration utilizes $O_2$, producing $CO_2$, and photosynthesis utilizes $CO_2$, producing $O_2$, it is difficult to separate the two processes. In fact, the photosynthetic rate is usually calculated by measuring $O_2$ diffusion values and then measuring, in the dark, $O_2$ consumed during respiration. This mechanism assumes that the $O_2$ consumed by dark respiration is the same as that consumed in the light, an inference that may not be correct.

Injury, disease, or exogenous chemicals can also affect the rate. It has been known for some time that wounding, pathogens, or mechanical stim-

*Table 14.1*
COMPARISON OF PHOTOSYNTHESIS AND RESPIRATION

| PHOTOSYNTHESIS | RESPIRATION |
|---|---|
| $CO_2$ and $H_2O$ used | Food, $O_2$, and $H_2O$ used |
| Food, $O_2$, and $H_2O$ produced | $CO_2$ and $H_2O$ produced |
| Radiant energy converted to chemically bound energy | Chemically bound energy converted to the readily usable energy of ATP or lost as heat |
| Occurs only in chlorenchyma cells | Occurs in all living cells |
| Occurs only in the light | Occurs in both light and darkness |
| Occurs in chloroplasts in eukaryotic cells | Occurs in cytoplasm and mitochondria in eukaryotic cells |
| Overall a reductive process | Overall an oxidative process |
| Rate, except for short periods, must exceed that of respiration if the organism is to survive | |

ulation will increase the respiration rate of affected cells. In addition, application of the so-called weed killers such as 2,4-D causes a rapid increase in the respiration rate which decreases only when the cells are close to death.

It might be informative to compare the two major chemical pathways of green plants, photosynthesis and respiration. Table 14.1 summarizes this information.

## SUMMARY

1. Food is an organic substance which is capable of being respired; thus, it includes only carbohydrates, lipids, and proteins.
2. One of the major roles of food in plants is to provide a source of energy.
3. Food is converted, via respiration, to a source of energy usable by the plant, ATP.
4. Respiration is an intracellular process, occurring in all living cells.
5. Respiration may occur either in the presence or absence of molecular oxygen.
6. If molecular oxygen is utilized, the process is referred to as aerobic respiration.
7. Aerobic respiration consists of three general series of reactions: the EMP pathway, the Krebs cycle, and electron transport and oxidative phosphorylation. It is in the latter series of reactions that molecular oxygen is utilized and most of the ATP formed.

8. In aerobic respiration, glucose, oxygen, and water form carbon dioxide and water

$$C_6H_{12}O_6 + 6H_2O \rightarrow 6CO_2 + 12H_2O$$

9. In the absence of molecular oxygen, the glucose may be respired via fermentation.
10. Fermentation yields a respirable substance, some ATP, and may or may not produce $CO_2$.
11. An alternate respiratory pathway is the pentose phosphate pathway. In it, a number of five-carbon compounds are formed, some $CO_2$ is produced, glucose-6-phosphate is regenerated, and the energy yield is almost as great as the EMP–Krebs cycle pathway.
12. Many of the intermediate compounds formed during respiration are important precursors of many organic compounds.
13. Normally, carbohydrates are the principal foods respired, but fats may also be respired. Proteins are less commonly respired.
14. Many factors affect respiration, including availability of substrate, type of substrate being respired, feedback mechanisms, various synthesis reactions utilizing respiratory intermediates, temperature, oxygen concentration, light, and injury, disease, or exogenous chemicals.
15. The rate of respiration during the daylight hours may be about one-tenth that of photosynthesis.

# Chapter 15

## ROLE OF FOODS IN PLANTS— II. ASSIMILATION, GROWTH, AND DIFFERENTIATION

I have attempted throughout our discourse concerning the biochemical activities of green plants to impress upon you that much of the information presented is hypothesis not truth. In this era of concern over the rights of various groups, chauvinism, and so forth, it is somewhat ironical to note that although our very existence depends upon green plants, in terms of biochemistry there is a definite bacteria and mammal chauvinism. Much of the information extracted from experiments with the above groups has simply been extended to include green plants as well. Such generalizations may help to simplify learning for the introductory botany student, but it also introduces serious misconceptions as to what we know about plants. We do know much about the chemistry of green plants, we suspect even more, but still more is yet to be known.

As in previous chapters, we will be discussing what investigators believe to be occurring. Obviously we cannot, nor would it be either practical or desirable to, present all the biochemical aspects of plants. Entire books have been written on photosynthesis, respiration, growth and development; in fact, books have been written just on specific aspects of these processes. Instead, I will attempt to present basic concepts about important or well-known aspects of plant growth and development, correct common misconceptions, and then deal in some depth with certain of these aspects.

## GROWTH

### *What Is Growth?*

Growth is an extremely complex process, and we do not even begin to understand how different reactions in a cell are correlated, how these reactions bring about specific growth phenomena, how reactions in one cell or organ correlate reactions in another cell or organ, or how certain reactions in one cell may be affected by reactions in another cell. As with most complex plant events, we do not even have a satisfactory description of growth. Intuitively, you and I both know what is meant when someone says, "Boy, my kids have really grown this summer!" In considering all the various attributes of plant growth, however, intuition may let us down. A plant may grow and yet not increase in size. A plant may grow and yet not have any increase in dry weight (weight after all water is removed); in fact, under certain conditions, a plant may actually *lose* dry weight yet have had an increase in length and thus has grown. The various parameters such as increase in length, increase in breadth, increase in numbers of cells, increase in size of cells, increase in weight (either dry or wet) all describe an aspect of growth yet separately may not manifest growth. I will make no attempt here to *define* growth. As we discuss the various phenomena of growth, I will attempt to make clear in what way I am using the term and hopefully you can develop a relatively clear *concept* of growth.

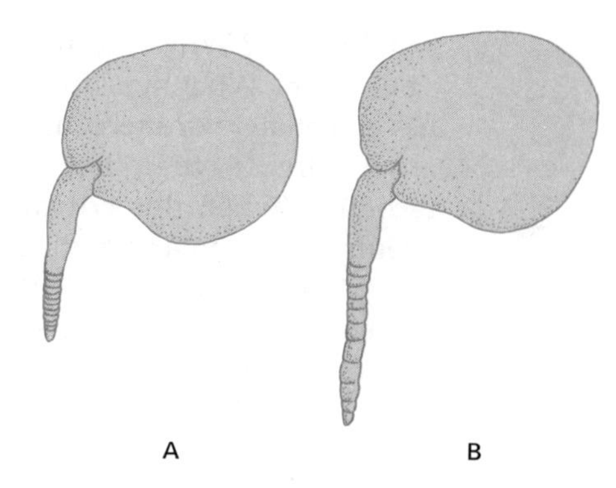


*Figure 15.1*
Region of growth in embryonic root of Broad Bean. *A.* At beginning; lines placed 1 mm apart. *B.* Forty-eight hours later.

Why separate the topics of growth and differentiation? Differentiation is an orderly series of changes; it may occur with or without growth. Often, however, differentiation cannot occur without growth; thus, they are combined as essentially a single process. It is admittedly quite artificial to separate the two phenomena; many of the processes that we will discuss as growth phenomena are really aspects of differentiation. But our major concern is not *which is what* but rather *where, how*, and to *what degree* these phenomena occur. You will want to recall a specific aspect of plant growth stated in an earlier chapter. Some plant organs exhibit *indeterminate growth;* that is, they continue to grow throughout the life of the organ. Other structures, however, exhibit *determinate growth;* they grow to a specific size and then stop.

## Where Does Growth Occur?

Growth is not uniform throughout the plant. As mentioned in earlier chapters, most growth is associated with regions called meristems. In most plant stems and roots, increase in length is due to the activity of *apical* meristems. Leaves, on the other hand, during one period in their development, grow pretty much all over, while at other times, they grow from localized meristem. Some leaves may elongate from activity of a *basal* meristem. *Intercalary* meristems may also contribute to an increase in length. These are normally located at the base of an internode and contribute to internodal elongation in stems. Growth in diameter of stems or roots is often associated with *lateral* meristems or *cambia*, although simple cell enlargement, *primary thickening* meristems, or other anomolous cambia may also be a factor. A typical example of the region of greatest root elongation is shown in Figure 15.1.

## What Happens When a Cell Grows?

One important event of growth is cell division. We have discussed previously the processes of mitosis and meiosis, so from a cytological standpoint, you already have some knowledge of these events. But what causes cell division? Why is it that cells of the vascular cambium all divide at the same time? What synchronizes these divisions? In other tissues, we may have a cell here or there, now or then divide. It has been stated that cambial cells divide when they reach some critical size. How does the nucleus recognize when this critical volume has been reached, and how are the events of cell enlargement so closely correlated that all the cambial cells reach that critical volume at the same time? Why is there an orderly progression of stages in nuclear division? Does one event somehow trigger the next? What causes the nucleus to double its DNA during late telophase, interphase, or early prophase of mitosis? What triggers cytokinesis, and why does it follow immediately nuclear division in some tissues, while it may be quite delayed in others? These events must be both self-regulated yet at the same time under control of genetic and environmental influences. As you can see, we have many questions concerning cell division but few answers.

Water uptake by osmosis is also an important aspect of growth. A particular type of plant hormone, called an auxin, causes an increase in plasticity of the cell wall; a reduction in wall pressure, resulting in decreased turgor pressure; and water movement into the cell, increasing the turgor pressure with a resulting irreversible expansion of the wall.

A third aspect of growth is an increase in protoplasmic substance, vacuolar contents, and cell wall materials. These increases involve one or both of two processes. Materials may diffuse into the cells from intercellular areas or from adjacent cells, or foods may be utilized in the production of these cell parts or both. The latter process is called *assimilation,* which originally referred specifically to the transformation of food into protoplasm. Most biologists, including myself, now prefer to broaden the concept of assimilation. Assimilation is the transformation of food into the substances of which cells are composed. These assimilates of plant cells represent some of our most important products of commerce. Cellulose of cell walls has hundreds of industrial uses such as in the manufacture of paper, rayon, cotton cloth, celluloid, collodion, cellophane, artificial rubber, charcoal, acetic acid (vinegar), and explosives. The pectic compounds of cell walls are important in making jellies, salad dressings, and various creams and emulsions, to mention a few. Plant pigments were, and still are, in certain parts of the world, important in making dyes. Resins and gums are used for manufacturing soaps, varnishes, certain medicines, and a variety of other products. Alkaloids are important sources of drugs. Vitamins are important health aids. Latex is the source of natural rubber. This list of plant products used by man is an extremely long one, and we have discussed them in Chapter 1. We should not overlook the importance of assimilates as food for man and other animals.

If these foods are to be assimilated, the products of photosynthesis are obviously converted or transformed into a variety of organic compounds. Some of these pathways have been discussed in Chapters 13 and 14, and we will not discuss them further.

# GROWTH PHENOMENA

Many of the so-called growth phenomena have already been discussed, to some degree or another, in previous chapters. In flowering plants, growth not only involves an increase in mass but also involves response to various external stimuli. Some of the growth phenomena we will be discussing involves changes in position of plant organs.

The bases for plant growth and development phenomena lie with either exogenous (or environmental) factors, endogenous factors, or both. Environmental factors include such aspects as *light, temperature, gravity, mechanical stimulation, sound, exogenous chemicals, electrical and magnetic stimulation, tidal and lunar influences,* and perhaps others. Endogenous factors include *hormones, enzymes, vitamins,* and *pigments.* A hormone is an organic substance produced in one cell or part of a plant but having its effect, at rather low concentrations, on a different cell or part of the plant. It is obviously necessary for a hormone to be transported in some way from its site of origin to its site of action.

Hormones have been implicated, or at least suggested, in almost all growth and differentiation phenomena. The generally recognized, major hormones include *auxins, gibberellins, cytokinins, abscisic acid, and ethylene,* although various other compounds may possess hormonal activity. I would like to briefly discuss some of the responses that hormones have been associated with. When considering this information, you will want to keep in mind that much of it was gathered by observing the results of exogenously applied hormones. These results are open to at least two questions: 1) Do endogenously formed hormones produce the same results as exogenously applied ones? 2) As the concentration of hormone applied was often considerably greater than that present normally in the cell, can it be reasonably assumed that the lower concentrations of endogenous hormones produce the same results?

## *Auxins*

Table 15.1 summarizes some of the known and suspected actions of auxin. It must be understood that auxin may not be the primary mechanism in each of the actions, and there is often an interaction between auxin and other hormones.

## *Gibberellins*

The gibberellins often interact with auxins and cytokinins, and often, their effect is to reverse the action of abscisic acid (ABA). Known and suspected activities of gibberellins are given in Table 15.2.

Note that many of the effects of gibberellin are similar to those of auxin. This has led some investigators to conclude that the major role of gibberellin is to simply stimulate auxin production, and it is then the auxin that

*Table 15.1*
**SUSPECTED ROLES OF AUXINS IN PLANT GROWTH AND DEVELOPMENT**

| | |
|---|---|
| Play part in phototropism | Increase membrane permeability to water |
| Play part in geotropism | Increase uptake of $Cl^-$, $K^+$, and $PO_4^{3-}$ ions |
| Stimulate cell elongation | Inhibit $NH_4^+$ uptake |
| Stimulate cell division in cambia and pericycle | Stimulate cytoplasmic streaming |
| Promotes elongation of shoot, hypocotyl, and coleoptile | Play part in xylem differentiation |
| May promote or inhibit root elongation | Play part in apical dominance |
| Play part in cell wall synthesis | Initiate roots on cuttings |
| Stimulate photosynthesis and respiration | Develop lateral roots |
| Stimulate photophosphorylation | Play part in fruit set |
| Stimulate RNA synthesis | Initiate flowers in some plants (may cause formation of only pistillate flowers) |
| Cause increase in certain enzymes | May cause parthenocarpy (production of seedless fruits) |
| Stimulate ethylene biosynthesis | Prevent abscission of leaves, flowers, and fruits |
| Cause change in membrane potential | |
| May cause some cells to become polyploid | |

*Table 15.2*
**SUSPECTED ROLES OF GIBBERELLINS IN PLANT GROWTH AND DEVELOPMENT**

| | |
|---|---|
| May stimulate auxin metabolism | Play part in apical dominance |
| Produce normalized growth in genetic dwarfs | Play part in fruit set |
| Stimulate cell elongation | May cause parthenocarpy |
| Stimulate cell division | Break dormancy in seeds and buds |
| May increase cell wall plasticity | Favor production of staminate flowers |
| Stimulate RNA synthesis | May substitute for vernalization in some plants |
| Induce synthesis of hydrolytic enzymes | Stimulate formation of polyribosomes |
| May affect permeability of cell membranes (stimulates leakage of $K^+$, $PO_4^{3-}$, $Mg^{2+}$ and $Cu^{2+}$ out of cells) | Stimulate ER membrane formation |
| Stimulate production of several nuclei in polyploid nuclei | |

produces the effect. This may be true in some instances, but it has been shown to be quite unlikely in others. Gibberellins also have some effects not produced by auxins.

## Cytokinins

Cytokinins have been shown to be absolutely essential for the growth of plant cells in culture. This would tend to indicate that it is essential for normal growth of cells *in vivo*. In common with gibberellin, cytokinins

*Table 15.3*
SUSPECTED ROLES OF CYTOKININS IN PLANT GROWTH AND DEVELOPMENT

| | |
|---|---|
| Promote cell division, cell elongation, or both | Break dormancy of seeds and buds |
| Stimulate auxin activity | Cause stomatal opening |
| Can reverse effect of auxin in apical dominance | Stimulate photosynthetic enzyme production |
| Delay senescence of leaves, cut flowers, fresh fruits and vegetables | Stimulate chlorophyll synthesis |
| | Decrease membrane permeability to $H_2O$ |
| | Inhibit lycopene synthesis |

may reverse the effect of ABA, and in common with both gibberellins and auxins, it exhibits multiple interactions. Table 15.3 summarizes some suspected activities of cytokinins.

## *Abscisic Acid*

Abscisic acid, formerly called dormin or abscisin II has many effects opposite those of the gibberellins and cytokinins. Abscisic acid seems to be the sole member of its class, as opposed to the many auxins, gibberellins, and cytokinins known to exist. This does not necessarily mean, however, that other, similar types of compounds do not exist in plants. They simply have not been recognized at this time. Table 15.4 summarizes some of the suspected activities of abscisic acid.

As you can see, the action of ABA is *generally* that of an inhibitor. Its actions are antagonistic to those of the so-called growth stimulators, IAA, GA, and cytokinin.

*Table 15.4*
SUSPECTED ROLES OF ABSCISIC ACID IN PLANT GROWTH AND DEVELOPMENT

| | |
|---|---|
| Promotes dormancy in seeds and buds | Generally inhibits shoot elongation |
| Promotes fruit abscission | May reverse GA effect of promotion of staminate flowers |
| May stimulate development of parthenocarpic fruit | May stimulate production of a senescence factor |
| Causes stomatal closure | Inhibits synthesis of hydrolytic enzymes |
| Generally inhibits shoot elongation | May induce a positive electropotential on the surface of membranes |
| May stimulate hypocotyl or epicotyl elongation in some plants | Inhibits leakage of $K^+$, $PO_4^{3-}$, $Mg^{2+}$, and $Ca^{2+}$ out of cells |
| May stimulate rooting of cuttings in some plants | Inhibits RNA synthesis |
| Reduces auxin transport | |
| Increases cold hardiness | |
| Increases membrane permeability to $H_2O$ | |

*Table 15.5*
**SUSPECTED ROLES OF ETHYLENE IN PLANT GROWTH AND DEVELOPMENT**

| | |
|---|---|
| Breaks dormancy of seeds and buds | Increases protein synthesis |
| Induces formation of roots on cuttings | May enhance auxin effects |
| Stimulates epinasty (the downward bending of leaves) | May block auxin transport |
| Interferes with geotrophic response | May regulate auxin biosynthesis |
| Stimulates ripening of fruits | May enhance respiration |
| Stimulates abscission of leaves and flower petals | May regulate sex expression in some plants |
| | Causes induction of certain enzymes |

## *Ethylene*

Unlike the other growth regulators, ethylene is a gas. There is some question as to whether or not ethylene is actually a hormone. It is produced to some extent in all cells and does not require directed transport. In other words, ethylene produced in the fruit acts on the fruit. Its production in lower plants, particularly fungi, would seem to be quite logically from ethyl alcohol. In higher plants, however, the pathway of ethylene synthesis is more complex, and may involve several different compounds, including ethyl alcohol. The most probable precursor would seem to be the amino acid methionine. Some of the proposed activities of ethylene are summarized in Table 15.5.

## *Florigen*

Florigen or the flowering hormone, represents one of a number of growth substances whose existence is postulated by observations that a stimulus is transmitted apparently from an induced to a noninduced region. This hypothetical regulator has not been isolated, and some investigators believe it to be two substances rather than a single substance.

A host of compounds have been isolated from plant tissue which act as either stimulators or inhibitors in various bioassays. But is is difficult to determine if these substances are actually hormones or, more importantly, if they have any real effect upon growth in the intact plant.

## *Enzymes and Vitamins*

Both enzymes and vitamins affect growth and development by their activities as catalysts. The phenotypic expressions noted in plants are, of course, determined largely by the kinds of chemical reactions occurring in the plants. Whether these reactions will occur is dependent largely upon the kind and concentration of enzymes present. However, certain enzymes will not function without a vitamin cofactor, thus certain vitamins are also important in enzymatic reactions.

*Table 15.6*
ACTIVITIES OF PLANT GROWTH AND DEVELOPMENT APPARENTLY INFLUENCED BY PHYTOCHROME SYSTEM

| | |
|---|---|
| Flowering | "Sleep movements" in leaves of certain plants |
| Seed germination | Chlorophyll formation |
| Chloroplast orientation | Carotenoid and flavonoid synthesis |
| Lateral bud initiation | Epidermal hair formation on hypocotyls |
| Bud dormancy | Membrane permeability (possible control of) |
| Spore germination | $NADP^+$ production (possible increase of) |
| Tuber formation | Plumular hook opening |
| Sex expression | Etiolation (increased stem elongation, poor leaf development, lack of chlorophyll) |
| Leaf expansion | |
| Growth of shoot and hypocotyl | |
| Auxin transport (possible regulation of) | |

## Pigments

Certain pigments are extremely important in plant growth and development. We have previously discussed the importance of the photosynthetic pigments in providing the chemically bound energy which is transformed into usable energy for all the growth activities, as well as providing directly assimilates for cell growth and differentiation.

Next to the photosynthetic pigments, probably the most siginificant pigment system in plants is the phytochrome system. The phytochrome system consists of two blue pigments, a red-absorbing form ($Pr_{660}$), and a far-red–absorbing form ($Pfr_{730}$). The $Pr_{660}$ is stated to be the inactive form, $Pfr_{730}$ the active form of the pigment, and the two are interconvertible. Table 15.6 summarizes some of the activities that the phytochrome system is believed to be involved with.

Pigments other than the phytochromes may also be important in plant growth and development. The carotenoids may be important in phototropic responses, perhaps acting as a light screen. Riboflavin, a pigment as well as a vitamin, may also be important, particularly in processes involving light absorption in the ultraviolet range, although it also has the blue absorption peak.

## Environmental Influences

Environment has its effect by probably mediating in some way the action of some or all the various hormones, enzymes, vitamins, and pigments. The most influential environmental factor is probably light or the absence of it. The processes influenced by light may be rhythmic, either circadian (occurring on an approximate 24-hour basis) or noncircadian, or they may be nonrhythmic. They may be directed (tropic) or nondirected (nastic). Some of the plant processes mediated by light are given in Table 15.7.

*Table 15.7*
**SOME HYPOTHESIZED PHOTOMORPHOGENETIC PROCESSES**

| | |
|---|---|
| Flowering (inhibits short-day plants, stimulates long-day plants) | Polar differentiation of rhizoids in germinating spores |
| Phototropism (bending of shoot or coleoptile towards the light) | Elongation of rhizoids |
| Photoreactivation | Conversion of fern protonema into prothallia |
| Chloroplast movement | Initiation, growth, and differentiation of dicot leaves |
| Chloroplast replication | Formation of palisade layers of leaves |
| Seed germination | Stimulation of opening and closing of *Mimosa* leaflets |
| Spore germination | Unfolding of leaves of grass seedlings |
| Anthocyanin production | Stimulation of cytoplasmic streaming |
| Plumular hook opening | |
| Temporary stimulation of grass coleoptile elongation | |
| Inhibition of stem, hypocotyl, and mesocotyl elongation | |

The response of shoots and roots to gravity is another well-known effect of environment. The production of zygomorphic (asymetrical) flowers may also be a response to gravity. Some fruit stalks will also show a gravity-induced response. A response to mechanical stimulation is shown by tendrils, certain insectivorous plants, and the plant *Mimosa.* Pollen tubes exhibit a response to chemicals or to water (humidity) as do roots. Some flowers seem to respond to differences in temperature, opening or closing as temperature varies. Plant growth, particularly of seedlings, seems to be affected by electrical fields. Some plant growth seems to be affected by differences, either in wavelength or intensity, of sound. As with hormonal factors, environmental factors are often interrelated in their effects on plant growth and development; thus, it is difficult to attribute the effect to any one factor.

You should now have some basic ideas concerning the influence of environmental and internal factors on plant growth and development. We will now go into more detail for those of you who are interested.

## *Growth Movements*

One aspect of plant growth that has intrigued man for a long time is that of growth movements. Charles Darwin, in 1880, reported on the bending of grass coleoptiles towards the light, and these pioneering studies led ultimately to the discovery of plant hormones. These growth movements are related normally to some environmental stimulus and may include *tropisms,* in which the direction of movement is related to the direction of the stimulus, *nastic movements,* in which the direction of movements is unrelated to the direction of the stimulus, and *rhythmic movements* which are periodic or time-related movements.

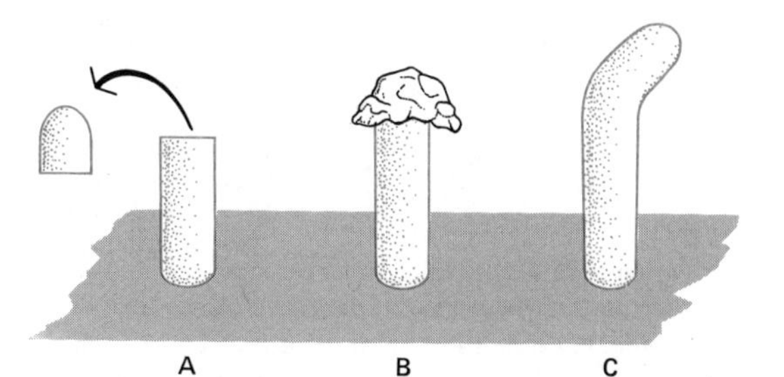


*Figure 15.2*
Experiments of Charles Darwin with grass coleoptiles; (A) tip removed, no response, (B) foil cap over tip, no response, (C) tip exposed to light, bending occurs.

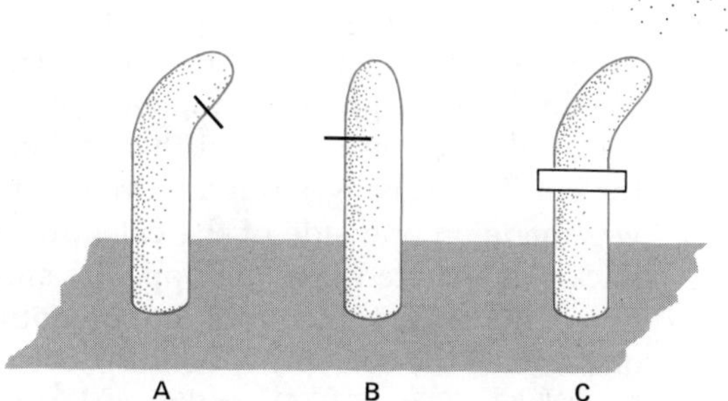


*Figure 15.3*
Experiments of Boysen-Jensen with Avena (oat) coleoptiles; (A) mica in side nearest light source, (B) mica in side away from light source, (C) tip separated from rest of coleoptile by block of gelatin.

In Chapter 2, we discussed the bending of a plant stem toward the light. Darwin (1880) discovered that a plant's ability for *phototropic response* resides in its tip. When he subjected a coleoptile with an intact tip to unidirectional light, the coleoptile bent toward that light source (Figure 15.2). When he removed the tip, the coleoptile no longer exhibited a response. He then placed a foil "hat" over the coleoptile tip, thus shielding it from light, and found the coleoptile exhibited no phototropic response. It behaved just as though the tip was not there.

Boysen-Jensen (1910–1913) modified Darwin's experiments somewhat. He inserted a sheet of mica about half way into the coleoptile near the tip; in the side toward the light source in one group of plants. He noted that when the mica was inserted into the side of the coleoptile nearest the light source, bending still occurred, but when it was inserted into the side away from the light source, bending did not occur. Boysen-Jensen found that bending would occur even if the tip was separated from the rest of the coleoptile by a block of gelatin (Figure 15.3). This suggested that a diffusi-

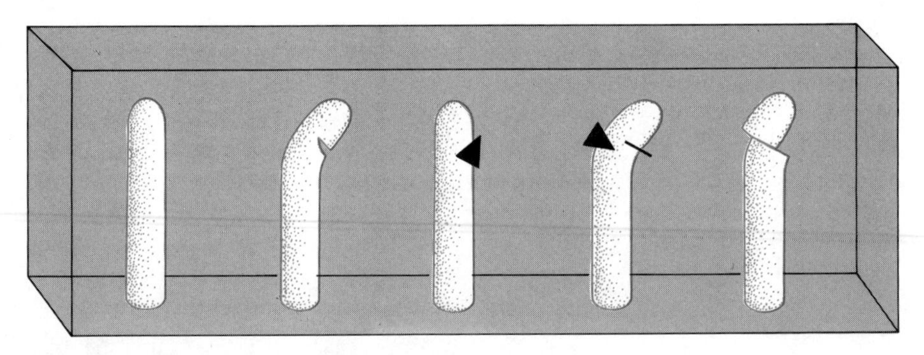

*Figure 15.4*
Experiments of Paal with *Avena* coleoptile; (A) control, (B) tissue removed from one side, (C) hole filled with gelation, (D) tip removed, replaced with gelatin on one side, mica on the other, (E) tip removed, replaced to one side.

ble substance was moving from the tip and that this substance was a growth *promoter* not a growth inhibitor.

Paal (1914–1918) of Hungary demonstrated essentially the same thing, but showed that light was not even necessary for bending to occur. If the coleoptiles were grown in the dark, no bending occurred, but if a piece of coleoptile was removed near the tip, bending occurred towards the side of the cut. If the cut was filled with gelatin, bending did not occur. If a cut was made on one side of the coleoptile and then filled with gelatin, and mica was inserted into the opposite side, bending occurred in the direction of the mica. If the tip of the coleoptile was cut off and then replaced to one side on the coleoptile, bending occurred away from the side of the coleoptile tip (Figure 15.4). Paal's work supported the conclusions of Boysen-Jensen but, in addition, indicated that light was not necessary for bending of coleoptiles. Thus, the tentative explanation of a diffusible growth-promoting substance seemed a reasonable one.

The actual existence of such a substance was demonstrated finally by Went in 1926. He reasoned that if such a chemical did exist, and if it was produced in the coleoptile tip and diffused downward from it, he should be able to collect the substance in a block of gelatin. He removed the tips from oat coleoptiles and placed a number of them on a block of gelatin. After one hour, he removed the tips and cut the gelatin into small squares. He then placed these squares of gelatin to one side of oat coleoptiles and noted that after only an hour's time, the coleoptiles were already exhibiting a negative curvature (Figure 15.5). After three hours time, the curvature was strongly negative. If plain gelatin was used little or no response was obtained, nor could he obtain any results with gelatin on which coleoptile rings, cut off just below the tip, had stood. Went stated that it was, " . . . evident that I had obtained the more or less hypothetical growth-regulators."

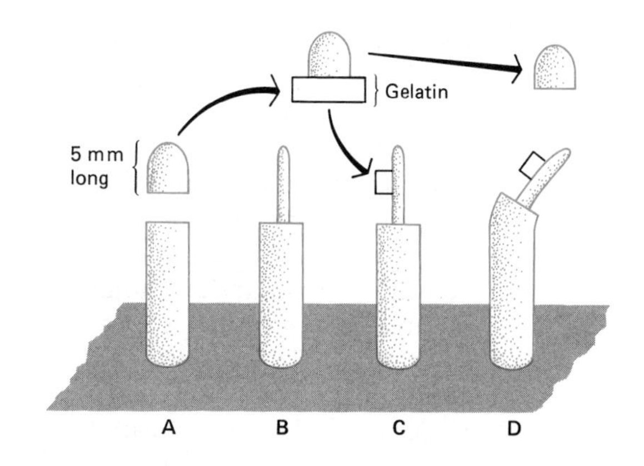


*Figure 15.5*
Went's experiments with *Avena* coleoptiles; (A) tip removed, placed on gelatin block, (B) primary leaf pulled out so its growth will not lift gelation block from surface, (C) gelatin block containing diffusible substance stuck on one side of coleoptile stump with melted gelatin, (D) after three hours, negative curvature measured.

Went also demonstrated the correlation with light by showing that more of the substance would diffuse into a block placed below the shaded side of a coleoptile than into one placed below the brightly lighted side. He called this substance *auxin.* It wasn't until 1934 that another Holland investigator, F. Kogl, isolated a substance from urine that produced the same response in coleoptiles as did Went's substance. Kogl's substance was chemically identified as *indoleacetic acid* (IAA), a substance which has since been isolated from many plant tissues. It may be somewhat surprising to you to learn that IAA has not been isolated from *Avena* (oat) coleoptiles. Although other growth regulators can cause the bending response, Went's auxin is believed to be IAA even though it cannot be isolated. The substance is apparently present in extremely minute amounts; it is estimated that it would take *20,000 tons* of coleoptile tips to produce *1 gram* of IAA. Obviously, the substance is active in extremely small amounts, a property of plant hormones in general.

In our considerations of the various growth phenomena, do not let the use of specific terminology bother you as it sometimes bothers the typical man-in-the-street. Botanists have not coined big words simply to bedevil the uninformed; however, I do think there is sometimes an unnecessary proliferation of terms. It is simply much easier to say "phototropism" than to say, "the bending of a plant organ toward or away from a unidirectional light source." To the informed, this single word tells everything that the long phrase does, and much more.

No matter which way you place a seed in the ground, when it germinates the roots will grow downward, the stem upward. If you place a plant on its side, the stem will begin to grow upward after a few hours. Such a growth response is termed *geotropism,* the response to gravity. This response is said to be *inductive,* in that the plant does not have to be stimulated continuously for a response to occur. Experiments have been performed subjecting plants to forces from over 58 *g* to less than 0.1 *g*. At the highest *g* force, the plants had to be subjected to the force for only five seconds to elicite a response. At the lowest *g* force, exposure time had to be increased to over an hour before the bending response was induced. These time periods, termed the *presentation time,* refer simply to the amount of time the plant organ must be stimulated to obtain a bending response, assuming the plant is returned to the usual conditions of gravity following this exposure. Thus, even though the plant is no longer being stimulated, bending will occur in the same fashion as it would have under continued stimulation. Some plants will show a geotropic response if stimulated continuously by a force as low as 0.005 to 0.001 *g*.

In Chapter 8, we discussed the influence of the root cap on the geotropic response. In corn, only the root cap need be removed for loss of response to occur. In the broad bean, however, both the root cap and apical meristem region must be removed before loss of response occurs.

Can the geotropic response also be attributed to auxin? Using radioactive IAA, Leopold was able to demonstrate that auxins are transported laterally in response to a gravitational field. But if we consider the example of a plant placed in a horizontal position, we would have to assume that the auxin will collect along the lower side and be fairly equally distributed along the root-shoot axis. Why then does only the shoot respond as we would expect? The lower side of the shoot, having the highest concentration of auxin, grows more rapidly than the upper side; thus, the shoot bends upward. In the root, the side having the most auxin elongates to a lesser degree than does the upper side having less auxin. The growth of root cells is obviously inhibited by the same concentration of auxin that will cause elongation of hypocotyl and stem cells. This explanation would seem logical enough, except no such asymmetrical distribution of auxin has been found in the root. It may be that the normal endogenous auxin level of the root is too small to be determined by present techniques. We might postulate the following mechanisms for response to gravity (Figure 15.6). Geotropic response can change with time. The flower stalk of the peanut, for example, is at first negatively geotropic but, following fertilization, becomes positively geotropic.

The response of a plant part to a mechanical stimulus is called *thigmotropism* or *haptotropism.* Tendrils, for example, illustrate a positive and sometimes quite rapid thigmotrophic response. Cells in the region touched contract, while those on the opposite side elongate greatly. Strangely enough, a uniform static pressure, no matter how strong, will not elicit a response, yet drawing a thread lightly across a tendril will elicit a response. It would seem that it is not the pressure but the rubbing that is essential. Associated with the twining habit of vines is a phenomenon called *circumnutation.* As the tip of a stem elongates, that tip traces a spiral type of motion, due probably to unequal growth in different areas of the tip. This phenomena will be exhibited whether the tip is in contact with a substrate or not.

A variety of tropic responses can be found in plants. In Chapter 8, we mentioned the concept of hydrotropism, or the growth of roots towards

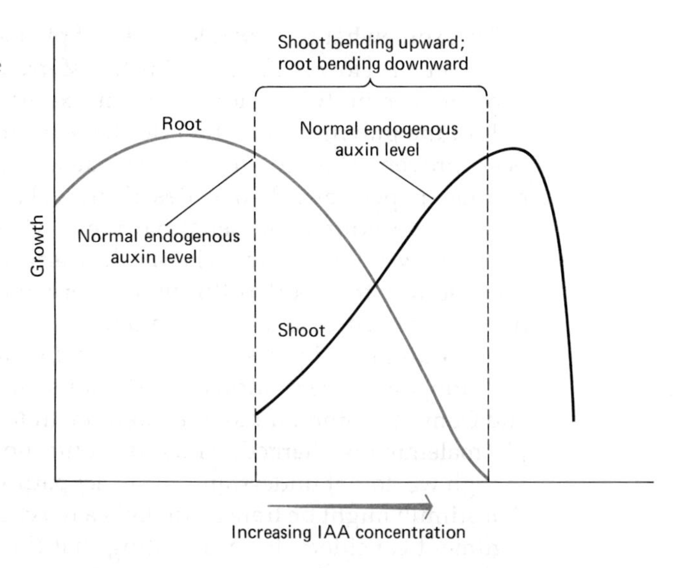


*Figure 15.6*
Postulated mechanism for response of stem and root to gravity and increased auxin concentration. The upward slope of the curve represents upward growth, the downward slope of the curve downward growth, with the leveling off of the curve indicating no response.

more moist regions. Response to chemicals *(chemotropism)*, wounding *(traumatotropism)*, and electrical fields *(galvanotropism)* are other tropic responses. Although we do not know the mechanisms of such responses, they may very well involve hormone mediation or variation of inorganic ion concentration, particularly calcium or potassium. Simply because we do not currently understand the mechanisms of certain plant growth phenomena, we should not automatically relegate these to the field of mysticism. On the other hand, we should not disregard totally certain explanations simply because they sound "far out" or are not within our understanding, at least not until we can come up with a more scientifically sound inference.

Nastic movements are quite common in many plants. Such movements may be stimulated by a variety of environmental conditions; thus, photonasty, geonasty, thigmonasty, chemonasty, and others are known. Many of these nastic responses are rhythmic and will be discussed in the next section.

An interesting and striking plant movement is that of the sensitive plant *Mimosa pudica.* Any stimulus, such as touch, wind, heat, or electricity that will cause a shock will cause a response termed a *seismonasty*. The reaction is rapid and may start 0.1 second after stimulation and progress rapidly from the point of stimulation. The reaction occurs in the expanded region at the base of the pedicels and petiole. These expanded regions called *pulvini*, react in two different ways: the pedicels undergo a shrinking of their upper sides, and the leaflets fold consequently upward; the petioles undergo a contraction of their lower sides so the whole leaf drops. In either case, there appears to be a rapid loss of water from the so-called motor cells of the pulvini.

This somewhat remarkable aspect of plants gives rise to two interesting questions: 1) How is the stimulus transferred from one part of the plant to another so rapidly? 2) How is water exported so rapidly from the motor cells of the pulvini? Many theories have been suggested but none have yet been proven. The Indian scientist, Jagadis Chandra Bose published numerous papers and books describing plant phenomena, many of his studies concerning the sensitive plant *Mimosa.* Bose declared that the stimulus was transported via an electrical impulse and further that this impulse was transported through a nervous system. Such statements do not seem too unreasonable when one considers that plant cells contain calcium, sodium, and potassium ions and are bathed in a solution of ions. In addition, the concentrations of these ions differ inside and outside the cell and from each other. It has been demonstrated rather conclusively that the plasmalemma is charged. Thus, it would not seem too far-fetched, even though we do not understand the exact pathway or mechanism, to assume that stimuli might be transmitted via a nervous pathway. But Bose hurt his argument considerably by insisting that the isolated vegetal nerve is indistinguishable from animal nerve. Plant anatomists and plant physiologists had studied many plants before Bose's time and have studied many others in the fifty-some years following this assertion, and no such nervous system has been found. This does not mean that some type of interconnected system for the rapid transmission of electrical impulses does not exist in plants, but it does indicate it is highly unlikely to be similar to that of animals.

The *thigmonastic* response that seems to interest most people is that of the insectivorous plants. The sundew, bladderwort, and venus flytrap all possess tactile hairs that, when stimulated, transmit an impulse which results in a closure of the trap (Figure 15.7). Contrary to what certain recent pseudoscientific offerings say, these plants will react to anything that stimulates the trigger mechanism, not just to so-called food organisms. The trapping mechanism of a Venus flytrap is interesting. It requires two successive stimuli before the leaf will spring shut. This can be accomplished by touching a single tactile hair twice or two adjacent hairs once. In addition, if one stimulates the tactile hairs immediately after the leaf has opened from a previous closure, the closing action will be somewhat damped, it will not close quite as rapidly as it did the first time. If one repeats the stimulus on the same leaf, it may not close at all. It is known that an electric potential is set up in the trap as it springs, so again the transport of an electrical impulse may be involved. In fact, the trap may be sprung by an electrical current.

There are about 450 species representing 15 genera of insectivorous plants. The trapping mechanisms exhibited by these species are classified into two groups, passive and active. In the former, the most common, referred to as a pitfall, is found in the so-called pitcher plants, *Heliamphora, Cephialotus,* and *Sarracenia,* in the cobra plant *Darlingtonia (Chrysamphora)*

*Figure 15.7*
Some insectivorous plants, demonstrating various trapping mechanisms. *A.* Pitcher plant. *B.* Bladderwort; SEM of bladder *C.* Sundew and Venus Fly Trap. *D.* Butterwort. [*A., C.,* and *D.* Courtesy of Carolina Biological Supply Company.]

*californica*, and in *Nepenthes.* The passive trap of *Genlisea*, found in the tropics of Africa and South America, has been described as a lobster pot. Another type of passive trap has been referred to as a bird lime or flypaper trap. Such plants, including the genus *Byblis* of Australia, *Drosophyllum*, and *herba piniera orvalhada* ("dewy pine") of Spain, secrete a sticky mucilage which serves to trap insects. The flypaper type mechanism may also be associated with an active mechanism in which growth movements of the plant aid in trapping the prey. Such is the case in the butterwort, *Pinguicula*, and the sundew, *Drosera.* Two other active mechanisms include the steel trap technique of the venus flytrap, *Dionaea*, mentioned earlier, and a similar trap in the small aquatic plant, *Aldrovanda.* The final type of active trapping mechanism has been called the mousetrap by F. E. Lloyd. Such a trap is characteristic of the bladderwort, *Utricularia* and *Biovularia* of Cuba and Eastern South America, and *Polypompholyx* of Australia.

Most of the insectivorous plants have some way of attracting their prey. In some, it is odor: *Sarracenia* is described as having the odor of violets, *Drosophyllum* as having a honeylike odor, and *Pinguicula* as having a fungus order. Others, such as *Nepenthes, Heliamphora, Cephalotus, Utricularia,* and *Darlingtonia* secrete a nectar or sugar solution which attracts insects to the trap. It is thought the attractive colors or bright fenestrations of *Sarracenia, Darlingtonia,* and *Cephalotus* may be an attractant, while the mucilage secretion of *Pinguicula, Drosera, Genlisea, Byblis,* and *Drosophyllum* may, in some instances, be both attractant and trap.

Regardless of the method of attracting prey or of trapping, digestion of the trapped insect will normally follow. In some instances, it requires only a few hours, in other instances several days. In some insectivorous plants, digestion of the prey is apparently through bacterial activity; in others, digestive enzymes are secreted by the plant.

It is essentially the active trapping mechanisms that we are interested in here. Such traps, once stimulated, respond through differential growth movements activating the trap. In bladderwort for example, the trapping mechanism is extremely complex and has been likened by F. E. Lloyd to an almost Rube Goldberg—like mousetrap. His description is interesting and I think worth repeating:

> Here . . . I present two models, in the form of mouse traps, designed *ad hoc,* to illustrate the way in which the trap of *Utricularia* has been and at present is thought to work.
>
> Two models are offered. One [see Figure 15.8] represents the mechanism of the trap as conceived by Cohn, Darwin and others. In this the door is a passive check valve, easily pushed inwards, but not outwards. In the model

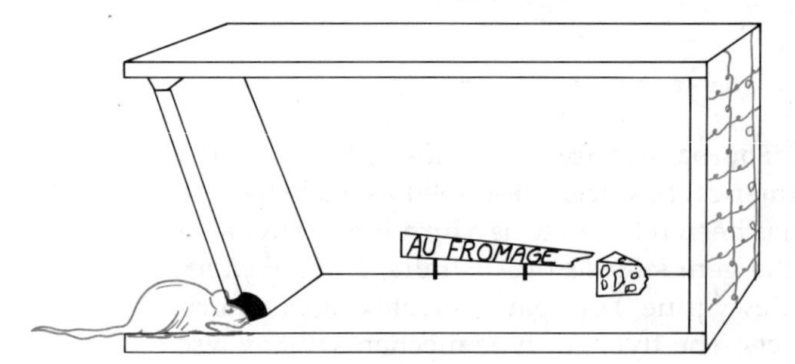


*Figure 15.8*
A mouse trap designed to embody the idea of the trapping mechanism of insectivorous plants held by Cohn and by Darwin and others for 15 years after them.

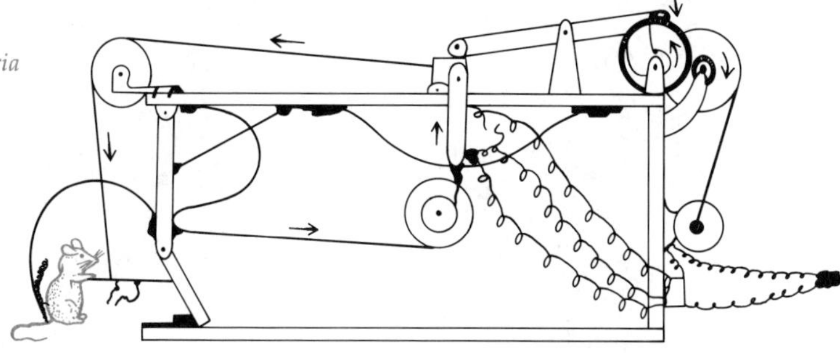

*Figure 15.9*
A mouse trap intended as a model embodying present ideas of the *Utricularia* trap as a mechanism.

a small hole in the bottom of the door allows the mouse to see the bait thus enhancing the effect of the lure by adding sight to smell. This model is an improvement on the *Utricularia* trap in having the bait on the inside. Its extreme simplicity is in contrast to that of the second model [see Figure 15.9] which affords an analog in which the complexity of the *Utricularia* trap as now understood is suggested without exaggeration.

. . . A captious reader may find difficulty in accepting the analogy as complete. I can say only that he would be right; but at least a purpose is served, to indicate that the *Utricularia* is a pretty complex bit of mechanism.

Some plants show growth phenomena which repeat over a definite time interval. These cyclic patterns may continue in the absence of an environmental stimulus; thus, they are some sort of timekeeping device. An example of such a rhythm can be observed in the different position assumed by leaves of certain plants at night and during the day. Some authors have considered such movements as photonastic, but they will continue even in the absence of light. The first record of these movements, to my knowledge, is a reference by Pliny in 77 A.D. The Swedish botanist Linnaeus, in 1751, coined the name *plant sleep* for these movements, and to this day, they are still referred to as *sleep* movements (Figure 15.10). An astronomer, DeMairan, found in 1727 that the leaves continued their cyclic opening and closing even in continuous darkness, and in 1835, de Candolle, found the movements continued in continuous light as well. De Candolle also found that the day-night positions could be reversed by keeping the plants in the dark during the day and placing them under artificial light at night. Thus, although light does not *cause* the movements, it certainly must *influence* them in some way. It is also noted that plants kept in continuous light shortened the time of their cycling by about one to one and a half hours. In addition, they would continue their sleep movements for only about 48 hours when kept either in continuous light or continuous darkness. They gradually damped out after this time.

In an attempt to determine the basis for these sleep movements, Semon (1905) published results of experiments with totally dark-germinated seeds, and those germinated in a light-dark cycle of 5, 5 and 24, 24 hours of

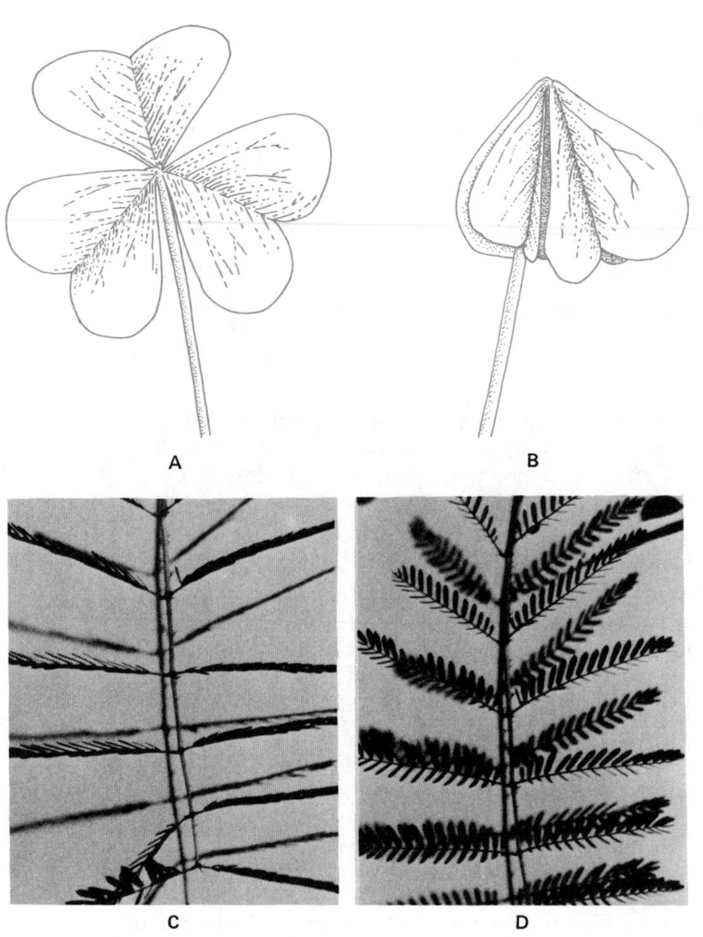


*Figure 15.10*
Leaves of the wood sorrel (*Oxalis acetosella*): *A.* Day position, *B.* Night position. Leaves of *Acacia lophantha:* *C.* night position, *D.* day position.

light and darkness. When the seedlings were placed in continuous light, they all showed the approximate 24-hour pattern. He concluded that these movements had to have a hereditary basis, not an environmental one. These types of rhythms, occurring on an approximate 24-hour basis, are called *circadian rhythms*, from the latin *circa* (''about'') and *diem* (''day'').

A most interesting organism is the luminescent alga, *Gonyaulax polyedra* (some species of *Gonyaulax* produce red tide). The ability of *Gonyaulax* to luminesce when stimulated is quite low to absent during the day, but great during the night, gradually diminishing towards morning. In addition, without stimulation, a glow will appear in a *Gonyaulax* culture about midnight, increase to its greatest intensity about four o'clock, and then decrease and disappear entirely by early to mid morning. The photosynthetic rate in *Gonyaulax* also exhibits a circadian rhythm. If the cells are exposed to a constant intensity of bright light during the day, the photosynthetic rate will gradually increase until noon, then gradually decrease. Still another circadian rhythm in *Gonyaulax* is associated with cell division. Almost all cell divisions occur in a 30 minute period in each 24 hours, this period being about 12 hours after the initiation of the dark period. The cells do not divide each day at this time, but divide, on the average, every second day. Many other types of circadian rhythms are known to exist in other organisms.

Another type of biological rhythm is *tidal rhythm*. Certain algae migrate upward as the tide falls, causing the mud of tidal flats to change color due, of course, to the pigmentation of the algae. The diatoms *Pleurosigma* and *Hantzschia amphioxys*, and the chrysomonad *Chromulina psammobia* exhibit such movement. When these organisms are brought into the laboratory, they continue to move up and down with the tidal period for a few days even though no tide is present. The alga *Euglena obtusa* was also reported to exhibit a tidal rhythm, but later investigations indicate that it was, in fact, a circadian rhythm, the rhythm exhibited by a number of diatoms of mud flats as well.

Certain marine algae, such as *Dictyota*, *Derbesia*, and *Halicystis*, liberate their spores or gametes at the time of spring tides, twice every lunar month. This suggests that the organisms are in some sort of a *semilunar rhythm*. However, such rhythms are hard to prove under laboratory conditions because of problems both in controlling environment conditions and in the length of time required for the study. There is no precise knowledge of the existence of semilunar or lunar rhythms at this time, but in *Dictyota*, at least, it is difficult to explain their behavior in any other way.

*Annual rhythms* might also logically be thought to occur. But, most of the phenomena considered to demonstrate an annual rhythm have been shown to be a photoperiodic response or simply a response to change in temperature. True annual rhythms are difficult to demonstrate; however, annual differences in growth have been demonstrated under carefully controlled conditions where light and temperature would not be a factor. Seeds have also been shown to demonstrate yearly maximum and minimum germination capacity.

Rhythms seemingly not triggered by the environment also exist. The beat of flagella in flagellated algae, protoplasmic flow in a slime mold, electric potential changes in plant cells or organs, certain biochemical processes, leaf movements, nutation or certain other growth aspects, all demonstrate a definite rhythm. Why and how do these rhythms occur? I have no idea, but if they are not controlled by environment, they must have some sort of endogenous control. Some biologists, perhaps overemphasizing the importance of nuclear transcription of genetic material, have stated that all rhythms must have their basis in such transcription. But there is little or no evidence to support such a statement. Other investigators have hypothesized that the basis for biological rhythms is in the nature of the cellular membranes.

The phenomenon of *photoperiodism* could be considered in the previous section as a rhythmic response or at least some sort of *biological clock.* The term was first used by Garner and Allard (1920) to describe the flowering response of certain varieties of soybeans and tobacco to the length of the light period.

Although the best known photoperiodic response is flowering, other responses are also thought to be correlated with the photoperiod. Often, temperature influence seems to be able to substitute for the photoperiodic effect, or the two influences interact producing a response.

Photoperiodism affects the elongation of buds and shoots, and this, in turn, corresponds to increased secondary growth and the production of spring wood in woody plants. Photoperiodism also affects dormancy of storage organs such as tubers, corms, and bulbs, cambial activity, and metabolism in shoot and root tissues. Renewal of growth in these regions is also attributed to photoperiodism. In addition, the development of reproductive structures in certain liverworts and mosses, winter hardening of certain dicots, dark fixation of $CO_2$, and the development of plantlets on *Bryophyllum* leaves have been attributed to photoperiodism.

The photoperiodic response appears to be mediated through the phytochrome system. The basic aspects of the system would seem to be the presence of blue pigment having two forms as mentioned earlier, a red-absorbing form ($Pr_{660}$), and a far-red–absorbing form ($Pfr_{730}$). This occurs because the conversion of $Pr_{660}$ to $Pfr_{730}$ requires considerably less energy than the conversion of $Pfr_{730}$ to $Pr_{660}$; thus, the equilibrium is far to the right. In darkness there is a slow conversion of $Pfr_{730}$ to $Pr_{660}$. The pigment is proteinaceous, and its primary action may be through control of membrane permeability, as there is some evidence that it is bound to the membrane. The various transformation of phytochrome are illustrated in Figure 15.11. It is not known with any certainty whether $Pfr_{730}$ acts directly producing the biological action, or whether $Pfr_{730}$ combines with some unknown factor X producing the biologically active form.

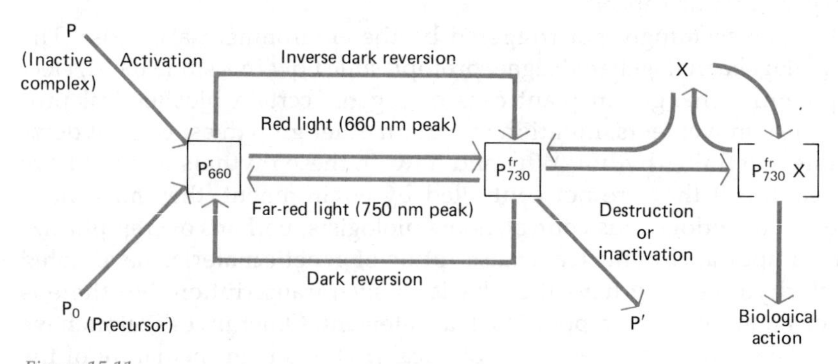


*Figure 15.11*
Differentiation of phytochrome. [From P. Rollin, *Phytochrome Control of Seed Germination,* Paris: Masson & Cie, Publishers, © 1970.]

When photoperiodism is analyzed in terms of a circadian rhythm, at least two actions of light can be distinguished. Light acts apparently as a signal, inducing or preventing a particular biological activity when it falls at a particular time in the circadian cycle. When light is given for several hours it resets the biological clock and initiates a new cycle. Both these functions of light seem to involve phytochrome.

The hour glass approach to phytochrome action proposes that the effective element in the inductive process is the duration of darkness. However, some light, usually several hours, is required if the inductive dark period is to be effective. Light may be necessary in the production of substrates for the dark reaction. The system may be coupled to the time-measuring reaction when a particular threshold value of $Pfr_{730}$ (or $Pfr_{730}X$) is reached. When the time measurement is completed, hormone production may begin. Hormone production seems to be highly sensitive to the presence of $Pfr_{730}$ and ceases on transfer to light.

Flowering in plants is an extremely complex event. As we mentioned earlier, the term *photoperiodism* resulted from studies of flowering plants. Those plants which flower under conditions of short days and long nights are termed *short-day plants* (SDP), while those which flower under conditions of long days and short nights are called *long-day plants* (LDP). In other plants, length of the photoperiod does not seem to be a factor; thus, flowering is not totally under photoperiodic control.

Grafting experiments, in which a floral induced plant is grafted to a noninduced plant, indicate that floral induction can be transferred. In such experiments, a noninduced plant held under conditions of noninduction to flowering, will undergo floral induction when grafted to a floral induced plant. To some investigators, this indicated that a single hormone is responsible for floral induction, thus leading to the *florigen* concept mentioned earlier. No such hormone has been isolated and purified.

In some plants, it was found that if the seed was not subjected to a period of cold temperature, the plant which developed from that seed would not flower or would flower at a much later time than usual. Such a requirement for cold treatment is known as *vernalization*. In some plants, vernalization alone suffices for floral induction, but in others subsequent exposure to inductive photoperiods is required. Melcher, in 1937, demonstrated that vernalized plants could transfer that condition to a nonvernalized plant through a graft. He suggested a hormone, which he named *vernalin*, might be involved. As with florigen, no such hormone has been isolated.

Considering then, that under certain conditions a plant will be induced to flower and under other conditions it will not, how can we explain such behavior? Perhaps two possible explanations occur to you; floral evocation may be brought about by the increase in concentration of one or more inducible substances, or by the decrease in concentration of an inhibitor substance, or perhaps both events occur simultaneously. What kind of evidence do we have to support one or the other of these hypotheses? The simplest way to describe it leads to a contradiction. In those plants having a phytochrome involvement in flowering, evidence seems to suggest that at least two separate photoperiodic processes, one favored by high and the other by low $Pfr_{730}$ levels, are required for flowering in both SDP and LDP. This also suggests two photoperiodic stimuli to floral evocation; thus, the case for a single florigen cannot be supported presently. But changes in concentration of ABA have also been noted furing floral evocation. As we mentioned earlier, ABA is a known inhibitor of many growth processes, and may also inhibit the flowering response.

Responses of different plants to a variety of chemical substances have yielded variable results. In general, it has been found that in short-day plants, flowering in long days or in subthreshold conditions can be evoked by increased concentration of sugars, GA, zeatin, kinetin, adenine, uricil, eosin, ABA, CCC, maleic hydrazine, TIBA, nicotine sulphate, or certain mineral salts. Among LDP, flowering in short days or subthreshold conditions has been evoked by application of auxins, GA, furfuryl alcohol, kinetin, vitamin E, and sucrose.

Evidence suggests that in short-day plants, long-day plants and in plants with dual photoperiod requirements, the day length, or length of the dark period, controls the production in leaves of one or more, probably more, transmissible substances with a controlling influence on flower evocation. Which of these actually limits evocation may depend on the plant and on the conditions under which it is growing. In some plants, the requirement for a positive stimulus may be dominant, in others the inhibitory effects may be dominant and limiting. There is no need to assume that the photoperiodic inhibitor specifically inhibits flower evocation, or that there is only one such substance. A compound which may inhibit evocation in one plant or under one set of conditions, may evoke flowering in another plant or under a different set of conditions. Thus, at this time we cannot state positively what causes a plant to cease vegetative growth at a stem tip, and instead produce a flower.

Most plants go through a period during which growth apparently ceases. These periods have been referred to as *rest*, *quiescence*, or *dormancy*. The three terms refer to somewhat different processes to those interested in seed biology and to those interested in fruit trees. It will add little or nothing to our understanding to argue about nuance, rather we should be interested in the biology of the matter. Dormancy can generally be considered as a developmental intermission occurring in buds, seeds, or the vegetative plant body.

Many temperate woody plants produce a specific type of bud called the *overwintering bud*. Whether or not a specialized overwintering bud is produced, there is normally a cessation of apical growth. Such dormancy may be under photoperiodic control, being induced by short days. The importance of the dark period in this cycle is illustrated by the fact that many buds can be brought out of dormancy by subjecting them to a small amount of light in the middle of their dark period. Buds can also be brought out of dormancy by subjecting them to long days, low temperature, or both.

Another explanation of bud dormancy assumes a hormonal basis. Growth inhibitors, particularly ABA, have been shown to increase as dormancy approaches, while growth stimulators such as auxins, gibberellins, and cytokinins decrease. The concentration of inhibitor declines throughout the dormant period, reaching its lowest level as dormancy passes. Growth stimulators such as auxin, gibberellin, and cytokinin increase as dormancy is passed. The actual significance of these growth stimulators in the breaking of dormancy is largely unknown. The exogenous application of auxin will not break dormancy, while exogenous application of gibberellin or kinetin will break dormancy, although often both must be applied in order to accomplish the action. Ethylene has also been shown to break bud dormancy in some instances. Some workers have suggested that bud dormancy and the breaking of dormancy represented a shift in respiratory efficiency, others have suggested that changes in the permeability of the plasmalemma to water (they become essentially impermeable) promotes dormancy, while still others suggest that the mechanisms for protein synthesis are lost from the shoot tip, thus promoting a cessation of growth.

One aspect of bud dormancy is the inhibition of lateral bud development through apical control. The apex of a shoot is able to communicate a distance of many centimeters thereby regulating not only the onset of growth but also the growth or the actual form of a lateral organ. This phenomenon is called *apical dominance*. The effect of apical dominance varies considerably among different plants, from being essentially absent in some to almost totally absent in others.

What causes apical dominance? How does the terminal bud exert its influence on the laterals? One way to investigate this aspect would be to decapitate the apex and study the changes that result.

On a cytological and biochemical level, an aspect of shoot tip decapitation is increased cell division and an increase in DNA synthesis. There is also a proliferation of ER, ribosomes, and storage material. A fibrous or membranous material appears in the vacuoles; none is seen in the vacuoles of inhibited buds. From an anatomical view, it was once considered that lateral bud inhibition was due to inhibition of vascular connections between the lateral bud and the stem. But release from apical dominance does not lead to increased vascular connections. One could also test the effect of various factors on the inhibited bud. Hormone effects might be one group of factors to be considered. Addition of exogenous cytokinins has been shown to increase cell division, increase DNA synthesis or increase thymidine uptake or both, increase the synthesis and methylation of fatty acids, and increase nitrogen availability in the lateral bud. In other words, it seems to mimic the effect of decapitation. However, cytokinins do not stimulate the growth of excised lateral buds, thus suggesting the cytokinins must interact with a stem factor.

Auxins have been implicated as inhibitory factors since the early work of Thimann and Skoog in 1933. If a ring of TIBA (an inhibitor of auxin transport) is placed around the stem just below the apex, the lateral buds will grow. It is postulated that auxin may act indirectly either by interacting with some factor from the roots (such as cytokinin) or by mobilizing essential substances away from the lateral bud towards the apex. It has been demonstrated that the apex is the site of nutrient accumulation in intact plants. If the apex is removed, the lateral buds become new sites for accumulation. It is possible to show a direct correlation between the IAA concentration at a particular node and the intensity of inhibition.

The results from exogenous application of gibberellins is variable. It may be that GA is an endogenous regulator whose effects are modulated by levels of other inhibitory compounds. Gibberellins might be involved in lateral shoot growth but may not be triggering the outgrowth of buds. In other words, GA may not induce growth of the lateral bud but may stimulate that growth after it has begun.

Abscisic acid has been implicated in lateral bud dormancy. Exogenous application of ABA has been shown to inhibit stem and bud growth. In stimulated lateral buds, ABA levels are less than one-tenth that of inhibited buds.

The effect of ethylene on apical dominance is not clear. Ethylene production is greatest at the nodes, and this production has been shown to be stimulated by IAA. Therefore, it has been suggested that apical dominance may be more directly due to ethylene, not to IAA per se. Bud growth has been shown to be inhibited by levels of ethylene, but removal of that gas did not cause a resumption of growth. Ethylene production from the stem decreased after decapitation, but production from the bud remained the same. Bud outgrowth induced by 0.1 ml kinetin was not inhibited by ethylene, even at concentrations as high as 1000 $\mu$liter/liter.

Environmental factors were also shown to affect apical dominance. As nitrogen concentration in the environment increased, there was a greater lateral bud growth. Increased availability of potassium also resulted in lateral bud outgrowth. It was shown that the greatest outgrowth of lateral buds occurred under conditions of decreased water stress. As the rate of transpiration increased, resulting in increased water stress, apical dominance was enhanced. Light was also shown to be a factor. As light intensity increased, apical dominance decreased. The effect of low light intensity on increased apical dominance is probably a secondary effect. Low light intensity will result in decreased carbohydrate production, thus limiting bud growth.

The dormancy of seeds has been investigated in somewhat more detail than that of buds. The environmental controls of seed dormancy are quite similar to those for bud dormancy, including light, temperature, growth substances, and mechanical treatments which improve diffusion of water and oxygen into the seed.

Dormancy of a seed may be correlated closely to the development of the rather hard, sometimes relatively thick, and essentially impervious seed coat. The breaking of dormancy in seeds is associated with some method by which the permeability of the seed coat is improved. This breaking of the seed coat barrier is called *scarification.* Scarification may be brought about in nature by several means including alternating warm and cold temperatures; movement of the seeds, by some means, across sand or rock particles; passage through the intestinal tract of animals; fire; microbial or fungal digestion.

In other instances, dormancy is apparently not coat imposed. This can be readily shown by excising the embryo from newly shed seed and subjecting it to cold treatment. If dormancy is a feature of the seed coat, the embryo might be expected to germinate following the cold treatment. This, in fact, does occur in many species, but in others, a period of time must pass before chilling will initiate germination. This period is called *after ripening.* Thus, some seeds have a coat-imposed dormancy, others an embryo-imposed dormancy, some have both.

Hormonal influences on the dormancy of seeds are essentially the same as those postulated for bud dormancy. It seems well established that seed dormancy is related intimately to the presence of the growth inhibitor ABA, while breaking of dormancy involves the growth stimulators gibberellin, cytokinin, ethylene, or some interaction between them.

## *Differentiation*

During the development of a plant from a single-celled zygote, various processes of differentiation take place. Cells, which are initially very similar to each other, become recognizably different from the cells that produced them and from neighboring cells having the same origin as themselves. It is these changes that constitute differentiation. As stated at the beginning of this chapter, it is not a random, haphazard type of process, but rather takes place in an orderly fashion. The simple enlargement of a cell represents a type of differentiation. Some cells never differentiate beyond this point while others, such as a xylem element, may pass through a series of differentiation events (Figure 15.12). What actually controls this orderly progression of development? The mechanisms are little understood, but as has been mentioned many times, they are influenced obviously by two major factors, genetics and environment. The interaction of these two factors are illustrated in Figure 15.13. A change in IV will result obviously in a change in V, and differences in I will result in differences in II, III, IV, and V. What a cell becomes is determined by its genetic potential, or competence to react, and the determination of micro- and macroenvironmental influences. We have previously discussed this aspect of toti-

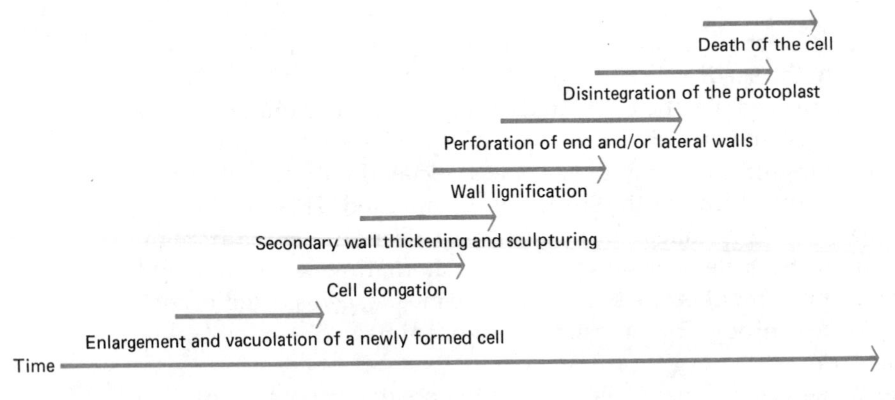


*Figure 15.12*
The series of events leading to certain maturity of a vessel segment. [From R. A. Popham, *Laboratory Manual for Plant Anatomy*, C. V. Mosby Co., St. Louis, 1966, p. 16.]

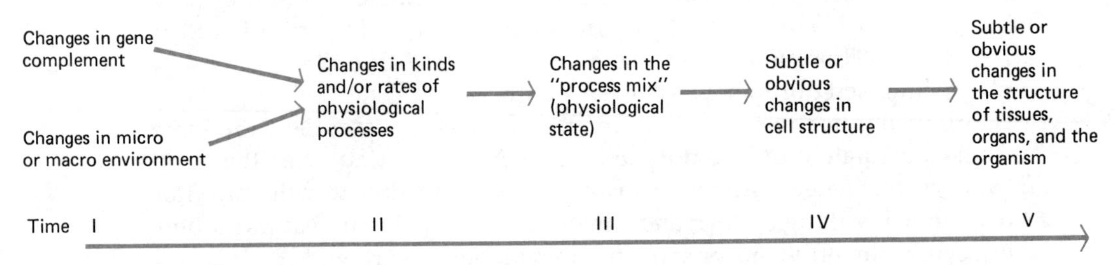


*Figure 15.13*
The influence of genetics and environment on the orderly processes of differentiation.

potency of plant cells and the effect of environment on the realization of this potential.

As you can probably deduce from the foregoing discussion, differentiation is a continuing process, thus a cell cannot "redifferentiate," "dedifferentiate," or undergo "regressive differentiation." The exact "physiological mix" can never be repeated nor be produced in reverse. The process of differentiation may reach a stage that is relatively long lasting for a particular cell or cells. This state is referred to as the stage of *physiological maturity* for that cell, which implies that, under the appropriate conditions, differentiation of that cell may continue. Maturity is not *certain* unless the cell is dead.

## Factors Affecting Differentiation

The genetic complement of cells may vary in cells of different parts of the same plant. For example, investigators have found that in cocklebur *(Xanthium)*, the diploid chromosome number of the embryo and shoot is ten but of the root, only eight. As a general rule, cells of meristematic regions, such as the apical meristem, pericycle, vascular and cork cambium, are usually diploid, while cells of the cortex may be tetraploid. In spinach, however, it was found that the cells of the cortex varied in ploidy according to their position in relation to the root tip. Cells at or near the root tip were primarily 2N, those 400 $\mu$m from the tip were generally 4N, those 500 to 600 $\mu$m from the tip were primarily 8N, while those 900 $\mu$m from the tip were primarily 16N. In tobacco, pith cells of the shoot apex were analyzed, and it was found that from 0 to 3.5 cm from the apex, 100 percent of the cells were 2N; 3.5 to 10.5 cm from the apex, 50 percent were 2N and 50 percent were 4N; from 15.5 to 20.5 cm from the apex, 70 percent of the cells were 4N, 16 percent were 8N, 9 percent were 2N, and 5 percent were aneuploid. In cycads, cells of the root tip contain 22 chromosomes. Cells away from the root tip divide by somatic reduction division producing cells with 11 chromosomes. These, in turn, divide producing some cells with six chromosomes and some with five chromosomes, which divide producing cells with two or cells with three chromosomes.

Simply because chromosome complements differ in different cells does not mean necessarily that these cells will be different morphologically. But the fact that certain genes may be present in a "double dose" might certainly result in increased concentrations of a particular enzyme which could, in fact, affect the morphology of a cell.

Barring mutation, the same genes will be present in every cell, with the possibility of the existence of ploidy; thus, differentiation cannot be under *direct* genetic control. It is obvious that a complex switching mechanism must exist for turning particular genes on or off at specific times. It is very possible that once a particular switch is thrown, its effect will throw another switch, etc., thus producing a series of physiological events resulting in a specific morphological expression. The switching mechanism may be an internal microenvironmental factor, such as a hormone, vitamin, histone, or some other relatively small molecule, or it may be some external environmental stimulus such as light, temperature, moisture, etc.

As we have just stated, plant growth and development is controlled by environmental influences, both internal and external. At an earlier time the importance of cell position was emphasized as the controlling factor in differentiation. Yet we know today that by simply altering an external environmental factor, morphological changes can be induced; thus, position cannot be the only factor involved. Through the work of Steward and his co-workers at Cornell, plant tissue culture became a feasible process. These techniques indicated that single cells were capable of developing into embryoids and eventually into entire plants. Not only could individual cells of the embryo produce entire plants, but so-called physiologically mature cells could act as zygotes. Thus, cells reveal apparently their totipotency only in isolation and not when part of tissue. Thus, the limitations on the capacity for development must be imposed by the environment, one aspect of which would be the position of the cell within the organism.

## *Some Specific Examples of Control of Differentiation*

The differentiation of sclereids may involve various factors. Position may be a factor. In some tissues, sclereids differentiate only near the surface; in others, only along the margin. Other experiments implicate osmotic pressure, while others implicate a *nearest neighbor* effect, with the differentiation of one cell as a sclereid stimulating an adjacent cell to do likewise. Hormonal factors may also be involved. There is some evidence that IAA may inhibit sclereid formation.

The differentiation of fibers presents an interesting phenomenon. It has been shown that in climbing tendrils, the fibers on the side of the stem adjacent to the supportive structure were not only more numerous, but they were also longer and thicker walled. The same effect could be produced, however, by simply rubbing one side of a tendril; thus, some type of thigmotropic influence was involved. It has also been found that treating a plant with gibberellin can induce the formation of fibers which are longer, wider, and thicker walled. This suggests the possibility that the thigmotropic stimulus may result in an increase in gibberellin in that portion of the plant.

Several environmental factors appear to affect epidermal cells. In some plants, such as *Elodea*, a cuticle forms over root cells only when in the light. If grown in the dark, no cuticle develops. In addition, dark-grown plants develop root hairs. When *Elodea* is grown in water, under the normal conditions of light and darkness, no root hairs develop. If grown in soil or mud, or as mentioned above, in the dark, root hairs do develop. This is thought to be associated with the absence of a cuticle in the dark.

In some plants, the root hairs develop as outgrowths of specialized epidermal cells called trichoblasts. In the aquatic plant *Hydrocharis*, it was found that these cells were polyploid, the amount of DNA in the nucleus increasing with distance from the root tip, up to about eight times that of hairless epidermal cells. Nucleolar volume was also noted to increase in trichoblasts of certain plants. In fact, even in plants not forming distinctive trichoblasts, there was an increase in nucleolar volume and a greater concentration of protein bodies in cells that would form hairs when compared with those that would not. Enzyme differences were also found in some trichoblasts, primarily in the levels of cytochrome oxidase and acid phosphatase. This difference, however, was found to vary according to distance away from the root tip, with no initial difference between the hair and hairless initials but with the hair initials being active 200 to 300 $\mu m$ behind the tip while the hairless initials were not. This suggests that the hairless initials lose, or are in someway inhibited in their ability to form, root hairs. Some trichoblasts occur in specific longitudinal rows along the root. This suggests a relationship with underlying tissues. In the white mustard, the trichoblasts are aligned between the underlying hypodermal cells, thus being associated with the intercellular space. There is some evidence that an inhibitor to root hair development may be transported from the stele through the cortical cells. Thus, those epidermal cells in direct contact with the hypodermal cells would be most inhibited. Epidermal hairs on aerial portions of the plant may form more prolifically during periods of water stress, although this is not the complete explanation as some aquatic plants have epidermal hairs on their leaves and stems as well.

There is evidence that the formation of xylem vessel members is controlled by a single gene. This is indicated by the existence of mutants in which the large metaxylem elements are lacking. Experimental evidence indicates that auxin concentration is of extreme importance in the differentiation of xylem elements and that there is an interaction between auxin and kinetin. Working with intact plants, it was shown that if a small portion of stem vein was removed, pith cells would differentiate into new xylem and phloem elements forming a new vein. If the leaves above the cut were removed, differentiation was greatly inhibited or lacking. Working with callus tissue in culture, if one cut a V-shaped notch in the callus and placed a shoot tip into that notch, a few days later cell differentiation occurs in the callus tissue about two to three cm below the notch. In 20 to 30 days, xylem will be differentiated, 50 days later both xylem and phloem will have differentiated, and by 60 days some cambial cells will have differentiated between the xylem and phloem. If the shoot tip is replaced by a block of agar impregnated with auxin and sucrose, similar results are obtained. When the auxin concentration was held constant but the sugar varied, it was found that with one to two percent sucrose, xylem only was formed; at two and one-half to three and one-half percent sucrose, xylem, phloem, and cambium was formed; at four percent sucrose, only phloem was formed.

In intact plants, phloem regeneration seemed to occur under somewhat different conditions than in the previous example of callus tissue. This is probably due to the fact that various gradients are already established in the stems of intact plants. Evidence indicated that auxin was the limiting factor in phloem regeneration, while sucrose had little or no affect (Table 15.8).

The logical consequence of differentiation is aging and eventual death. Thus, *senescence* can be thought of as simply a stage or level of differentiation. Senescence may be associated with individual cells, tissues, organs, or whole plants. We are all familiar with the dropping off of leaves, flower petals, and fruits, a process called *abscission.* In biennial herbs, the whole aboveground vegetative plant dies the first year, but the roots remain alive and from them the flowering plant is produced the second year, following which the whole plant dies. In perennial herbs, the aboveground portion of the plant dies each year. In perennial deciduous woody plants, the leaves die, the flowers die, the fruit may fall off, but the whole aboveground plant does not die each year. But when you consider that most of

*Table 15.8*
**THE EFFECTS OF IAA AND SUCROSE ON PHLOEM REGENERATION IN PLANTS OF *COLEUS* WITHOUT SHOOT ORGANS**

| TREATMENT | MEAN NUMBER OF STRANDS REGENERATED IN 5 DAYS |
|---|---|
| Controls (intact plants) | 34.6 ± 2.8 |
| All shoot organs removed | |
| Plain lanolin | 9.1 ± 1.7 |
| 0.1% IAA in lanolin | 42.8 ± 1.4 |
| 1% IAA in lanolin | 64.1 ± 11.4 |
| Water | 13.2 ± 5.2 |
| Water + 0.025% sulphanilamide | 11.2 ± 0.9 |
| 20 g/liter sucrose + 0.025% sulphanilamide | 14.6 ± 3.7 |

*Source:* From LaMotte and Jacobs, "The Role of Auxin in Phloem Regeneration in *Coleus Internodes*," *Devel. Biol.*, 1963, 8:80–98.

the aboveground portion is wood, that wood is xylem tissue, and that xylem elements are nonliving, then most of a woody plant is actually composed of nonliving cells. The process of senescence is thus very important but unfortunately little understood.

The patterns of senescence vary greatly. In the differentiation of a xylem element, the wall becomes thickened, the protoplast undergoes dissolution, and then the end walls are digested, leaving the side walls intact. In other cells, only the protoplast undergoes dissolution, while in still others, the entire cell may be digested. What controls the specificity of these patterns is quite unknown.

Studies seem to indicate that as senescence of a cell occurs, there is an increase in the respiration rate, an increase in hydrolytic and phosphorolytic enzymes, an accompanied decrease in synthestases and RNA synthesis, and a shift in the concentration and kinds of hormones present. If a tissue region or organ is undergoing senescence, the hydrolytic breakdown products are exported out of the region and translocated into the younger regions of the plants.

The possible causes of senescence are several, none of which seem to cover adequately all situations. Molisch (1928) suggested that senescence of vegetative parts of the plants occurred as a result of nutrient depletion during flower and fruit formation. He noted that some plants die very shortly after fruit production, but if the flowers of these plants are removed, senescence could be delayed. In the case of plants that flower only once, termed *monocarpic* plants, this observation would seem to be plausible. In fact, some plants such as the century plant *(Agave)* or bamboo may live vegetatively for many years, then flower, produce fruits, and die. But many other plants flower and produce fruit every year, and certain parts of the plant can senesce without the death of the entire plant. Obviously, the explanation of Molisch cannot apply to plants in general.

Leopold (1959) noted that the presence of young flowers in soybeans resulted in the senescence of the leaves, even though these flowers were not large enough to cause a depletion of nutrients. He also noted that the ripening of fruits could also cause senescence, even though translocation of nutrients into the fruits had essentially ceased. He found that senescence occurred in plants producing staminate flowers only, and the removal of either the staminate or pistillate flowers could delay senescence. He suggested that perhaps the reproductive organs produced some inhibitor substance that is translocated out of these parts into the vegetative parts where they induce senescence. He also found that in a seedling, removal of the epicotyl and cotyledonary bud delayed senescence of the cotyledons. Thus, both the vegetative and reproductive buds can be associated with senescence in other parts of the plant.

A British biochemist, M. Dixon, suggested that senescence is due to chemical changes which compound the normal instability of enzymes. It has been noted that assimilatory mechanisms decrease at the onset of senescence. This may be due to the presence of inhibitors or could be due to the cells loss of ability to produce these enzymes. It is known, for example, that certain hormones can affect the gene-directed synthesis of mRNA. The senescence-delaying effect of cytokinin or auxin might be due to their stimulation of RNA synthesis, while the effect of ABA in promoting senescence may be due to its effect of inhibiting RNA synthesis. Even this explanation has its problems, as either auxin, gibberellin, or cytokinin may retard senescence in some plants, but not in others, while in still others several of these growth factors are required before senescence will be delayed. On the other hand, in some plants ABA seems to promote senescence, in others, ethylene; and the effects of these two substances may differ in different organs of the same plant.

Following senescence in certain organs, they simply fall off. This process of abscission is most notable in leaves, flowers, and fruits. The abscission of these plant parts does not occur simply because the parts are dead, but rather a specific zone of cell division occurs across the base of the organ. Abscission usually occurs in this zone, although in some plants, abscission occurs without the production of this so-called *abscission zone*. During abscission, hydrolytic enzymes are produced that catalyze the digestion of the middle lamellae between these cells, and the cells simply fall apart, leaving only the vascular system intact (Figure 15.14). Wind, or some other mechanical force, causes this thin strand to break normally and the organ falls. Preceding, or soon following, this event, cell divisions occur in the stump producing a corky layer which covers the wound, preventing water loss and entry of pathogenic organisms.

The physiological aspects of leaf abscission seem to be correlated with both external environmental factors and hormonal factors. The affect of external factors are difficult to correlate with leaf abscission. There is some evidence that a photoperiodic effect might be involved, but abscission has been shown to occur under conditions of high light intensity,

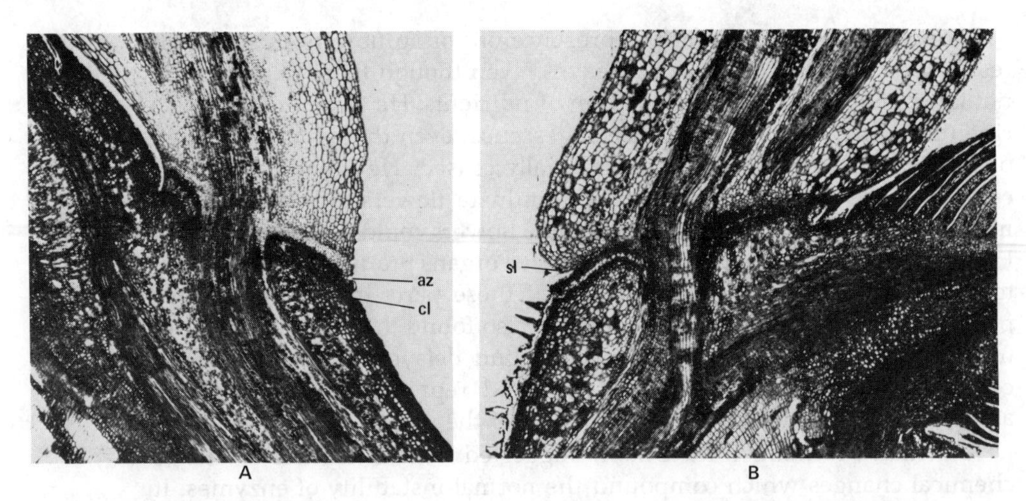


*Figure 15.14*
Leaf abscission. *A. Ulmus americana.* Note the abscission zone (*az*) and the suberized layer beneath it (*cl*). *B. Acer* sp. Note the development of a separation layer (*sl*) just above the abscission zone.

darkness, dryness, wetness, high temperature, low temperature, removal of the blade, zinc, nitrogen, or calcium deficiencies, disease, high ethylene concentrations, or high ABA concentrations. The application of exogenous auxin may either retard or stimulate leaf abscission. If the auxin is applied distal to the abscission zone (such as on the leaf blade) it tends to retard abscission, if applied proximal to the abscission zone (such as on the shoot above the leaf) it tends to accelerate abscission. If the auxin concentration is high enough, it inhibits abscission regardless of direction. The role of auxin in abscission would seem to be that of maintaining ongoing physiological functions. It may also act by mobilizing nutrients.

Cytokinins are considered to be the most effective retarders of plant senescence. They have been shown to retard chlorophyll loss, prolong peptidase, ribonuclease, and deoxyribonuclease activity, retard the decrease in auxin levels, and suppress the usual increase in oxygen levels in senescent leaves. Synergistic effects between cytokinins and auxins are common.

Although there is considerable evidence that gibberellin tends to promote fruit set, thus tending to prevent abscission of fruit, there is little evidence that GA plays a major role in retarding leaf abscission. In fact, most evidence seems to point to an abscission-promoting effect of GA.

Most current evidence seems to indicate that ethylene promotes leaf abscission by initiating cell separation in the abscission zone through enzyme induction. But this ability depends largely upon the sensitivity of this tissue to ethylene. Sensitivity seems to be based primarily upon auxin concentrations in the abscission zone, if it is high, ethylene will have no effect, if it is low, ethylene will promote abscission. Ethylene apparently acts in the intact leaf by altering the synthesis, conjugation, or destruction of auxin in the blade. Ethylene also appears to reduce auxin transport in the venous tissues of the blade, thus reducing the supply of auxin to the petiole. When one considers that the abscission zone develops across the base of the petiole, some combination of reduced auxin synthesis, enhanced auxin destruction, or reduced auxin transports would all contribute to a reduction of auxin efflux from the blade to the petiole, thus promoting leaf abscission.

Another compound that seems to be associated with the promotion of leaf abscission is ABA. Abscisic acid was found to be 20 times as high in strawberry leaves collected in September as in those collected in June. Abscisic acid is a less effective defoliant of young leaves than of old leaves, but in warmer climates, it is relatively effective. Chlorophyll degradation in leaves is accelerated by ABA, and RNA and protein syntheses are reduced. Thus, ABA acts both as a senescence factor and a promotor of abscission. Abscisic acid apparently promotes abscission through stimulation of pectinase and cellulase synthesis. It may also act by blocking assimilatory enzyme synthesis.

## SUMMARY

1. The concept of plant growth has many aspects, including an increase in length, increase in breadth, increase in numbers of cells, increase in size of cells, and increase in weight.
2. Differentiation is an orderly series of changes and may occur with or without growth.
3. Growth is associated normally with regions of the plant called meristems or cambia.
4. Important events associated with growth include cell division, water uptake, and assimilation.
5. Plant growth is influenced by a number of factors, both environmental and genetic.
6. Environmental factors influencing plant growth may be either or both external or internal. External factors include the affects of light, temperature, gravity, mechanical stimulation, sound, electrical and magnetic stimulation, tidal and lunar influences, application of exogenous chemicals, and others. Internal or endogenous factors include hormones, enzymes, vitamins and pigments.
7. The major classes of hormones include auxins, gibberellins, cytokinins, abscisic acid and ethylene.
8. In general, auxins, gibberellins and cytokinins are growth stimulators, abscisic acid is a growth inhibitor, but these effects depend upon the plant, the concentration of the hormone, and the environment.

9. An important pigment is phytochrome. It has two forms, a red-absorbing form ($Pr_{660}$) and a far-red–absorbing form ($Pfr_{730}$).
10. Light, or absence of it, has a great variety of influences on plant growth. Light has its effect on growth in both a quantitative and a qualitative fashion. Light is involved in flowering (or inhibition of it), phototropism, photoreactivation, the movement of chloroplasts, replication of chloroplasts, seed germination, spore germination, anthocyanin production, plumular hook opening, cytoplasmic streaming, temporary stimulation of grass coleoptile elongation, polar differentiation of rhizoids in germinating spores, elongation of rhizoids, conversion of fern protonema into a prothallium, initiation, growth and differentiation in dicot leaves, formation of palisade layer in some leaves, stimulation of opening and closing in *Mimosa* leaflets, the unfolding of leaves in grass seedlings, and the inhibition of stem, hypocotyl and mesocotyl elongation.
11. Growth movements may be tropisms (direction of movement directly toward or away from source of stimulus) or nasties (movement not directed).
12. Examples of tropic responses include phototropism, geotropism, thigmotropism or haptotropism, chemotropism, traumatotropism, and galvanotropism.
13. Nastic movements include seismonasty and thigmonasty.
14. Some growth phenomena show a definite repeatable pattern. Such cyclic patterns are termed natural rhythms and may be diurnal (daily), semilunar, tidal, annual, or other.
15. The photoperiodic response seems to be mediated through the phytochromes. The phytochrome system has two pigments, a red-absorbing form ($Pr_{660}$) and a far-red–absorbing form ($Pfr_{730}$). In the dark, $Pr_{660}$ predominates. During the day, $Pr_{660}$ undergoes a slow conversion to $Pfr_{730}$. The conversion of $Pfr_{730}$ to $Pr_{660}$ in the dark can be interrupted by a flash of far-red light.
16. Dormancy is an important feature of buds and seeds. It may be dependent upon either environmental or hormonal aspects or both.
17. Differentiation of cells, tissues, and organs is based upon both genetics and environment. Cells are totipotent, not predestinated, thus maturity of a cell is not certain unless that cell is dead.
18. Polyploidy may be an important genetic factor affecting growth.
19. The senescence (old age) of cells is generally indicated by an increased respiration rate, increased hydrolytic and phosphorolytic enzymes, a decrease in synthetases and RNA synthesis, and a shift in concentration and kinds of hormones.

# Chapter 16

## PLANT REPRODUCTION

Reproduction is considered to be a basic and essential aspect of survival for a species. As mentioned earlier, a method for reproduction would have evolved very early in the development of living organisms. As you recall, reproduction involves the ability to replicate the genetic material of the organism (the DNA and RNA), and then transmit that material by some mechanism into new organisms. Thus, reproduction results in the multiplication of organisms. It is during the replication and transmission process that introduction of new genetic characters or new combinations can occur, thus introducing variation into the offspring. As you recall from previous discussion, the synapsis of chromosomes during meiosis I and the subsequent recombination which occurs during sexual reproduction represents a major source of such variation in a species.

Basically, we will discuss reproduction in plants as occurring in three general ways: vegetative, in which an ordinary vegetative cell or groups of vegetative cells produce a new organism similar to the one from which it was propagated; asexual, in which a specialized cell or group of specialized cells produce a new organism similar to the one from which it was propagated; and sexual, in which a series of events occur leading to fusion of gametes and the formation of a new organism having new combinations of genetic material. The first two mechanisms for plant reproduction are frequently classed together by some botanists into the general category of asexual reproduction. The offspring produced by asexual mechanisms from a single organism, being in most cases genetically similar to each other, and asexually producing more organisms like themselves, represent a *clone*. We will not attempt to discuss all of the variability encountered in reproductive mechanisms of plants, but will attempt to discuss some of the more common or more important of these.

## VEGETATIVE REPRODUCTION

As you recall from previous discussion, vegetative reproduction is the principal and, in many cases, the only means of reproduction in the nonvascular plants, especially in the Monera. But even in the flowering plants, vegetative reproduction is a common and important method of propagation and occurs naturally or is often aided by man. In some flowering plants, vegetative reproduction is the only means of propagation.

In the Monera, *fission* was the principal means of reproduction, consisting merely of the splitting of the cells by some means (Figure 7.2). In colonial or filamentous organisms, and in true multicellular organisms as well, *fragmentation*, the simple breaking apart or breaking off of pieces of the plant body is common. *Mitosis* and *cytokinesis* is common to all eukaryotic organisms, either in multiplication of the cell or in multiplication of the organism. In multinucleate organisms, *budding* may be an important mechanism. It occurs simply by a segregation of protoplasm by an invag-

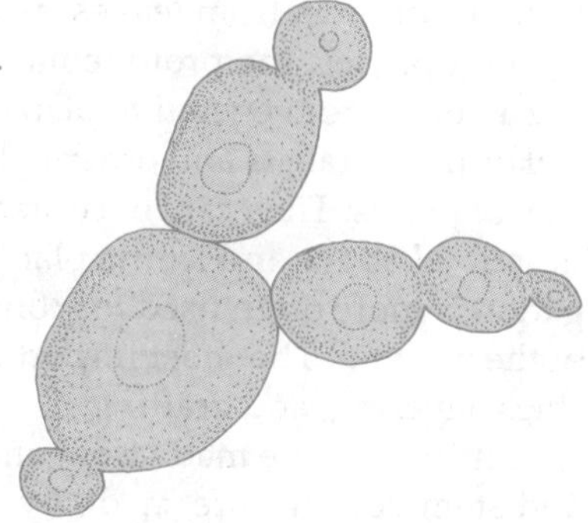
*Figure 16.1*
Budding in yeast.

ination of the plasma membrane and wall, and the eventual separation of the bud from the parent organism (Figure 16.1). You recall that within the Bryophyta, budlike structures called *gemmae* are produced. In the lichens, clusters of fungal hyphae and algal cells form reproductive bodies called *soredia.*

In the flowering plants especially, vegetative propagation may occur from any of the three principal organs. *Roots* may develop adventitious buds and produce new shoots. Leaves may produce adventitious roots and buds, either from their tip, margin, or base. Stems, both aerial and underground, form adventitious buds and roots. Thus new plants may be propagated from upright portions of aerial stems, from aboveground horizontal stems *(runners),* and from parts of underground stems *(rhizomes, tubers, corms,* and *bulbs).* A common example of propagation from an aerial stem occurs when a long slender stem droops and bends so as its tip touches the ground. When this occurs, adventitious roots may develop from a node close to the stem tip, with the bud developing into an upright stem. Such a mechanism is termed *tip layering* (Figure 16.2). Tip layering

*Figure 16.2*
Simple, common, or tip layering. After the roots have formed, the stems are cut off below the roots and the new plants set out.

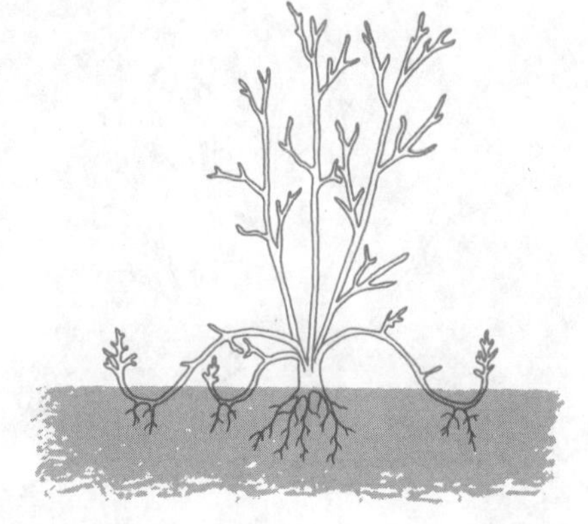

is often common from leaves, as in the walking fern. You will recall that in the Bryophytes, the protonema develops adventitious buds, thus producing a number of vegetative plants from a single protonema.

Horticulturalists and others play a major role in the vegetative propagation of plants. They can, of course, simply make use of the natural methods described above and let the plants do the propagation. But artificial propagation is more often used for convenience and to increase the effectiveness of the method. The most important of these artificial methods include cuttings, layering, and grafting.

Cuttings can be made from any vegetative organ: root, stem or leaf. Root and stem cuttings are made by simply cutting the organ into pieces and placing them in water or a rooting medium such as moist sand. Sometimes such cuttings are treated with a dilute mixture of auxin. Stem cuttings are the type used most commonly, as root cuttings can only be made from those species that form adventitious buds. Leaf cuttings are made by simply immersing the petiole in water or rooting medium.

Layering involves covering a portion of the organ with layers of a moist substance. This is done principally with stems. *Mound layering* utilizes the principal observed in tip layering. A branch is simply placed on the ground and portions of it covered with moist soil (Figure 16.3). After adventitious roots have developed, the branch on either side is simply cut and the new plants dug up and transplanted. In *air layering*, a branch is first cut about half way through and from this cut a second cut is made about three inches upward through the center of the stem. The cut surfaces are then generally powdered lightly with a growth hormone and moist sphagnum moss inserted between the cut surfaces and wrapped entirely around the cut region of the branch. The ball of sphagnum is then wrapped with a sheet of polyethylene and tied top and bottom so as to retain the moisture in the spaghnum (Figure 16.4). The polyethylene does not prevent air from entering, however. After adventitious roots have developed, the branch is cut off and planted.

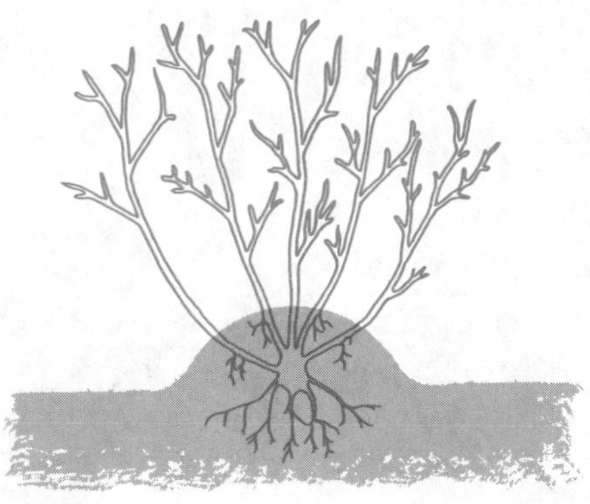

*Figure 16.3*
Mound layering. The plant is pruned back, its base covered with soil, and the shoots develop roots. These shoots are then separated from the parent plant and set out.

*Figure 16.4*
Air-layering. [Photograph by Robert G. Trumbull, III.]

Grafting is simply a modified method of propagation by cuttings. The cutting is called the *scion*, the plant to which it is grafted, the *stock*. Typically the stock produces only the root system and the lower portion of the stem; the scion, the shoot system. In making a graft, the stock and scion are cut in such a way that the two dovetail (Figure 16.5). Grafting has an advantage in that a high yielding but low disease resistant plant can be grafted onto a plant that is more resistant to disease but lower yielding. Thus, one can produce a more disease resistant, high-yielding plant. For a good graft it is usually necessary for the cambia of stem and scion to be in contact. Successful grafts have been made with monocots, which of course do not have a vascular cambium, but such grafts are successful less than 10 percent of the time. Many of the dwarf varieties of fruit trees purchased at your local nursery are the results of grafts. Some plants have the effect of dwarfing others. Thus, a peach scion grafted to an apricot stock will be dwarfed; pear scions grafted to quince will develop into dwarf pear trees.

Another type of grafting is known as budding. In this technique only the bud of the wanted variety is sliced off, being certain to make the cut so some cambium is removed with the bud. The bark of the stock is cut about three inches around the stem, then sliced vertically for several inches, and the bark partly peeled back. The bud is placed in the peeled area, the bark placed back around it and tied with waxed stem so that the bud is held in place (Figure 16.6).

Occasionally, a bud produced in the graft region will have some of its cells contributed by the stock, some by the scion. Thus, the shoot developing from such a bud will have characters of both stock and scion. Such a shoot is known as a *graft chimera*. A chimera is any region that is composed of areas of genetically different cells.

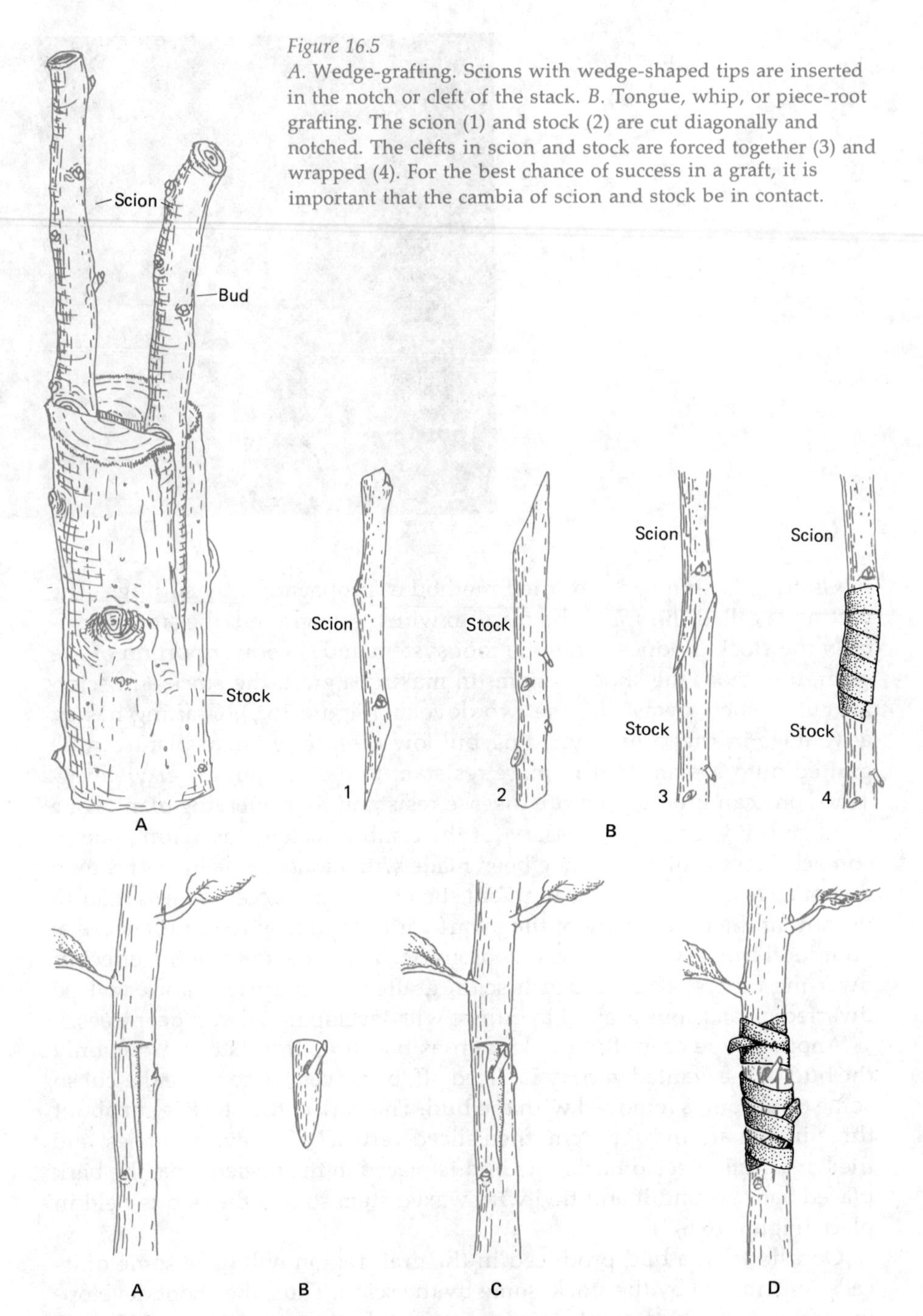


*Figure 16.5*
*A.* Wedge-grafting. Scions with wedge-shaped tips are inserted in the notch or cleft of the stack. *B.* Tongue, whip, or piece-root grafting. The scion (1) and stock (2) are cut diagonally and notched. The clefts in scion and stock are forced together (3) and wrapped (4). For the best chance of success in a graft, it is important that the cambia of scion and stock be in contact.

*Figure 16.6*
Budding. *A.* The growing stock is cut to receive a bud. *B.* A bud with a shield-shaped section of bark. *C.* The bud is inserted in the T-shaped cut in the stock. *D.* The bud is wrapped to protect it and hold it in place.

## ASEXUAL REPRODUCTION

Asexual reproduction involves the development of specialized reproductive bodies by the vegetative plant, with those specialized structures producing other vegetative plants similar to the parent. The most common means of asexual reproduction is by *spores.* Motile spores called *zoospores* are produced in all the algal groups with the exception of the blue-green and red algae. In the blue-green algae, and many of the true algae as well, a variety of nonmotile spores may be produced. The *akinete* has its wall fused with that of the vegetative cell in which it developed, the *hypnospore* is simply a thick-walled akinete, the *aplanospore* has the spore wall separate from that of the vegetative cell. Short sections of filament called *hormogonia* may also be produced (Fig. 7.3). As we mentioned earlier, the *heterocysts* in blue-green algae may function as a spore. In the red algae, spores are produced in a specialized structure called the monosporangium, with the spore being a *monospore.* In the ascomycete fungi, *conidiospores* are produced. Conidia are typically produced in chains from a modified hyphal tip (Figure 16.7). In the chrysophytes, *cysts* or *statocysts* may be formed. Normally, with the exception of hormogonia, all these reproductive bodies look distinctly different from the organism in which they are formed. But in the green algae, reproductive bodies called *autospores* or *autocolonies* are produced. These form, in the case of the former, from single-celled organisms; in the case of the latter, from colonial organisms. They look like miniature copies of the organisms that produced them.

In the plant kingdom, other events occur that are processes of asexual reproduction. Some of these have been discussed earlier. The Italian plant morphologist Flávio Resende describes these events in the following way. He states, "If a non-fertilized gamete germinates and so substitutes the zygote producing a sporophyte generation, we have then what is called *parthenogenesis.* If, however, the same non-fertilized gamete germinates and produces a gametophyte generation the process is designated *apomictosis.*" He continues, "*Apogamy* refers to the process of the substitution of

*Figure 16.7*
Some types of conidiospores bearing conidia. *A. Aspergillus B. Penicillium.*

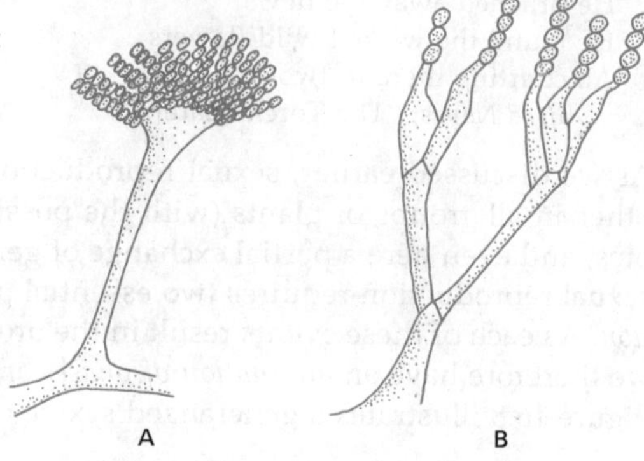

the zygote for a vegetative cell of the gametophyte generation and *apospory* the substitution of the gonospore [Note: gonospore is the product of meiosis] for a vegetative cell of the sporophyte generation." Resende further states, "When the gonospore loses its function as a transition spore (the cell initiating the gametophyte generation), and produces again the sporophyte from which it was derived, the alternation of generations is eliminated from the life cycle." He calls such an event *apogonosis*.

Resende's concepts tend to lose something in translation but are excellent ideas in that they clearly delimit the terms. Most American morphologists would characterize parthenogenesis as simply the germination of an egg cell resulting in production of a new individual, apogamy as the formation of an organism without gametic union, and apospory as the formation of a gametophyte other than from spores. Such concepts do leave room for misconception. Parthenogenesis could be a form of apogamy, and both parthenogenesis and apogamy could be construed as resulting in the formation of either a gametophyte or sporophyte (i.e., organism). I would agree with Resende that the terms should have a more limited concept. Parthenogenesis should be considered as the formation of a *gametophyte* from an unfertilized egg or even from an antherozoid or sperm cell (although the latter would rarely occur). Apogamy should refer to the production of a sporophyte from a *vegetative* cell of the gametophyte (thus eliminating the gamete from consideration). Apospory, according to most plant morphologists, would be the production of a gametophyte from a *vegetative* cell of the sporophyte.

Regardless of how broad or how definitive we wish to be in our concepts of these events, the processes serve as a clear focus to our concept of sporophyte and gametophyte. It obviously matters not whether the sporophyte or gametophyte is haploid, diploid, or polyploid; what *does* matter is the product: the sporophyte, produces spores, the gametophyte, gametes.

## SEXUAL REPRODUCTION

So beautiful, bright and early
He brushed away the dews
He found the wicked wild-flowers
All courting there in twos.
*Alfred Noyes*, "The Torch Bearers"

As we discussed earlier, sexual reproduction occurs in some form or another in all groups of plants (with the possible exception of the prokaryotes, and even here a partial exchange of genetic material can occur). True sexual reproduction requires two essential processes, *meiosis* and *fertilization*. As each of these events result in the production of different products, we therefore have an *alternation of phases or generations* as a consequence. Figure 16.8 illustrates a generalized sexual cycle. As we can see, there are

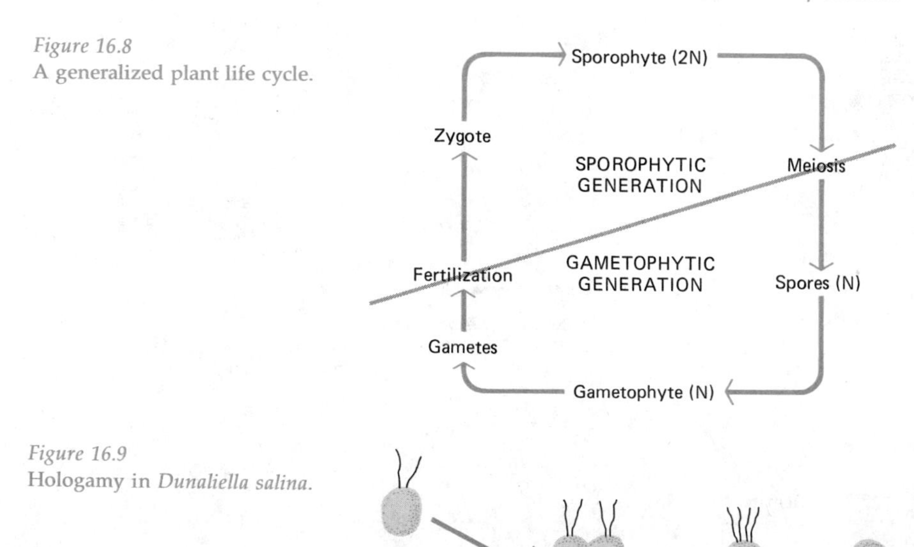


*Figure 16.8*
A generalized plant life cycle.

*Figure 16.9*
Hologamy in *Dunaliella salina.*

two different stages, sporophyte and gametophyte, as one progresses from zygote to zygote. A number of biologists refer to these alternating phases as an alternation of generations, others may refer to them as an alternation of phases. Which is right? Think for a moment about the concept of a generation. You of course are of one generation, your parents are of a different generation. In this sense, a generation is from zygote to zygote. But is consideration of human reproduction applicable to plants? Before we attempt to answer these questions, let us look at a few life cycles of different plants, keeping in mind that all are simply variations on the general scheme shown in Fig. 16.8.

The simplest type of sexual reproduction involving meiosis and fertilization occurs in the green algae (Figure 16.9). This type of life cycle involves the fusion of morphologically similar gametes. So long as a habitat or culture tube contains a single strain of *Dunaliella* (a *Chlamydomonas*-like organism lacking a cell wall and found in brackish water), only asexual or vegetative reproduction occurs. But if a physiologically different mating type is present (generally these mating types are designated + and −), the vegetative cells will begin to clump in a very gametelike fashion (Figure 16.9B), and in fact some of these cells will begin to fuse. The fusion product is a quadriflagellate zygote (Figure 16.9C). After a short time, the zygote becomes quiescent, the flagella are absorbed, and a thick wall forms; the latter character being responsible for the name *zygospore* for this structure (Figure 16.9D). The zygospore nucleus undergoes meiosis forming four

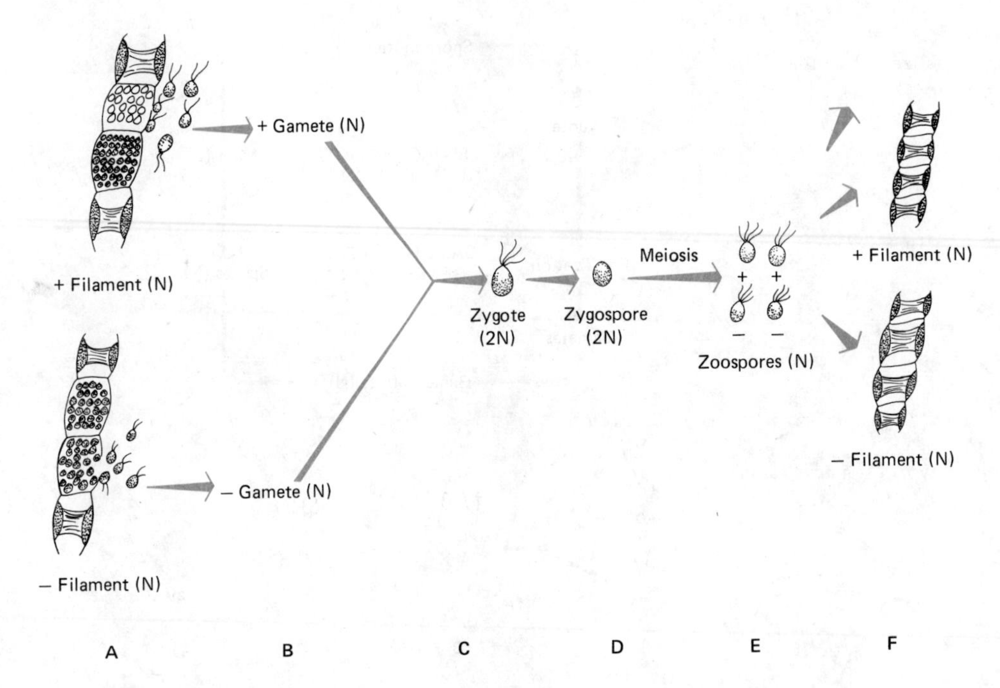


*Figure 16.10*
Isogamy in *Ulothrix* sp.

haploid nuclei, with accompanying reorganization of the zygospore cytoplasm, and resultant *zoospores* liberated upon the breakdown of the zygospore wall. However, unlike the general type of life cycle, these spores do not produce the gametophyte, they *are* the gametophyte. Thus, we have a very simple kind of reproductive cycle, with the products of meiosis actually serving as gametes for the next fertilization cycle. This modified type of isogamous, sexual reproduction is called *hologamy*, as the whole cell fuses. There is no true sporophyte either in this type of life cycle as meiosis occurs within the modified zygote *(zygotic meiosis)*.

In the filamentous green alga *Ulothrix*, a more typical kind of isogamous sexual reproduction occurs (Figure 16.10). Any cell of the vegetative filament, except for the basal *holdfast* (or anchoring cell), can undergo mitotic division and compartmentalization of its contents forming up to 64 small biflagellate gametes (Figure 16.10B). In some species, gametes formed in different cells of the same filament may fuse; in other species, gametes from different filaments fuse. The former species are termed *homothallic*, the latter *heterothallic*. The resulting zygote develops a thick wall forming a zygospore. Eventually, meiosis occurs in the zygospores, and four quadriflagellate zoospores are released (Figure 16.10E). Under favorable conditions, the zoospores will come to rest, anterior end down, on a substrate, and from these, new vegetative filaments arise. In this type of life cycle we clearly have a gamete-producing plant (the gametophyte), but meiosis is still zygotic. Clearly there is an alternation between diploidy and haploidy, with the zygote (or zygospore) being the only diploid structure.

In both *Chlamydomonas* and *Ulothrix*, species are present which produce anisogametes, that is the gametes are both motile but different in size. Meiosis is zygotic and the life cycle, in general, is similar to that described for *Ulothrix*.

Oogamy also exists in the green algae, being present in a number of genera. The genus *Oedogonium* can serve as an example (Figure 16.11). In this genus, small multiflagellate sperm unite with a large nonmotile egg. The resultant zygote forms a thick wall and is frequently referred to as an oospore. Meiosis occurs within the oospore producing four multiflagellate zoospores, each of which develop into a new *Oedogonium* filament. Meiosis is again zygotic.

A slightly different kind of cycle occurs in some species of the green alga *Cladophora*. The isomorphic gametes unite forming a zygote which develops into a diploid free-living plant. Certain cells of this plant undergo meiosis producing quadriflagellate zoospores. The zoospores consist of two physiological types, with one type producing a + mating-type gametophyte, the other a − mating-type gametophyte, but in both cases, a free-living haploid plant is formed. Certain cells of these plants undergo mitosis producing, respectively, + and − biflagellate gametes which fuse forming the zygote, and the cycle begins again. Here we clearly have two different plants, one which produces spores and one which produces gametes. Meiosis does not occur in the zygote as with the species discussed thus far but occurs in the sporophyte (sporic meiosis). But the gametophyte and sporophyte look alike; they have the same size and morphology. Such species of *Cladophora*, and other plants in which sporophyte and gametophyte are similar, are said to exhibit an *alternation of isomorphic generations*.

In the siphonous marine alga *Derbesia*, quite a different situation exists. As in *Cladophora*, the zygote produces a free-living sporophyte on which sporangia develop. Within these sporangia meiosis occurs producing spores. The spores give rise to a free-living gametophyte. The gametophyte mitotically produces anisogamous gametes which fuse forming a zygote. But in this instance, the sporophytes and gametophytes are quite dissimilar. So much so they were identified as different genera until their life cycle became known. *Derbesia* is thus an example of *alternation of heteromorphic generations*.

As we mentioned in Chapter 7, *Chara* is an unusual plant for an alga. Its reproduction reflects this also. The life cycle of *Chara* has not been clearly elucidated. But it appears that meiosis is zygotic, with the simple product of that division producing through germination a filamentous structure called a *protonema*. The protonema produces a shoot initial and a rhizoid initial. The shoot initial consists of a dome-shaped apical cell which cuts

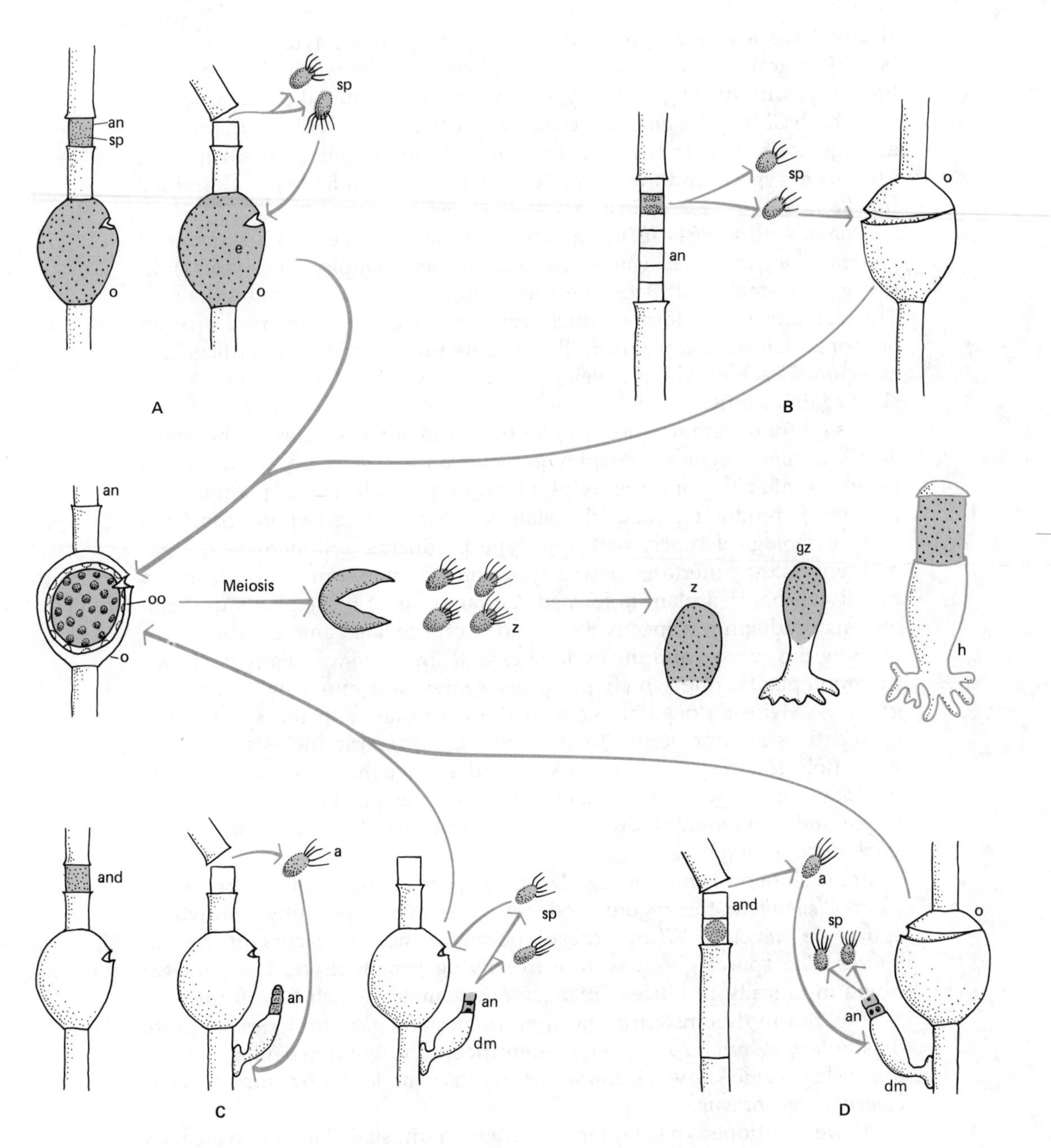


*Figure 16.11*
Oogamy in *Oedogonium* sp.: *A.* homothallic, *B.* heterothallic, *C.* gynandrosporous, *D.* idioandrosporous. *An,* antheridium; *sp,* sperm; *o,* oogonium; *e,* egg; *oo,* oospore (zygote); *z,* zoospore; *gz,* germinating zoospore; *h,* holdfast; *and,* androsporangium; *a,* androspore; *dm,* dwarf male.

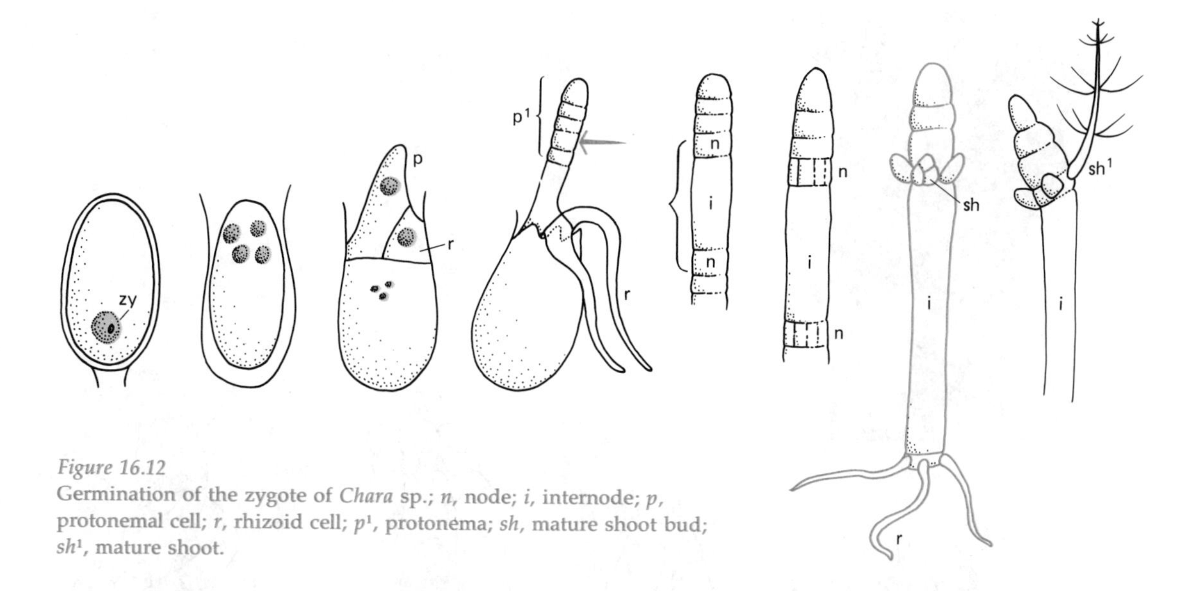


*Figure 16.12*
Germination of the zygote of *Chara* sp.; *n*, node; *i*, internode; *p*, protonemal cell; *r*, rhizoid cell; *p*¹, protonema; *sh*, mature shoot bud; *sh*¹, mature shoot.

off derivatives in a transverse direction. Each of the derivatives divide forming a nodal initial and an internodal initial (Figure 16.12). The internodal initial elongates considerably, while the nodal initial divides in various ways, forming a ring of nodal cells, the whorled branches, and the sex organs.

The sex organs of the Charophyta consist of a more or less spherical *antheridium* and a tubular, somewhat elongated, *oogonium*. These structures are sometimes called the *globule* and *nucule*, respectively. The ontogeny of the sex organs is shown in Figure 16.13.

At maturity, the antheridium consists of a chain of colorless cells called the *antheridial filament*, each of which produces a single biflagellated sperm (antherozoid). The antheridial filaments are borne from cells called the *capitulum*, which in turn are attached to a cell designated the *manubrium*. The manubria, in turn, are attached to epidermis-like cells which form the surface layer of the antheridium. These are termed the *shield cells* (Fig. 16.14).

The mature oogonium consists of five outer, spirally elongated, tubular cells, the *tube cells*, which are delineated at their apices into one or two rows of cells called the *corona*, or *crown cells*. These outer cells enclose the egg cell borne on a short stalk of vegetative cells (Figure 16.15).

Following fertilization, the zygote develops a thick wall and the resultant oospore remains dormant for a period of time. Meiosis occurs apparently within the oospore producing four haploid nuclei, three of which disintegrate, with the single functional nucleus producing the protonema of the gametophyte.

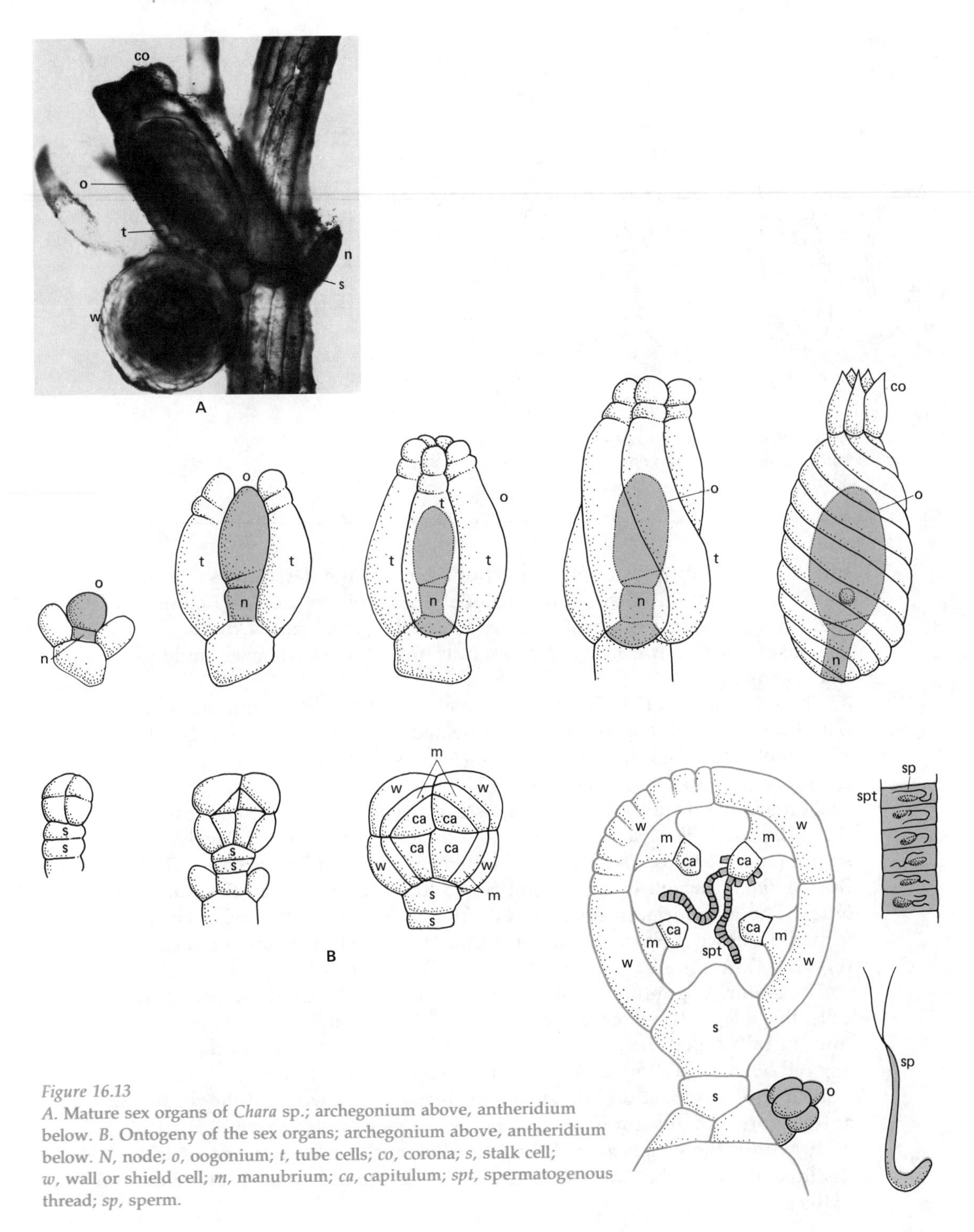


*Figure 16.13*
*A.* Mature sex organs of *Chara* sp.; archegonium above, antheridium below. *B.* Ontogeny of the sex organs; archegonium above, antheridium below. *N*, node; *o*, oogonium; *t*, tube cells; *co*, corona; *s*, stalk cell; *w*, wall or shield cell; *m*, manubrium; *ca*, capitulum; *spt*, spermatogenous thread; *sp*, sperm.

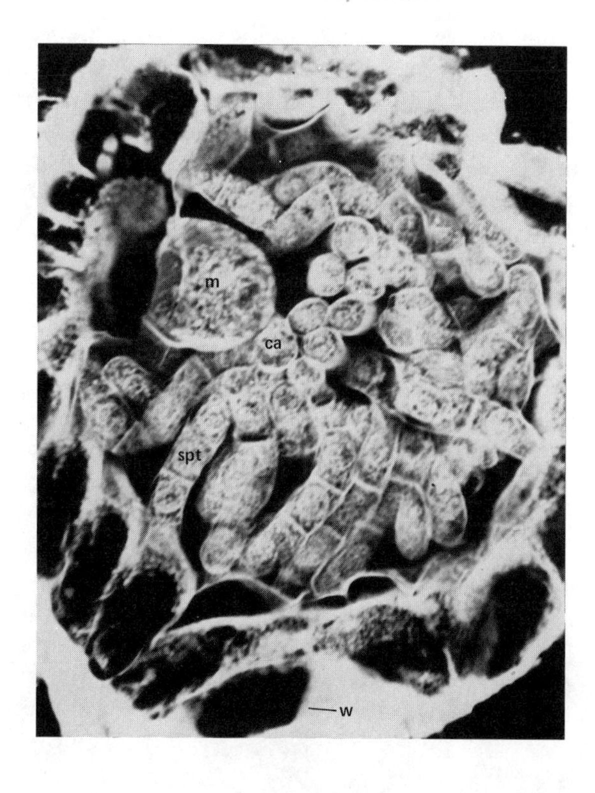


*Figure 16.14*
Section through a *Chara* antheridium; *w*, wall cell; *m*, manubrium; *ca*, capitulum; *spt*, spermatogenous thread. [Photograph by David R. Williams.]

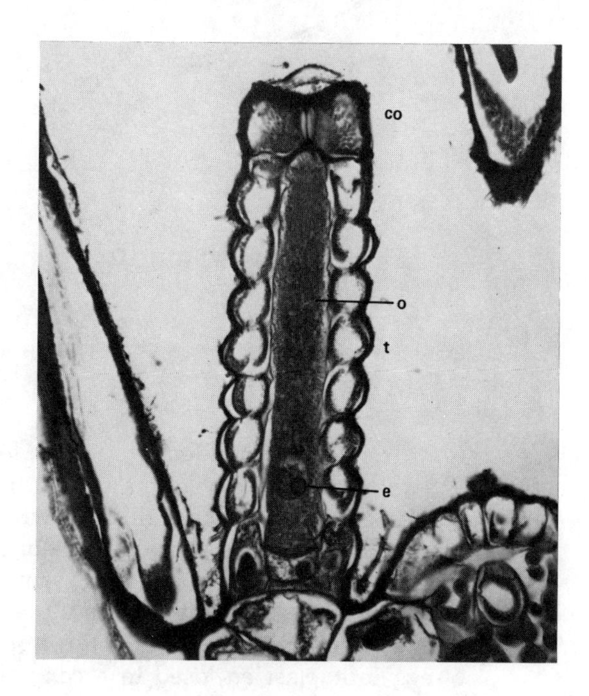


*Figure 16.15*
Section through a *Chara* archegonium; *o*, oogonium; *t*, tube cell; *co*, corona; *e*, egg.

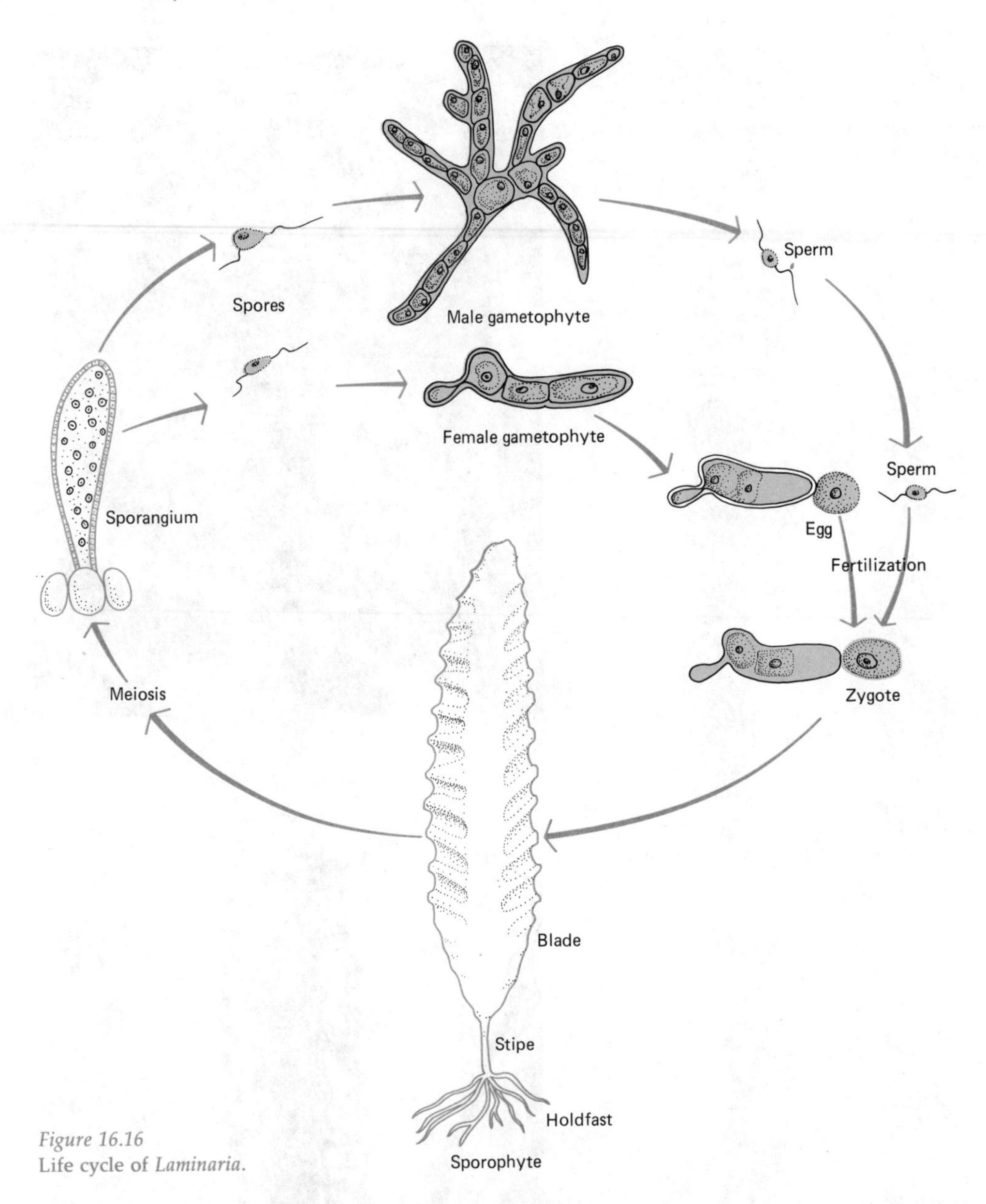


*Figure 16.16*
Life cycle of *Laminaria.*

The use of the older terms globule and nucule reflects the thinking by some writers that the terms antheridium and oogonium should be restricted specifically to those cells from which the sperm and the egg are derived directly, thus arguing that the sex organs of the Charophyta are really unicellular. On the other hand, the same individuals would argue that the sex organs in the Bryophyta are multicellular (Figures 7.32, 7.35 and 7.36).

Harold Bold has written: "It is not clear to the writer, however, why the egg protoplast enclosed in a cell wall and surrounded by tube cells. . .

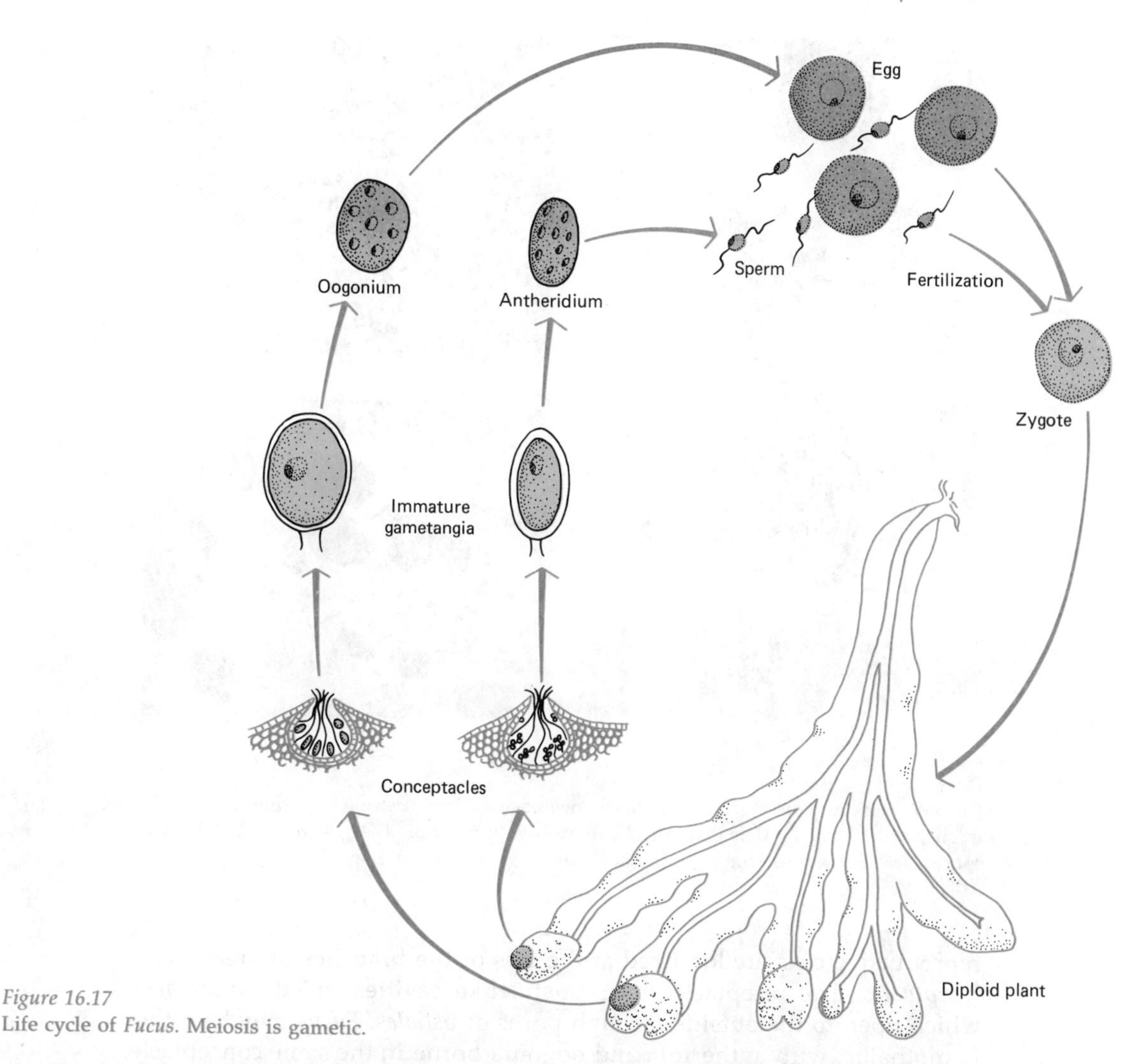


*Figure 16.17*
Life cycle of *Fucus*. Meiosis is gametic.

should all together be considered unicellular, while the egg protoplast of a liverwort or moss, enclosed in its cell wall and surrounded by venter, neck canal, and neck cells, should be considered multicellular. Even if they are not homologous, both are apparently multicellular organs." Can you know which is correct in this instance? Look at the way the structures originate.

The brown alga, *Laminaria*, illustrates the type of life cycle most characteristic of its group; an alternation of heteromorphic generations (as shown earlier for *Derbesia*). The diploid sporophyte does develop specialized spore-producing structures, termed *unilocular zoosporangia* (*unilocular* referring to the fact that the sporangium contains a single chamber, *zoosporangia* referring to the fact that motile spores are produced). Otherwise the life cycle is similar to that already discussed (Figure 16.16). The life cycle of the brown alga *Fucus* is quite different (Figure 16.17). The

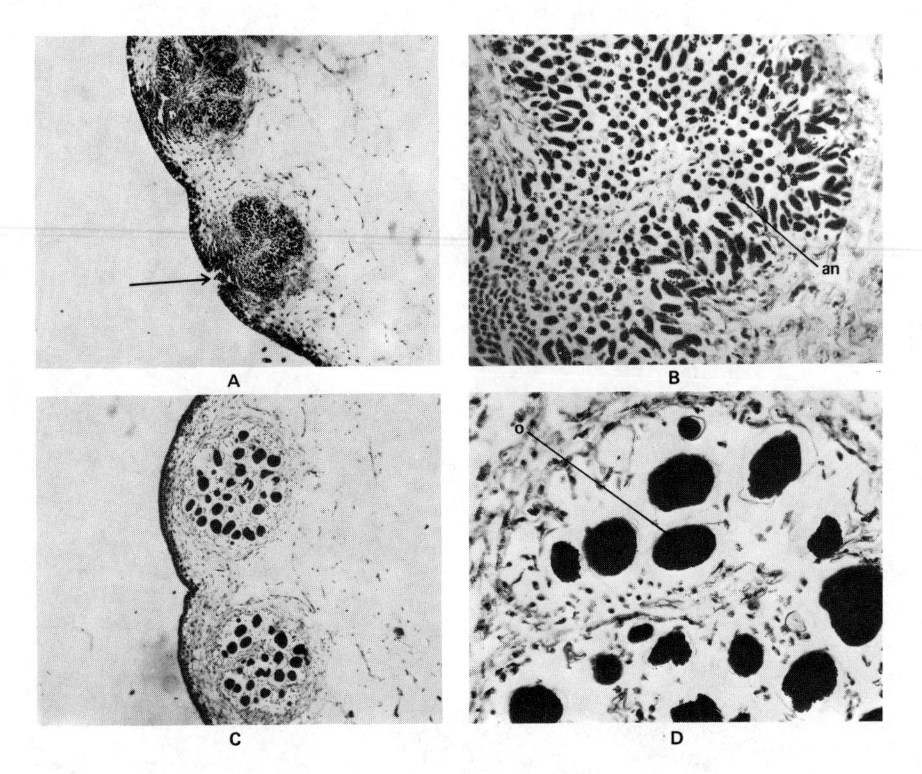


*Figure 16.18*
*Fucus* receptacles, cross-section. *A*. Male conceptacles. Note opening or ostium (*arrow*) in one conceptacle. *B*. Male conceptacle, more highly magnified; *an*, antheridia. *C*. Female conceptacles. *D*. Female conceptacle, more highly magnified; *o*, oogonium.

reproductive cells are localized at the tips of the branches in areas called *receptacles*. The receptacles bear pustulelike cavities called *conceptacles* which open to the outside through pores or *ostioles*. *Fucus* may be either homothallic, with antheridia and oogonia borne in the same conceptacle, or they may be heterothallic, with antheridial conceptacles borne on one plant, oogonial conceptacles borne on a different plant (Figure 16.18). Meiosis occurs during the formation of antheridia and oogonia, followed immediately by mitosis, producing in the former 64 laterally biflagellate sperm, in the latter, 8 nonmotile eggs. At maturity, both sperm and eggs are discharged into the water where fertilization occurs. The resulting diploid nucleus does not become thick walled nor does it enter a period of dormancy but rather germinates forming a new *Fucus* plant. Meiosis in this organism is interpreted to be *gametic meiosis* (that is, results in the formation of gametes), and the four nuclei produced at the conclusion of meiosis are not spore nuclei but simply a stage in the development of gametes.

Life cycles in the red algae can be quite complex, as characterized by

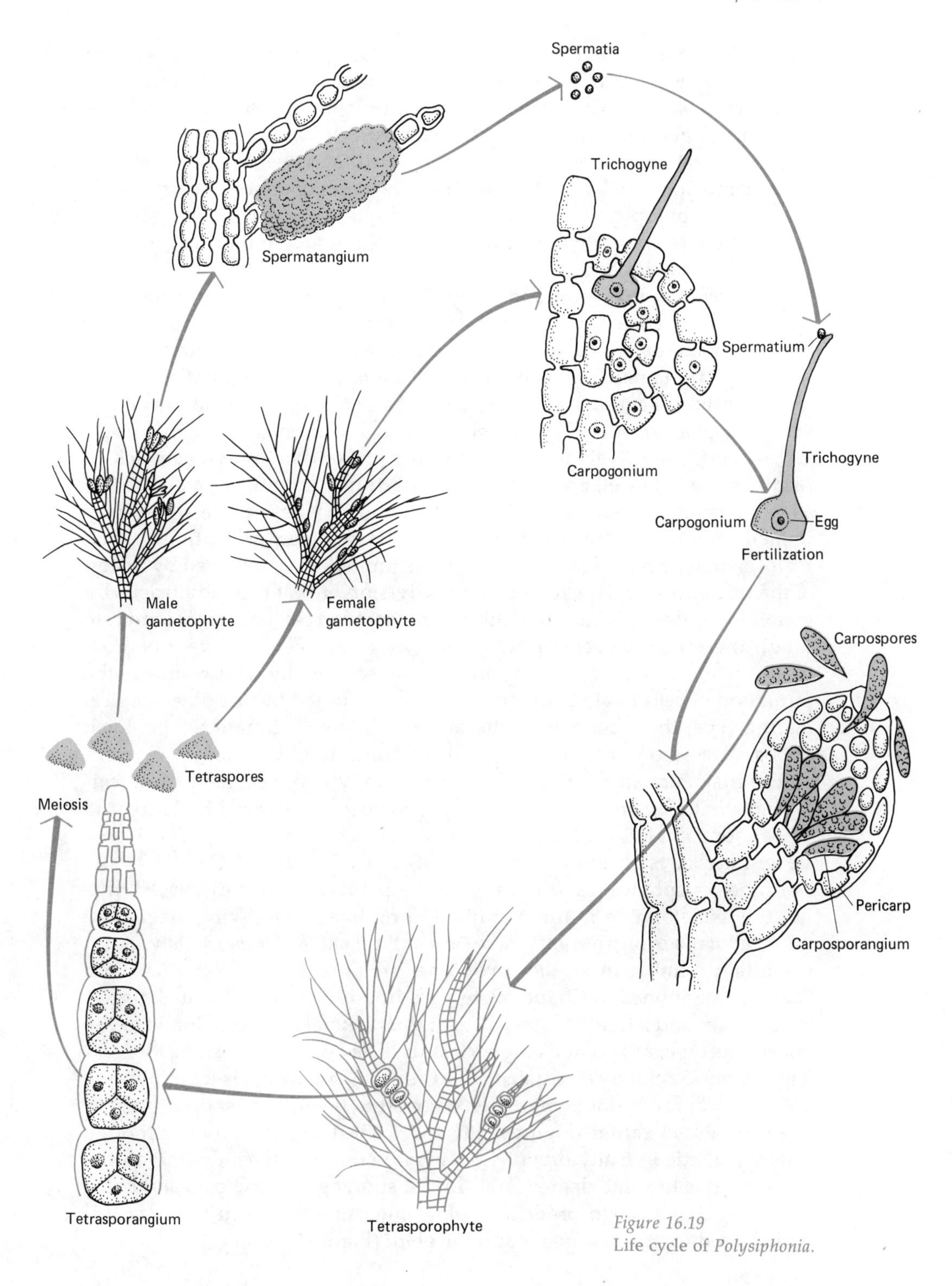


*Figure 16.19*
Life cycle of *Polysiphonia.*

that of *Polysiphonia* (Figure 16.19). The female reproductive organ is called the *carpogonium* and is essentially an oogonium with an apical protuberance, the *trichogyne*. The male sex organs are called *spermatangia* and produce nonflagellated male gametes, termed *spermatia*. Postfertilization stages are rather complicated in their detail, and we will not consider them here. However, note that two different types of spores are produced, hence two sporophytes. The carposporophyte is borne on the gametophyte, the tetrasporophyte is a free-living plant and similar morphologically to the gametophyte.

Life cycles in the fungi and fungallike groups are similar in many respects to those already mentioned. Meiosis may be zygotic or sporic, there may be an alternation of isomorphic generations, but more commonly, the diploid phase is much reduced. It might be important to mention at this time that three stages are recognized during sexual reproduction in eukaryotes: 1) plasmogamy, 2) karyogamy and 3) chromosomogamy. These stages refer respectively to 1) fusion of the protoplasm of two gametes, 2) fusion of their nuclei, and 3) fusion of portions of homologous chromosomes during synapsis in prophase I of meiosis. These three stages may immediately follow each other in some of those plants exhibiting zygotic meiosis, in which the fusion of gametes is immediately followed by fusion of the nuclei, which is followed by the division of that diploid nucleus by meiosis. In other plants, particularly the ascomycete and basidiomycete fungi, there may be considerable separation between the event of plasmogamy and that of karyogamy and chromosomogamy. This results in the formation of cells having two nuclei, one contributed by one physiological mating type, the second from the other physiological mating type. Such cells are referred to as *dikaryons*, and the condition, *dikaryotic*.

The rust, *Puccinia graminis*, illustrates a life cycle typical in its complexity with that of many rusts. As the rust requires two hosts, barberry and wheat, for completion of its life cycle, it is termed *heteroecious*. The life cycle for *Puccinia* is given in Figure 16.20. (See also Figure 7.27).

In the bryophytes, as with most of the plants we have discussed thus far, the gametophyte is the dominant, long-lived, free-living stage. The multicellular sex organs, *antheridia* and *archegonia*, are superficially, if not essentially, similar in all the bryophytes (refer to Figures 7.32, 7.35, and 7.36). As mentioned in Chapter 7, fertilization is effected by a small flagellated sperm and a large nonmotile egg; thus, sexual reproduction is of the oogamous type, as it is in all higher plants. The events of plasmogamy and karyogamy are closely associated, and the resultant zygote is a diploid cell. The embryonic development of the zygote into the mature sporophyte has been discussed earlier (Figures 7.41 and 7.45). The sporophyte, although photosynthetic in many bryophytes, is none the less dependent upon the gametophyte to some degree. Within the sporangium, meiosis and chromosomogamy occur to produce haploid meiospores, which upon germination produce the new gametophytic plant (Figure 16.21).

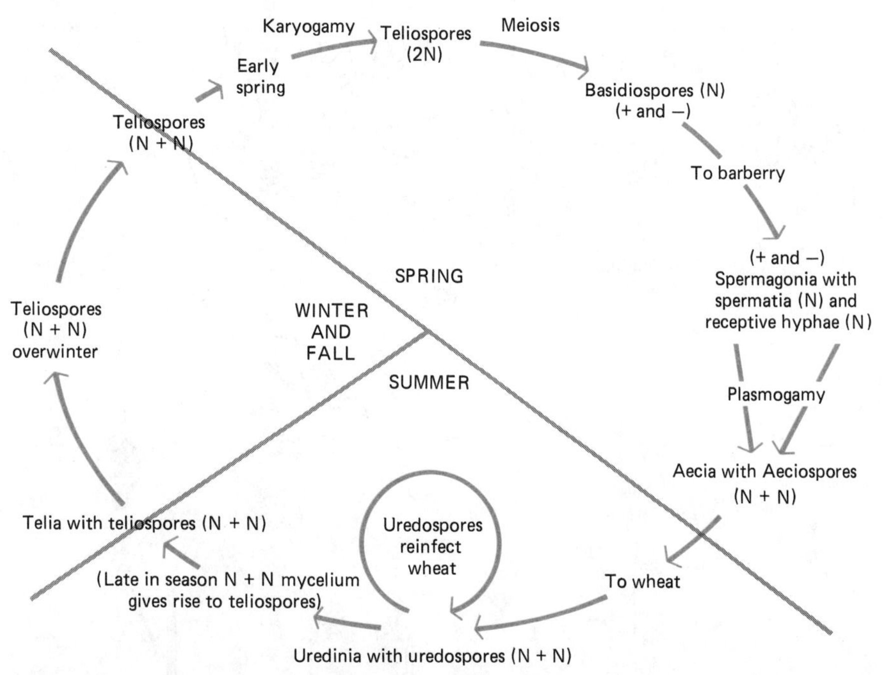


*Figure 16.20*
Life cycle of the wheat rust, *Puccinia graminis.*

In the ferns, we find a heteromorphic alternation of generations, both of which as free living. The sporophyte, however, is the more dominant, larger phase. The reproductive organs of ferns have been illustrated previously in Figure 11.21 and 11.22. A fern life cycle is shown in Figure 16.22.

In the gymnosperms, and even more so in the angiosperms, the sporophyte is the dominant, long-lived, free-living stage. The gametophyte is reduced essentially to a one-celled stage which is retained within a sporophytic structure. The gametophytes are nongreen, thus completely parasitic upon the sporophyte. The gymnosperm and angiosperm life cycles are shown, respectively, in Figure 16.23 and 16.24.

We have now considered a variety of types of life cycles, all simply modifications of the generalized cycle shown in Figure 16.8. Let us come back to our original question. Have we been discussing an alternation of *phases* or an alternation of *generations?* As we pointed out earlier, some botanists consider all the cycles we discussed to represent an alternation of phases with the exception, perhaps, of those in which no obvious spores were produced but rather in which the gametes were produced meiotically as in animals. Their argument is that a generation represents a period from the production of one zygote to the production of the next zygote; in other

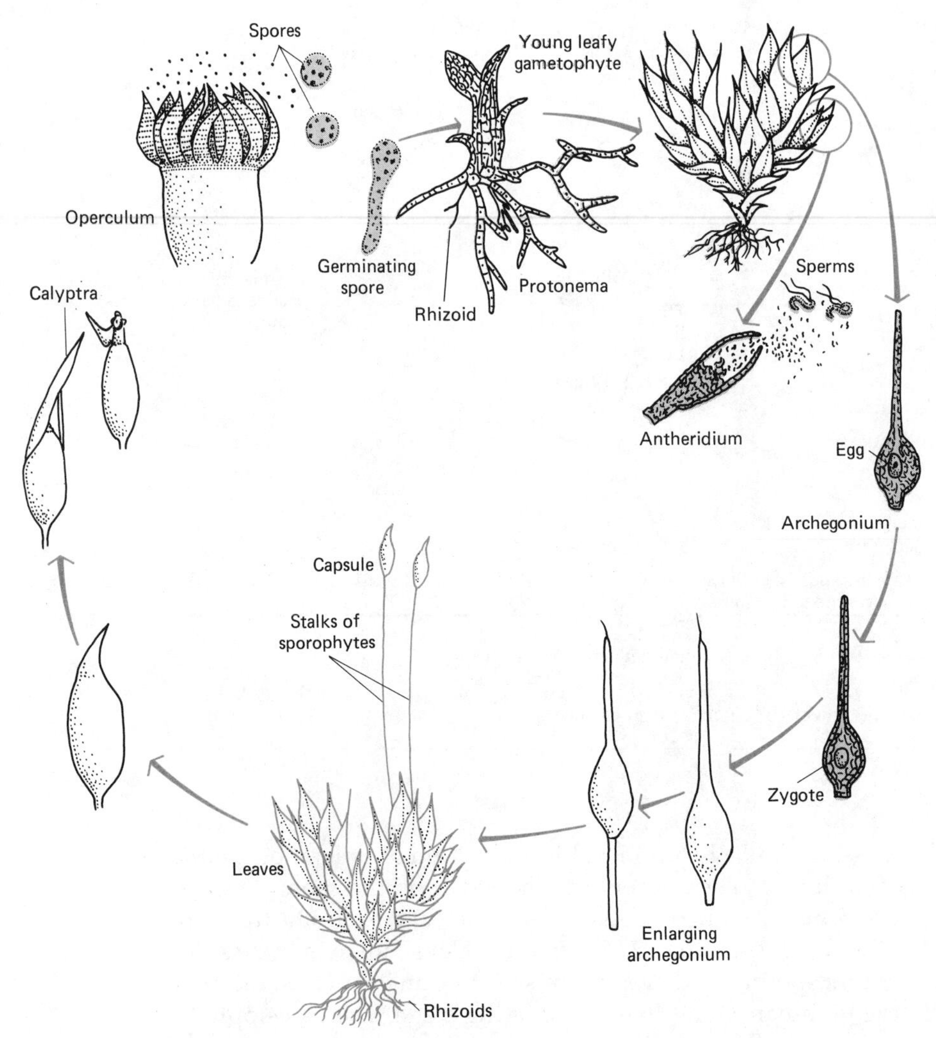


*Figure 16.21*
A representative Bryophyte life cycle. The sporophyte is epiphytic (in some cases, parasitic) upon the gametophyte.

words, a new generation is not formed until there is an offspring produced by fusion of gametes. In the plant life cycles we have been discussing, this only happens once. Therefore, we simply have two different phases of a single generation, thus an *alternation of phases.* Other botanists point out the word generation comes from "to generate," that is to produce something. The consequence of meiosis produces something, the spore; the consequence of fertilization produces something, the zygote. Those two products are different; hence, the phrase *alternation of generations* is ap-

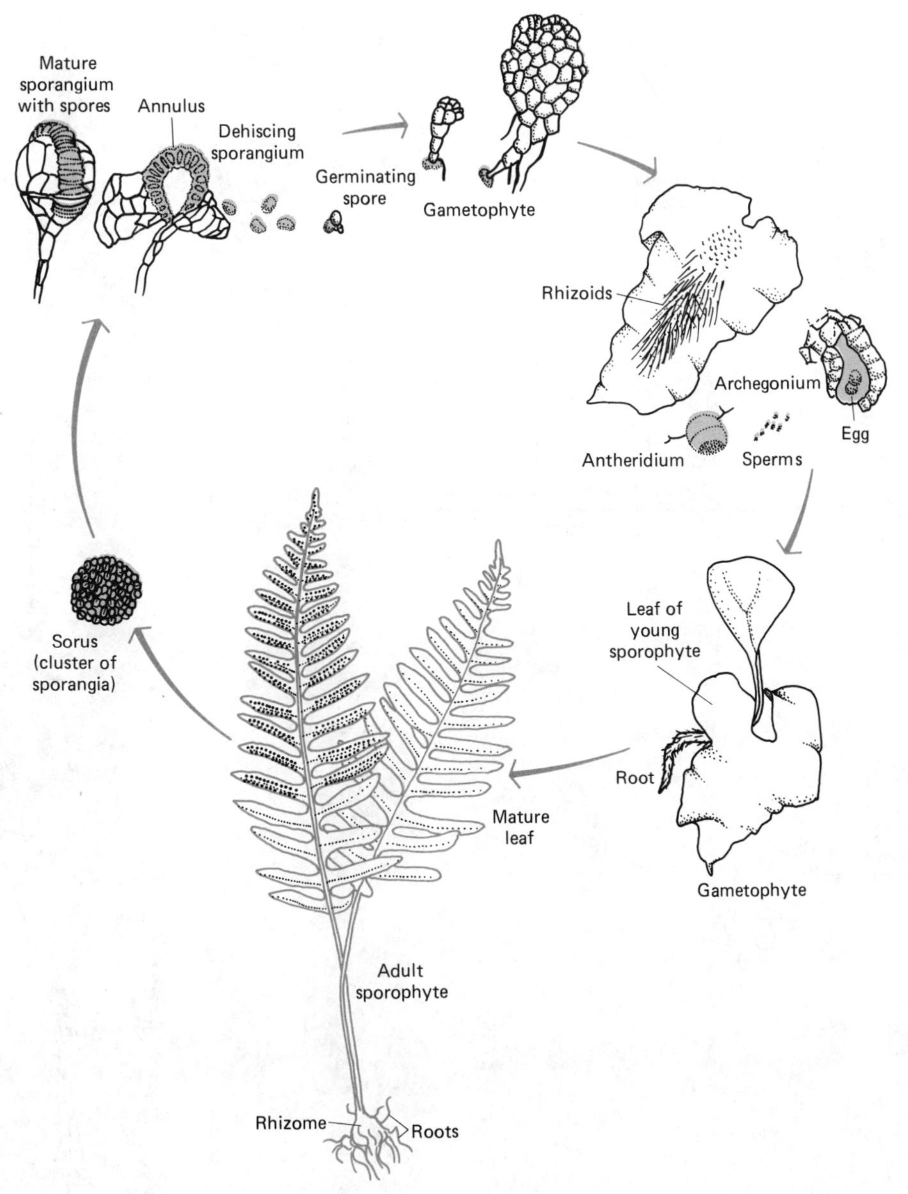


*Figure 16.22*
A representative fern life cycle. Both sporophyte and gametophyte are free-living plants.

propriate. Bold has recognized that the products generated by zygotic meiosis and by sporic meiosis are different, thus he has modified the concept somewhat by referring to the former instance as a *cellular alternation of generations*, the latter a *morphological alternation of generations*. Which do

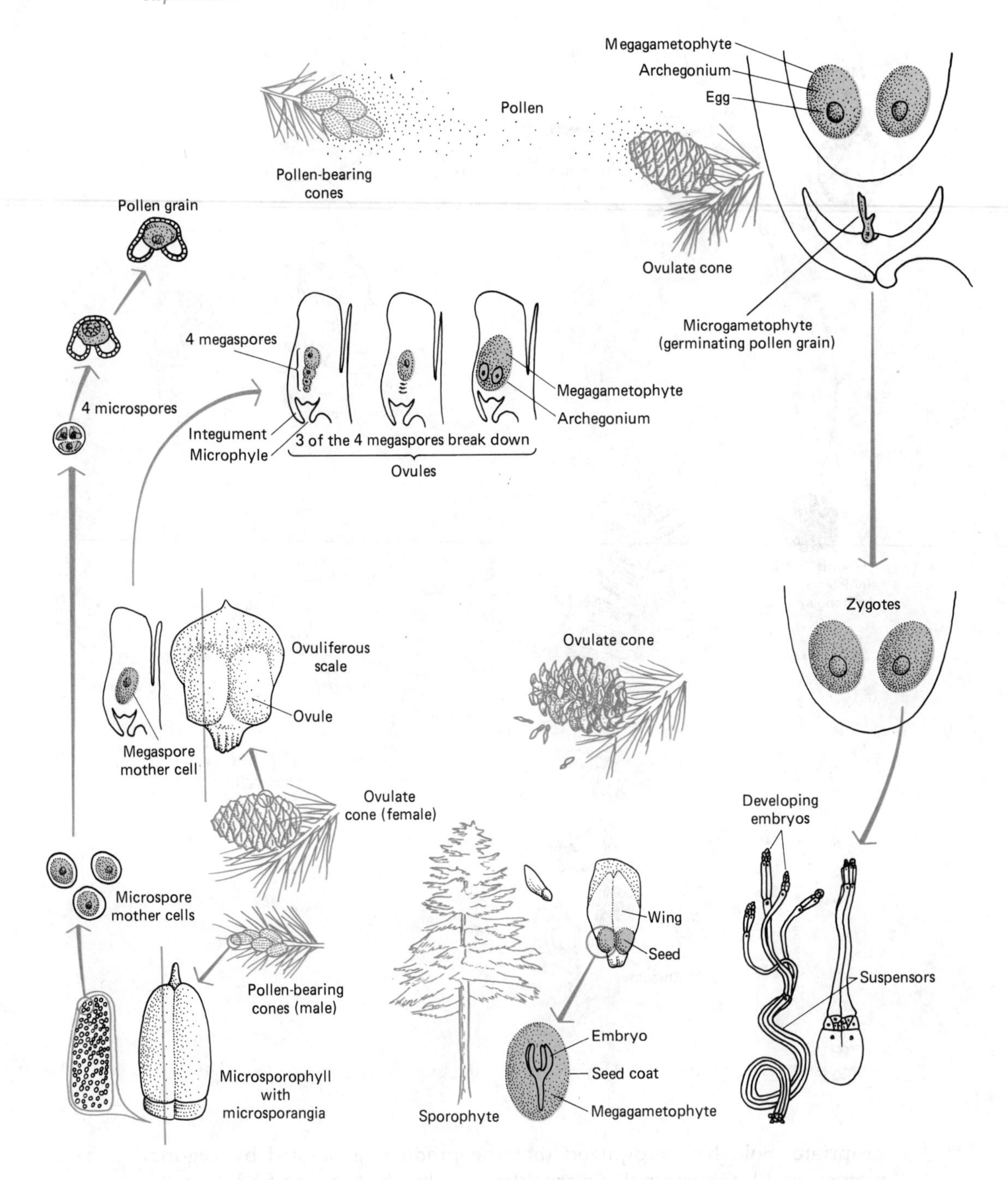


*Figure 16.23*
Life cycle of a pine. The gametophyte is much reduced and parasitic upon the sporophyte.

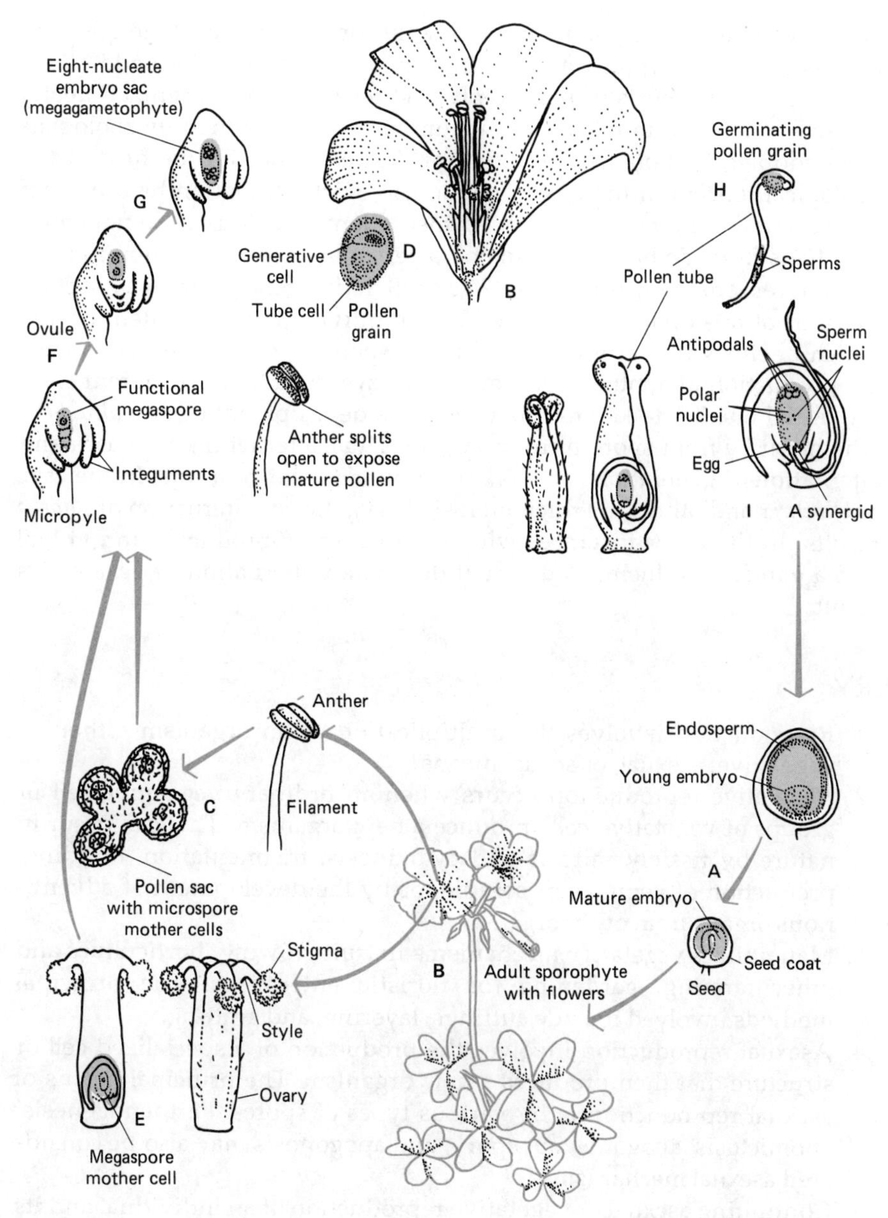


*Figure 16.24*
The life cycle of an angiosperm.

you agree with, or are you now thoroughly confused? The whole question obviously hinges upon one's definition of generation. Resende (1968) points out that aspects of plant reproduction were not recognized nearly so early as that for animals, thus, a number of terms coined by zoologists were adopted by botanists. We do not, however, need to be held to the zoological concept of the term. For the zoologist, copulation had to occur before a new generation could be formed. Plants, obviously, do not copulate. Thus Resende has chosen to characterize the term generation as used in plant reproduction as follows: "We shall define generation in a life cycle as a part of this cycle characterized by an individual, independent or not, which begins with germination of a cell, spore or zygote, and ends with the production of another cell type, either zygote or spore, different from the initial one." The key to this concept is development of an *individual*, whether it be from spore or from zygote. Thus, alternation of phases and alternation of generations are *both valid* but *different* concepts. In the case of *Ulothrix* and others in which meiosis is zygotic, an alternation of *phases* results. In the other instances where both a spore-producing individual and a gamete-producing individual occurs, an alternation of *generations* results.

## SUMMARY

1. Reproduction involves the multiplication of an organism either by vegetative, asexual, or sexual means.
2. Vegetative reproduction occurs when an ordinary vegetative cell or groups of vegetative cells produce a new organism. This can occur in nature by fission, mitosis and cytokinesis, fragmentation, budding, production of gemmae or soredia, or by the development of adventitious organs from other organs.
3. Man utilizes vegetative mechanisms in fruit growing, horticulture and other farming, gardening, or floristic enterprises. The principal methods involved include cuttings, layering, and grafting.
4. Asexual reproduction involves the production of a specialized cell or structure that then produces a new organism. The principal means of asexual reproduction is by various types of spores. Parthenogenesis, apomictosis, apogamy, apospory, and apogonosis may also be considered asexual mechanisms.
5. Continuing asexual or vegetative reproduction of an individual and its dependents produces a group of genetically similar plants called a clone.
6. Sexual reproducion involves the generation of a new individual either through the fusion of gametes or through meiospores.
7. Gametic fusion can be isogamous, anisogamous, or oogamous.
8. Meiosis can be either zygotic or sporic. It is rarely gametic in plants, although that is the principal mode of animals.

9. The sexual cycle includes normally a spore-producing individual, the sporophyte, and a gamete-producing individual, the gametophyte.
10. In the life cycle of many plants, the above two individuals alternate with each other in the plant's life cycle. This is termed alternation of generations.
11. In some plants, there is alternation only between the gametophyte and the zygote. This is termed alternation of phases.
12. In both alternation of generations and alternation of phases, three events must occur: 1) plasmogamy, 2) karyogamy, and 3) chromosomogamy. These three events may occur at widely separated times, but they must occur if the cycle is a sexual one.
13. In plants we can follow the evolution of the sporophyte from a much reduced, dependent stage to the dominant, long-lived independent stage we find in seed plants.

# Chapter 17

## EVOLUTION OF PLANTS

# EVOLUTION OF PLANT DIVERSITY—POSSIBLE MECHANISMS

In Chapters 7 and 11, we introduced you to some of the considerable diversity that exists among organisms known commonly as plants. We discussed in Chapter 3, the possible mode of origin of a living cell. If we assume that such a cell arose only once and all other living things evolved from it, we are adopting a monophyletic (*mono*, "one"; *phyle*, "clan") approach. However, it seems likely that if conditions were appropriate for the formation of *one* living entity, it would seem logical to assume the *others* could have formed as well, with each entity possessing its own unique DNA. Thus, we are assuming that life was different *from the beginning*, and that from these differing entities, other living things evolved. This would be a polyphyletic (*poly*, "many"; *phyle*, "clan") origin. Most botanists are more inclined to accept the latter explanation.

## *Spontaneous Generation*

Ideas concerning the origin of plants and animals, particularly the tiny ones, were the basis for controversy among scientists for many hundreds of years, particularly during the seventeenth, eighteenth and nineteenth centuries. Aristotle, in speaking of the origin of worms, wrote:

> Some of them are produced from similar animals. . . Others do not originate in animals of the same species, but their production is spontaneous, for some of them spring from the dew which falls upon the plants. . . . Some originate in rotten mud and dung; the hair of animals, or in their flesh or excrements, whether ejected, or still existing in the body, as those which are called helminths.

Many scientists believed in *spontaneous generation:* living things arising spontaneously from nonliving. Van Helmont believed he saw rats generated spontaneously from bran and old rags. Buffon, Lamarck, and Needham were also believers, and the great English physician William Harvey at least partially believed in spontaneous generation. Even Francesco Redi, who helped to disprove that worms arose from rotten meat, believed that intestinal worms and the gall flies arose spontaneously.

Redi was the court physician to the Grand Duke of Tuscany. Most scientific research of that time was done by individuals who were independently wealthy and had an interest in science or were under the patronage of a member of the aristocracy as was Redi. Redi conducted a number of experiments concerning the appearance of maggots (fly larvae) in rotten meat, and in 1668 published a book titled *Experiments on the Generation of Insects.* In it, he stated the hypothesis on which his experiments were based:

> It being thus, as I have said, the dictum of ancients and moderns, and the popular belief, that the putrescence of a dead body, or the filth of any sort of decayed matter engenders worms; and being desirous of tracing the truth in the case, I made the following experiments . . .

Redi then went on to describe carefully how worms developed on three dead snakes he had placed in an open box, how these worms were transformed into egg-shaped objects (pupae), and then the emergence of adult flies from these egg-shaped cases. He used the flesh of a number of different animals, all with the same results. He noted that flies seemed to hover over the decaying meat, and those flies were the same kind that later developed from the egg-shaped objects he had seen. He wondered if there might possibly be a cause and effect relationship. He stated that such a possible relationship was not a novel idea with him, as even the nonscientists were well aware of it:

> A fact . . . well known to hunters and butchers, who protect their meats in summer from filth by covering them with white cloths. Hence, Homer, in the nineteenth book of the Illiad, had good reason to say that Achilles feared lest the flies would breed worms in the wounds of Patroclus, whilst he was preparing to take vengeance on Hector.

How would you test the hypothesis that perhaps the flies were the cause of the worms which developed in rotten meat? Probably, the same way Redi did. He simply closed and sealed four containers of meat, while leaving four others open. Sure enough, although the meat rotted in both sets of containers, only the open vessels had evidence of worms. However, Redi foresaw possible criticism of his experiment. The vitalistic theory, discussed in Chapter 2, was widely held in his day, and many believed the *vital force* to be in the air. Thus, by closing the vessel to the air, Redi had excluded the vital force and obviously many would contend that life could not develop. How would you get around that problem? Redi simply repeated the experiment using fine netting to cover one set of vessels rather than sealing them. He reported:

> I never saw any worms in the meat, though many were to be seen moving about on the net-covered frame. These attracted by the odor of meat, succeeded at last in penetrating the fine meshes and would have entered the vase had I not speedily removed them.

Redi stated that he saw some adult flies deposit their "worms" on the net while others "dropped them in the air before reaching the net." We do not know whether Redi was allowing his imagination to get the better of him at this point or not. Most of his report shows the result of careful observation. Perhaps, Redi was referring to eggs and not actually the larva when he used the word *worm*, although some flies are ovoviviparous and bear live young rather than lay eggs. In any case, what Redi was observing would be very tiny, although visible to an acute observer. Many investigators have reported that Redi did not use a microscope, but he states elsewhere in his report that the abdomen of the fly he was looking at bore an occasional hair "as shown by the microscope. . . ."

Unfortunately, Redi's experiments could not always be repeated by other investigators, so the controversy continued. As more scientists began to have access to and use the microscope, further evidence seemed to pile up on the side of those believing in spontaneous generation. Surely, those tiny microorganisms seen in hay infusions arose through transmutation of nonliving material within the infusion.

As the Church was one of, if not the strongest, political and philosophical force at that time, it is not surprising that it entered the argument. Oparin (1957) stated that Thomas Aquinas, St. Basil, and St. Augustine were all believers in spontaneous generation. Yet some theologians believed the concept to deny special creation of living things by God, a belief held especially during the eighteenth Century. Two churchmen who represented the opposite poles in this regard were two Catholic priests, Lazzaro Spallanzani and John T. Needham. Needham believed that a vital force was inherent in every organic particle. He carried out a variety of experiments to test his hypothesis. Of one he reported:

> I took a quantity of mutton gravy hot from the fire and shut it up in a phial closed with a cork so well masticated that my precautions amounted to as much as if I had sealed my phial hermetically. I thus excluded the exterior air that it might not be said my moving bodies drew their origin from insects or eggs floating in the atmosphere. I neglected no precaution even so far as to heat violently in hot ashes the body of the phial that anything existed even in that little portion of air which filled up the neck it might be destroyed and lose its productive faculty.

But Needham found that within several days the vessel teemed with microorganisms. He conducted many other such experiments, using a variety of infusions, with the same results. He inferred that his experiments proved nonliving materials were transformed to living organisms. George L. Leclerc, Comte de Buffon, a contemporary of Needham supported these findings and concluded that "germs of life" existed in all of nature.

Spallanzani challenged Needham's views. He believed Needham's results were the consequence of insufficient heating of the vessels and set up his own experiments. Spallanzani used infusions similar to those of Needham but sealed them hermetically and subjected them to prolonged boiling. He found that microorganisms did not appear.

Needham's response was quite predictable:

> Nothing remains to be done save to speak of Spallanzani's last experiment which he himself believes to be the only one in his entire treatise that appears to have some force against my ideas. He hermetically sealed 19 vessels filled with different vegetable substances and he boiled them, thus closed, for the period of an hour. But, from the method of treatment by which he has tortured his vegetable infusions, it is plain that he has greatly weakened, or perhaps entirely destroyed, the vegetative force of the infused substances. And, not only this, he has, by the exhalations and by the intensity of the fire, entirely spoiled the small amount of air that remained in the empty part of his vessels. Consequently, it is not surprising that his infusions, thus treated, gave no sign of life. This is as it should have been.

> Here then, in a few words, is my last proposition, and the result of all my work: let him renew his experiments, using substances sufficiently cooked to destroy all the supposed germs that one may believe to be attached to the substances themselves or to the interior walls of the vessel, or floating in the air within the vessel. Let him seal his vessel hermetically, leaving within them certain amount of undisturbed air. Let him then plunge the vessels into boiling water for several minutes, the time which is necessary to harden a hen's egg and to kill the germs. In a word, let him take all the precautions that he wishes, provided that he seek only to destroy the supposed foreign germs which come from the outside. I reply that he will always find these microscopic living creatures in number sufficient to prove the correctness of my ideas.

Essentially, Needham was saying, do things my way and you will get the same results that I did, which of course would be true. Spallanzani did repeat his experiments but of course was unable to convince Needham or others that held the same belief. The idea of a vital force in the air was strengthened by the experiments of the chemist Gay-Lussac, who showed that a liquid subjected to prolonged boiling and hermetically sealed, lacked oxygen. He further showed that oxygen was vital to life.

This was the situation in 1864, each side dogmatically claiming their viewpoint to be correct and neither able to prove that point to the satisfaction of the other. In an attempt to throw additional light on the subject, the French Academy of Science offered a prize for: "An endeavor by means of careful experiments to throw new light on the question of spontaneous generation." A professor by the name of Louis Pasteur decided to enter the fray, much against the advice of his colleagues, and at his own expense. Pasteur used a sugar and yeast-water solution as a fermentable substance, boiled it, then allowed air which had passed through a red hot platinum tube to enter the flask. No microorganisms developed. By heating the air Pasteur was thus subjected to the same arguments that he had destroyed the "vital force" in the air. He modified his experiment by placing a cotton plug, which was full of dust previously filtered from the air, in a small tube and placing the whole between the platinum tube and the flask. By means of a stopcock he was able to close off the platinum tube, which he heated, and evacuate the flask and tube containing the cotton. Following evacuation he then allowed the heated air to slowly pass over, through and around the tube of cotton and into the flask. No growth occurred in the flask. He then tilted the flask, allowing the little tube carrying the dust-laden cotton plug to slip down into the flask. Within 24 to 48 hours, growth had occurred in the flask. Pasteur believed he had clearly shown that heating the air had not damaged it, and the air would support life if the life was present to begin with.

But Pasteur's primary opponent in the contest, Felix Archimede Pouchet, was not convinced and set out to discredit the experiments. He sterilized flasks of fermentable substances and sealed them hermetically in a manner similar to Pasteur. He found, as Pasteur had found, that no growth occurred. He then carried his flasks up a mountain, and near the edge of a glacier at 6,000 feet elevation, he opened them, allowed them to remain open for a short time, and resealed them. As the air at the edge of a glacier was, in his opinion, *pure* air without air-borne microorganisms, if growth did develop in the flask, it would have to be caused by the air, not by the microorganisms in the air. Growth did develop, and Pouchet believed he had proved his point.

Pasteur responded by doing the same experiment, but with a little dash of showmanship. He simply opened 19 flasks in his lecture hall, resealed them, walked outside and open 18 others underneath some trees. After the same length of time as indoors, he resealed them. Only 4 of the 19 opened in the lecture hall subsequently contained microorganisms, while growth occurred in 16 of the 18 opened outdoors. Thus, said Pasteur, it was not the air but rather microorganisms in the air that may or may not cause the growth. With this, Pouchet withdrew from the competition and the prize was awarded to Pasteur. The awarding of the prize to Pasteur pretty well laid to rest the idea of spontaneous generation.

In a little quirk of fate, we now know that both Pouchet and Pasteur were right. If you recall our earlier discussion in Chapter 7 concerning the bacteria, we pointed out that some can form endospores, and these endospores are highly resistant to heating. Once heat has been withdrawn, they germinate. The pertinent aspects of this situation were stated long afterwards by one of Pasteur's assistants:

> The battle was won, for Pasteur was sure of his experiments. . . . Had anyone told us that this brilliant victory amounted to nothing, he would have surprised us very much. Nevertheless, such was the case. Pasteur was right; Pouchet, Joly, and Mussett (Pouchet's collaborators)—were right also, and if, instead of withdrawing, they had repeated their experiments, they would have embarrassed the Commission very much, and Pasteur would not have known how to reply to them.
>
> It is in reality quite true that if one opens, at any point whatsoever on the globe, flasks filled with decoction of hay as Pouchet did, it often happens that all the flasks become clouded and filled with the living organisms. In other words, with this infusion, the experiments of Pasteur with *yeast water* do not succeed. . . . The fact is that the germs already exist in the hay infusion. They have remained inert as long as the flask, sealed during the boiling, remains devoid of air. They develop when the air enters, thanks to its oxygen. But, Pasteur did not yet know this result.

If living things did not arise spontaneously, how did they come into being? This was to be the question asked following Pasteur's work. For some, the question was believed to be unanswerable, for others it reinforced the idea of special creation. In other words, if we discount spontaneous generation of life, then we must accept the fact that life began by direct creation. When confronted by fossil evidence and other kinds of evidence that life has not *always* existed on earth, some theorize that it was transported to earth from some other planet or other galaxy. As stated by a French botanist: "The vegetation of the earth had beginning and will have an end, but the vegetation of the universe, like the universe itself, is eternal."

## Evolution

Thus, the groundwork was laid for the consideration of how the diversity of life came to be. The Greek, Xenophones, seems to have been the first to recognize fossils as such and to realize that the sea formerly covered much of the land. Another Greek, Heraclitus (510 - 450 B.C.), stated that, "Struggle is life," and, "All is flux"; ideas that we state somewhat differently now, but the concepts of which are basic to modern thought on evolution. Aristotle also had ideas on the subject, ideas developed from the teachings of Thales and Anaximander. Aristotle stated clearly but then rejected, the theory of the survival of the fittest as an explanation of evolution of adaptive structures. The Frenchman Lamarck, noting fossil evidence, did not consider species immutable, that is, unchanging, but he could not produce plausible mechanisms that would effect this change. As you might expect, any theories counter to those of special creation would encounter resistance from the Church and in fact are still being resisted by many of the more fundamentalist sects. However, St. Augustine, in the twelfth century, argued against the error of searching the scriptures for laws of nature:

> It very often happens that there is some question as to earth or the sky, or the other elements of this world . . . respecting which one who is not a Christian has knowledge derived from most certain reasoning or observation, and it is very disgraceful and mischievous and of all things to be carefully avoided, that a Christian speaking of such matters as being according to the Christian Scriptures, should be heard by an unbeliever talking such nonsense that the unbeliever perceiving him to be as wide from the mark as east from west, can hardly restrain himself from laughing.

From all the evidence, philosophical discussion, and arguments, it was up to Charles Darwin and his contemporary, Alfred Wallace, to finally propose, not only the idea of evolution (which was not new), but a proposed mechanism, natural selection. Little was known about inheritance or the facts of reproduction in Darwin's time, and in fact, Darwin, like Lamarck, believed that acquired traits could be inherited. In any case, Darwin's famous book opened the modern era of research on evolution. We now know that natural selection is only one force acting upon diversity. These forces of change have been identified as 1) mutation, 2) recombination, 3) natural selection, and 4) gene flow.

Mutation is an inheritable change in the genome of the individual. This may be brought about by addition, loss, or change in the chemical nature of a gene or allele. The altering of the position of a gene on the chromosome by breaking and rejoining of chromosomes is also considered to be a mutation.

Recombination has been spoken of several times before. This is, as mentioned earlier, a reorganization of alleles on homologous chromosomes brought about through a reciprocal exchange of chromosome parts during synapsis and crossing over.

Natural selection was described by Lerner (1958) as, "nonrandom differential reproduction of genotypes." Natural selection entails both aspects of mutation and effect of environment on the phenotype, and is a force that has its effect on phenotype not genotype.

Gene flow simply refers to the fact that different populations of a species have many genes in common yet have genes which are different. Through the migration of particular organisms into or out of a particular population, and the subsequent gain or loss of their particular genome in the breeding population, gene frequencies in the population will change. Particularly unique genes may be gained by the population or lost to it. Thus, gene flow refers simply to the movement of genes into or out of a population via migration of individuals. Natural selection is influenced by environment, mutations, and recombination, and in turn affects phenotype. Gene flow has its effect in determining what genomes will be present in a population for natural selection to effect.

Let us look at these forces of change in a little more detail. As discussed in Chapter 6, Mendel's work pointed out the existence of dominant and recessive genes for a character, and the number of terms, combinations, and character differences have a mathematical derivation. Let us look at a typical monohybrid cross again. If we have an individual which is a homozygous dominant for a particular phenotype *(AA)* and that individual breeds with one homozygous recessive for that same character *(aa)*, then the offspring, as you know, will all look like the dominant parent but will genotypically be *Aa*. When this generation interbreeds we will get the typical Mendelian 3:1 phenotypic ratio, comprised of $1/4$ *AA*, $1/2$ *Aa*, $1/4$ *aa* (Figure 17.1). If the $F_2$ population freely and randomly interbreeds, what would be the proportions of *AA*, *Aa* and *aa* in the $F_3$? We would first have to determine the number of different crosses that could occur. There would be 3 (the number of different genotypes) raised to the second power (2 = number of alleles); $3^2 = 9$ different crosses. These would be *AA* × *AA*, *AA* × *Aa*, *Aa* × *AA*, *AA* × *aa*, *aa* × *AA*, *Aa* × *Aa*, *Aa* × *aa*, *aa* × *Aa*, and *aa* × *aa*. The results of these crosses are shown in Figure 17.1.

How do we calculate the number of each genotype that will appear in the $F_3$? Simply on the basis of probability. In the first cross, for example, between *Aa* ×*Aa*, only $1/4$ of our $F_2$ progeny is *AA*. Thus, the probability of one *AA* breeding with another *AA* is $1/4 \times 1/4 = 1/16$. As all the $F_3$ progeny of such a cross will be AA, the contribution of this cross to the total number of AA in the $F_3$ would be $4/4 \times 1/16 = 1/16$. Likewise in the cross between *AA* × *Aa*, the proportion of *AA* in the $F_2$ is $1/4$, that of *Aa* $1/2$, thus, the prob-

*Figure 17.1*
A cross between heterozygotes with free and random interbreeding of the $F_3$.

$F_2$

*Aa* X *Aa*

| Gametes | *A* | *a* |
|---|---|---|
| *A* | *AA* | *Aa* |
| *a* | *Aa* | *aa* |

$F_3$

*AA* X *AA*

| Gametes | *A* | *A* |
|---|---|---|
| *A* | *AA* | *AA* |
| *A* | *AA* | *AA* |

*AA* X *Aa*

| Gametes | *A* | *A* |
|---|---|---|
| *A* | *AA* | *AA* |
| *a* | *Aa* | *Aa* |

*Aa* X *AA*

| Gametes | *A* | *a* |
|---|---|---|
| *A* | *AA* | *Aa* |
| *A* | *AA* | *Aa* |

*AA* X *aa*

| Gametes | *A* | *A* |
|---|---|---|
| *a* | *Aa* | *Aa* |
| *a* | *Aa* | *Aa* |

*aa* X *AA*

| Gametes | *a* | *a* |
|---|---|---|
| *A* | *Aa* | *Aa* |
| *A* | *Aa* | *Aa* |

*Aa* X *Aa*

| Gametes | *A* | *a* |
|---|---|---|
| *A* | *AA* | *Aa* |
| *a* | *Aa* | *aa* |

*Aa* X *aa*

| Gametes | *A* | *a* |
|---|---|---|
| *a* | *Aa* | *aa* |
| *a* | *Aa* | *aa* |

*aa* X *AA*

| Gametes | *a* | *a* |
|---|---|---|
| *A* | *Aa* | *Aa* |
| *A* | *Aa* | *Aa* |

*aa* X *aa*

| Gametes | *a* | *a* |
|---|---|---|
| *a* | *aa* | *aa* |
| *a* | *aa* | *aa* |

ability of this cross occurring is $1/4 \times 1/2 = 1/8$. From such a cross, one-half of the offspring will be *AA;* thus, this cross will contribute to the total number of *AA* in the $F_3$, $1/2 \times 1/8 = 1/16$. These calculations have all been made for you if Figure 17.1. If you add up each value for *AA, Aa,* and *aa,* you find the totals are $1/4$*AA*, $1/2$*Aa*, $1/4$*aa*, exactly as it was in the $F_2$. If you allowed the $F_3$ to freely and randomly interbreed, the ratios of *AA* to *Aa* to *aa* in the $F_4$ would be $1/4 : 1/2 : 1/4$, just as before. In a large population in which there is random mating and an absence of the forces of change we discussed earlier, the original proportions of dominant alleles to recessive alleles will remain constant from generation to generation. This aspect was pointed out in 1908 by G. H. Hardy, an English mathematician, and G. Weinberg, a German physician, and is now known as the *Hardy-Weinberg Law.*

The Hardy-Weinberg law is usually stated in algebraic terms. If we let the frequency of $A = p$, the frequency of $a = q$; the frequency of these two alleles, assuming they are the only alleles for a particular gene in a gene pool, would have to equal one; that is, $p + q = 1$. Knowing this, how then do we find the relative proportions of *AA, Aa,* and *aa* as we found above, the hard way, working it out for each cross? We know that the *A* allele can be in either the male or female gamete; thus $A \times A = A^2$. In the *Aa* individuals, we have an *a*♂ × *A*☿, and an *a*☿ × *A*♂ or 2*Aa*. Finally, *aa* can be either male or female; thus, $a \times a = a^2$. If we convert these to our algebraic terms, we have $p^2 + 2pq + q^2$ which converts, if you recall the binomial theorem, to $(p + q)^2$. If, as we stated above, $p + q = 1$, then $(p + q)^2 = 1$. So if one-half of our gene pool is $A$, the other one-half $a$, the proportion of $AA$ will be 0.25, just what Mendel said it would be and what we found earlier. That is, $A^2 + 2Aa + a^2 = (0.50)^2 + 2\ (0.50)\ (0.50) + (0.50)^2 = (0.25) + 2\ (0.25) + (0.25) = 0.25 + 0.50 + 0.25 = 1$.

But the forces of change discussed earlier will affect that ratio. Populations in nature simply are not in Hardy-Weinberg equilibrium. If they were, there would be no evolution.

Mutations do arise from time to time in a population. The effect of some mutations is obvious, for others the effect is so slight it is not even noticeable. Most mutations are more or less deleterious, some even result in death of the organism. The latter are referred to as *lethal mutations.*

If most of the viable mutations are not beneficial to the organism and in fact are harmful to some degree or another, how can we postulate that evolution is compounded of them? The answer is a relatively simple one; a mutation that may be of no benefit or even harmful to the organism in a particular environment, may become useful or even essential if that environment changes. Some mutations are unique, occurring in only a single organism at a single point in time. Such mutations, simply on the basis of probability, are generally not propagated in the population. Mutations which are not unique are called recurrent mutations and occur with a measurable frequency of $10^{-4}$ to $18^{-8}$ mutations per generation.

In commenting on mutations, the great geneticist Theodosius Dobzhansky stated:

> Most mutants that arise in any species are, in effect, degenerative changes; but some, perhaps a small minority, may be beneficial in some environments. If the environment were forever constant, a species might conceivably reach a summit of adaptedness and ultimately suppress the mutation process. But the environment is never constant; it varies not only from place to place but from time to time. If no mutations occur in a species, it can no longer become adapted to changes and is headed for eventual extinction. Mutation is the price that organisms pay for survival. They do not possess a miraculous ability to produce only useful mutations where and when needed. Mutations arise at random, regardless of whether they will be useful at the moment, or ever; nevertheless, they make the species rich in adaptive possibilities.

In other words, mutations occur, and as a consequence, the organism may be better able to adapt; the organism does not adapt by making mutations.

What is the effect of mutations on Hardy-Weinberg equilibrium? This effect can be assessed mathematically, keeping in mind that an allele can not only mutate to a different allele but a second mutation can occur to change the new allele back to the old one. The latter instance is called *back mutation.*

If we let the rate of mutation $= \mu$, the rate of back mutation $= v$, the original allele $= A_1$, the mutated allele $= A_2$, the initial gene frequency of $A_1 = p_0$, and the initial gene frequency of $A_2 = q_0$, then the number of mutations is the product of rate and the initial frequency. The change in frequency of gene $A_1$ in one generation is $\mu \times p_0$. But gene $A_2$ will mutate back to $A_1$ at a rate $v \times q_0$. Thus, the frequency of the mutated gene is $\Delta q = \mu \times p_0 - v \times q_0$. As $A_2$ increases, there are fewer $A_1$ left to mutate and more available to back mutate. Eventually an equilibrium will be reached, at which point no further frequency changes will take place as the result of mutation alone: $p \times \mu = q \times v$, or $p/q = v/\mu$ and $q = \mu/\mu + v$. In the absence of selection, the frequency of gene $A_1$ depends exclusively on the rate at which it mutates to $A_2$ and on the rate $A_2$ mutates back to $A_1$. These rates are very low, thus changes that mutation alone introduces into a population are very small. If back mutation does in fact occur, a gene can never become established at 100 percent in a population. In the absence of both selection and back mutation, it would take a minimum of 5,000 to a maximum of 50 million generations (depending on whether the rate is $10^{-4}$ or $10^{-8}$) to replace one-half of the $A_1$ genes by $A_2$ genes in a population. In three years, a bacterium could produce 50,000 generations, but in a pine tree it would take approximately 1 million years to produce that many generations.

The effect of recombination can also be mathematically determined. If we let $g =$ the number of diploid genotypes and $r =$ the number of alleles at any one locus, then $g = r(r+1)/2$. If we have two different loci, on different chromosomes, what would be the total number of possible genotypes for the two genes considered together, if each locus has the same number of alleles? It would be $g_A + g_B = r(r+1)/2 \times r(r+1)/2 = r(r+1)^2/2$. For example, if we had two alleles at each of two loci, $r = 2$, and $g_A + b_B = 2(2+1)^2/2 = 9$.

The number of loci in a genome has been estimated to be in the order of $10^4$. If three alleles per locus is assumed, the number of theoretically possible recombinant genotypes is $6^{(10)4} = 6^{40}$. This number is only theoretical, as the genes are arranged in chromosomes, and thus the loci on that chromosome are not free to assort independently at meiosis. Crossing over can break up linkage groups, but usually does not occur with any great frequency (between 1 to 3 per generation). Also, although some loci have 100 or more alleles, many others are monomorphic (have no alleles).

Natural selection accounts for the number of offspring that will reach reproducing age and thus limits the type of genome that will be passed to future generations. The effect of environment may cause some phenotypes to increase, some to decrease, while others will show no net change. Those phenotypes that over several generations are seen to decrease, are said to be *acted against* by natural selection. Selection may tend to stabilize the genome of a population by eliminating extreme types, it may be disruptive in that two or more extreme types increase at the expense of intermediate forms, or it may cause a directional change in the genome, that is a shift in the direction of a particular characteristic.

The introduction of new genes into a population from adjacent populations (gene flow) would of course have an effect on Hardy-Weinberg equilibrium, particularly if such gene flow was fairly extensive. In fact, gene flow has been postulated by some investigators to be an important factor in genetic variability. But Levin and Kerster (1976) conclude from their studies of migration statistics that "the local gene pool is little dependent on migration for the maintenance of genetic diversity." They based their conclusion on two factors: 1) migration rates in plants are very low and are almost exclusively between neighboring populations, which tend to have a correlated gene pool anyway, and 2) populations seem to retain a high level of generic polymorphism in spite of vanishingly small inputs of extraneous genes. Research has indicated that many individuals in a population are frequently heterozygous at numerous loci. This research has been extended to include electrophoretic studies of the various loci encoding for structural proteins, and again extensive heterogeneity was found both within and between populations.

If gene flow is to be essentially discounted as a factor for introducing or maintaining genetic variability, what then should be considered? Most investigators consider the primary force to be mutation and introgressive hybridization. The latter simply refers to a backcross of an offspring to its parent. We will discuss this aspect in a little more detail later.

We mentioned earlier that natural selection has its effect on the phenotype. Phenotypic variability in populations may occur as the result of geographical isolation. Habitats are discontinuous, and even in populations that are relatively close to each other, gene flow is quite limited. Thus, in spatially separated populations, each population has responded essentially to the selective aspects of its environment in a different way. In the field, growing under natural conditions, plants may be phenotypically different species; that is, they look different enough that most taxonomists would identify them as different species. But simply because these plants are phenotypically different, does this mean they must also be genotypically different? How can we find this out? Two general methods have been used. The plants can simply be grown together under uniform environments. It is assumed that if in fact the plants are genetically different, they will remain phenotypically different from each other when grown in a uniform environment. The second method assumes that different species should not be able to interbreed to produce fertile hybrids. One can simply cross pollinate the two species and see if fertile hybrids are produced.

Phenotypic characters may be related with geography in general; other factors being equal, northern plants are smaller than southern plants, western plants are larger than eastern plants. The effect of altitude generally mimics latitude. In other words, the effect of going up a mountain would be similar to that of traveling from south to north.

In most instances, geographically isolated populations were found to be genetically different. Such difference in a species, developed as a result of geographical isolation and selection, are referred to as *ecotypes*.

Although the capacity to change as the result of environment is not inherited directly, the capacity to respond phenotypically to environmental changes is. Thus it is under control of natural selection. Some species seem to be highly variable and able to grow under a wide variety of environmental conditions, while others require quite specific conditions, and even slight changes in the environment will cause their decline. Flowering plants tend more or less to be in the latter group, blue-green algae in the former. In fact, investigations by Francis Drouet and others have shown a surprising potential of blue-green algae. If one takes a particular species of blue-green algae and subjects it to a variety of different environmental conditions: light quality, light quantity, differing temperatures, differing concentrations of nitrates, and so forth, not only will the species generally survive, but it may assume the morphological form of a different species under each environmental condition. Thus, Drouet concludes that differing organisms that have formerly been called different species, and, in many instances, even different genera, are actually genetically similar and simply represent different forms of the same species. These observations have really knocked blue-green algae taxonomy into a cocked hat. In any case, the blue-green algae represent perhaps an extreme in what is termed *genetic plasticity*, the ability of an organism to exhibit varying phenotypic expression with the same genotype.

The seeming lack of genetic plasticity in higher plants may not only be a factor of the genome. It has been shown that many plants can adapt quite well to a new environment if no competition exists. But once competition is introduced, they fail.

We have been talking quite a bit about a species, but do you know just what a species is? If they look different, they must be different species, right? But even leaves on the same tree may look different, and certainly red maple trees do not all look alike, yet we call them the same species. Is it just a matter of degree; do they have to look quite a bit different? Certain plants grown in differing environments may look considerably different, yet grown in the same environment they look quite similar, interbred, and produce fertile offspring. Then, is it only a matter of hybridization? To the zoologist, the concept of a species has traditionally been that of groups of individuals which breed and produce fertile offspring among themselves but not between themselves and other groups. But we can find plants that do not hybridize in nature, they have somewhat different genomes, yet can be artificially hybridized producing at least some fertile offspring. Do we no longer call them species? Why did these species not hybridize in nature? One reason may have been geographical isolation as we mentioned earlier, but sometimes species can occur side by side and yet not hybridize. This might be due to differences in their time of flowering, a difference in their pollination system, or a variety of other factors called *biological isolation*. Although there is no single explanation that will cover all aspects of what a plant species is, in general, it is considered to be groups of actually or potentially interbreeding natural populations, which are reproductively isolated from other such groups.

How are new species formed? The process by which this occurs is called speciation. There are two major modes: *phyletic speciation* and *true speciation.* In phyletic speciation, there is a transformation of one species into another with time. This obviously does not result in an increase in the number of species, one is lost, one is gained. In true speciation, an ancestral species gives rise to one or more new species without losing its own identity. In the *allopatric theory of speciation,* two steps are required: 1) reproductive isolation brought about by physical (geographical) separation, and 2) the independent evolution of these reproductively isolated populations. This kind of separation has lead to what is referred to as evolutionary radiation of plants on islands that have been separated for eons of time.

The theory of *continental drift* has also played an important role in speciation, or lack of it, in areas that are actually widely separated today. In the late sixteen hundreds, Francis Bacon pointed out that there seemed to be a fit between the outline of the coasts of Europe and Africa with that of North and South America. It was not until 1915 that Alfred Wegener, a German geologist, introduced the theory of continental drift. He envisioned the continents as rafts of lighter rock floating on heavier, semifluid rock. He believed that at one time a supercontinent existed, composed of our present continents (later named Pangaea). During the Triassic, it began to split into two large land masses, Gondwanaland in the Southern hemisphere, Laurasia in the northern hemisphere. Toward the end of the Mesozoic, these masses became further separated into South America, Africa, India, Antarctica, and Australia. North America, however, was still connected to the great Eurasian land mass. During the Cenozoic Era, North America split off and drifted away and India became attached to the Eurasian continent. Mid-oceanic ridges mark the regions of separation between the continents (Figure 17.2). The theory of continental drift was laughed at by many scientists, and it was not until the 1960s that it began to receive serious attention. But thirty years before that, the French botanist Henri Gaussen recognized the theory as a possible explanation for the widely scattered distribution of cycads in both North and South America, South Africa, Malaysia, the Caribbean, and Australia. It also helped explain the puzzling, discontinuous distribution of certain fossil species.

In general, four steps are given in the speciation process: 1) the separation of the original gene pool, 2) the independent evolution of those gene pools, 3) secondary merger of the gene pools through hybridization, and 4) competition between new gene pools. The results of such competition can have one of three outcomes. One species may simply outcompete the other; thus the non-overlap of species in a particular area is restored. Two species may, however, divide the environment in such a way that no real competition occurs, but the breeding barrier remains and thus they exist

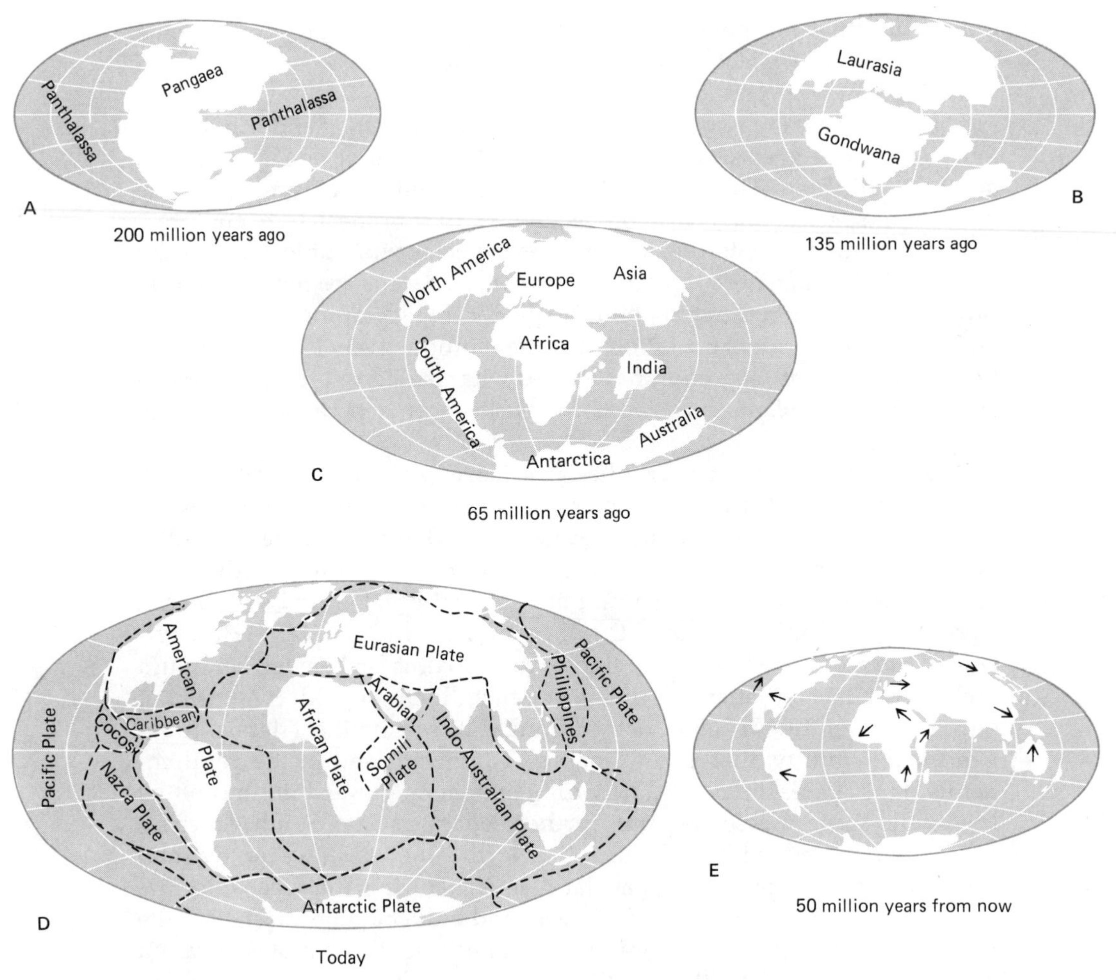


*Figure 17.2*
Continental drift. *A.* The supercontinent Pangaea. This is postulated to have existed during the Triassic Period. *B.* Break-up of Pangaea into Laurasia and Gondwana; postulated to have occurred during the Cretaceous Period. *C.* Further separation during the Tertiary Period. *D.* The position of the continents today. *E.* Projected position of the continents in 50 million years.

as distinct species. A third outcome would be that the two populations hybridize forming a single interfertile population, thus masking the effect of the original speciation.

In regard to competition, Gause has stated that, "No two forms can share exactly the same environmental requirements for an indefinite period of time. Eventually one form will replace the other." This has become known as the *Gause exclusion principle.* Another principle is called *character displacement* and involves the concept that plants that are most dissimilar in their requirements can grow where a minimum number of plants of another population can grow. Thus, the species which occupy an area of overlap will be most dissimilar to one another, while those not in an area of overlap may be more similar.

The separation of the original gene pool and subsequent independent evolution of that gene pool in speciation often involves hybridization. Even if hybridization between two species is rare in nature, these hybrids may be quite important because of the way they recombine the parental characters. The new genetic combination may be better suited to a changing environment, may be able to occupy areas of the habitat where neither parent would grow, or may be better adapted to occupying areas of intergrading habitat. The evolutionary role of the hybrid depends of course on the effect of its genetic composition on that of the parental population. As the number of hybrids are often very few, or in some cases unique, opportunity for the hybrid to cross with other hybrids is rare. Thus, backcrosses of hybrids to one or both parental species are extremely important. This process is known as introgressive hybridization.

By such backcrossing, the hybridizing species retain their own distinct characters and usually become part of one or the other of their parent populations. In some instances, introgressive hybridization does not occur, and the hybrids may become established as a distinct population with characteristics more or less intermediate between the two parental populations. Both of these instances require the production of a fertile hybrid. But as we mentioned earlier, crosses between different species often produce an infertile hybrid. Does this mean that the plant will simply die out and the new recombinant not be a factor? Not necessarily. Sterile hybrids can reproduce themselves in a variety of ways. Apomixis (the origin of new individuals without nuclear or cellular fission), principally by vegetative propagation, is quite prevalent, and if this can be combined with an occasional hybridization, a whole series of related species can result. Due to the mode of vegetative propagation, members of a particular habitat may in fact be members of the same clone. This makes them well adapted to their particular environment and generally able to outcompete most other plants. The Kentucky bluegrass, *Poa pratensis*, is an excellent example of such an apomictic system.

Rather than being a slow, evolving process, speciation can be abrupt. One such instance is the development of polyploidy. One way in which this might occur is through apospory. If a diploid cell of the sporophyte gives rise directly to the gametophyte, then the gametophyte will also be diploid. If gametes are formed in the usual way, by mitosis, then the gametes will also be diploid and fusion of those gametes will produce a tetraploid sporophyte. Another mechanism involves hybridization. If two species having nonhomologous chromosomes cross, the hybrid offspring will generally be sterile because of its inability to undergo synapse or pairing of homologous chromosomes during meiosis. If, however, those nonhomologous chromosomes somehow duplicate, then each chromosome will have a homologue, although twice as many chromosomes will be present at synapse and meiosis can proceed normally. Regardless of the way the polyploid is formed, it acquires instant reproductive isolation.

Catastrophic selection is also a factor in abrupt speciation. Such speciation is brought about by environmental stress or *mutator genotypes* (genotypes that produce a high frequency of mutation). They cause a drastic chromosomal rearrangement in the plant. If such changes are not lethal to the plant, and such a plant can somehow become detached from its population, it can then establish a new population with new chromosomal arrangement.

In our foregoing discussion, we have essentially assumed that given reproductive isolation, and given time, speciation will occur. However, two species of sycamore, *Platanus orientalis* and *P. occidentalis* are both morphologically and physiologically distinct from each other. Fossil evidence indicates that they have been isolated from each other at least from the Tertiary period (1 million or more years ago). Yet artificial hybrids were formed. They are now widely cultivated as an ornamental, and the hybrids are vigorous and highly fertile. This indicates that no drastic change in the order of the genes has occurred in the parent species over all those years of isolation.

## EVOLUTION OF PLANT DIVERSITY—POSSIBLE PATHWAYS AND INTERRELATIONSHIPS

In discussing how different groups of plants may have arisen, we can provide no answers, only theories. We have previously discussed the possible origin of the prokaryotic cell. It has been a tacit assumption that it is the closest to the first cell. But Reanney (1974) postulates that the genome of existing prokaryotes represents regressive evolution, an interesting but not widely held hypothesis. The evolution of the eukaryotic cell, as we have already mentioned, is open to considerable question. We could postulate a step by step evolution, or if we subscribe to the symbiotic theory we do not have to worry about how mitochondria and chloroplasts evolved. All of a sudden, they just jumped into cells. This is facetious of course, but the theory, if one accepts it, does tend to limit thinking on the question of the evolution of these structures because it postulates they were already preformed. Regardless of how the eukaryotic cell formed, and again I do not think one needs to hypothesize that only one was formed, the various eukaryotic organisms evolved from it in some manner. We could of course state that the primitive eukaryotic cell had a chloroplast from the beginning, and from it the green plants evolved. A different primitive eukaryotic cell may have lacked chloroplasts, and from it protists and animals were derived, with those protists which are now green deriving their chloroplasts secondarily. We could even postulate a different primitive ancestral, eukaryotic cell type for each division, although this is probably highly unlikely as some divisions do show close affinities with each other.

Many different schemes have been proposed for the evolution of the various plant groups and certain of these differ quite considerably from each other. Assuming the same information is available to each individual investigator, why would they come up with different pathways of evolution? They simply interpret that information differently. What are the aspects we can look at in attempting to derive a phylogenetic system? We could look at fossil data and at extant plants. Fossil data give us limited information as many gaps seem to exist and not all of a particular plant may be fossilized, with the parts missing being critical ones. But some valuable evidence has been gained from paleobotanical studies. What can we look at in extant plants? Their morphology and anatomy is classically looked at. But which is more important, the life cycle, mode of cell division, sexual reproductive mechanisms, or certain anatomical aspects? Is one more important than others? This, of course, demands a value judgment, and not all investigators will have the same opinion. What about pigmentation, production of oxygen during photosynthesis, presence of certain respiratory cytochromes, cell wall characteristics, nature of the storage product, presence or absence of motile stages, nature of that motility, and characteristics of organs of motility? All of these are aspects to be considered. Which are the more important in showing phylogenetic affinity? It is simply a matter of judgment. I do not think anyone can say with certainty. More recently biochemical studies have provided additional data. Cell wall constituents, both primary and secondary; pigmentation, including the presence of flavonoids; biochemical pathways to synthesis of various secondary compounds—all have provided additional information.

In general, I believe one must consider that those aspects of an organism providing distinct advantages would, under the influence of selective pressure, probably evolve into various, and unrelated, groups of plants. The characters on which one must base phylogenetic considerations must be those that would seem to provide no particular advantage or may in fact be somewhat selected against; the theory being that these probably would not have arisen many times. Thus, if different groups of plants possess these characteristics we might consider they had a common origin. But how do we know what characteristics may or may not have been slightly selected against during the time organisms were evolving? Obviously, a variety of climatic conditions existed during the millions of years that the different groups of plants were evolving and becoming distinct one from another. What kind of selective influence these different conditions might have had on a particular trait can only be guessed.

So what is the phylogenetic relationships between extant groups? In many cases it is pure supposition. For many years it has been stated that the terrestrial higher, green plants had their origin from the green algae because of similarity between their chlorophyll and carotenoid pigments, cellulose in the cell wall, starch as a storage product, 9 + 2 flagellar structure, alternation of generations, and so on. This supposition led to two alternative theories for how this event occurred. One was the idea of *antithetic alternation of generations.* This theory interprets the sporophyte as a new structure, arising from the prolonging of meiosis in the zygote. As a consequence, the zygote underwent a few mitotic divisions producing some sterile vegetative tissue. If the event of meiosis was prolonged or put off long enough, a sporophyte as we find it in higher plants would have resulted. But after it was discovered that some algae already had a sporophyte and it was identical in appearance to the gametophyte (alternation of isomorphic generations), it was proposed that the sporophyte was not really a new structure but simply a gametophyte modified to produce spores. This was called the *homologous theory.* When attempting to describe these two differing theories, eminent botanists will describe the homologous theory as, ". . . the postponement of the reduction division from the zygote to the tetrasporangium . . ." (Fritsch), or, ". . . derived from haploid organisms simply by suppression of meiosis at the time of germination of the zygote" (Stebbins). Other prominent botanists in discussing the antithetic theory states, ". . . the sporophyte . . . has been interpolated between successive gametophyte generations because of a delay in meiosis" (Bold), or ". . . the interpolation of a growth phase between syngamy and meiosis. . ."(Wahl). The arguments sound pretty much alike. In that case, why have botanists been arguing for years which theory was correct? Primarily, the argument has been whether the sporophyte was an immediately full-sized isomorphic one, with heteromorphic generations being derived from it, or whether the sporophyte simply developed gradually into a heteromorphic one.

The main thrust of both of these theories was in relating the green algae to the mosses and the mosses to the other higher plants. There is beginning to be a good body of evidence available which would deny both of these associations. The bryophytes are probably a polyphyletic group, but one of those ancestors was not a green algae. Nor is there any evidence that the vascular plants developed from a bryophyte ancestor. In fact, most of the evidence is to the contrary. How did the bryophytes evolve? We do not know, but we could postulate that a common, primitive ancestor produced both the green algae and the bryophytes, with each group separating very early during their evolutionary development. We could also postulate a polyphyletic origin: that the various groups of green algae and bryophytes arose from entirely different ancestral types and that they simply became more similar during evolutionary development. The first of these suggestions, that the groups arose from a common ancestor, separated, but then evolved along similar lines is called *parallel evolution.* The second concept, that they evolved from different ancestors but then became more similar is called *convergent evolution.* Of course, even groups that look quite distinct from each other may have had a common ancestor. The origin of groups from a common ancestor with the subsequent evolution of very dissimilar characteristics is called *divergent evolution.*

## *Evolution of Flowering Plants*

Although we can draw few definite conclusions concerning the evolution of the so-called lower plants, we do have better information concerning the evolution of flowering plants. Fossil angiosperms have been identified from the lower Cretaceous, some 125 million years ago. They seem to appear rather suddenly in the fossil record, an event the paleobotanist Axelrod attributes to their place of origin. He suggests that evolution of the angiosperms may have occurred away from the lowland basins, the areas where fossils are usually found. Thus, the earliest angiosperms simply were not fossilized. As evolution of the flower occurred and mechanisms for dispersal of pollen and seed became more efficient, the angiosperms became established in the lowland areas. By this time, the characteristics of the group were fairly well established.

Angiosperm evolution is postulated to have progressed through an extinct group called the Rhyniophytina. This group contains several known fossil members, including one first discovered by two Englishmen, Kidston and Lang, from the Old Red sandstone belts near Rhynie, Scotland. The plant was called *Rhynia*, in honor of its place of discovery (Figure 17.3). From the *Rhynia*-type plant, another fossil type *Psilophyton princeps*

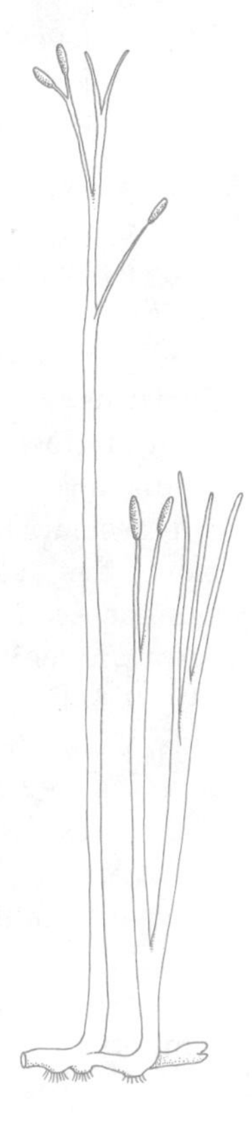

*Figure 17.3*
Diagrammatic reconstruction of *Rhynia major*. [Redrawn from Kidston and Lang, *Trans. Roy. Soc. Edinb.*, vol. 52, part IV, 1921.]

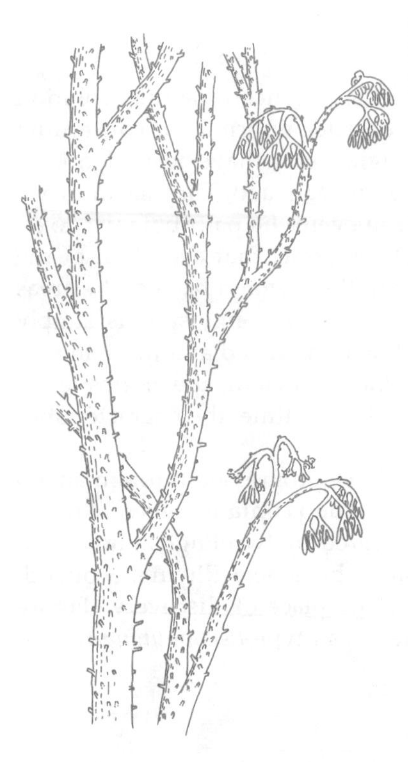

*Figure 17.4*
*Psilophyton princeps* Dawson. Restoration by Hueber (1967). [After Andrews and Kasper.]

(Figure 17.4), may have evolved. A *Psilophyton*-type plant is suggested as the progenitor of the extant groups that include *Equisetum*, *Psilotum*, and the true ferns. In addition, it is also suggested as the progenitor of an extinct group called the progymnosperms. From the progymnosperms supposedly arose the gnetums, conifers, ginkgos, cycads, seed ferns, and angiosperms.

The angiosperms were suggested as originating as an offshoot from the line that also produced the seed ferns, or pteridosperms (so-called because of the resemblance of their leaves to those of ferns; Figure 17.5). Other botanists have suggested that the angiosperms may have evolved directly from the seed ferns. The fossil evidence of stages between the pteridosperm progenitor and the protoangiosperm is not known.

In 1934, Dr. Albert Smith collected a strange looking plant in the Fiji Islands, which he could not identify. Later, it was found to be similar to a flowering tree collected in 1941 by Otto Degener. The great botanist I. W. Bailey classified the plant as a member of the dicot order Ranales, and named it *Degeneria*. The microsporophylls are broad and flat and the carpel is very leaflike during its development, becoming an inwardly folded, three-veined sporophyll in which the two rows of ovules are borne some distance from the edge of the megasporophyll. It is thought that early members of the magnoliaceae (a family within the order Ranales) may

*Figure 17.5*
Reconstruction of the seed fern *Medullosa noei.* [Redrawn from W. N. Stewart and T. Delevoryas, *Bot. Rev.* 22:45, 1956.]

have looked very much like *Degeneria.* Because of this evidence, plus the fact that the treelike habit was well developed in the pteridosperms, it is thought that the protoangiosperms were treelike in appearance, with the present herbaceous habit in both dicots and monocots being a derived, and therefore a more modern, characteristic. Both the modern monocots and dicots are thought to have evolved from some magnoliaceous ancestor.

The occurrence of fossil gymnosperms and angiosperms in regions where they no longer grow, and the occurrence of particular species in various, distant parts of the world presented botanists with a considerable problem. This was a particular enigma for plants having seeds too heavy, too dense, and too large to have been distributed by wind, water, or animal vectors. How did such plants obtain such widely scattered distribution? Continental drift, discussed earlier, may help to explain it.

## *Evolution of the Flower*

We have previously discussed the parts of a flower. How did such an organ evolve? We can only surmise this from a comparative study of modern forms because flowers, as you might expect, are rarely found in the fossil record, and those that are found are generally poorly preserved.

The most primitive carpel is considered to be simply a leaflike blade folded lengthwise, enclosing a number of ovules attached to the inner surface. The margins are covered with stigmatic hairs and the edges are not sealed. As that type of carpel evolved, its margins became sealed, and the stigmatic region became localized at the apex as a "crest." Eventually the upper portion of the carpel became elongated into a style, with the stigmatic part occupying a small region at its tip. Such a progression would result in the formation of a simple ovulary. The origin of carpels in close proximity to each other and the subsequent partial or complete fusion of these would result in formation of a compound ovulary (Figure 17.6).

The stamens of primitive flowers are likewise flattened leaflike structures with microsporangia borne in their upper or lower surfaces. The modern stamen may have evolved from such a structure through differentiation of the blade into a slender stalk bearing the microsporangia at its tip. It is also possible that the stamens of many plants are not homologous to leaves at all but rather represent the modification of a branch tip bearing terminal sporangia. Stamens may also become fused, and some have become secondarily sterile, being modified into nectaries, the glands secreting sugary nectar that attracts various pollinators.

The perianth may simply consist of modified leaves: the sepals look very leaflike, and the petals may represent modified sepals, or the petals may be modified stamens. Petal fusion has occurred in a number of different flowering plants.

In general, the primitive flower is considered to have many parts (indefinite in number), and consists of an elongated shoot having the four flower parts arranged on it in a spiral fashion, a superior ovulary, and radial symmetry. The evolution of diversity of flower types illustrates an excellent example of coevolution. It is assumed that the earliest angiosperms, like various groups of gymnosperms, were wind pollinated. Drops of sticky sap were undoubtedly present on the stigmatic area in which pollen grains would become trapped. Insects, particularly beetles it is thought, visiting the plant to feed on sap and resin exuding from it would have encountered the sap and pollen grains of the ovule. Once they had discovered this source of food, they probably would have returned to it and to other structures of the same type, thus picking up and carrying a little pollen from plant to plant. Such a mechanism for pollination would probably have had some selective advantage over wind pollination. Those plants that developed substances that would be more attractive to insects, such as nectaries, would have a selective advantage over those less attrac-

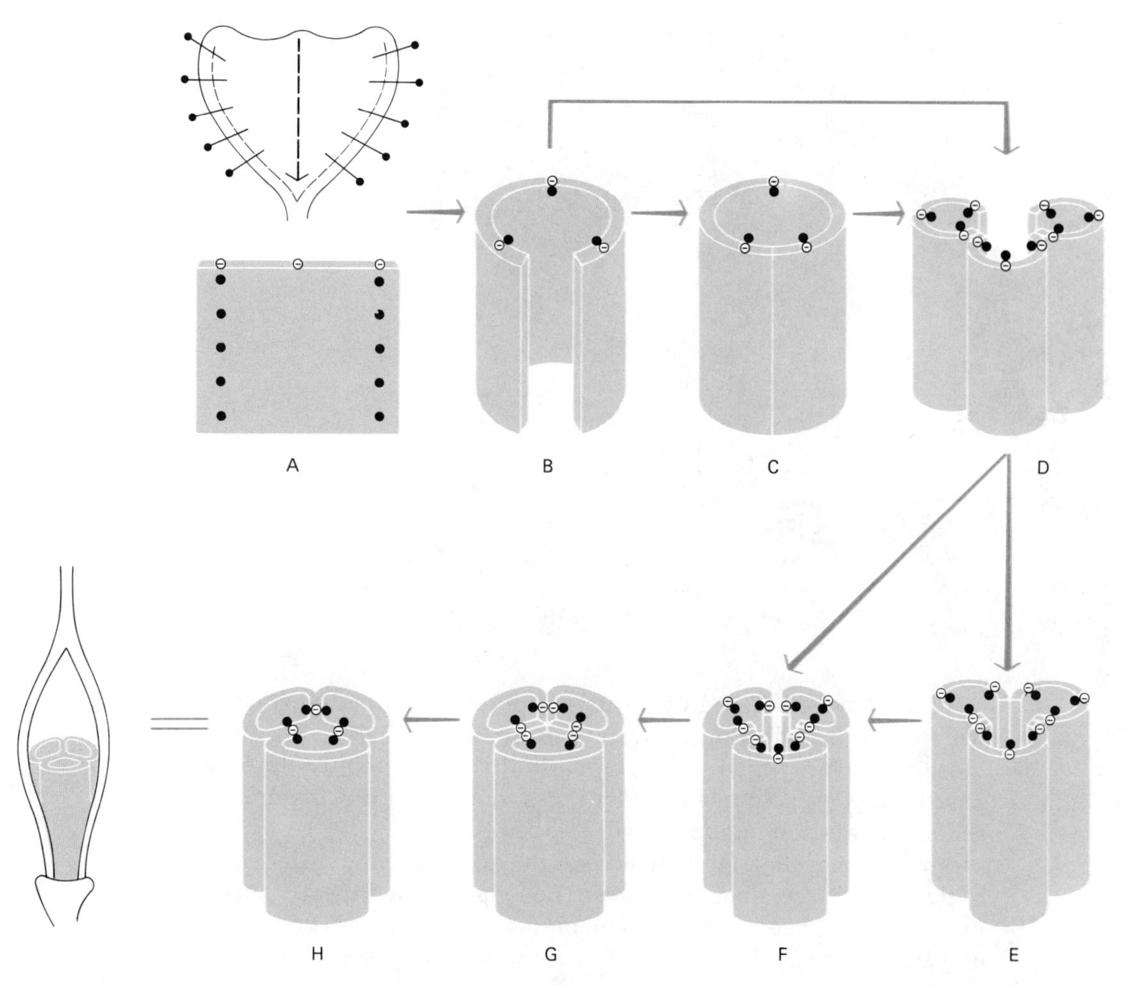


*Figure 17.6*
A theory on the evolutionary development of the compound ovulary. *A.* Carpel with submarginal ovules. *B.* and *C.* Ovulary of a single carpel formed by infolding and fusion of margins. *E.* to *H.* Ovularies formed by fusion of separate carpels.

tive to insects. As some insects were just as likely to eat the ovule as sample the sap, those plants which developed methods resulting in protection of the ovule, such as having it enclosed within the carpel, would have a distinct reproductive advantage. Another obvious advantage to the plant would be having both sex organs in the same flower. The insect visiting the ovule or ovulary would pick up some pollen at the same time, thus pollination would not necessarily depend simply on the chance transfer of pollen already on the ovule or stigmatic area.

*Figure 17.7*
An example of a beetle-pollinated flower. [From a color photograph by Dee Ann Houston.]

As we have discussed previously, cross fertilization has greater potential for the introduction of variability than does self-fertilization; thus, those flowers in which self-incompatibility developed would necessarily have to be cross pollinated. Under these conditions, the numerous genetic self-sterility factors supposedly developed.

As floral morphology changed and the nectaries became less accessible, or perhaps inaccessible, to beetles, the long-tongued insects such as butterflies, moths, bees, and wasps made their appearance. Thus, coevolution of plants and insects developed. For as the various insects evolved, they in turn exerted a selective influence on floral morphology, resulting, in time, in still greater diversification in floral types.

Figure 17.7 shows an example of a beetle-pollinated flower. In general, such flowers are not showy but have very distinctive odors, as beetles do not depend much on visual sense. These odors are not the sweet types of odors associated with bee, butterfly, or moth-pollinated flowers. They are dung type or fetid, spicy, or fruity type odors. Most have their ovules buried, thus they are not as likely to be eaten by the beetles. Flies also act as pollinators. An example of a fly-pollinated flower is shown in Figure 17.8.

Bee flowers are often quite showy, as bees do perceive colors, although not in the same way we do. They cannot distinguish between certain colors such as yellow, yellow-green, and orange. They see red as black, and they do perceive ultraviolet markings. Special markings may be present in some flowers that lead the way to the location of the nectar. These are termed *honey guides.* Such honey guides may be in visible color or in ultra-

*Figure 17.8*
An example of a fly-pollinated flower. [From a color photograph by Dee Ann Houston.]

violet. Some bee or wasp-pollinated flowers have evolved distinctive morphological variability. Many are zygomorphic, with the enlarged lip petal providing a landing platform for the insect. The flowers of the orchid genus *Ophrys* are modified in such a way as to mimic both the form and odor of a female bee (Figure 17.9). The male bee, deceived by this impersonation, goes from flower to flower attempting to copulate with them. Pollination occurs incidentally as the bee moves from flower to flower performing the mating movement. Still other orchids mimic insects in form and odor. In these instances, the insect does not attempt to mate with the flower but rather recognizes it as an intruder. In an attempt to protect its home range, it attacks the flower trying to drive it away. In the ensuing "battle" pollen is knocked off onto the insect, and deposited on the next flower the insect battles.

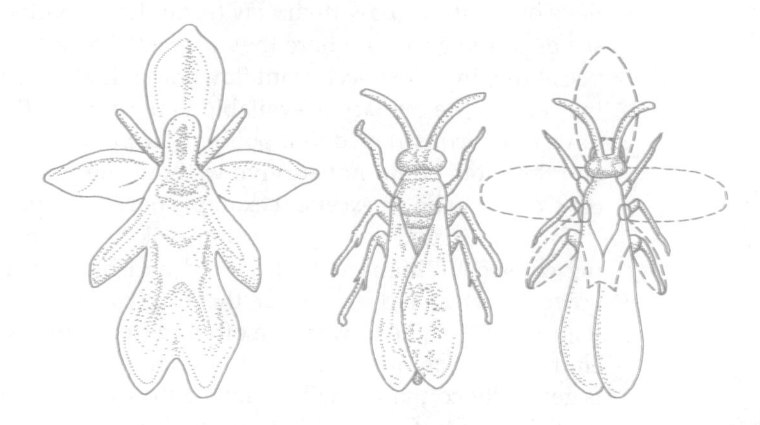

*Figure 17.9*
*Ophrys insectifera.* This orchid has a lip that resembles the female *Argogorytes mystaceus.* A male wasp, attempting copulation, picks up pollen.

*Figure 17.10*
The *Yucca* flower. This flower is pollinated by a night-flying pronuba moth. [From a Kodachrome photograph. Courtesy of the Carolina Biological Supply Co.]

Moth- and butterfly-pollinated flowers, like bee-pollinated flowers, generally emit a strong sweet scent. As moths are nocturnal in their habits, the odor is strongly emitted only at night. The flowers pollinated by nocturnal insects are generally white, pale yellow, or pink. At least some species of butterflies are able to perceive reds, and many butterfly flowers are orange or red in color. The nectary of moth and butterfly flowers is typically found at the base of a tubular corolla; thus, only the long-tongued insects are able to get to it. One of the most unusual insect-flower relationships exists between the yucca plant and the pronuba moth (Figure 17.10). When the yucca flower opens, it is visited at night by the moth, who collects the pollen, rolls it into a tight little ball and carries it in her mouth parts to another flower. There she pierces the ovulary wall of the flower with her ovipositor, lays eggs among the ovules, then packs the mass of pollen into the hollow style. The moth larva and seeds develop simultaneously, with the larva feeding on the yucca seeds. It is estimated that about 20 percent of the seeds may be eaten. When the larvae are fully developed, they gnaw their way through the ovulary wall and lower themselves to the ground where they pupate. The adult emerges from the pupae at the time the yucca plant flowers again. In the meantime, the remaining seed in the ovulary is available to perpetuate the species. As the yucca flower is so constructed that pollination could not occur without the moth, and the moth could not survive without the yucca plant, this situation is considered to be an excellent example of coevolution.

Bird-pollinated flowers are often strikingly colored, often with contrasting color combinations such as red and blue, or red and green. All red flowers, particularly those of the tropics, are generally bird-pollinated (Figure 17.11). The flowers are often rather strong and heavy textured and emit copious amounts of nectar (which usually has little odor). The nectar is generally contained within long tubes which only the birds with their long slender beaks are able to reach. It would seem that the red color, in-

*Figure 17.11*
An example of a bird-pollinated flower. [Courtesy of U.S. Department of Agriculture.]

*Figure 17.12*
An example of a bat-pollinated flower. [Courtesy of U.S. Department of Agriculture.]

visible to most insects, the lack of odor, and the protection of the nectar serve to ensure that insects will not visit such plants. If they did, such copious nectar would surely more than satisfy their need, and they would not likely move on to other flowers. Thus, no cross pollination would result. The birds, however, with their high metabolic rate, require much more nectar to satisfy their requirements, thus move from flower to flower.

Bat-pollinated flowers (Figure 17.12) are generally similar to bird-pollinated flowers in many respects. However, bats are nocturnal; thus, the flowers generally open at night and are not brightly colored but rather are white or pale-yellow.

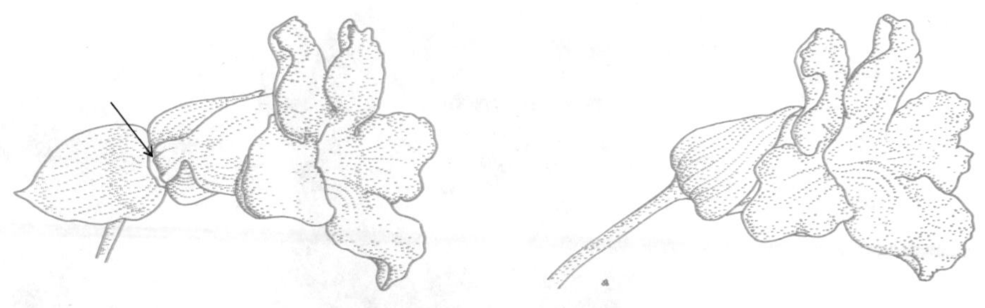

*Figure 17.13*
*Thunbergia* sp. flower. Note the extrafloral nectaries (*arrow*). These are visited by ants that serve to protect the flower from nectar thieves.

Another unusual flower-insect relationship exists in the flowers of *Thunbergia*. Some insects, such as the robber bees, do not go into the flower for nectar but drill through the base of the corolla to the nectaries. *Thunbergia* has accessory nectaries externally at the base of the flower, and these are visited by ants (Figure 17.13). The ants consume the nectar while at the same time discourage visits by the robber bees, thus protecting the flower and insuring pollination by the normal insect vectors.

Wind-pollinated angiosperms are thought to be secondarily evolved. Their flowers are characterized by emergent stamens producing small, light, nonadherent pollen grains. The stigma are also generally emergent and often have feathery outgrowths or branches which serve to interrupt the wind-blown pollen grains (Figure 17.14). Generally, wind-pollinated species occur in groups as the method is very inefficient and nearly all the pollen will be deposited within 100 meters of its source.

## SYNOPSIS

In our discussions of evolutionary methods, mechanisms, and pathways, we made a number of points, introduced considerable theory, but drew few conclusions. The main thrust of our discussion was that plant speciation did not occur through spontaneous generation but rather evolved. If we direct our attention to how life evolved rather than how species evolved, as we discussed it in Chapter 3, to discount the idea of spontaneous generation would leave us in quite a quandary. George Wald, in the August, 1954 edition of *Scientific American* perhaps states it best:

> I think a scientist has no choice but to approach the origin of life through a hypothesis of spontaneous generation. What the controversy reviewed above (spontaneous generation of species) showed to be untenable is only the belief that living organisms arise spontaneously under present conditions. We have now to face a somewhat different problem; how organisms may have arisen spontaneously under different conditions in some former period, granted that they do so no longer.

*Figure 17.14*
An example of a wind-pollinated flower. [From a Kodachrome by the Carolina Biological Supply Co.]

Thus, we have come full circle. A theory, widely accepted at one time, which hundreds of years of effort by many scientists, angry discourse, and reams of pages of print finally disproved, is now once again accepted. It is accepted, however, with qualification and in an entirely different light.

In considering any viewpoint concerning the evolution of plants, we should heed the warning of Dobzhansky:

> Theories that ascribe evolution to "urges" and "telefinalisms" imply that there is some kind of predestination about the whole business, that evolution has produced nothing more than was potentially present at the beginning of life. The modern evolutionists believe that, on the contrary, evolution is a creative response of the living matter to the challenges of the environment. The role of the environment is to provide opportunities for biological inventions. Evolution is due neither to chance nor to design; it is due to a natural creative process.

## SUMMARY

1. The origin of different groups of plants was probably polyphyletic.
2. At one time, the origin of new species was considered to be due to spontaneous generation.

3. Considerable argument and conflicting opinions existed concerning spontaneous generation until the idea was laid to rest through the work of Louis Pasteur.
4. The idea that species may have evolved existed in some form from the time of the Greeks, but it was not until Darwin and Wallace suggested a mechanism for evolution, natural selection, that any widespread acceptance of the theory was evident.
5. The forces of change have been identified as mutation, natural selection, recombination, and gene flow.
6. The forces above effect the equilibrium of alleles in a population; this equilibrium being known as Hardy-Weinberg equilibrium.
7. The effect of mutations is deleterious generally, often not noticeable immediately, and has generally only a slight effect on the Hardy-Weinberg equilibrium.
8. Recombination does not change the ratio of the alleles in a population, but only the ways in which they are combined; thus, it has its effect on phenotype.
9. Natural selection also has its effect on phenotype, and may be stabilizing, disruptive, or directional in its effect.
10. The introduction of new genes into a population by gene flow has probably little effect on diversity.
11. Geographical isolation may be an important factor in speciation, resulting in the formation of various ecotypes.
12. Some species show considerable genetic plasticity, thus show considerable phenotypic variations although genotypically similar.
13. A plant species has been described as a group of actually or potentially interbreeding natural populations which are reproductively isolated from other such groups.
14. New species may be formed by phyletic speciation or by true speciation.
15. Phyletic speciation involves the change of one species into a different one.
16. True speciation involves the separation of gene pools and the independent evolution of those gene pools.
17. One way in which gene pools may have become separated is by continental drift.
18. Some species can become established in a new area only if there is a lack of competition. The effect of competition on some plants is stated in the Gause exclusion principle that no two forms can share exactly the same environmental requirements for an indefinite period of time.
19. Hybridization is extremely important is introducing variability into a population, particularly that of introgressive hybridization.
20. Abrupt speciation may occur through polyploidy or catastrophic selection.
21. Pathways of evolution are inferred on the basis of paleobotanical, morphological, anatomical, and biochemical information.

22. The evolution of characters in a plant group could have come about through parallel, convergent, or divergent evolution.
23. It is suggested that the seed plants evolved from a fossil group called the progymnosperms which, in turn, had evolved from a *Psilophyton*-type plant.
24. The flowering plants evolved supposedly from a group known as the pteridosperms, or seed ferns, or from a common ancestor to that group.
25. The most primitive angiosperm group is supposedly the magnoliaceae family of the order Ranales.
26. Both the monocots and dicots supposedly had their origin from a Magnoliacean-type ancestor, with the herbaceous members of each representing derived groups.
27. It is proposed that the angiosperm flower coevolved with certain insect groups.

# Chapter 18

## PLANT ECOLOGY

Throughout this text we have been discussing the effect of various ecological parameters on plant form and function. We now want to look at those aspects which comprise plant ecology in a more integrated way.

The word ecology was actually first proposed by the zoologist Reiter in 1865. He combined the Greek words *oikos* ("house" or "home") and *logos* ("study" or "discourse"). One year later Haeckel, another zoologist, defined ecology in its modern sense as "the body of knowledge concerning the economy of nature—the investigation of the total relations of the animal to its inorganic and organic environment." We now know that the term plant could be substituted for animal in Haeckel's description and be equally as appropriate.

Aspects of plant ecology have their origins back in 1305 when de Crescentius became the first to recognize the existence of competition among plants. King, in 1685, was the first to describe the concept of succession, a concept which gained strong support following the work of Warming in the late 1800s, Cowles (1899, 1901), Tansley (1911), and Clements (1904, 1916). The relationship of soil, vegetation, and climate was first developed by the Russian Dokuchaev in the late 1800s and by the American Coffey in 1912.

Plant ecology as a science essentially had its beginnings in plant geography when early taxonomists such as Humboldt, de Candolle, Engler, Gray, and Kerner described the distribution of plants. Such description quite naturally led to questions as to why plants were so distributed. Natural history studies have also been a primary developmental source of ecology.

The basic approaches of modern ecology attempt to answer questions dealing with one or more levels of study: 1) the organism, 2) population, 3) community, and 4) ecosystem. Up to this point, our interest has been with the individual organism and aspects affecting its morphology, anatomy, physiological processes, growth, development, and others. Population ecology considers the interactions associated with a number of organisms of the same species, while community ecology deals with the interactions and interrelationships of populations. Ecosystem ecology, on the other hand, attempts to include all aspects of the interactions of all organisms in a given area with both the biotic and abiotic elements of their environment.

## PLANT GEOGRAPHY

As mentioned above, one of the first aspects of plants to interest plant ecologists was that vegetation of different parts of the earth was different. The Austrian botanist Anton Kerner von Marilann stated in 1863, "Every plant has its place, its time, its function, and its meaning—In every zone the plants are gathered into definite groups which appear either as developing or as finished communities but never transgress the orderly structure and correct composition of their kind." This statement exemplifies the teleological approach taken by many early plant ecologists and unfortunately an approach utilized by many modern ecologists as well. Such personification discouraged investigations into cause and effect relationships but rather led to consideration of different aspects. Geographers drew maps of the world showing the aerial extent of each of the world vegetation types, and indeed, the world seemed to be divided up into neat parcels called tropical forest, temperate forest, needle leaf forest, evergreen hardwood forest and savanna, steppe and half desert, heath, dry desert, tundra and cold woodland, cold desert, and others. Each of these groups of vegetation seemed to have their own particular form; thus, they were called *formations.* In a general way, the formations appeared to succeed each other on a latitudinal basis, and this pattern seemed to some botanists to be repeated as one traveled up a mountain. Thus was established the generalization that one would encounter the same plant formations in the same order traveling up a mountain as would be encountered if one simply traveled northward.

From this point of focus, a whole series of questions seemed relevant. Why should this be? Why should large areas of the land have vegetation of a special form? Why, as stated by Kerner, should every plant have its place and the different formations "respect each others territory?"

The noted plant taxonomist, Alphonse de Candolle, was the first to seriously consider the answers to these questions. While gathering plants from all over the world, he became aware of the ecological problem of formations. To him, it seemed obvious that the answer lay with some aspect of weather; rainfall and temperature appearing to be most critical. On this basis, he postulated five plant groups: megatherms, xerophiles, mesotherms, microtherms, and hekistotherms (most to least heat required, respectively). De Candolle's system, based as it was on climatic elements, soon became the basis for mapping climate. In other words, vegetation maps were used to form climatic maps.

So each formation has its own units of form. This is known. This can be observed. But why is it so? In 1934, Raunkiaer devised a system for classifying life forms based upon the assumption that climate is the controlling influence. If this is so, Raunkiaer reasoned that plants must have evolved some mechanism for existing under the extremes of climatic conditions of the region. He reasoned further that the perennating bud, that structure by which the plant "renews" itself periodically, would be the most obvious structure to look at. As a result of his studies, Raunkiaer erected classes or categories of life forms, each defined by the relative exposure of its perennating bud. Raunkiaer expected that the composition of life forms in each formation would be characteristic, that each would reveal its own *life form spectrum,* and so it was. Each of the formations named by de Candolle could also be described by a life form spectrum.

What are the various associations recognized today? It depends upon who is doing the classifying. Whittaker recognizes over 30 different formation types.

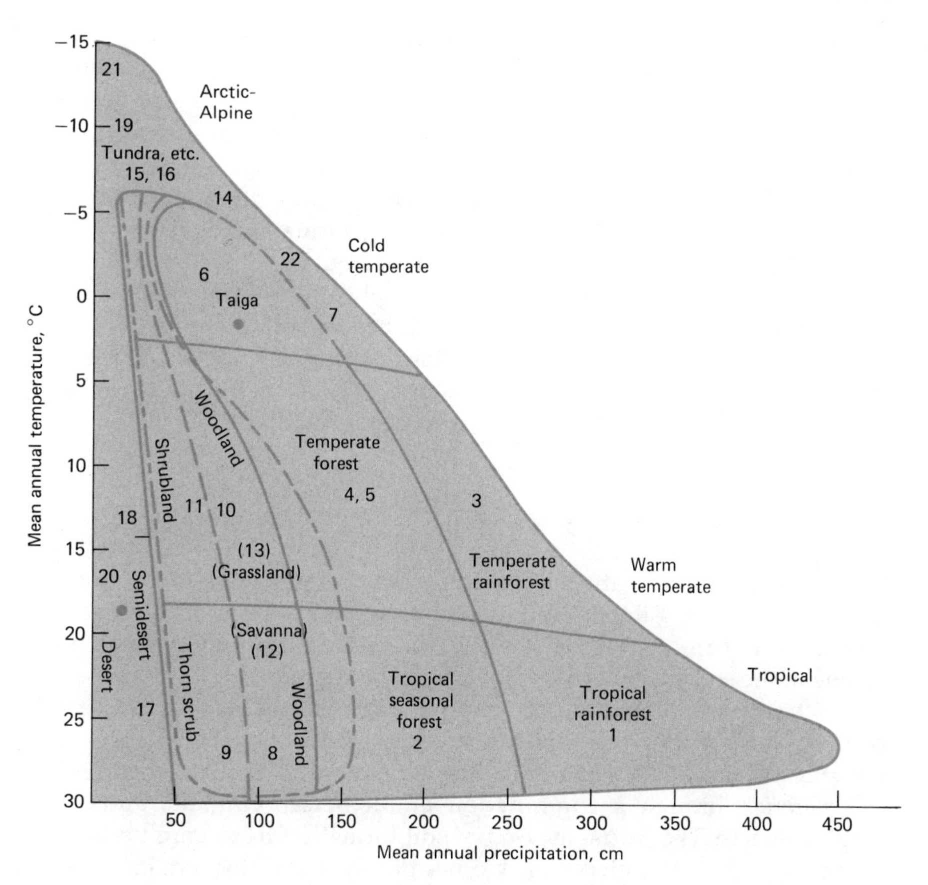


*Figure 18.1*
A pattern of world biome-types in relation to climatic humidity and temperature. Boundaries between types are approximate. In climates between forest and desert, maritime versus continental climates, soil effects, and fire effects can shift the balance between woodland, shrubland, and grassland types. The dot-and-dash line encloses a wide range of environments in which either grassland, or one of the types dominated by woody plants, may form the prevailing vegetation in different areas. (1) Tropical rain forest; (2) tropical seasonal forests; (3) temperate rain forests; (4) temperate deciduous forests; (5) temperate evergreen forests; (6) taiga; (7) elfinwoods; (8) tropical broadleaf woodlands; (9) thornwoods; (10) temperate woodlands; (11) temperate shrublands; (12) savannas; (13) temperate grasslands; (14) alpine shrublands; (15) alpine grasslands; (16) tundras; (17) warm semidesert shrubs; (18) cool semideserts; (19) arctic-alpine semideserts; (20) true deserts; (21) arctic-alpine deserts; (22) hydric. [From R. H. Whittaker, *Communities and Ecosystems.* Copyright © 1975, Macmillan Publishing Co., Inc., New York.]

Such formation types can be arranged into a pattern in relation to the major climatic gradients; rainfall and temperature, mentioned earlier as being major influences (Figure 18.1). But the whole idea of discrete geographical areas of vegetation, whether latitudinal or altitudinal, is a gross oversimplification. The boundaries are purely man made, and natural communities or formations do not sharply abut each other except under unusual circumstances. Rather, they intergrade continuously, sometimes over hundreds of miles. Neither are the various growth forms adapted similarly in different continents. Under certain climatic conditions, forests may be formed in one continent while only grasslands or shrublands may develop on others. More recent studies have indicated that climate is not the only factor responsible for determining formation. Soil conditions, frequent burning, and other factors may be equally important. Thus, the formation as a unit of plant structure is simply a conception of man. It has some usefulness in a very general way, it has served to stimulate further research on plant distribution, but it does not indicate the specific global distribution of plant growth forms, nor does it give any insight to the more important aspects of interactions within each formation.

## SOILS

Both climate and vegetation, linked as we indicated above, collaborate in forming the soil. Climate influences vegetation which influences soil which, in turn, influences vegetation. This cycle, continuing through time, results in the soil characteristic for a particular region.

Soil has been defined as the weathered superficial layer of the earth's crust containing living organisms and the products of their decomposition. Soils are formed through the long-term modification of *parent geological material* through a combination of biological, climatic, and topographic effects. The initial aspects of soil formation from bare bedrock is termed *weathering,* and is, for a time, purely a physiochemical process (Table 18.1). The process rapidly becomes biogenic as organisms such as blue-green algae, lichens, and other pioneering species begin to grow on the weathered material, and organic substances begin to incorporate into the crusts of weathered rock debris (see Figure 7.29).

The interaction of various soil-forming factors results in the differentiation of a variety of soil types. These are defined by the nature of the mineral matrix, the vertical distribution of organic matter, and the movement and redistribution of various inorganic constituents. If we dig a pit in the soil mantle, a succession of layers or *horizons* will be exposed. The Soil Survey Staff of the U.S. Department of Agriculture (1960) has defined a soil horizon as "a layer which is approximately parallel to the soil surface and that has properties produced by soil forming processes but that are unlike those of adjoining layers." Horizons can be classified in regard to both position and constitution. Five main groups of horizons have been identified and are designated *O, A, B, C,* and *R.* The *O* horizon is the organic horizon formed on the surface and dominated by fresh or partially decomposed organic matter. The *A* horizon formed either at or adjacent to the surface and is a mineral horizon enriched with either or both organic mat-

*Table 18.1*
EXAMPLES OF THE VARIOUS WEATHERING PROCESSES

| PHYSICAL WEATHERING | CHEMICAL WEATHERING |
| --- | --- |
| **WETTING-DRYING** Disruption of layer lattice minerals which swell on wetting | **HYDRATION** Reversible change of hematite to limonite which is accompanied by swelling and so disrupts cementation of sandstones and other rocks $Fe_2O_3 \rightleftharpoons Fe_2O_3 \cdot 3H_2O$ |
| **HEATING-COOLING** Disruption of heterogeneous crystalline rocks in which inclusions have differential coefficients of thermal expansion. Surface flaking of large boulders, particularly in arid climates, due to sun heating | **HYDROLYSIS** Silicate breakdown $K_2Al_2Si_6O_{16} \rightarrow Al_2O_3 \cdot 2SiO_2 \cdot 2H_2O$. K and surplus Si are washed away in solution |
| **FREEZING** Disruption of porous, lamellar, or vesicular rocks by frost shatter due to expansion of water during freezing | **OXIDATION-REDUCTION** $Fe^{3+} \rightleftharpoons Fe^{2+}$ causes disruption of cementation as $Fe^{2+}$ is much more soluble than $Fe^{3+}$ |
| **GLACIATION** Physical erosion by grinding process | **CARBONATION** $CaCO_3 \rightleftharpoons Ca(HCO_3)_2$ leads to solution loss of limestone or disruption of $CaCO_3$ cemented rocks as the hydrogen carbonate is more soluble than the carbonate |
| **SOLUTION** Removal of more mobile components such as Ca, $SO_4$, Cl | **CHELATION** Essentially a consequence of biochemical activity, various metals being dissolved as chelates with organic products of plant and microorganism activity |
| **SAND BLASTING** Erosion of upstanding rocks in arid, desert, conditions | |

*Source:* From J. R. Etherington, *Environment and Plant Ecology*, New York. Wiley, 1975, p. 58.

ter and downward loss of soluble salts, clay, iron, or aluminum and the consequent enrichment with silica or other resistant minerals. The *B* horizones are formed below the surface and have one or more of the following features: 1) enrichment with inwashed clay, iron, aluminum, manganese, or organic matter, 2) residual enrichment with sesquioxides or silicate clays, 3) sesquioxide coatings of mineral grains sufficient to give differential color to the layer, or 4) alteration of the original rock material to give silicate clays or oxides under conditions where 1, 2, and 3 do not apply. The C horizons are mineral layers below the *B* layer, excluding true bedrock, while the *R* horizon is bedrock.

Soil classification is greatly influenced by the work of the Russian Dokuchaev (1900), and the classical United States system was based largely upon it. Soils were classified into orders, suborders, great groups, subgroups, families, and series. It is the great groups that are most often referred to. These include tundra, podsol, grey-brown podsolic, red-yellow podsolic, lateritic, chernozem, chestnut, brown, desert, and others.

The characteristic of each great group is dependent, in part, upon the effects of climate. Under extremely cold conditions, soils undergo a type of development referred to as *gleization*. In the more temperate regions, soil-forming processes are of a type called *podsolization*, while in the subtropical and tropical regions, *laterization* is the characteristic weathering process. Under arid conditions, a characteristic process called *calcification* will occur.

In artic regions, where the soils are cool, continually wet but not saline, a sticky, compact, structureless layer develops at the bottom of the *B* horizon. This layer, called the *glei*, is blue gray in color because of the essentially anaerobic reduction of iron compounds. As a consequence of this layer, accumulation of peaty material at the surface occurs, and because of expansion and contraction of the soil water on freezing and thawing, a mixing of the soil horizons results. Such conditions characterize the tundra soils.

In cool, moist climates, particularly where coniferous forests predominate, the slow decomposition of the leaf litter and other organic debris results in the production of varying amounts of acids. Thus, rainwater percolating through the soil is acid and dissolves out the free carbonates and adsorbed basic ions, a process termed *decalcification and leaching*. This results in an increase in soil acidity, which in turn brings about leaching of iron and aluminum compounds, leaving the relatively insoluble silica behind in the *A* horizon. Thus, podsol soils are highly acid, of low fertility, with an ashy-gray *A* horizon of siliceous sand and dark-brown *B* horizon of accumulated iron oxides leached out of the *A* horizon.

The gray-brown podsolic soils develop under slightly warmer conditions than do the podsols, and typically under deciduous forests of maple, beech,and others. This type of vegetation results in the return of more bases to the soil. The warmer climate results in greater decomposition of leaf litter and other organic materials. Thus, the soil is more fertile, and the irons and aluminums are not leached out of the *A* horizon to the degree they are in podsol soils. Mixing of the *A* and *B* horizons may occur as the result of earthworm activity, and the soil may form specific structural aggregates.

Under conditions of warm temperature and high rainfall, silica is converted to silicic acid and lost by leaching, leaving a residuum of iron and aluminum oxides. Such a process, called *laterization*, results in the formation of reddish-colored soils. These soils are extremely low in organic content due to the high rate of decomposition. In the Southern part of the United States, both laterization and podsolization occur to some degree producing a red-yellow soil, low in fertility, lacking calcium and phosphorus, having more iron and aluminum in the *A* horizon than typical podsols but less than in tropical soils. They also have less silica in the *A* horizon than the typical podsol but more than found in true lateritic soils of the tropics.

In grasslands and desert regions, precipitation is insufficient to leach calcium and magnesium from the *A* horizon, and percolation waters are in intimate contact with organic substances which tends to adsorb the cations as they are released. Ions of calcium and magnesium which are not adsorbed will leach only a short distance before they are precipitated. Thus, a dark-colored *A* horizon of extreme depth and of high organic content is formed. This horizon is saturated with calcium carbonate; thus, the pH rarely falls below 7. It typically rests directly on a highly calcareous *C* horizon. At some level, there will be a zone of calcium carbonate deposition. Such soils include the *chernozem*, *chestnut*, and *brown* soils. In the more southern portions of the grasslands, where weak laterization has occurred, soils will have a slightly reddish tinge.

The desert soils typically have a very low organic content. As leaching is minimal, the soil will be calcareous, or under certain conditions, saline.

When one considers the conditions under which soil formation occurred, the influence of man can begin to be appreciated. The forests were cleared, the grasslands plowed. Wind and water carried away much of the organic-rich topsoil, and the farmer made certain it was not returned by carrying off the produce of the land. Under such conditions, soil is not a renewable resource.

## PRINCIPLES OF ECOSYSTEMS

Geography and soils are two aspects of habitat, and these play a basic role in determining the kinds of plants present in a specific area. On the other hand, the kinds of plants present will determine many characteristics of the habitat. Thus, the plants present in a given place (the *community*) and the environment of that region are inseparable, a concept recognized by the Greek philosopher Hippocrates but developed as a basic ecological concept by the Russian, V. V. Dokuchaev, in 1894. This idea of interaction of communities with each other and with their environment has led to development of the *ecosystem* concept in ecology. The ecosystem can be visualized as consisting of four basic components: 1) abiotic substances, 2) producers, 3) consumers, and 4) decomposers. A second way of viewing the ecosystem is to consider it to consist of the following: 1) abiotic elements (environment), 2) biotic elements (producers, consumers, decomposers), 3) energy input and utilization, and 4) nutrient input and cycling. The relationships between, and among, these various components could be illustrated schematically as in Figure 18.2.

### *The Abiotic Elements*

The abiotic elements comprise the totality of physical factors and chemical substances which interact with the biotic elements. These factors modify or regulate aspects of plant growth and development, dispersal, and reproduction.

How do these basic abiotic substances influence plant growth? Obviously, the basic requirements vary with the species and with the situation. In 1840 the German botanist and physiologist Justin Liebig observed

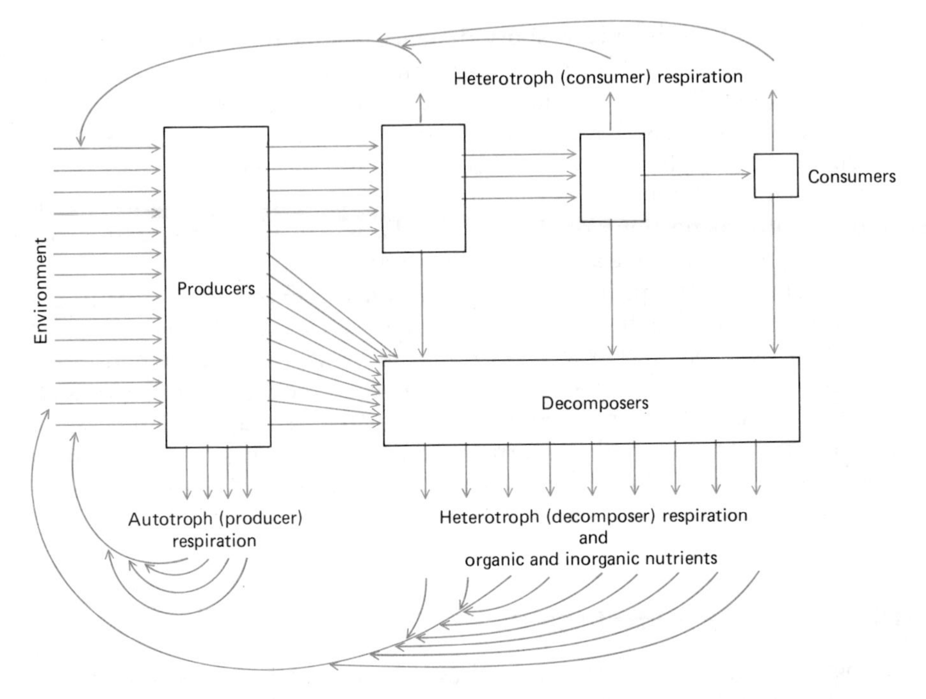


*Figure 18.2*
Relationships between and among the basic ecosystem elements.

that crop yield was dependent on the amount of nutrient present in the least quantity. His statement that "growth of a plant is dependent on the amount of foodstuff which is presented to it in minimum quantity" has since come to be known as *Liebig's law of the minimum.* The statement has been extended to include factors other than mineral nutrients by Blackman (1905), Taylor (1934), and others.

Subsequent work since Liebig's time has shown that certain modifications or constraints must be applied to Liebig's law if it is to have any practical usefulness. The first is recognition of the fact that Liebig's law is applicable only under steady-state conditions, that is, when the inflow of energy and materials balances outflow. But due primarily to human influence, steady-state conditions are rarely found in most temperate ecosystems. Under transient-state conditions, there is no theoretical basis for any one-factor hypothesis.

A second important consideration is factor interaction. The presence of one factor may modify the rate of utilization of another. Under certain conditions, an organism may be able to substitute a closely related substance for one deficient in the environment.

Not only may too little of something regulate plant growth, but also too much. Thus, organisms have both an ecological minimum and maximum.

This idea is embodied in *Shelford's* (1913) *law of tolerance*, which holds that each abiotic factor has a maximum and a minimum level for each constituent population of a given ecosystem. Between these points lies a range known as the *limits of tolerance*. As you might expect, organisms differ in their range of tolerance, and organisms having a wide range of tolerance for one factor may have a narrow range of tolerance for another. Organisms most likely to be widely distributed are those having a wide range of tolerance for all factors. As with Liebig's law, factor interaction may modify the limits of tolerance. Some plants may actually live under much less than optimum conditions with regard to a particular physical factor. In such instances, it is often found that other factors have greater importance. In other instances, population interactions (predators, competition, and so on) seem to be involved.

Interactions between the abiotic and biotic elements of an ecosystem can be understood, in part, by considering the cyclic aspects of key elements or factors. Such cycles, diagrammatically presented, aid in our understanding of these relationships, but it must be remembered that they are gross oversimplifications of what actually occurs in nature.

Carbon is a basic constituent in all organic compounds. Figure 18.3 portrays the general pathway of carbon in an ecosystem. Carbon exists in the

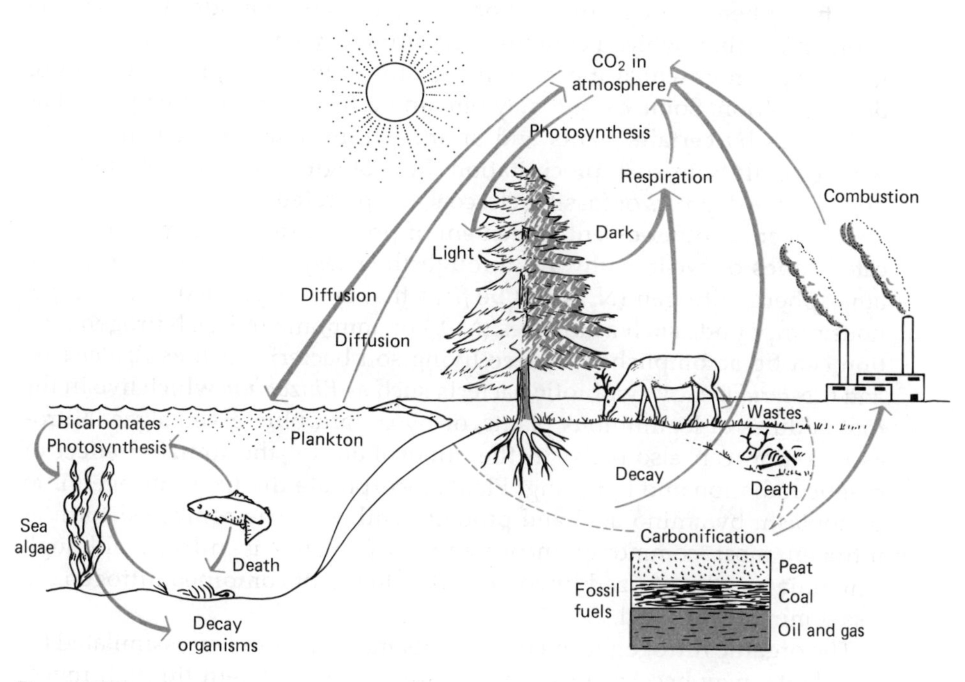


*Figure 18.3*
The carbon cycle. [Redrawn from R. L. Smith, *Ecology and Field Biology*, Harper & Row, Publishers. Copyright © 1966.]

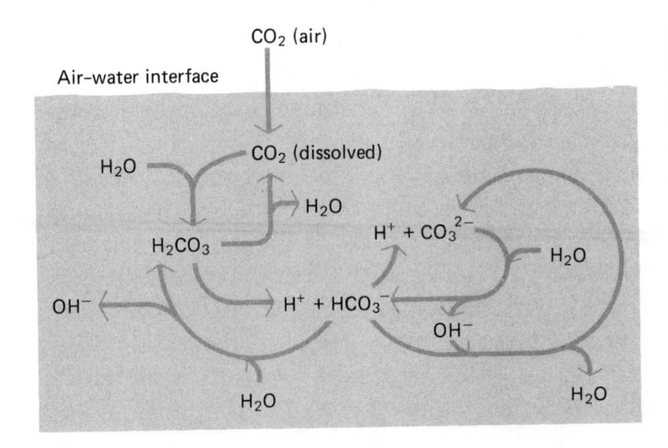


*Figure 18.4*
Relationship between carbon dioxide, carbonic acid, bicarbonate, and carbonate in freshwater systems.

atmosphere primarily as carbon dioxide and, in water, as free $CO_2$ or bound in bicarbonates and carbonates. The relationship between $CO_2$, carbonic acid ($H_2CO_3$), bicarbonate ($HCO_3^-$), and carbonate ($CO_3^{2-}$) is shown in Figure 18.4. Carbon is incorporated into plant protoplasm, as you recall, in the process of photosynthesis. From plants, organic carbon may go into animals by ingestion, where it goes through various stages of digestion and assimilation. From both plants and animals, $CO_2$ reenters the atmosphere through respiration of organic compounds. That portion of organic carbon which is not respired ultimately passes into dead organic material, from which it can return to atmospheric $CO_2$ by oxidation or decomposition. Some carbon may remain bound by being precipitated as carbonates by certain plants and animals. Limestone or marl may ultimately result from such precipitation. In geologic time, some bound carbon became deposits of fossil fuels, coal and petroleum.

Nitrogen, another essential element of protoplasm, has a more complicated series of cyclic pathways through the ecosystem (Figure 18.5). The atmospheric nitrogen ($N_2$) must be *fixed* that is incorporated into nitrogenous compounds such as nitrates ($NO_3$) or ammonia ($NH_3$). Nitrogen fixation can be accomplished by free-living soil bacteria such as *Azotobacter* and *Closteridium*, by symbiotic bacteria such as *Rhizobium* which live in the root nodules of leguminous plants, or by certain blue-green algae. Nitrogen fixation can also occur in the atmosphere by the ionizing effect of cosmic radiation and lightning. Plants incorporate the fixed nitrogen into protoplasm by amino acid and protein synthesis. The combined form of nitrogen most commonly incorporated by plants is nitrate, although ammonia may be utilized by some. Other forms of combined nitrogen are less commonly utilized.

The organic nitrogen formed by plants may be stored or assimilated by the plants, may become incorporated into animal protein through ingestion and assimilation, and will ultimately be dissociated through death and decomposition by bacterial or fungal activity.

During decomposition, ammonia is produced from amino acids by the

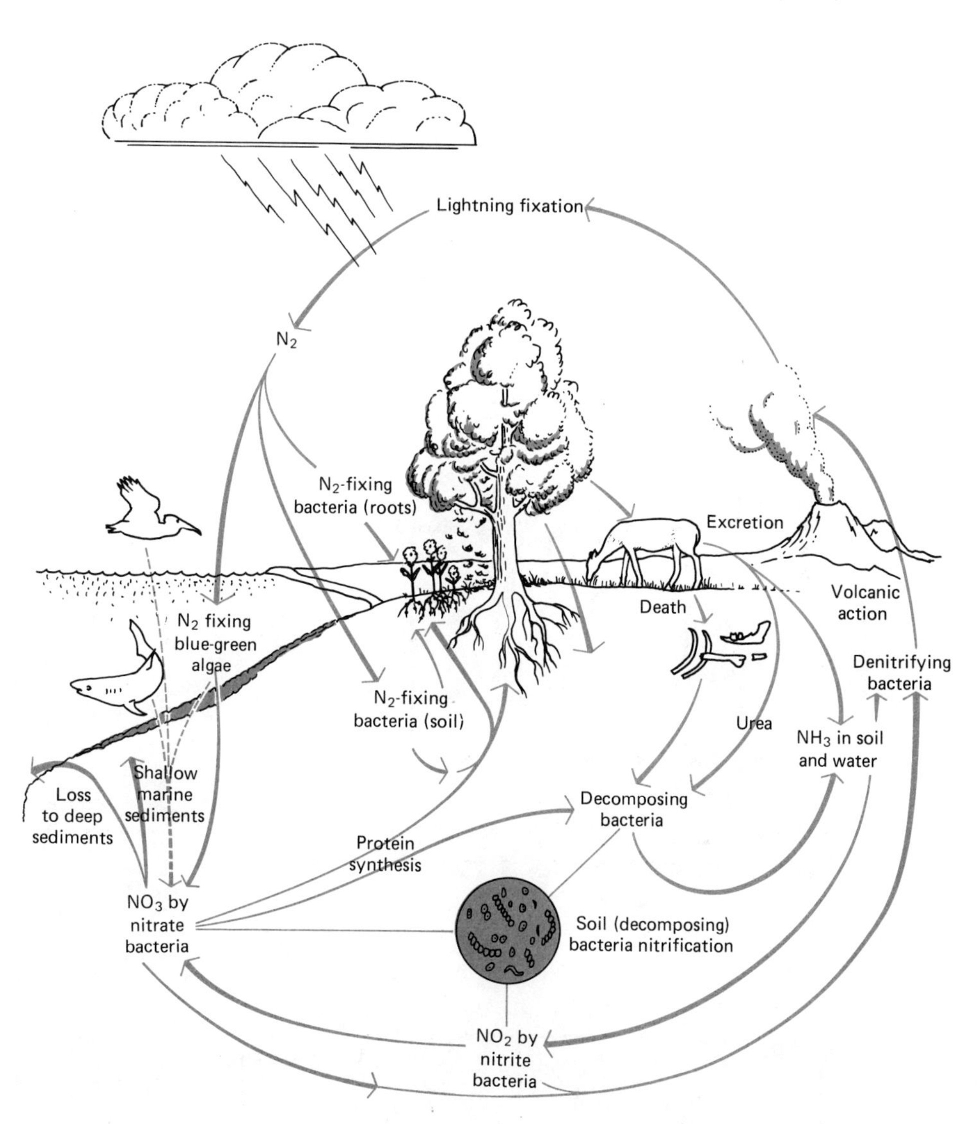


*Figure 18.5*
The nitrogen cycle. [Redrawn from R. L. Smith, *Ecology and Field Biology*. Copyright © 1966 by Harper & Row, Publishers, New York.]

action of ammonifying bacteria such as *Pseudomonas, Proteus,* and others. The ammonia is generally converted into nitrite ($NO_2$) by nitrite bacteria such as *Nitrosomonas,* and into nitrate from nitrite by nitrate bacteria such as *Nitrobacter.* Nitrogen is returned to free nitrogen by the action of denitrifying bacteria such as *Pseudomonas, Thiobacillus,* and others.

Phosphorus is less abundant in the ecosystem than nitrogen (a ratio of 1:23), but is relatively more abundant in living organisms. Soluble phosphates are assimilated by plants in protein synthesis, in enzymes, in phosphorylated organic compounds, and many others (Figure 18.6).

These phosphorus compounds may enter animals by ingestion and assimilation, and both plants and animals may be acted upon by organisms of decay and decomposition. Phosphorus entering bone, becoming precipitated to form phosphate rocks, or found in guano represent relatively insoluble forms. These relatively insoluble forms of phosphorus are grad-

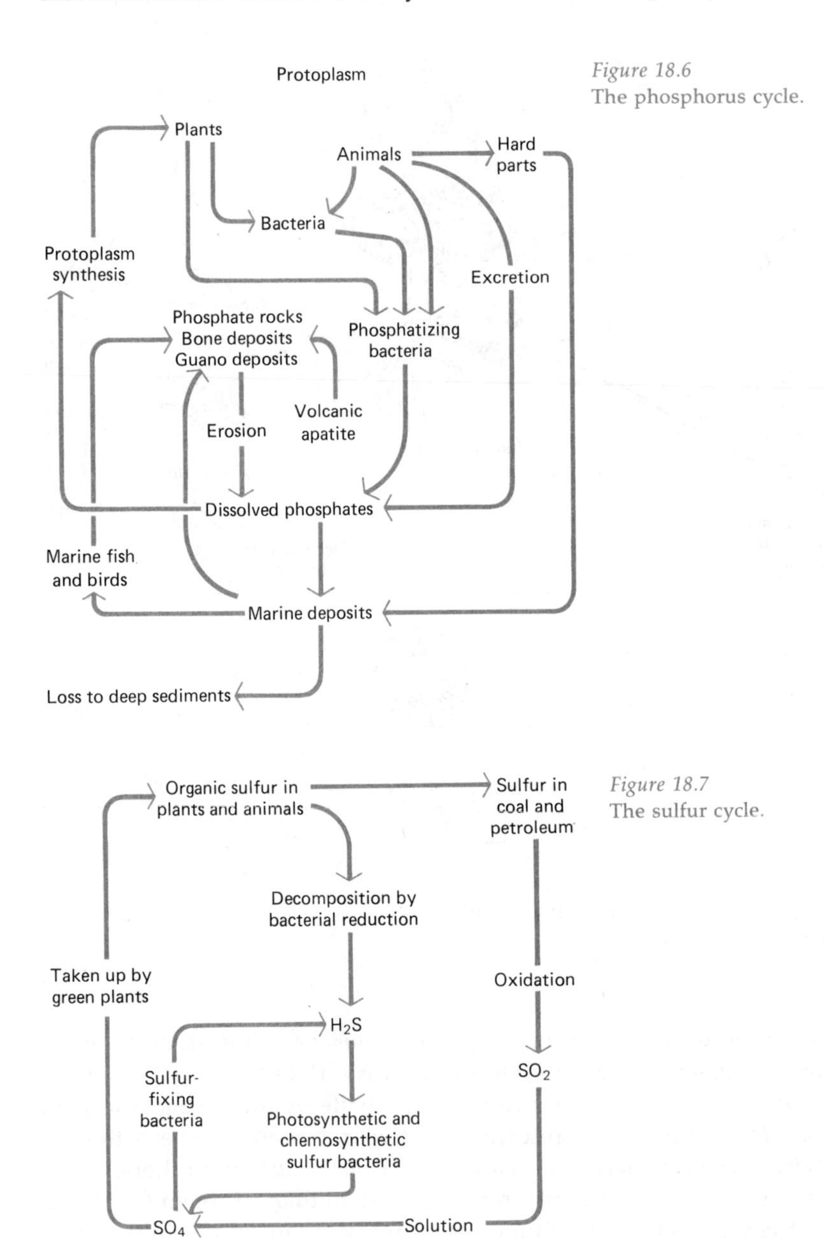


*Figure 18.6*
The phosphorus cycle.

*Figure 18.7*
The sulfur cycle.

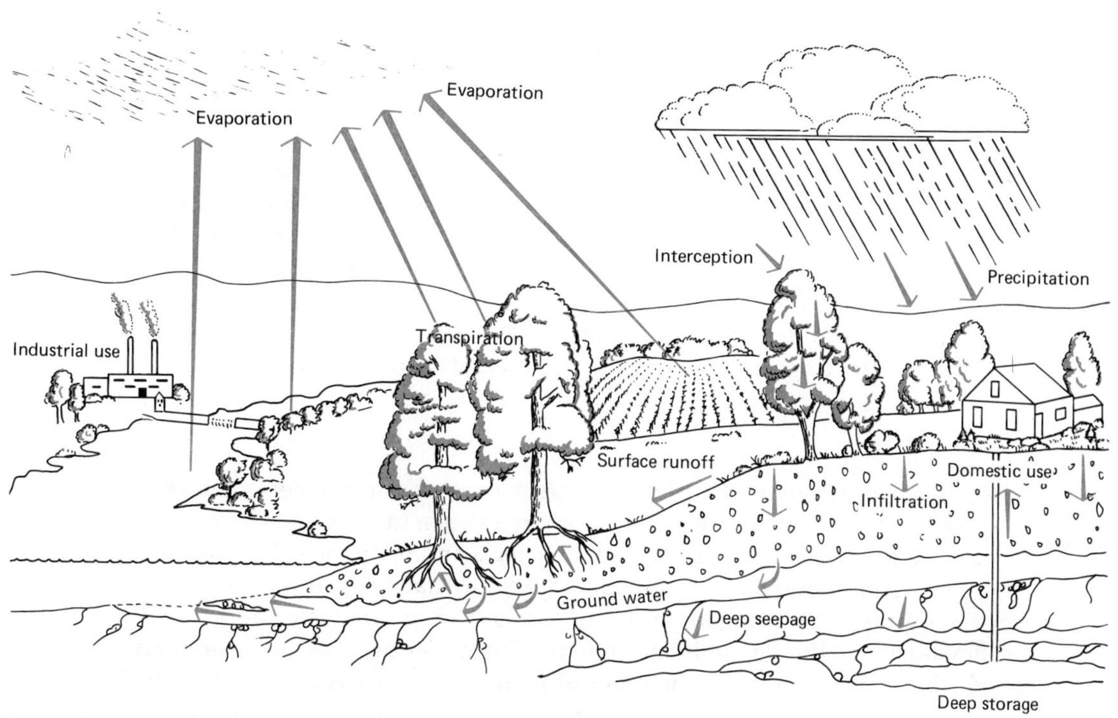


*Figure 18.8*
The water cycle. [Redrawn from R. L. Smith, *Ecology and Field Biology.* Copyright © 1966 by Harper & Row, Publishers, New York.]

ually eroded and some of the phosphorus returns to the ecosystem as soluble phosphate. Some of this phosphorus may be artificially recovered through mining of guano and crushing of bone to form bone meal. Of those phosphates entering the ocean, some is deposited in shallow sediments and a part is lost to the deep sediments.

Phosphates may be artificially introduced to an ecosystem through phosphate detergents and crop fertilization. In countries where artificial fertilizers are scarce and expensive, phosphate loss from soils is a critical problem. This problem is often compounded by social and agricultural practices, such as burning of dried cow dung as fuel and complete removal of crops from the land.

Sulfur, while essential in protein synthesis, is rarely if ever limiting to plant growth. But excess sulfur can be toxic to both plants and animals. In nature, sulfur exists in elemental form and as hydrogen sulfide ($H_2S$), sulfites ($SO_2$), and sulfates ($SO_4$). Organic sulfur in plants and animals is decomposed by bacterial action to $H_2S$, which is further oxidized to sulfates by sulfur-oxidizing bacteria (Figure 18.7).

The water cycle (Figure 18.8), like the $CO_2$ cycle, is being affected by man on a global scale. Atmospheric $H_2O$ is a relatively small component of the total cycle, representing 0.13 geograms (1 geogram $= 10^{20}$ grams). Evaporation from the oceans accounts for 3.8 geograms of water, while 3.4 geograms are returned to the oceans by precipitation. The situation on land is reversed, with 0.6 geograms of water evaporating, much of this from plant surfaces through transpiration, while 1.0 geogram is returned through precipitation. Circulating ground waters contribute 2.5 geograms, with the recharge rate of ground waters being 0.8 geograms annually (1.0 geogram precipitation − 0.2 geograms run off). An estimated 0.25 geograms of water are in freshwater lakes and streams. Human intervention is rapidly decreasing the hydrological budget of terrestrial ecosystems through two major events: 1) the construction of more reservoirs thus increasing water surface area and hence the evaporation rate, and 2) an increase in the rate of runoff due to agriculture, mining, forestry, and other human activities.

## The Biotic Elements

As stated previously, the biotic elements consist of producer organisms, consumer organisms, and decomposer organisms. In general, we can characterize producer organisms as those bacteria and plants which synthesize organic compounds. They are *autotrophic* in that they utilize inorganic compounds, producing organic materials and protoplasm from them. Obviously, all life depends upon them. Consumer organisms are animals which utilize the organic materials of plants, either directly or indirectly. Consumers are unable to produce organic compounds from inorganic raw materials; thus, they are said to be *heterotrophic.* Decomposer organisms are bacteria and fungi which degrade organic compounds. As they utilize dead organic material as a food source, their nutrition is said to be *saprophytic.* Saprophytic nutrition is simply a specialized case of heterotrophic nutrition.

The basic aspect of production is the process of *photosynthesis.* Photosynthesis requires carbon, nitrogen, phosphorus, and other specific minerals in addition to water. Oxygen and hydrogen, bound in compounds such as enzymes are thus also important to photosynthesis. There is indeed interaction between the abiotic cycles discussed earlier and the biotic elements. *Chemosynthesis* is another mechanism by which organic compounds can be synthesized from inorganic substances. Chemosynthetic bacteria obtain energy by the oxidation of ammonia to nitrites, nitrites to nitrates, sulfides to sulfur, or ferrous to ferric ions, to cite some examples.

The term *productivity* is used in discussing the rate of energy production. Thus, a number of different parameters of productivity and respiration exist. *Gross primary production* (GPP) refers to the amount of carbon fixed per unit area per unit time. *Net primary production* (NPP) refers to gross primary production minus plant respiration. *Autotrophic respiration* ($R_A$) refers to the respiration of green plants (GPP − NPP), while *heterotrophic respiration* ($R_H$) refers to the respiration of animals and decomposers. *Ecosystem respiration* ($R_E$) is thus calculated as $R_A + R_H$. *Net ecosys-*

*Table 18.2*
COMPARATIVE PRODUCTIVITY (IN g CARBON/m² yr) OF FOUR ECOSYSTEMS

| ECOSYSTEM PRODUCTIVITY PARAMETER | MESIC FOREST* | XERIC FOREST** | PRAIRIE† | TUNDRA‡ |
|---|---|---|---|---|
| Gross primary production (GPP) | 1620 | 1320 | 635 | 240 |
| Autotrophic respiration ($R_A$) | 940 | 680 | 215 | 120 |
| Net primary production (NPP) | 680 | 640 | 420 | 120 |
| Heterotrophic respiration ($R_H$) | 520 | 370 | 271 | 108 |
| Net ecosystem production (NEP) | 160 | 270 | 149 | 12 |
| Ecosystem respiration ($R_E$) | 1460 | 1050 | 486 | 228 |
| Production efficiency ($R_A$/GPP) | 0.58 | 0.52 | 0.34 | 0.50 |
| Effective production (NPP/GPP) | 0.42 | 0.48 | 0.66 | 0.50 |
| Ecosystem productivity (NEP/GPP) | 0.10 | 0.20 | 0.23 | 0.05 |

*Source:* After Reichle, 1975.

*Early successional deciduous forest (*Liriodendron tulipifera*) on alluvial soil. (After Reichle et al., 1973.)

***Quercus* and *Pinus* forest on sandy soil. (After Woodwell and Botkin, 1970.)

†US/IBP Grasslands, Biome program, personal communication of raw data-calculations and interpretation by D. E. Reichle. (Refinements after Andrews et al., 1974.)

‡Adapted from the data of P. C. Miller, L. L. Tieszen, P. I. Coyne, and J. J. Kelly in Bowen, ed., *Tundra Biome Symposium,* 1972.

*tem production* (NEP) can thus be calculated as GPP − $R_E$. Other parameters are shown in Table 18.2.

Following primary production, the organic compounds of plants may 1) be metabolized within the plant itself, 2) be stored and later consumed by a herbivore, or 3) enter a cycle of decomposition resulting in the production of inorganic compounds. As you recall, these aspects were a part of most of the biogeochemical cycles discussed previously.

Consumers may be conveniently classified into three categories: 1) primary consumers, or herbivores, 2) secondary consumers, or carnivores, and 3) multilevel consumers, or omnivores. Primary consumers are animals that feed directly upon primary producers. Such organisms are often referred to as grazers. Secondary consumers are predators or carnivores that feed upon the herbivores. There may be several levels of carnivores; that is, certain predators may feed upon other predators. We could envision a situation in which a water scorpion which feeds upon crustacea (herbivores) may be eaten by a frog, which in turn is eaten by a fish, which is eaten by a larger fish, with the latter finally eaten by an osprey. The water scorpion would be termed the secondary consumer, with the frog being a tertiary consumer, the smaller fish a quaternary consumer, and so on. This brings us to the concept of trophic levels and food chains which we will be discussing shortly.

Multilevel consumers, or omnivores, refer to those animals that feed on both plants and animals. Scavengers are a special kind of consumer in that they feed on dead organic material. But unlike the decomposers, scavengers are holophytic feeders; that is, they ingest the organic matter rather than secreting hydrolytic enzymes into the environment to effect digestion. Some scavengers feed solely on dead animals, some solely on dead plant material, while others may feed on either.

Another special group of consumers are the parasites. All are heterotrophic, but not all are holophytic. Some may be saprophytic. Plant parasites feed directly on plants and thus are herbivores; animal parasites derive their nutrition from animals and thus are carnivores. The parasite differs from the predator only in the fact that it does not eat the host, although occasionally it may kill it.

Decomposition involves the digestion of complex organic materials to simpler ones. Normally, through a series of such decomposition processes, insoluble organic substances are converted to soluble compounds. During the process, the decomposer obtains food, and soluble nutrients are produced which can be utilized by plants. Of course, all organisms are decomposers in a sense, in that they respire, decomposing simple sugars into simpler organic compounds or to $CO_2$ and $H_2O$. We do not classify such organisms as decomposers, however, because decomposition is not their principal role in the ecosystem.

Decomposers respire as do living cells of all organisms. Some respire by fermentative pathways, others aerobically. As a result of these processes, most decomposers produce by-products which inhibit their own group. Thus, no single bacterium or fungus performs the complete range of decomposition. Through the action of one decomposer, conditions are created which are optimum for growth of a second decomposer and so on. Some decomposers have very specific chemical functions, others can exist under a wide range of conditions.

During the process of decomposition, many organisms produce metabolic by-products which have a regulatory influence on the growth of other organisms. Such chemical substances are termed *ectocrines.* Antibiotics are well-known ectocrines. We will be discussing some of these compounds, called *allelochemics* by Whittaker (1970) (rather than ectocrines), in a later section on interspecific interactions.

## *Energy Flow and Trophic Structure*

The ultimate source of all energy in a natural ecosystem is solar radiation. At sea level, solar radiation averages 15,000 calories per square meter per minute. Assuming 10 hours of sunlight per day, this would total 9 million calories per square meter per day. This total daily solar energy is equivalent to the energy in 684 billion tons of coal. This is sufficient to supply 1 million watts of light to each acre of ground. If this solar energy could be harnessed, the amount of solar energy striking the surface of the United States every 20 minutes is sufficient to meet the country's entire power needs for one year (Southwick, 1976).

Of the solar energy striking the earth's surface, 98 percent is reflected. Of the two percent absorbed, only about one-half is in the wavelengths utilized in photosynthesis. Thus, the ecological efficiency of terrestrial green plants is generally one percent or less. In aquatic ecosystems, this figure becomes even lower, averaging about 0.18 percent in the ocean.

The series from plants to other organisms through which transfer of energy occurs constitutes *food chains.* The parts of the food chain constitute the *trophic levels.* A trophic level refers to those organisms which obtain their food in the same general way. It is a basic law of thermodynamics that when energy is transferred between the various trophic levels, some will be lost due to conversion to heat. As a general rule, the amount of energy available to the next trophic level from its predecessor is approximately ten percent. Therefore, in a given ecosystem, there are fewer and fewer organisms that can be supported at each trophic level. This forms what is commonly known as the *Eltonian pyramid,* after the British ecologist Charles Elton who first emphasized this relationship.

A simple food chain is shown in Figure 18.9. Such representations are a gross oversimplification of what actually occurs under natural conditions. Food chain relationships multiply and diversify to form complex *foodwebs.* The relationships between populations can be expressed schematically by construction of diagrams such as Figure 18.10.

Energy flows in a natural community have been difficult to determine due to the extreme complexity of the foodweb. However, by selecting a relatively simple example and ignoring many populations, some generalizations may be expressed (Figure 18.11). In this example, about one percent of the total solar energy available is converted into plant tissue. The dominant herbivores, the meadow mice, consume only two percent of this available energy. Some 10 to 20 percent of the potential energy in the con-

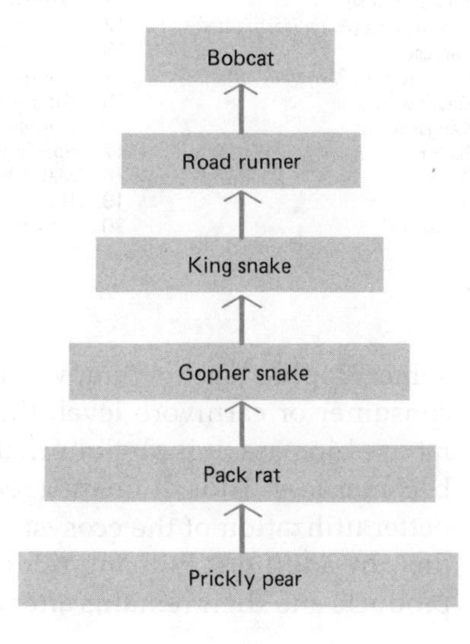


*Figure 18.9*
One food chain from chaparral in southern California. The boxes only represent trophic levels and are not proportional to the number of animals present at each level. [Redrawn from A. S. Boughey, *Fundamental Ecology.* Copyright © 1971, International Textbook Company, Scranton, Pa.]

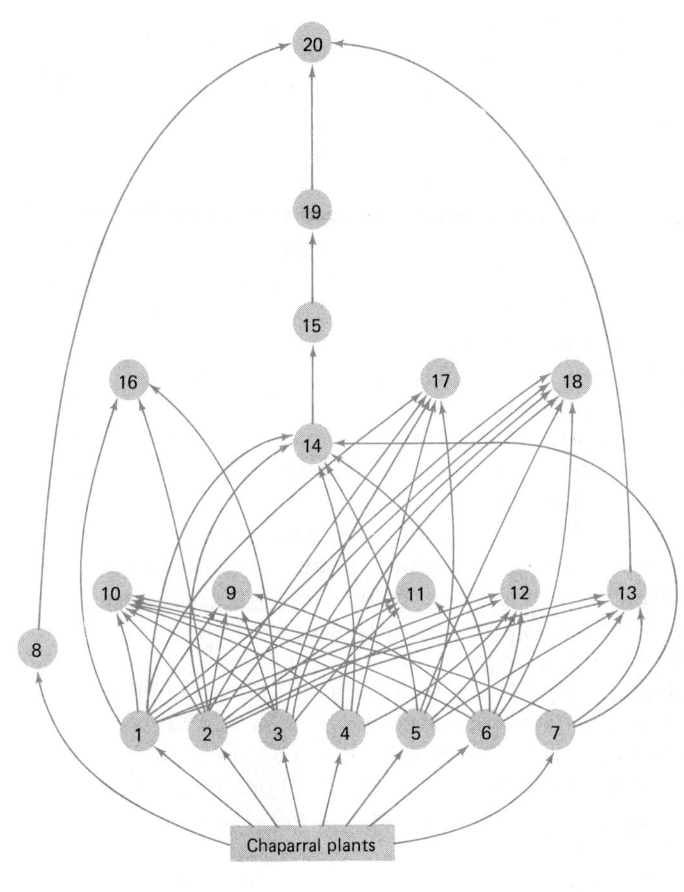


1. Pocket mouse
2. Deer mouse (*P. californianus*)
3. Harvest mouse
4. Deer mouse (*P. maniculatus*)
5. Meadow mouse
6. Kangaroo rat
7. Pack rat
8. Quail
9. Raccoon
10. Opossum
11. Striped skunk
12. Weasel
13. Red fox
14. Gopher snake
15. King snake
16. Horned owl
17. Sparrow hawk
18. Marsh hawk
19. Road runner
20. Bobcat

*Figure 18.10*
The basic foodweb of the chaparral ecosystem. Deliberately omitted from this is a shrub-mule–deer-cougar food chain, which is now sometimes absent because of human disturbance of this ecosystem. Earlier hunters from middle Pleistocene times also are believed to have disrupted food chains by exterminating such animals as sloths, camels, and horses and thus causing their predators, such as "lion" and saber-toothed cat, to also pass to extinction. The chaparral ecosystem we have today, even in areas referred to as "undisturbed," has been vastly and irreversibly changed by approximately 20,000 years of human occupation. Omitted from this diagram entirely are insects. Every animal in this food web probably eats one or more insect species at some stage. [Redrawn from A. S. Boughey, *Fundamental Ecology*. Copyright © 1971, International Textbook Company, Scranton, Pa.]

sumer trophic level is removed by herbivorous insects. At the secondary consumer or carnivore level, the weasels utilize about 30 percent of the mouse biomass. It is obvious that at each trophic level the available energy becomes less. Thus, humans (generally the top trophic level) could derive better utilization of the ecosystem through a vegetarian rather than a meat diet. In addition, humans return little to the ecosystem, as their waste products and their remains after death are generally deposited elsewhere.

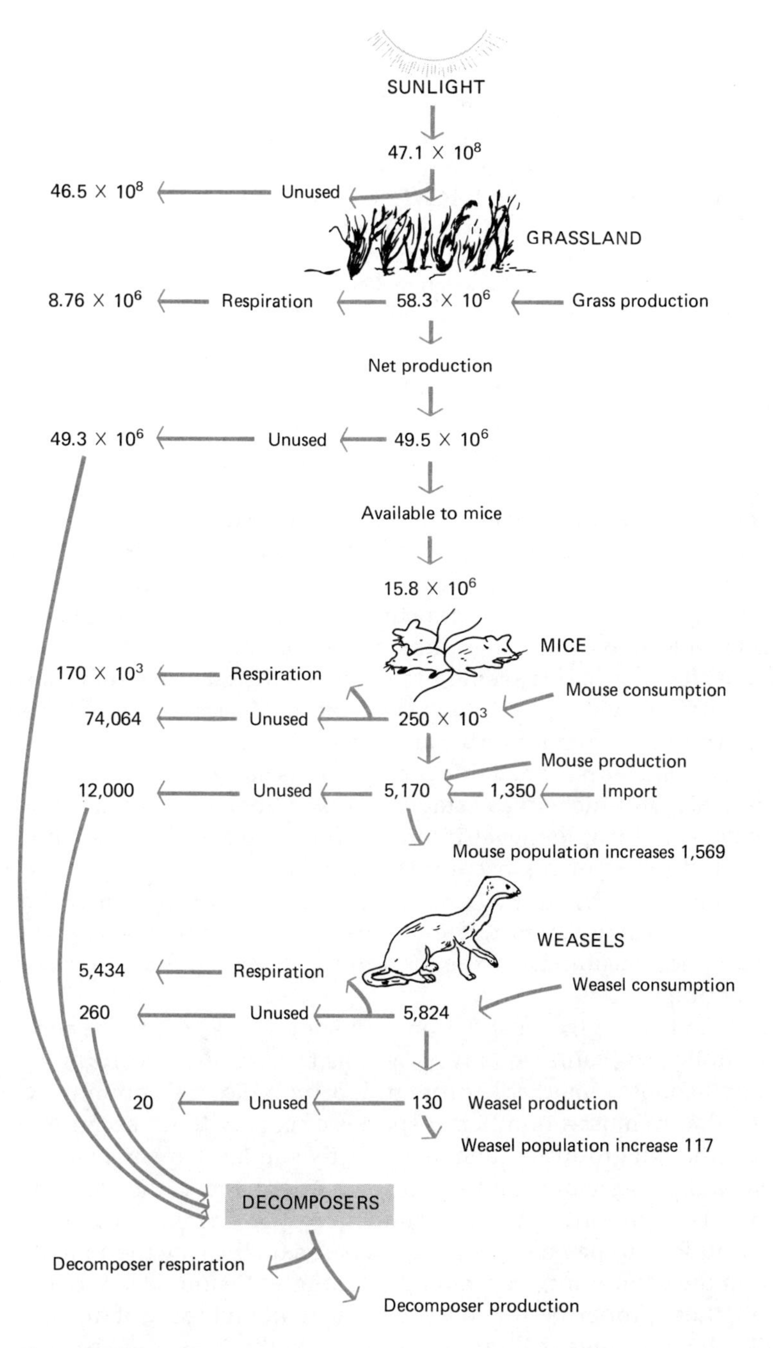


*Figure 18.11*
Energy budget for a simple natural food chain in a degraded forest ecosystem in southern Michigan. Figures express energy values in calories/m²/year. [Redrawn from F. B. Golley, in *Ecological Monographs* 30(2):187–200, 1960.]

# THE CONCEPT OF NICHE

And NUH is the letter I use to spell NUTCHES
Who live in small caves known as NITCHES for NUTCHES
These NUTCHES have troubles, the biggest of which is
The fact that there are many more NUTCHES than NITCHES
Each NUTCH in a NITCH knows that some other NUTCH
Would like to move into his NITCH very much
So each NUTCH in a NITCH has to watch that small NITCH
Or NUTCHES who haven't got NITCHES will SNITCH
*Dr. Seuss*

This verse exemplifies the way an ecological term has crept into everyday thought and speech. Unfortunately, Seuss, as many other nonscientists and even some biologists, uses the term niche as equivalent to habitat. It is not.

The zoologist Charles Elton was interested in the feeding habits of animals and, as a consequence of his studies, established the Eltonian concept of "the pyramid of numbers." Elton found that animals of different trophic levels were likely to be of distinctly different sizes, with the large animals being far less abundant than the smaller ones on which they feed. In other words, feeding habits seemed to reflect all the adaptive qualities of an animal; its behavior, physiology, and mechanical abilities. These qualities in turn fit the animal for its particular role in life; what might be termed its *profession*. Elton termed this profession its *niche*.

Thus, the concept of niche was established by a zoologist based upon feeding behavior and trophic level interaction of animals. Plants do not "feed," and most green plants occupy the same trophic level; thus, the pyramid of numbers described by Elton would not seem appropriate to the discussion of plant ecology. In fact, one can read several plant ecology textbooks without once encountering the term niche; several others only superficially mention it.

But if one is to believe the postulations of Pielou (1974), the niche concept can be applied to plants equally as well as to animals. According to Pielou, two of the most remarkable properties of an ecological community are 1) the fact that in most a number of species coexist without excluding one another, although groups of them make fairly similar demands on the environment and presumably compete to some extent and 2) the fact that some species are represented by an abundance of individuals whereas other species, judged to be no less successful based on the way their members persist in the community, are rare. According to Pielou, it is the desire to explain these properties that has led to the modern theory of niche.

What does the term mean? Elton described it as the functional role of the organism in its community. Many people, as exemplified by the Seuss writing, seem to mistakingly equate niche with habitat. The modern interpretation (Hutchinson, 1957) refers to niche as "an *n*-dimensional hypervolume bounded by limits imposed by the environment, including both abiotic and biotic parameters." Hutchinson describes this hypervolume as the *fundamental niche* of the species. A fundamental niche of a species is composed of two disjunct subsets: all conditions in which the organism could succeed but for the fact that it is excluded by competitors; all conditions in which the organism does succeed in fact. The latter case is called the *realized niche* of the species. According to the niche theory, it would be impossible for two (or more) species to coexist indefinitely in the same area if their niches are identical. This statement is known as the *principle of competitive exclusion* or the *Gause principle*, based upon the observation of the Russian biologist G. F. Gause. The Gause principle raises two questions: 1) how dissimilar must two niches be for several species to occur in the same area and 2) what actually delimits a species?

## Niche Diversification

Niche diversification is found far more often among animals than among plants. For example, three ichneumonid wasp species of the genus *Megarhyssa, M. atrata, M. macrurus*, and *M. greene*, were found to occur together in beech-maple forests in Michigan (Heatwole and David, 1965). All were found to parasitize a single host insect, the wood wasp *Tremax columba*, in whose larvae they lay their eggs. The host is the grub, wood borer stage found in dead logs and stumps. Each of the three parasitic wasp species differ in length of their ovipositors. A female will lay eggs only in larvae which her ovipositor just reaches when inserted at right angles to the wood surface. As the larvae are found at various depths in the wood, any particular larva would be the target for only one species of *Megarhyssa* wasp. Therefore, each wasp, although functionally similar and occupying the same area, parasitizes a different segment of the host population.

Two varieties of flax *(Linum usitatissimum)*, one the fiber flax, the other oil seed flax, were grown together. Not only did they coexist, but a mixed planting yielded more dry weight of plant material per unit area of soil than did either variety grown alone. Later studies indicated that the two varieties grew at different rates, thus together they made much more efficient use of incident light than was possible when growth was synchronized.

Many aspects of plants and niche diversification relate to the fact that many plant species have several recognizably different forms. Such species are termed *polymorphic*. Labeling of these different kinds of variants is complicated, and several different sets of terminology have developed. Turesson noted that in many species of plants, *genetic variants* have evolved that are adapted to different environments. These he called *ecotypes*. Gilmour and Gregor (1939) devised a set of terms based upon use of the suffix *-deme*, a term used to connote a group of individuals within a species. Prefixes were added to this suffix to create terms having very specific meanings, namely, *topodeme*, group occurring in a single, specified region; *ecodeme*, group occurring in a specific kind of habitat and not necessarily confined to a single region; *genodeme*, genetically related group known to differ genetically from other such groups; *gamodeme*, group able to interbreed and close enough to each other to do so; *topoecodeme*, group occurring in a specific habitat and region; *genoecodeme*, group forming a genetically distinct ecodeme (the ecotype of Turesson); *hologamodeme*, group capable of interbreeding though not necessarily close enough in space to do so; and *clinodeme*, group belonging to one of a series of related groups exhibiting clinal (continuous) variation.

There are various studies which illustrate the relationship between niche and polymorph. In one of these, the grass *Agrostis tenuis* was found growing in a valley containing a small, derelict copper mine. One genoecodeme of the grass is tolerant of the ore-contaminated soil, the other is not and thus grows only on uncontaminated soil. In this example, a single species of organism is associated with two niches and has two genoecodemes to occupy the niches.

Bird's-foot trefoil growing in England was found to be a dimorphic species in that some plants contain a cyanogenic glycoside, others do not. These two variants are different genodemes. The cyanogenic plants are selectively favored by virtue of their toxicity to herbivores such as slugs, snails, and voles. The acyanogenic genodeme persists because of greater fertility and a faster rate of growth, thus having a selective advantage in areas where herbivores are sparse. In this situation, only a single niche is involved. The same herbivores are present, differing only in the degree of density, but a single plant species with *two* genodemes occupies that niche.

## INTERSPECIFIC INTERACTIONS

In general, there are three effects one species may have on another: no significant interaction, beneficial interaction, or detrimental interaction. As we are discussing two-species interactions, a combination of interactions result. These interactions have commonly been assigned the symbols 0, +, and − respectively. Thus, the combinations 00, −−, ++, +0, −0, and +− result. Three of these (++, −−, and +−) have been subdivided resulting in nine important interactions shown in Table 18.3. These are *neutralism*, *mutual inhibition type competition*, *resource use type competition*, *amensalism*, *parasitism*, *predation*, *commensalism*, *protocooperation*, and *mutualism*. For a given pair of species, the interaction may change; that is, they may exhibit parasitism at one time, commensalism at another, and neutralism at still another. In general, the interactions may be classified as *negative* and *positive*.

*Table 18.3*
**INTERACTIONS IN A TWO-SPECIES POPULATION**

| TYPE OF INTERACTION* | EFFECT ON SPECIES** 1 | 2 | GENERAL NATURE OF INTERACTION |
|---|---|---|---|
| 1. Neutralism | 0 | 0 | Neither population affects the other |
| 2. Competition: direct interference type | − | − | Direct inhibition of each species by the other |
| 3. Competition: resource use type | − | − | Indirect inhibition when common resource is in short supply |
| 4. Amensalism | − | 0 | Population 1 inhibited, 2 not affected |
| 5. Parasitism | + | − | Population 1, the parasite, generally, smaller than 2, the host |
| 6. Predation | + | − | Population 1, the predator, generally larger than 2, the prey |
| 7. Commensalism | + | 0 | Population 1, the commensal, benefits while 2, the host, is not affected |
| 8. Protocooperation | + | + | Interaction favorable to both but not obligatory |
| 9. Mutualism | + | + | Interaction favorable to both and obligatory |

0 No significant effect
+ Growth, survival, or other population attribute benefited
− Growth, survival, or other population attribute inhibited

*Source:* After E. P. Odum, *Fundamentals of Ecology*, 3rd ed., Philadelphia: Saunders, 1971, p. 211.

*Types 2 through 4 can be classed as negative interactions, types 7 through 9 as positive interactions, and 5 and 6 as both.

## Negative Interactions

Interspecific competition, either direct or indirect, is one of the negative interactions of which a great deal has been written. In its broadest sense, competition refers to the interaction of two organisms for the same resource, be it space, food, or other. We have previously discussed the tendency for competition to bring about an ecological separation of closely related or otherwise similar species—the competitive exclusion principle or Gause principle.

Demonstration of competition in the field is often difficult, as competition effects are hard to separate from other possible causes of negative correlation. Thus, much of the detailed work on competition has been carried out under controlled conditions. In addition, essentially all of the classical examples of competition relate to animal populations. However, competition between plant populations under natural conditions has been established. Often, these effects are dependent upon competition for soil microsites during germination or for a specific soil mineral nutrient, commonly nitrogen.

Competition is well developed in aquatic ecosystems. Harper (1961) demonstrated the existence of competition under experimental conditions

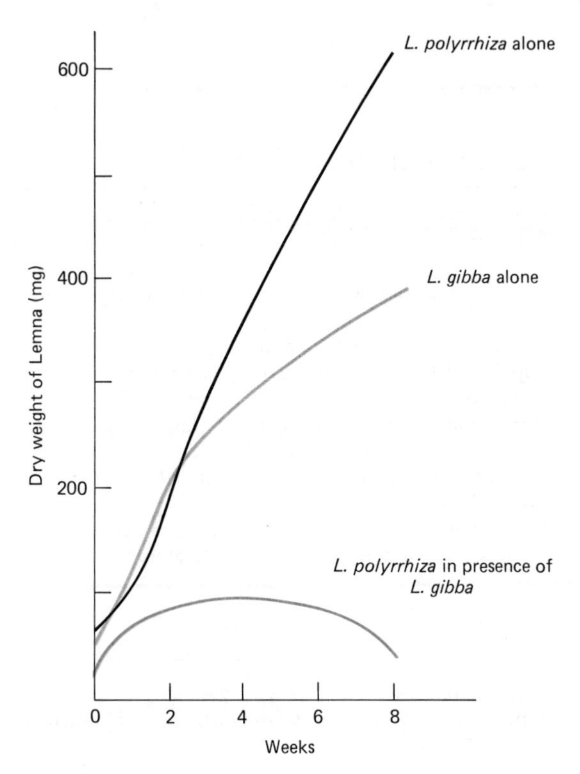


*Figure 18.12*
Competition between two species of duckweed. [Redrawn from A. S. Boughey, *Fundamental Ecology*. Copyright © 1971 by International Textbook Company, Scranton, Pa.]

in plants. Using two species of duckweed, it was demonstrated that growth was drastically reduced when the two were grown together (Figure 18.12). Competition within phytoplankton communities under natural conditions has been demonstrated. When nitrate and phosphate concentrations are relatively high, green algae and chrysophytes predominate. As the amount of nitrates decrease, blue-green algae replace the chrysophytes and greens. Apparently, nitrogen becomes a limiting factor, and under such conditions, nitrogen-fixing blue-greens are better competitors.

Other negative interactions involve predation, parasitism, and antibiosis. Predation and parasitism are quite similar from an ecological standpoint, although parasites and predators do exhibit differences in size, at least at the extremes. Parasites also have a higher biotic potential than do predators and are often more specialized in structure, metabolism, host specificity, and life history. The so-called parasitic insects are somewhat intermediate between parasite and predator. They consume the entire prey as do predators, yet have high host specificity and the high biotic potential of parasites.

Many examples of parasitism of plants could be discussed. A classic example might be that of the chestnut blight in America. Prior to 1904, the

American chestnut was an important member of the eastern North American forest, living in coexistence with its particular parasites, diseases, and predators. Likewise, the oriental chestnut in China was coexisting with its particular parasites, and so on, one of which was the fungus *Endothia parasitica.* In 1904, this fungus was accidentally introduced into the United States. The American chestnut proved to be completely unresistant to this new parasite, and by 1952, essentially all the large chestnuts had been killed.

Several examples of insect predator-parasites are also available. In the late 1800s, the prickly pear cactus *(Opuntia)* was introduced to Australia from the United States. It quickly escaped cultivation and, by 1925, covered some 60 million acres of range land. Because of its dense growth, the range was ruined for grazing. Australian scientists searched hard for an answer to the problem. Eventually, entomologists found the answer. They sent people to America to collect insects that fed on prickly pear. These collectors found and brought back a number of likely candidates, among them the moth *Cactoblastis cactorum.* The moth was reared in captivity and its eggs placed on the wild prickly pear cactus. The eggs hatched and the caterpillars thrived. They ate every cactus within reach, then pupated, and, as a full grown moth, they flew to other cacti to lay their eggs and produce a new generation of caterpillars. Thus, the wave spread, and within a few years the entire 60 million acres was once again grassland.

This does not mean, however, that the prickly pear no longer exists in Australia. There are a few scattered survivors, and these seem to exist in a dynamic equilibrium with surviving *Cactoblastis cactorum.* Thus, the number of prickly pear cactus in Australia is now relatively constant, regulated in fact by the insect in a sort of predator-prey relationship.

Another plant that became a pest in pastures was the Klamath weed or St. John's wort *(Hypericum perforatum)* in California. The beetle *Chrysolina quadrigemina* was brought in to control the weed, which it did, essentially eliminating it from the open pasture land. But the week is still very common in the shady woods surrounding the fields, either because the habitat is unfavorable to the beetle or the beetle can not readily find the plants. A botanist studying the distribution of St. John's wort in California today, and unfamiliar with the history of the plant, would surely come to the conclusion that the distribution of the plant was due to habitat influence; St. John's wort is obviously a shade-loving, woodland species. In this instance we know that conclusion to be invalid, but how many other plant population studies may be likewise influenced by unknown animal or insect predators?

In England, the effect of predators was noted in that species diversity in pastures grazed by sheep was shown to increase. The reason for this seemed clearly to be that certain plants were more palatable to grazing sheep than others thus were eliminated, while less palatable species replaced them.

The effect of a herbivore in controlling plant numbers was demonstrated dramatically in Britain in 1954. The virus myxomatosis, a fatal disease of rabbits, struck in that year and virtually eliminated all rabbits, where previously tens of thousands had existed. The year after the rabbits died, there appeared 33 species of plants on a tiny island off Wales that had not previously been seen on the island in over 300 years. Obviously, the seeds for such plants had to have come from the mainland, but there is no reason to believe that this had not happened every year over the past 300. In previous years, rabbits had simply eaten the young seedlings as quickly as they appeared.

While herbivores would clearly influence plant distribution and number, at least in some instances, plants can influence the development of other plants by their addition of chemical substances to the environment. The fact that some plants seem antagonistic to others was noted as early as 1832 by de Candolle.

These chemical interactions of plants may be stimulatory as well as antagonistic, both cases classed as *allelochemics* by Whittaker (1970). Allelochemics responsible for chemical antagonism of one species by another is termed *allelopathy*. Allelopathics are apparently quite variable, both in their mode of action and their mode of excretion into the environment.

The effect of allelopathy may be to vary the rate of plant succession or determine species composition within one or several seral stages. Allelopathy may influence succession by 1) suppression of an established species by an invading species, 2) suppression of an invading species by an established one, 3) elimination of a species due to its own autotoxicity, 4) inhibition of soil organisms, and 5) influencing the sequence of species.

Allelopathic effects have been demonstrated in old field succession in Oklahoma and by the inhibition of herbs and grasses under certain shrubs. The latter case is demonstrated by the bare zone clear of other plants around California chaparral.

Still another control of population growth is fire. Its effect may be negative or positive. Fire has been a factor in shaping forests and grasslands for centuries. It has been stated by Muller (1929) that in southeastern Europe:

> Virgin forests require fires for fullest fruitation just as the Phoenix arose only from fire. There is, in short, no virgin forest that has not resulted from fires; for fires are as "natural" as the forest itself.

The effects of fires are several. Fire obviously affects the soil. The organic matter that burns is slash (branches and other debris left on the ground after logging operations), living moss, and the humus layer. These, if unburned, have little effect on the fertility of the soil because they decompose very slowly. After burning, the nutrient status may be excellent for a long period of time. Burning may result in a great loss of total nitrogen from the site, but increased mineralized nitrogen, the latter of great consequence. Several studies have shown that the amount of nitrogen leached from a site after burning is small. Other minerals such as magnesium, and especially calcium, are tightly bound to soil particles thus are little affected by burning. Burning may be detrimental if the humic layer is very thin. If the burning is too intense, the entire humic layer may be consumed. Burning may also affect allelopathic agents in the soil, destroying heat labile toxins produced by certain dominants in the community.

Fire also affects soil organisms. Algal pioneers typically appear on burned sites, while soil bacteria may show no effect, be completely destroyed, or actually increase in numbers. The effect on bacteria is a result of depth and intensity of the fire, temperature during burning, soil moisture, and soil chemistry. Actinomycetes seem less affected by heat and drying than are the bacteria. In the lower soil levels, actinomycetes may actually increase in the years following a fire, possibly because of downward leaching of ash minerals. Some fungi are found only on burned sites. This may be due to nitrification, sterilization of soil by heat, the effect of heat on spore germination, the formation of nutrient substances due to heating, the chemical properties of the ash, and altered biotic competition. The main factors appear to be temperature, acidity, and nutrient source. The effect of fire on plant pathogens is variable; some are destroyed or reduced in numbers, while others actually increase; fire opens portals of entry on trees or stimulates prolific growth of the host plant, thus causing spread of the pathogen. In the root rot fungus *Rhizina undulata*, germination of ascospores is enhanced by heat and mycelial growth of the fungus stimulated by the heated extract of pine roots. Thus, the serious spread of root rot in Scotch pine or Mungho pine may follow burning.

The effect of fire on soil fauna is difficult to ascertain, due to varying sampling methods, habitat, time of study, and intensity of the fire. However, two generalizations seem apparent: 1) the effect of fire on soil fauna is greater in forests than in grasslands and 2) with the exception of mesofaunal species (mites and collembolans) and spiders, population reductions do not seem to be directly caused by the heat of the fire but rather by the resultant change in the environment. Population decline appears to be the result of a transition to xeric condition, along with the lack of food and the occurrence of greater temperature fluctuations due to removal of cover.

The effect of fire on grasslands is variable, depending upon the intensity of the fire, the thickness of the humus, dryness of the soil, fertility of the soil, and other factors. In general, fires may be important in the maintenance of certain grasslands.

Fires in forests may be an important factor in the determination of species composition, particularly between hardwood species and conifers. Conifers are largely fire resistant and some in fact require fire for opening the cone and deseminating of the seed. Hardwoods, on the other hand, are usually fire sensitive. In addition, pine seedlings, because of their short, slow growing, vertical roots, survive best in mineral soil or on reduced, burned humus one inch or less in depth, while the roots of hardwood species are able to penetrate deep, unburned litter.

We will close this section by quoting the great naturalist John Muir, writing about Yellowstone National Park in 1901. In reading this passage, we are quite cognizant of the fact that John Muir probably did not know a teleological statement when he saw one and probably could not have cared less.

> The lodgepole pine, which, though . . . thin-skinned . . . and . . . easily killed by fire, takes pains to store up its seeds in firmly closed cones, and holds them from 3–9 years, so that, let the fire come when it may, it is ready to die and ready to live again in a new generation. For when the killing fires have devoured the leaves and thin resinous bark, many of the cones, only scorched, open as soon as the smoke clears away; the hoarded store of seeds is sown broadcast on the cleared ground, and a new growth immediately springs up triumphant out of the ashes. Therefore, this tree not only holds its ground, but extends its conquests after every fire.

## *Positive Interactions*

Positive interactions include commensalism, cooperation, and mutualism. Such interactions are most commonly demonstrated in animals but do exist for plants as well.

Mutualism is demonstrated by several examples. A classic one is said to be the lichen, consisting of both an algal and a fungal component (see our earlier discussion on Chapter 7). Similar relationships exist between algae and certain invertebrates. The nodule growth of nitrogen-fixing bacteria on the roots of legumes is still another well-known example. Perhaps less well known are the ectotrophic, endotrophic, or peritrophic fungi that form mycorrhizal structures either inside or outside the roots of pines, oaks, beech, and other trees. The term *mycorrhiza* refers to the combination of fungal hyphae and root. If the hyphae grow between cells of the cortex and extend outward from the surface of the root, they are termed *ectotrophic.* If the hyphae grow within cells of the root, they are termed *endotrophic.* If the fungal hyphae simply grow around the root without actually penetrating it, they are termed *peritrophic.* The fungi apparently decompose insoluble nutrients to soluble ones which can diffuse into the root cells. The fungi are apparently supplied with photosynthates from the plant. Examples of animal-plant mutualism have been discussed earlier in Chapter 17. These include the yucca moth-yucca plant relationship in pollination, and the ant-acacia system.

Among commensals are epiphytes such as the well-known orchids, Spanish moss, and others. The epiphytes depend upon the host for support only. All are capable of photosynthesis.

As stated earlier, if the relationship is not obligatory, it is termed cooperation. It is sometimes difficult to determine when the interactions are obligatory to the two participants and thus truely mutualistic or when they simply represent cooperation. The ant-acacia example may represent the latter situation.

## PLANT COMMUNITIES

It has been recognized since the time of Theophrastus that plants are grouped into communities. Each plant community is the product of interaction between two phenomena, the same two phenomena that we have mentioned throughout this text as affecting plant activities. These are genetic aspects, that is, the *genetic plasticity* or *ecologic amplitude* of the plants comprising the flora, and the effects of environment, that is, its hetero- or homogeneity.

The fact that one considers various kinds of communities in an ecosystem presumes that some sort of biologic homogeneity in terms of species composition and community structure exists within a specific area. Thus, a community has boundaries in space. Where two different communities come into contact they intergrade. This area of intergradation, whether extremely narrow or extending for many miles, is called an *ecotone.* A specific community is generally characterized by its *dominant* and *constant* species; within an ecotone, a shift in dominance usually occurs.

A community also has boundaries in time. Geologic evidence indicates that at one time or another every part of the land surface has been denuded of its vegetation. On these denuded areas, plant communities develop, at first perhaps extremely simple in composition, but changing with time. This change from one type of community to another within a specific area is called *succession.*

The natural replacement of one community by another through time was easily observed in many areas of the earth, thus recognized early. Theophrastus wrote of the sequence of vegetation on fields and flood plains, and in 1685, W. Kind outlined plant succession in mire communities. The Frenchman Buffon also discussed successional aspects, and Kerner (1863) described, in a qualitative way, plant succession on burned forest lands, sand dunes, and other habitats of central Europe. In fact, in many of the European botanical writings since 1870, the idea of classifying vegetation by successional relationship was inherent. This work is exemplified by E. Warming's *Oecology of Plants: An Introduction to the Study of Plant Communities* (1909), which was translated into English by Groom and Balfour.

The idea of succession was not generally accepted by American botanists until the work of Cowles (1899) on Lake Michigan sand dunes, and the later voluminous and vigorously defended work of Frederick Clements (1916). From a deductive standpoint, Cowles work, and that of other scientists on other habitats, appeared to support the view that a successional series seemed to result in a community of plants that was apparently self-duplicating and perpetual. The community which gains essentially permanent occupancy of the habitat was named the *climax* community. Circumstantial evidence seemed to suggest that given enough time, the climax of a succession would be the same, although begun on different sites. All of these successions supposedly proceed by pioneering new sites. Such succession is termed *primary*.

Much of Clements work, on the other hand, dealt with old field succession and other types of succession occurring on disturbed land. This is land upon which a climax had once been present, and upon which seral stages occur leading to replacement of that climax community. This type of succession is called *secondary*. Clements work and his philosophy was brought together in his book titled *Succession* (1916), and as the result of his writings, his persuasive presentations and the zeal of his students, the Clementian point of view *was* American plant ecology for many years to follow. Clements looked upon the climax community as a *superorganism;* one which was born, grew, matured, and under certain conditions, eventually died. On the first page of his book he states:

> The developmental study of vegetation necessarily rests upon the assumption that the unit or climax formation is an organic entity (Research Methods, 199). As an organism, the formation arises, grows, matures, and dies. Its response to the habitat is shown in processes or functions and in structures which are the record as well as the result of these functions. Furthermore, each climax formation is able to reproduce itself, repeating with essential fidelity the stages of its development. The life-history of a formation is a complex but definite process, comparable in its chief features with the life-history of an individual plant.

Although widely accepted, Clements views were vulnerable. Many studies seemed to indicate that not one but several different climaxes were possible within the same general area; thus arose the *polyclimax theory*. Gleason (1917, 1926), disagreed with the whole concept of the community as a superorganism and the ordered succession of plant communities to climax. He stated that one could explain all successions and communities as no more than the randomly acquired assortment of plants suited to a particular habitat. With time, this number would naturally increase.

The Clementian theory of succession, although criticized from the outset, was so forcefully presented that it attracted thousands of disciples and ruled ecological thought for many years. Along with the theory came many general assumptions, some unfortunately still held today by many plant ecologists. Among these assumptions, or hypotheses, were that successions proceed to 1) maximum ecological efficiency, 2) maximum biomass, 3) efficiently managed information, and 4) stability. Let us examine each of those hypotheses.

Lindeman's (1942) classic paper on trophic dynamics discusses ecological efficiency admirably. As the unoccupied site obviously bears no vegetation at all, productivity must be zero; thus, efficiency is also zero. As the first pioneers come in, production begins but at first is necessarily low. With each successive successional level, productivity of the site increases, although the energy received from the sun remains essentially the same. Ecological efficiency, then, must be increasing as succession proceeds.

However, productivity, and therefore efficiency, cannot continue to rise indefinitely; it must be naturally limited, and what better place for the natural limit on ecological efficiency than the climax community? Thus, the climax community should maintain maximum efficiency. But a growing body of data indicates that early successional stages may be highly productive and in fact may have the highest productivity. There may be actually a trailing off of productivity in later successional stages (Kira and Shidei, 1967; Odum, 1969). In terrestrial ecosystems, the explanation seems to be a genetic one: competition is greatest in youth, thus the plants have evolved their greatest efficiency when young. The older plants are not so efficient. By competition, we are referring to the interference between two populations at the same trophic level. In an aquatic ecosystem, where the same general decline in mature communities is seen, the explanation seems to involve secretion of growth inhibitors, or ectocrines, and the decline in mineral nutrients in the system due to active uptake by the plant communities. Thus, the predictions of Lindeman do not appear to be realized in fact.

One of the more obvious attributes of succession in a terrestrial ecosystem is the general tendency towards an increase in size. Thus, perhaps the real determining principle in plant successions is a progression toward maximum biomass that the habitat can support. If both living and dead organic material produced by a community is included, it seems fair to say that communities do achieve their greatest biomass at maturity. When a system reaches the point that gross primary productivity and respiration are equal, the accumulation of biomass is at an end, and the system may then be called mature.

In the early stages of successions, there is a progressive increase in diversity, an increase that may or may not continue through to the so-called climax stage. Actually, the so-called climax stage may have fewer species than those stages preceeding it, due to ecological dominance of one or several species. But in many instances, each successional stage contains a larger number of species. The ecosystem becomes increasingly complex, and in the view of some, this complexity results in greater organization and increased stability. The so-called information content of the system is increased during succession. What is information? The accumulation of species, the massing of organic molecules, increase in vertical structure all represent information to the system, and this information, in turn, directs the accumulation of more information. Thus, the system becomes self-regulating and self-determining. According to Ramon Margalef (1958, 1963, 1968), ecosystems evolve through succession, achieving the most efficient maintenance of structure, the most efficient management of information, at maturity.

Significant to this hypothesis is the observation that biomass increases fastest in early successional stages. The young ecosystem is seemingly storing up information (biomass) for the future. This is obviously a very teleological way of looking at it and from our earlier discussions we should be very suspect of it. But according to Margalef, not only does it store information for its own future but the evolving ecosystem exports information to neighboring ecosystems, such as export from the plankton to the benthos in an aquatic system.

Margalef expounded on his theories in a classical paper (1963) called, "On certain unifying principles in ecology." He expressed his views on the basis of thermodynamics and in mathematical terms through information theory. To many this approach was appealing, but close examination indicates his observations were not particularly profound. They were really nothing more than a restatement of the concept of succession in mathematical terms. As succession starts with little or no mass and little or no complexity, it must necessarily increase in both mass and complexity. Such an increase requires work, such work results in the organization of molecules, such organization results in decreased entropy within the system.

Margalef's hypothesis seems to treat the ecosystem as a functional biological unit, a unit thought to develop by organizing itself in such a way that information is conserved and managed. Herein lies its greatest fault. Ecosystems cannot be treated like species, ecosystems cannot evolve, they cannot be selected for or against. Only individuals can evolve, and the concept of ecosystem is simply that, a concept. In addition, the extreme teleology utilized by Margalef is quite unacceptable.

In the above discussion, we alluded to successions leading to increased stability. What is stability? If we define stability as the absence of species turnover and population fluctuations, then stability increases tautologically with succession. In other words, succession implies stability. Such an idea is inherent in our concept of succession, which is defined as occurring when the composition of the community is changing, and stopping when the composition of the community is not changing. Another way of looking at stability would be to regard it as resistance to perturbation. This is most conveniently measured by the speed with which a community returns to its original state after a temporary disturbance. From this viewpoint, stability decreases through succession; disturb early succession and it becomes early succession, disturb a so-called climax community and it too becomes early succession. It takes the latter a long time to return to its previously undisturbed state.

May (1973) devoted an entire book to a discussion of the relationship between diversity and stability. He found that in a wide range of mathematical models complexity resulted in instability. In general, the more diverse a community, the more stringent and exotic are the restrictions on the dynamic behavior of additional species if their addition is to result in a more stable community or one that is stable at all. May's analysis does not suggest that diversity and dynamic stability increase with succession, but in fact offers a considerable argument that complex systems are not inherently stable. Futuyma (1973) has also argued that complex ecological systems are intrinsically fragile and fossil evidence has indicated more extinctions in complex communities than in simpler ones.

Thus, stability, in the naive sense of absence of change, increases trivially during succession, but dynamic stability, the speed with which a community rebounds following a temporary disturbance, decreases stringently during succession. Competition among successional species favors dispersal, self-replacement, and longevity, those characteristics that lower or destroy dynamic stability.

Henry S. Horn (1974) has placed the argument squarely in the light of modern environmental concerns. He states:

> Conservationists and advocates of wilderness preserves have often cited the conventional generalization that diversity conveys stability, arguing that diverse natural communities should be conserved for their stabilizing influence on the simple artificial communities invented by man. One could equally argue that if complex systems were inherently stable, they should need no protection. The opposite view, that diverse communities and complex communities are inherently fragile, is a much more powerful reason for their requiring protection from human disturbance, perhaps even from well-intentioned management.

## SUMMARY

1. The study of ecology involves the investigation of the relations of organisms to other organisms and to their environment.
2. Plant ecology as a science had its beginning in plant geography.
3. Levels of study include 1) organism, 2) population, 3) community, and 4) ecosystem.
4. Early plant ecologists looked at the world distribution of plants and, on the basis of those studies, constructed world vegetation types.
5. These vegetation types were called formations.
6. Formations appeared to succeed each other as one traveled from the equator northward, with the same general pattern appearing as one traveled up a mountain.
7. Early plant ecologists believed that control of plant formations was climatic, primarily by temperature and moisture.
8. The idea of discrete boundaries between plant formations is not consistent with fact, rather formations continually intergrade.
9. Various local factors such as soil, burning, predation, and others may result in formations other than those generally occurring or expected to occur in a certain region.
10. Soil development is linked to both climate and vegetation.
11. Soil is formed as the result of both weathering and biogenic activity.

12. Soil may be classified as vertical layers called horizons.
13. Soils may also be classified on the basis of their chemical and biological properties, with these aspects based on the soil-forming activities of gleization, podsolization, lateralization, and calcification.
14. Gleization involves the formation of a sticky, compact, structureless layer at the bottom of the *B* horizon and the consequent accumulation of a peaty layer above it.
15. Podsolization involves the decalcification and leaching of iron and aluminum ore *A* horizons.
16. Lateralization results in the removal of silica from the *A* horizons and the formation of iron and aluminum oxides in the horizon, along with considerable loss of organic content.
17. Calcification involves the formation of a calcareous upper layer of soil.
18. The habitat plays a role in determining the makeup of the plant community and the plant community, in turn, determines the characteristics of the habitat.
19. The realization of community and environment interaction has led to the ecosystem concept in ecology.
20. The ecosystem can be viewed as consisting of biotic and abiotic elements.
21. The former consist of producers, consumers, and decomposers.
22. The abiotic elements represent the environmental aspects.
23. Energy and nutrient input and utilization may also be viewed as an ecosystem component.
24. Abiotic regulation is exerted, in part, by the limiting inorganic nutrient. This concept is stated in Liebig's "law of the minimum."
25. Liebig's law is appropriate only under steady-state conditions and without factor interaction.
26. Abiotic regulation is also exerted, in part, by the limits of tolerance. This concept is embodied in Shelford's "law of tolerance." This law is also subject to modification.
27. Interaction of the biotic and abiotic elements of the ecosystem can be seen, in part, through study of cyclic aspects such as water, oxygen, carbon dioxide, nitrogen, phosphorus, sulfur, and others.
28. Autotrophic organisms are those which produce their own food, either by photosynthesis or chemosynthesis.
29. Heterotrophic organisms are not capable of producing their own food but must obtain it through digestion or ingestion of green plant material.
30. The two principal types of heterotrophs are holophytic, which ingest their food whole, and saprophytic, which obtain their food through extracellular digestion.
31. Some heterotrophs are consumers. These consume other plants and animals as food.
32. Some heterotrophs are decomposers. These decompose dead plants and animals as a source of food.

33. Producers and consumers occupy various levels of a food chain or food web. While producers occupy the bottom level only, consumers are multilevel and may be herbivores, carnivores, or omnivores. Some consumers are termed parasites.
34. During the process of decomposition, soluble nutrients are formed and metabolic by-products having a regulatory effect on the growth of plants may be produced.
35. Only about 10 percent of the available sunlight is actually utilized by plants, and as the energy is transferred from trophic level to trophic level, approximately 90 percent of that energy is lost at each step.
36. The concept of niche was a zoological concept established to describe the organism's role in its environment.
37. Recent theories of niche are based upon the desire to explain certain plant interactions within the community.
38. Certain plants, while seeming to have similar roles in the community, still coexist; while another plant which may appear to have a quite specific role may exist only rarely or incidentally.
39. The explanation of the above situations may lie in niche diversification.
40. Species of plants may occupy what appears to be a similar niche due to polymorphic affects.
41. Polymorphism refers to the presence of different forms of a single species.
42. Plant interactions and niche exclusion have traditionally been viewed from the standpoint of competition.
43. Interspecific interactions include neutralism, mutual inhibition type competition, resource use type competition, amensalism, parasitism, predation, commensalism, protocooperation, and mutualism.
44. Neutralism denotes an interaction in which neither population affects the other.
45. Mutual inhibition type competition defines an interaction in which there is a direct inhibition of each species by the other.
46. Resource use type competition defines an interaction in which there is indirect inhibition when a common resource is in short supply.
47. Amensalism denotes an interaction in which one population is inhibited, the other is not affected.
48. Parasitism denotes an interaction in which one population, the parasite, benefits, and the second population, the host, is inhibited. Generally, the parasite is smaller than the host.
49. Predation denotes an interaction in which one population, the predator, benefits, and the second population, the prey, is inhibited. The predator is usually larger than the prey.
50. Commensalism denotes an interaction in which one population benefits, the second is unaffected.
51. Protocooperation denotes an interaction favorable to both population but not obligatory to either.

52. Mutualism denotes an interaction favorable to both populations and obligatory to both.
53. Interspecific interactions may be classified as either negative or positive.
54. Competitive exclusion is often cited as a factor bringing about an ecological separation of closely related or otherwise similar species.
55. Grazing may be very important in determining community structure, and differences in plant communities previously attributed to competitive exclusion may in fact be due to selective grazing or predation by animals.
56. Allelopathic effects may also determine plant community structure.
57. Allelopathy refers to the chemical antagonism of one species by another.
58. A further influence on plant community structure may be fire. Its effect may be negative or positive.
59. Fire has an affect on soil, soil organisms and various other fauna which may directly or indirectly affect plant communities.
60. Fire may affect the plant community directly due to the differential effect of burning on plant species.
61. Fire may actually be necessary for the continued existence of certain species.
62. Positive interactions include mycorrhizal relationships between fungal hyphae and tree roots.
63. The specific makeup of a plant community is dependent upon the ecologic amplitude of the plants comprising the flora and the heterogeneity of the environment.
64. An area of intergradation of the two different plant communities is called an ecotone.
65. Plant communities may change through time, a process called succession.
66. Succession may be primary or secondary depending upon whether it begins in a newly formed habitat or in a disturbed one.
67. Succession is generally thought to proceed to an apparently stable, self-duplicating community termed the climax community.
68. The plant ecologist Clements viewed the climax community as a superorganism and the process of succession as being specifically structured.
69. The current view of succession is that it indeed does occur but is certainly not ordered nor does it result in a climax community having properties generally associated with organisms.
70. Successions apparently do not proceed toward maximum ecological efficiency, are not directed toward more efficient management of information, and do not result in an increase in the dynamic stability of the system.

# APPENDIX I
# A FEW BASIC CHEMICAL CONCEPTS

Many chemical substances are composed of molecules. A molecule might be described as the smallest unit of these substances that still retain all the characteristics of the substance. Molecules contain an inherent energy called *kinetic energy* which keeps them constantly in motion. Obviously, this motion of molecules is quite different in different substances, with gases having the most, solids the least. Molecules, in turn, are composed of subunits called *atoms*, and it is these atoms that we are generally concerned with when we speak of chemical reactions. If a substance is composed of a single kind of atom, we call it an *element* (Table AI.1); if composed of different atoms in a fixed ratio, a *compound*. The atom contains still smaller subunits, three of which we shall be concerned with: the *electron (e)*, *proton (p)* and *neutron (n)*. The electron bears a *negative* electrical charge, the proton a *positive* charge, and the neutron is electrically *neutral.*

*Table AI.1*
**SOME OF THE ELEMENTS FOUND IN PLANTS**

| ELEMENT | SYMBOL | COMMON VALENCE NUMBERS |
|---|---|---|
| Aluminum | Al | +3 |
| Boron | B | +3 |
| Calcium | Ca | +2 |
| Carbon | C | −4, +4 |
| Chlorine | Cl | −1 |
| Copper | Cu | +1, +2 |
| Hydrogen | H | +1 |
| Iron | Fe | +2, +3 |
| Magnesium | Mg | +2 |
| Manganese | Mn | +2, +3 |
| Molybdenum | Mo | +3, +5 |
| Nitrogen | N | −3, +5 |
| Oxygen | O | −2 |
| Phosphorus | P | −5 |
| Potassium | K | +1 |
| Silicon | Si | +4 |
| Sodium | Na | +1 |
| Sulfur | S | +6, −2 |
| Zinc | Zn | +2 |

Both the proton and neutron occupy a small space in the center of the atom, with the electrons moving in a very prescribed, orbicular fashion around this central portion. If the number of protons and electrons are equal, the atom will possess no net charge and thus be neutral. If the atom loses an electron, it will become positively charged; if it gains an electron, negatively charged.

$$Na \rightarrow Na^+ + e^-$$

$$Cl + e^- \rightarrow Cl^-$$

Such charged atoms are referred to as *ions*. The letters Na and Cl are the *chemical symbols* for sodium and chlorine, respectively.

Some atoms of the same element may have differing numbers of neutrons. For example, oxygen may have either 15, 17, or 18 neutrons. These atoms would be written as $^{16}O$, $^{17}O$, $^{18}O$ and are called *isotopes*. Certain isotopes may emit penetrating rays and thus be *radioactive*.

The electrons occupy fixed orbitals about the core of the atom, with each orbital containing a maximum of two electrons and having a specific energy. Orbitals make up shells which can contain a specified maximum number of electrons, with that number being two for shell one. The shells beyond one are composed of subshells. Shell two is composed of subshell 2*s* containing a maximum of two electrons, and subshell 2*p* with a maximum of six electrons. Shell 3 has three subshells: 3*s* with a maximum of two electrons, 3*p* with a maximum of six electrons, and 3*d* with a maximum of ten electrons. Shell four has 4 subshells: 4*s*, 4*p*, 4*d*, and 4*f*, with a maximum of 2, 6, 10, and 14 electrons, respectively. The outermost shell of electrons in any atom is called the *valence shell* and determines the chemical behavior of the element. It is this outermost shell that is involved in chemical bonding because of the tendency of atoms to either gain, lose, or share electrons so as to reach the maximum number for the shell. This is not always so however, and some atoms can combine in several different radios and thus have several *valence states*. (See Table AI.1).

## CHEMICAL BONDING

As mentioned earlier, atoms can combine in various ways forming compounds. In the reaction between an atom of sodium (Na) and chlorine (Cl) (Figure AI.1), the outermost electron of Na is lost, thus forming a new outermost shell, the second, which has a complete set of 8 electrons. Similarly, if Cl, which has 7 electrons in its outer shell, were to gain an electron, it would reach the stable configuration of 8 electrons. As the result of this exchange, Na, having lost an electron, becomes a positively charged ion, and Cl, having gained an electron, becomes a negatively charged ion. Thus, these two ions will become bond together by the considerable electrostatic force between ions of opposite charge. This force creates what is

Figure AI.1

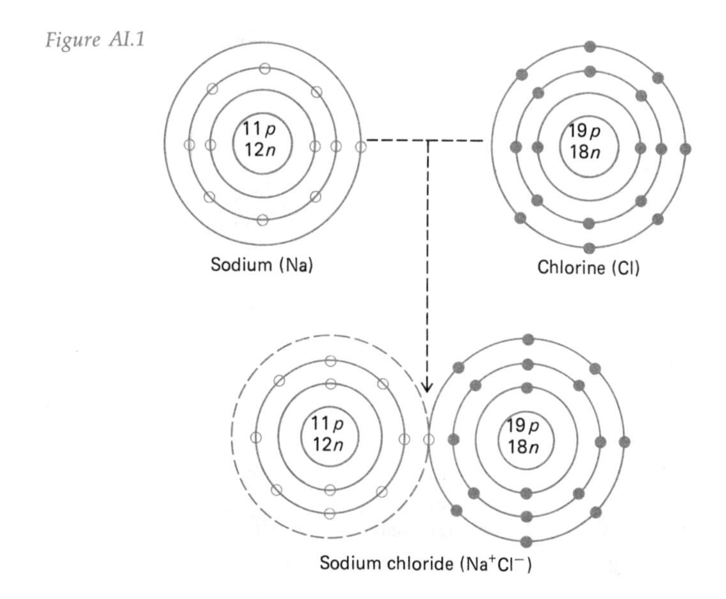


known as an *ionic bond.* Energy is required to remove the electron from sodium but is liberated when the electron is added to chlorine. The compounds formed by ionic bonds are called *ionic compounds,* and such compounds will normally dissolve in water. The molecules of the compound separate (or dissociate) into the component electrically charged particles. The positively charged ion is called a *cation,* the negatively charged ion an *anion.*

Some atoms reach a stable electron configuration, not by gain or loss of electrons, but by *sharing* with another atom or atoms. The shared electrons orbit about both nuclei and act as though they are in the valence shell of each atom. Such bonds are termed *covalent bonds* and may be formed by the sharing of a single pair of electrons, two pairs of electrons (double covalent bond), or three pairs of electrons (triple covalent bond). An example of covalent bonding is exhibited in ammonia (Figure AI.2). As you can see, ammonia is formed by the combining of three hydrogen atoms by single covalent bonds to a single nitrogen atom. Such covalent bonds are somewhat weaker than ionic bonds.

Some covalent bonds are described as *polar,* others as *nonpolar.* If the two atoms forming the bond have the same electronegativity, then the electrons will be shared equally by the two atoms and the bond will be nonpolar. But when different atoms are involved in forming a covalent bond, one is likely to be more electronegative than the other. In ammonia, for example, the electrons are bound more closely to the nitrogen than to the hydrogen, and thus, nitrogen will bear a *slight* negative charge and the hydrogens will bear *slight* positive charges. (These slight charges are symbolized by the Greek letter delta as $\delta^+$ and $\delta^-$.) Such a covalent bond is polar.

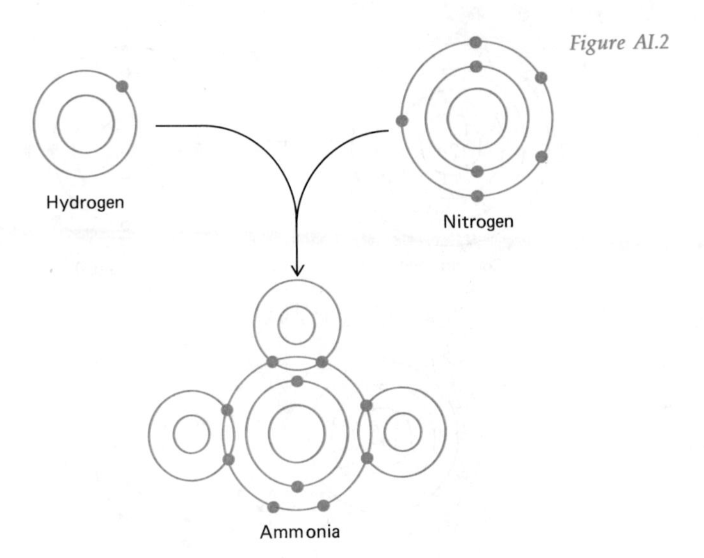


*Figure AI.2*

The formation of slight charges in certain covalently bound molecules is of considerable importance in biology. As these forces are generally associated with $H^{\delta+}$, $O^{\delta-}$, and $N^{\delta-}$, the slightly positively charged hydrogen of a compound is essentially a naked proton, and as protons are very small, an adjacent molecule can approach very close to it. Under these conditions, an $O^{\delta-}$ or $N^{\delta-}$ of one molecule can form an electrostatic force of attraction between the oppositely charged $H^{\delta+}$ of a second molecule (or the folded back portion of the same molecule), thus forming a type of weak ionic bond. As these bonds are always associated with hydrogen, they are termed *hydrogen bonds.* They are about five percent as strong as carbon-carbon, carbon-oxygen, and carbon-nitrogen covalent bonds.

# CHEMICAL REACTIONS

## *Chemical Equations*

All known chemical reactions that occur in plants can at least be *summarized* by means of a *chemical equation.* A chemical equation is simply a statement showing which atoms react, how many of them react, and what product or products are formed as the result of that reaction. For example, water is composed of hydrogen and oxygen. The hydrogen molecule is composed of two atoms of hydrogen and thus is written as $H_2$; the oxygen molecule is composed of two oxygen atoms ($O_2$). Water, however, is composed of only one atom of oxygen, but two atoms of hydrogen. Thus, the equation $H_2 + O_2 \rightarrow H_2O$ would not be properly balanced as we have one more oxygen reacting than is accounted for in the product. This could be easily corrected by having two molecules of water as the product: $H_2 + O_2$

$\rightarrow 2H_2O$. Now we have two oxygen atoms reacting and two formed. Unfortunately, we now also have four hydrogen atoms accounted for in the product but only two reacting. This can be corrected very easily by having two hydrogen molecules reacting with the single oxygen molecule: $2H_2 + O_2 \rightarrow 2H_2O$. Now we have four hydrogen on the left side of the equation, four on the right side, and two oxygen molecules on the left side of the equation, two on the right. The equation is now balanced, in line with the requirement that the total mass of the reactants must equal the total mass of the products. You will note that the equation is balanced in terms of the number of atoms, not the number of molecules. The number of atoms is written as a subscript after the chemical symbol for the element, while the number of molecules is placed before the chemical formula for that molecule. The total number of atoms is thus calculated quite simply by multiplying the number of molecules present times the number of atoms in each element composing that molecule.

## *Oxidation-Reduction*

An important type of reaction in organisms is that involving oxidation and reduction. The original meaning of the term *oxidation* referred to the *addition of oxygen* to a compound, with the compound *losing* the *oxygen* being *reduced.* Today, oxidation, in a chemical context, refers to the *loss of electrons*, with the molecule or atom which *gains* those electrons being reduced. We also realize that loss of the electron through loss of a hydrogen atom will accomplish the same purpose; thus, if a molecule loses a hydrogen it will be oxidized, if it gains a hydrogen, reduced. In our previous diagram showing the formation of common table salt, the equation may be written as follows:

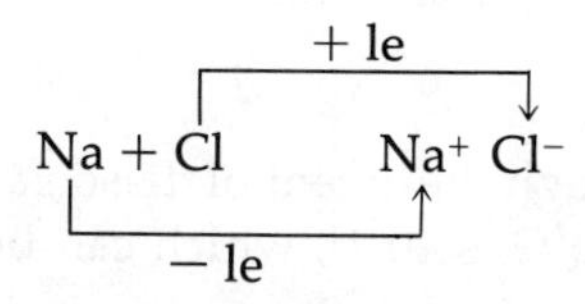


In this equation, both reactants are initially neutral, but in the reaction, sodium loses an electron (is oxidized) while chlorine gains that electron (is reduced). As you can see, the oxidation of one substance in a reaction must necessarily be accompanied by the reduction of another, and the number of electrons lost by one must equal the number gained by the other.

## *Acids and Bases*

Several hundred years ago certain compounds were noted to have properties in common: they would dissolve metals, and even when highly diluted, would be sour to the taste. These were called, *acids.* Another group of compounds was noted to rarely dissolve metals (only a few special ones) and when diluted were brackish to the taste. These were called *bases.* Somewhat later it was recognized that acids are compounds that dissociate yielding hydrogen ions, ($H^+$) while bases are compounds that dissociate yielding hydroxyl ions ($OH^-$).

(an acid) $HCl \rightarrow H^+ + Cl^-$
HYDROCHLORIC ACID

(a base) $NaOH \rightarrow Na^+ + OH^-$
SODIUM HYDROXIDE

But, the above description of an acid is an oversimplification. Actually, the *pure* compounds do not have the properties that we associate with acids. These properties develop, however, when the compounds are dissolved in water. The chemical reaction involves the transfer of a proton from the acid to the solvent (water).

$$HCl + H_2O \rightarrow H_3O^+ + Cl^-$$

It is the *hydronium ions* ($H_3O^+$) that give water solutions of these compounds their "acidic" properties.

Acids and bases also have the property that when they combine they neutralize one another forming a salt.

$$H_3O^+Cl^- + Na^+OH^- \rightarrow 2H_2O + Na^+Cl^-$$

The above product is a water solution which is neither acidic nor basic, but contains, in fact, only sodium chloride (common table salt). Only strong acids and bases will completely dissociate as shown in the preceding equation. Weak acids and bases only partly dissociate in solution.

As water can act as both an acid and a base, we might suppose that it would contain small amounts of hydronium and hydroxide ions.

$$2H_2O \rightarrow H_3O^+ + OH^-$$

Measurements have shown this to be true, although the extent of dissociation is low. The value of this dissociation at 25°C is $10^{-14}$, which can be expressed as:

$$[H_3O^+][OH^-] = 10^{-14}$$

The brackets in this expression signify that the concentrations are expressed in molarity (the number of moles of a compound dissolved in 1 liter of solution). This low dissociation constant indicates that water solutions never contain only $H_3O^+$ or $OH^-$, but always both ions. The relative proportions of the two determine the acidity or basicity of the solution.

If a 0.1 mole (M) solution of HCl is prepared by adding hydrochloric acid to water, the hydrochloric acid generates 0.1 mole of hydronium ion. As the amount of hydronium ions present in purewater is very small (0.0000001 M), the final concentration of the hydronium ion in the solution is, for all practical purposes, 0.1 M. If this is true, then the concentration of the hydroxyl ion must be $10^{-13}$ M:

$$[OH^-] = \frac{10^{-14}}{[H_3O^+]} = \frac{10^{-14}}{10^{-1}} = 10^{-13}$$

As the ratio of hydronium to hydroxide ions is very large ($10^{-1}/10^{-13} = 10^{12}$), only the acidic properties of such a solution need be considered for most purposes. Thus, the hydronium ion concentrations are a measure of the acidity of the solutions. A more convenient way has been devised for expressing these numbers, the *pH unit.* The pH of a solution is the negative logarithm of the hydronium ion concentration: pH = -log $H_3O^+$. Thus in the following:

$$[H_3O^+] = 1 \times 10^{-4};\ pH = -\log 1 \times 10^{-4} = -1(-4) = 4$$
$$[H_3O^+] = 1 \times 10^{-7};\ pH = -\log 1 \times 10^{-7} = -1(-7) = 7$$
$$[H_3O^+] = 1 \times 10^{-13};\ pH = -\log 1 \times 10^{-13} = -1(-13) = 13$$

Even more complex examples are handled in the same way:

$$[H_3O^+] = 3 \times 10^{-3};\ pH = -\log 3 \times 10^{-3} = -1(\log 3 \times 10^{-3}) = -1(.48 - 3) = 2.52$$

With the pH system one can refer quantitatively to the acidity of a solution; acidic solutions have a pH less than 7, basic solutions have a pH greater than 7.

### Reaction Rates

A final consideration concerns the rate of a chemical reaction. Certain inorganic ions, when added together, may react very rapidly forming a product. Large organic molecules, on the other hand, react very slowly or essentially not at all. Why is this so? For one thing, a large molecule may have only a single reactive site, and thus, collision of these molecules may not occur in such a fashion as to bring these reactive sites together. A second reason is that large molecules very often require that they be given a considerable amount of energy in order to overcome the repulsion of the electrons on the two reactants. This energy requirement is termed the *energy of activation.* If the energy of activation can be lowered, the rate of reaction can be increased. Certain substances capable of lowering the energy of activation of a reaction are termed *catalysts.* It is a special property of catalysts that although they participate in the reaction, they are neither changed nor consumed in the reaction.

## ORGANIC COMPOUNDS

The chemistry of organic compounds is essentially the chemistry of carbon compounds. The carbon-carbon covalent bond, which can be single or multiple, is a stable and strong bond, thus making possible the formation of long chains of carbon. These chains can either be open or closed, and the strength of the bond, along with the high covalency of carbon, makes possible the formation of a vast number of both simple and extremely complex molecules. But the carbon chain itself is relatively inert; thus, the properties of an organic compound are generally due to relatively small groups attached to the carbon chain, termed *functional groups.*

The simplest of the organic compounds are the *hydrocarbons*, which, as the name implies, are composed only of hydrogen and carbon. Not many biological compounds are pure hydrocarbons, but two common ones are methane ($CH_4$) and ethylene ($CH_2{=}CH_2$). We will refer to enthylene when we discuss plant hormones. Most biological compounds contain oxygen in addition to carbon and hydrogen, while others also contain nitrogen, sulfur, or phosphorus.

## Functional Groups

As the functional groups are so important in biological reactions, it might be well to become familiar with a few of the common ones. These groups can best be studied by relating them to some of the principal types of organic compounds found in cells.

Carbohydrates are compounds which contain carbon, hydrogen, and oxygen in the general ratio ($CH_2O$) or a modification of that ratio. To be classified as a carbohydrate, a compound must possess two specific properties: 1. it must have an aldehyde

```
    H
    |
R—C=O
```

or ketone

```
    O
    ‖
R—C—R
```

group, where R signifies the rest of the organic molecule exclusive of the functional group, and 2. it must have two or more primary or secondary alcohol groups. In primary alcohols, the —OH group is attached to a carbon atom which has only *one* carbon atom attached *directly* to it, while secondary alcohols have their —OH group attached to a carbon atom which has *two* carbon atoms attached directly to it.

Carbohydrates are characterized by several of the more important functional groups: the —OH or *hydroxyl* group, the

```
  |
—C=O
```

or *carbonyl* group, the

```
    H
    |
R—C=O
```

or *aldehyde* group, and the

$$R-\overset{\overset{\large O}{\|}}{C}-R$$

or *ketone* group. Some carbohydrates are characterized by the functional group

$$R-\overset{\overset{\large OH}{|}}{C}=O,$$

the *carboxyl* group, and are known as *carboxylic acids.* Such acids are also incorporated in lipids as well. If the hydroxyl portion of a carboxyl group is removed, a functional group

$$R-\overset{\overset{\large O}{\|}}{C}-$$

an *acyl* group, is formed.

The lipids, as a class of compounds, defy rigorous definition, but we can state that in general they contain carbon, hydrogen, and oxygen, usually in a different ratio than in carbohydrates, and may contain, in addition, phosphorus or nitrogen. The functional groups of lipids are similar to those of carbohydrates.

Proteins are usually very large molecules composed of a chain or chains of amino acids. They contain the elements carbon, hydrogen, oxygen, nitrogen, and usually sulfur. Some proteins may also contain phosphorus. In addition to carboxyl groups, several other important functional groups are associated with proteins. These include the —SH or *sulfhydryl* group, the $-NH_2$ or primary *amino* group, and the —NH or *amide* group.

## MAJOR CLASSES OF BIOLOGICAL COMPOUNDS

### *Carbohydrates*

Carbohydrates are a major source of energy to the cell, the starting compounds for the synthesis of many other types of compounds, part of the structural framework of the cell, and components of glycolipids, glycoproteins, and nucleic acids. We mentioned earlier the basic carbon, hydrogen, oxygen nature of carbohydrates, although certain carbohydrates can contain phosphorus and nitrogen as well.

The most fundamental group of carbohydrates are termed *monosaccharides.* These are compounds which do not hydrolyze yielding simpler compounds. Monosaccharides commonly consist of from three to seven carbons. The simplest way to illustrate the molecular structure is to show the compound as a straight chain, as illustrated for glucose and fructose.

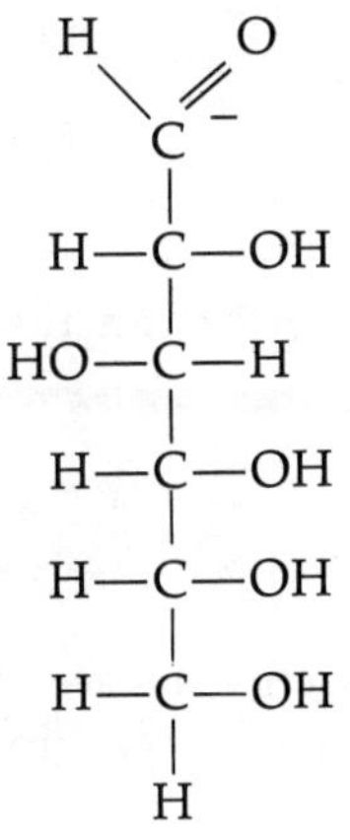

```
  H     O
   \   //
     C
     |
 H—C—OH
     |
HO—C—H
     |
 H—C—OH
     |
 H—C—OH
     |
 H—C—OH
     |
     H
```

GLUCOSE

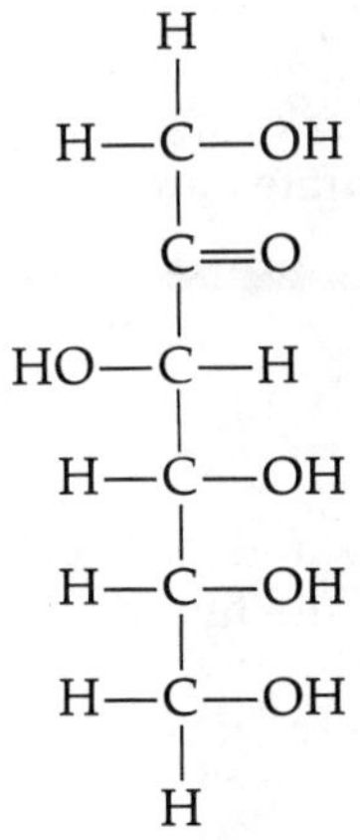

```
     H
     |
 H—C—OH
     |
    C=O
     |
HO—C—H
     |
 H—C—OH
     |
 H—C—OH
     |
 H—C—OH
     |
     H
```

FRUCTOSE

However, there is considerable evidence that these compounds do not exist linearly in the cell but, rather, form a ring structure as illustrated below.

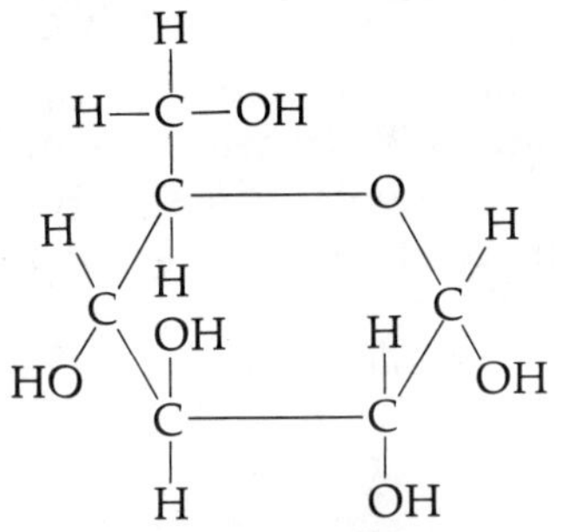

GLUCOSE

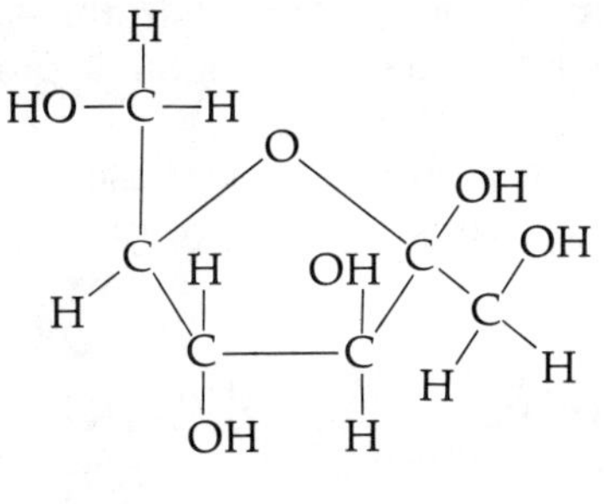

FRUCTOSE

As you can see, glucose and fructose have the same general formula: $C_6H_{12}O_6$. They differ in that glucose is an aldehyde, fructose a ketone. In addition, glucose is a *pyranose* (six-membered ring) while fructose is a *furanose* (five-membered ring).

Two molecules of a monosaccharide can combine, with the elimination of a water molecule, forming a *disaccharide*. If our previous examples, glucose and fructose, were to combine, a very common disaccharide, *sucrose* would result. Sucrose can easily be broken back down to glucose and fructose by the addition of a water molecule, a process called *hydrolysis*.

SUCROSE

Both glucose and fructose, having the reactive aldehyde and ketone groups, will act to reduce certain other compounds. Thus, they are called *reducing sugars.* Sucrose, however, is formed through the combining of the aldehyde and ketone group with the elimination of $H_2O$; thus, it is incapable of reducing other compounds. It is known as a *nonreducing sugar.*

When monosaccharides polymerize, they form *polysaccharides.* Thus, polysaccharides are simply repeating units of monosaccharides, with an $H_2O$ being lost for each pair of monosaccharides which unite. Most polysaccharides have relatively high molecular weights and are thus not water soluble as are monosaccharides and disaccharides. Common polysaccharides include *starch*, the major storage form of carbohydrates, and *cellulose*, the major component of cell walls.

## *Lipids*

As mentioned earlier, it is somewhat difficult to characterize lipids. Most lipids are completely insoluble in water but are soluble in polar solvents such as alcohol, acetone, and benzene. Some are important as reserve food materials *(fats* and *oils)*, some are structural components of membranes *(phospholipids* and *glycolipids)*, while others are exuded through the cell wall forming a coating over the surface *(waxes).*

A simple fat is formed by esterfication. Ester formation results when an alcohol and an acid are mixed together. In the case of fats, the alcohol is a secondary alcohol *glycerol*, the acids are termed *fatty acids.* Fatty acids are long chains of carbon and hydrogen, with the only oxygen being that associated with the carboxyl group. If each carbon has single covalent bonds and thus has two hydrogen attached to it, the fatty acid is said to be *saturated.* If a double covalent bond exists between two of the carbons in the chain, thus resulting in those two carbons each having only a single hydrogen attached to them, the fatty acid is said to be *unsaturated.* The general reaction for the formation of a fat would be as follows:

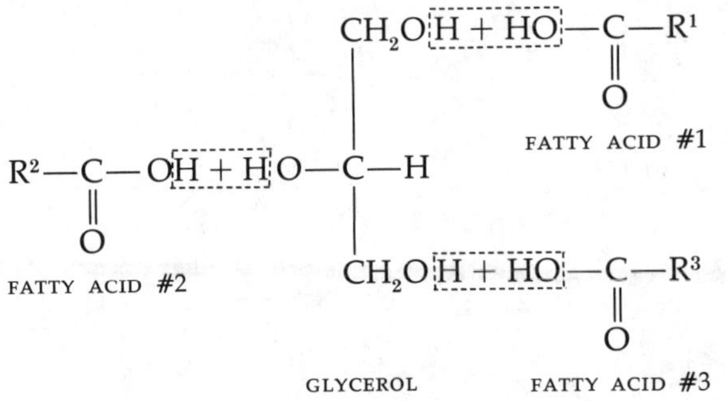

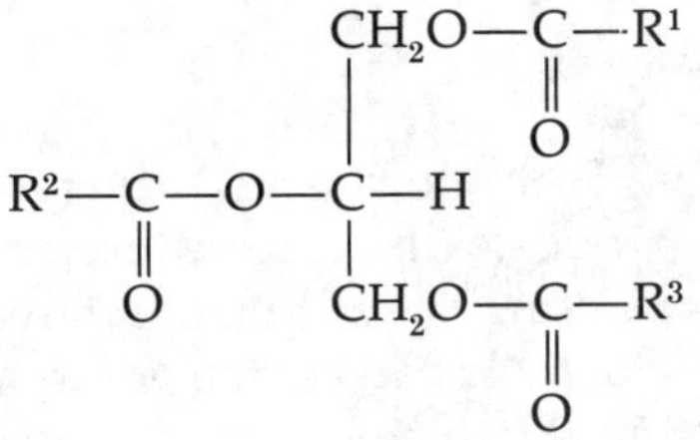

```
                         CH2O[H + HO]—C—R1
                          |           ‖
                          |           O
                          |
                          |      FATTY ACID #1
R2—C—O[H + H]O—C—H
   ‖              |
   O              |
                  |
FATTY ACID #2    CH2O[H + HO]—C—R3
                              ‖
                              O

             GLYCEROL        FATTY ACID #3
```

```
             CH2O—C—R1
              |   ‖
              |   O
R2—C—O—C—H
   ‖      |
   O     CH2O—C—R3
              ‖
              O
```

A FAT (TRIGLYCERIDE)

You can see that for each esterfication a water molecule is formed. A given fat usually contains three different fatty acids, some may contain two or, rarely, three similar fatty acids. A fat is usually a solid at room temperature, while an oil is a liquid at room temperature. The melting point is influenced by both length of the carbon chain of the fatty acid and degree of saturation. The melting point increases with length of carbon chain and with increased saturation; thus, fats are typically fully saturated, oils have a degree of unsaturation.

Other important lipid-related compounds include a diverse group called the isoprenoids. These include such compounds as *steroids* (ergosterol, cholesterol, and others), *carotenoids* (pigments), *essential oils* (oils having an essence or odor, such as *turpentine*) *rubber*, and the *gibberellins* (hormones).

## *Proteins*

Proteins are formed by the polymerization of amino acids. Plants contain some 20 different amino acids, the simplest of which is glycine.

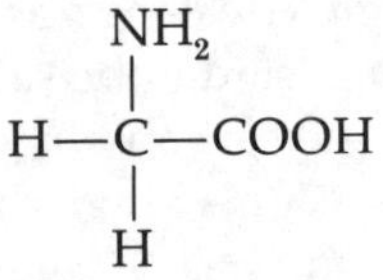

```
   NH2
   |
H—C—COOH
   |
   H
```

Other amino acids are formed by simply adding a different R group to the amino-containing carbon and by the addition of other amino or carboxyl groups. In the example shown above (glycine) the R group is simply represented by an H.

In solution, amino acids generally exist in ionized form.

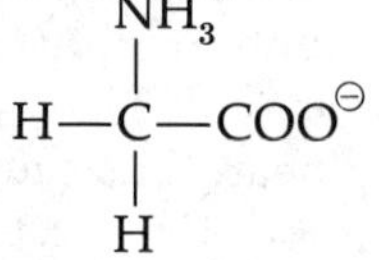

$$\begin{array}{c} NH_3^{\oplus} \\ | \\ H{-}C{-}COO^{\ominus} \\ | \\ H \end{array}$$

Such an ionized form is called a *dipolar ion* or, in German, *zwitterion* because of the two different charges. This ion has no *net* charge as the positive and negative charges are balanced. But if one changes the pH of the solution, by addition of an acid or base, the ion will then assume a charge.

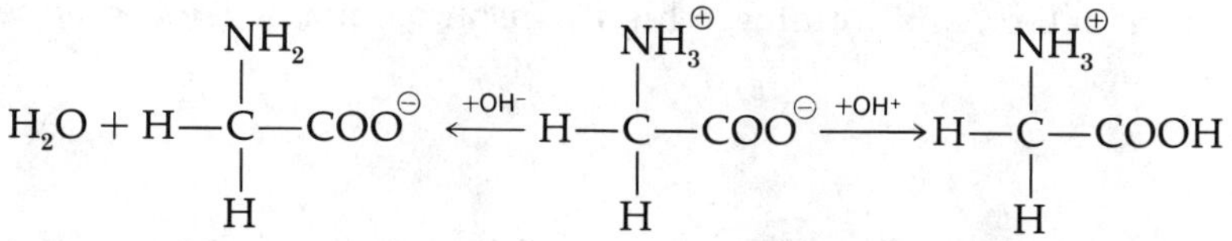

$$H_2O + \begin{array}{c} NH_2 \\ | \\ H{-}C{-}COO^{\ominus} \\ | \\ H \end{array} \xleftarrow{+OH^-} \begin{array}{c} NH_3^{\oplus} \\ | \\ H{-}C{-}COO^{\ominus} \\ | \\ H \end{array} \xrightarrow{+OH^+} \begin{array}{c} NH_3^{\oplus} \\ | \\ H{-}C{-}COOH \\ | \\ H \end{array}$$

As you can see, the addition of an acid causes the molecule to be positively charged, addition of a base causes it to be negatively charged. The pH of a solution at which there is no net charge on the amino acid is known as the *isoelectric point* for that amino acid.

An understanding of this property of amino acids, and thus of proteins, is extremely important. The activity of *enzymes* (organic catalysts), which are wholly or in part protein, will also be controlled by pH. Thus, an enzyme will have only a certain pH range in which it will be functional.

When amino acids combine forming protein, they do so through linkage of the carboxyl group of one with the amino group of the other, with the elimination of one water molecule.

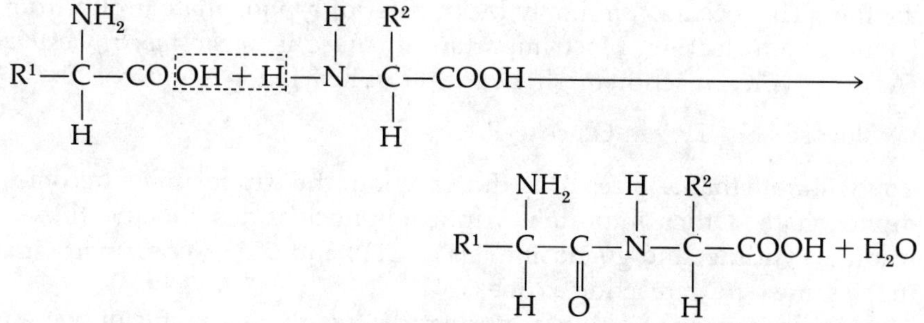

$$\begin{array}{c} NH_2 \\ | \\ R^1{-}C{-}CO \\ | \\ H \end{array}\!\!\boxed{OH + H}\!\!\begin{array}{c} H \quad R^2 \\ | \quad\; | \\ {-}N{-}C{-}COOH \\ \quad\;\; | \\ \quad\;\; H \end{array} \longrightarrow$$

$$\begin{array}{c} NH_2 \quad\;\; H \quad R^2 \\ | \qquad\quad | \quad\; | \\ R^1{-}C{-}C{-}N{-}C{-}COOH + H_2O \\ | \quad \| \qquad\;\; | \\ H \quad O \qquad\;\; H \end{array}$$

A DIPEPTIDE

The union of the amino and carboxyl groups form an *amide bond*, with this amide linkage between two amino acids termed a *peptide bond*. The resulting group is called a *peptide*. Thus, proteins are formed by a series of peptide linkages, and the *polypeptide chain* is a major part of the protein molecule. Coming off this chain will be numerous R groups as side chains.

The protein molecules can then become very large and can fold, bend, and coil in various ways. As you can see from the structural formula for the peptide, nitrogen, oxygen, and hydrogen stick out from the backbone of the molecule; thus, the coiled, folded or bent structure of the protein can be maintained by the formation of cross-linkages through hydrogen-bonding.

These hydrogen bonds can be rather easily broken by heat, thus causing the protein structure to be disrupted. When such dissociation occurs in enzymes, it leads to loss of activity. This inactivation is known as *denaturation.* Most of you have observed the result of drastic protein dissociation when you cook an egg. As the egg is heated up, the egg white, composed of a type of protein called albumen, will become dissociated, with the protein molecules becoming long and fibrous, rather than highly folded, piling together, and forming a highly insoluble mass. This is termed coagulation.

## *Nucleic Acids*

As the name implies, these molecules were associated with the nucleus. But we now know they are not found exclusively there. They are also associated with cytoplasmic organelles: ribosomes, mitochondria, and chloroplasts.

We discussed earlier (Chapter 6) the structure of nucleic acids being composed of nitrogenous bases, sugar, and phosphate. Nucleic acids are simply high molecular weight polymers of nucleotide units.

The genetic importance of the nucleic acids RNA and DNA has been emphasized in Chapter 6, but nucleotides have other biological importance in addition to their incorporation into these two compounds. Many metabolic reactions involve the transfer of energy from one molecule to another. This occurs commonly by transfer of a phosphate group from a triphosphonucleotide. Most important of these is *adenosine triphosphate* (ATP). A typical reaction might be the phosphorylation of glucose.

$$\text{Glucose} + \text{ATP} \rightarrow \text{Glucose-P} + \text{ADP}$$

You will note that as a result of this reaction, the triphosphate becomes a diphosphate. Other important triphosphonucleotides include those of uridine, cytidine, and guanosine (UTP, CTP, and GTP). These participate in the same type of reaction as does ATP.

Another important group of nucleotides are those capable of being reversibly oxidized or reduced. Thus, they are capable of serving as either oxidizing or reducing agents and are found to be involved with essentially every oxidation and reduction reaction occurring within living cells. An important example of such a compound is *nicotinamide adenine dinucleotide* (NAD). It undergoes the following type of reaction:

$$NAD^+ + H^+ + 2e^- \rightarrow NADH$$

(oxidized form) (reduced form)

Other nucleotides may become part of a coenzyme, such as coenzyme A, or become a portion of a vitamin, such as vitamin $B_{12}$.

# APPENDIX II
# A GEOLOGIC TIME TABLE*

| | | Subdivisions based on strata/*time*: Systems/*Periods* | Subdivisions based on strata/*time*: Series/*Epochs* | Radiometric dates (millions of years ago) | Outstanding events in evolution of living things |
|---|---|---|---|---|---|
| PHANEROZOIC | CENZOIC | Quaternary | Recent or Holocene Pleistocene | | *Homo sapiens* |
| | | | | 2? | |
| | | Tertiary | Pliocene | | Later hominids |
| | | | | 6 | |
| | | | Miocene | | |
| | | | | 22 | |
| | | | Oligocene | | Primitive hominids<br>Grasses; grazing mammals |
| | | | | 36 | |
| | | | Eocene | | Primitive horses |
| | | | | 58 | |
| | | | Paleocene | | Spreading of mammals<br>Dinosaurs extinct |
| | | | | 63 | |
| | MESOZOIC | Cretaceous | Many | | Flowering plants |
| | | | | 145 | |
| | | Jurassic | | | Climax of dinosaurs<br>Birds |
| | | | | 210 | |
| | | Triassic | | | Conifers, cycads, primitive mammals<br>Dinosaurs |
| | | | | 255 | |
| | PALEOZOIC | Permian | | | Mammallike reptiles |
| | | | | 280 | |
| | | Pennsylvanian (upper Carboniferous) | | | Coal forests, insects, amphibians, reptiles |
| | | | | 320 | |
| | | Mississippian (Lower Carboniferous | | | |
| | | | | 360 | |
| | | Devonian | | | Amphibians |
| | | | | 415 | |
| | | Silurian | | | Land plants and land animals |
| | | | | 465 | |
| | | Ordovician | | | Primitive fishes |
| | | | | 520 | |
| | | Cambrian | | | Marine animals abundant |
| | | | | 580 | |
| PRECAMBRIAN (Mainly igneous and metamorphic rocks; no worldwide subdivisions.) | | | | | Primitive marine animals |
| | | | | 1000 | Green algae |
| | | | | 2000 | |
| | | | | 3000 | Bacteria; blue-green algae |

*The geologic column, major worldwide subdivisions, selected dates, and events in the evolution of life. (Dates are best estimates, after R. L. Armstrong, 1971, unpublished.)

# INDEX